TECHNOLOGIES FOR THE WIRELESS FUTURE

TECHNOLOGIES FOR THE WIRELESS FUTURE

Wireless World Research Forum (WWRF)

Volume 2

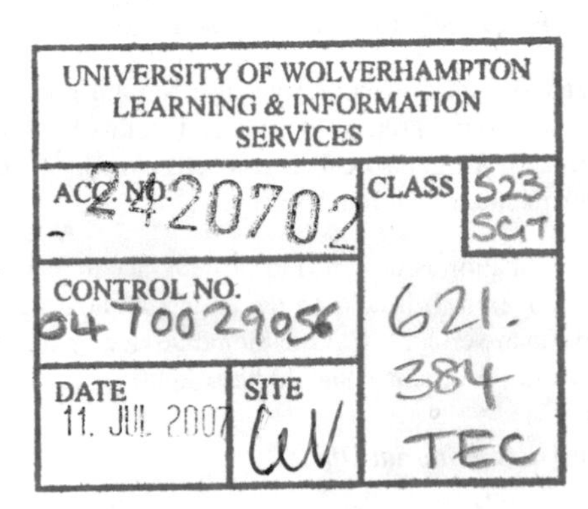

Edited by
Rahim Tafazolli
The University of Surrey, UK

Main Contributors

Mikko Uusitalo
WWRF chair 2004–, Nokia, Finland

Angela Sasse
WWRF WG1 chair 2004–2005, University College London, UK

Stefan Arbanowski
WWRF WG2 chair 2004–2005, Fraunhofer Fokus, Germany

David Falconer
WWRF WG4 chair 2004–2005, Carleton University, Canada

Gerhard Fettweis
WWRF WG5 chair 2004–2005, University of Dresden, Germany

Panagiotis Demestichas
WWRF WG6 chair 2004–, University of Piraeus, Greece

Mario Hoffmann
WWRF SIG2 chair 2004–, Fraunhofer, Germany

Amardeo Sarma
WWRF SIG3 chair 2004–, NEC, Germany

John Wiley & Sons, Ltd

Published in 2006 by John Wiley & Sons Ltd, The Atrium, Southern Gate, Chichester,
West Sussex PO19 8SQ, England

Telephone (+44) 1243 779777

Email (for orders and customer service enquiries): cs-books@wiley.co.uk
Visit our Home Page on www.wiley.com

Other Wiley Editorial Offices

John Wiley & Sons Inc., 111 River Street, Hoboken, NJ 07030, USA

Jossey-Bass, 989 Market Street, San Francisco, CA 94103-1741, USA

Wiley-VCH Verlag GmbH, Boschstr. 12, D-69469 Weinheim, Germany

John Wiley & Sons Australia Ltd, 42 McDougall Street, Milton, Queensland 4064, Australia

John Wiley & Sons (Asia) Pte Ltd, 2 Clementi Loop #02-01, Jin Xing Distripark, Singapore 129809

John Wiley & Sons Canada Ltd, 22 Worcester Road, Etobicoke, Ontario, Canada M9W 1L1

Wiley also publishes its books in a variety of electronic formats. Some content that appears in print may not be available in electronic books.

British Library Cataloguing in Publication Data

A catalogue record for this book is available from the British Library

ISBN-13 978-0-470-02905-3 (HB)
ISBN-10 0-470-02905-6 (HB)

Typeset in 10/12pt Times by Laserwords Private Limited, Chennai, India.
Printed and bound in Great Britain by Antony Rowe Ltd, Chippenham, Wiltshire.
This book is printed on acid-free paper responsibly manufactured from sustainable forestry in which at least two trees are planted for each one used for paper production.

Contents

List of Figures

List of Tables

List of Tables

List of Contributors

Chapter 1 editors	Mikko Uusitalo (Nokia)
Authors	Sudhir Dixit (Nokia)
	George Dimitrakopoulos and Panagiotis Demestichas (University of Piraeus)
	Sung Y. Kim and Byung K. Yi (LGE)
	Larry Swanson (Intel)
Chapter 2 editor	Mikko Uusitalo (Nokia)
Authors	Mikko Uusitalo (Nokia)
Chapter 3 editors	Angela Sasse (University College London, UK)
Authors	James Stewart (University of Edinburgh, UK)
	Andy Aftelak (Motorola, UK)
	Hans Nelissen (Vodafone, NL)
	Jae-Young Ahn (ETRI, Korea)
	Axel Steinhage (Infineon Technologies, Germany)
	Maria Farrugia (Vodafone Group R&D, UK)
	David Pollington (Vodafone Group R&D, UK)
Chapter 4 editors	Stefan Arbanowski (Fraunhofer FOKUS, Germany) and Wolfgang Kellerer (DoCoMo Euro-Labs, Germany)
Authors	Stefan Gessler (NEC Europe, Germany)
	Mika Kemettinen (Nokia, Finland)
	Michael Lipka (Siemens, Germany)
	Kimmo Raatikainen (High Intensity Interval Training (HIIT)/University of Helsinki/Nokia, Finland)
	Olaf Droegehorn (University of Kassel, Germany)
	Klaus David (University of Kassel, Germany)
	François Carrez (Alcatel CIT, France)
	Heikki Helin (TeliaSonera, Finland)
	Sasu Tarkoma (HIIT/University Helsinki, Finland)
	Herma Van Kranenburg (Telematica Instituut, The Netherlands)
	Roch Glitho (Ericsson Canada/Concordia University, Canada)
	Seppo Heikkinen (Elisa, Finland)
	Miquel Martin (NEC Europe, Germany)

Nicola Blefari Melazzi (Università degli Studi di Roma – Tor Vegata, Italy)
Jukka Salo, Vilho Räisänen (Nokia, Finland)
Stephan Steglich (TU Berlin, Germany)
Harold Teunissen (Lucent Technologies, Netherlands)

Chapter 5 editors Mario Hoffmann (Fraunhofer SIT), Christos Xenakis, Stauraleni Kontopoulou (University of Athens), Markus Eisenhauer (Fraunhofer FIT), Seppo Heikkinen (Elisa R&D), Antonio Pescape (University of Naples) and Hu Wang (Huawei)

Authors Mario Hoffmann (Fraunhofer SIT), Christos Xenakis, Stauraleni Kontopoulou (University of Athens), Markus Eisenhauer (Fraunhofer FIT), Seppo Heikkinen (Elisa R&D), Antonio Pescape (University of Naples) and Hu Wang (Huawei)

Chapter 6 editors David Falconer (Carleton University), Angeliki Alexiou (Lucent Technologies), Stefan Kaiser (DoCoMo Euro-Labs), Martin Haardt (Ilmenau University of Technology) and Tommi Jämsä (Elektrobit Testing Ltd)

Authors Reiner S. Thomä, Marko Milojevic (Technische Universität Ilmenau)
Bernard H. Fleury, Jørgen Bach Andersen, Patrick C. F. Eggers, Jesper Ø. Nielsen, István Z. Kovács (Aalborg University, Denmark)
Juha Ylitalo, Pekka Kyösti, Jukka-Pekka Nuutinen, Xiongwen Zhao (Elektrobit Testing, Finland)
Daniel Baum, Azadeh Ettefagh (ETH Zürich, Switzerland)
Moshe Ran (Holon Academic Institute of Technology, Israel)
Kimmo Kalliola, Terhi Rautiainen (Nokia Research Center, Finland)
Dean Kitchener (Nortel Networks, UK)
Mats Bengtsson, Per Zetterberg (Royal Institute of Technology – KTH, Sweden)
Marcos Katz (Samsung Electronics, Korea)
Matti Hämäläinen, Markku Juntti, Tadashi Matsumoto, Juha Ylitalo (University of Oulu/CWC, Finland)
Nicolai Czink (Vienna University of Technology, Austria)

Chapter 7 editors

Gerhard Fettweis, Ernesto Zimmermann (TU Dresden), Ben Allen (King's College London), Dominic C. O'Brien (University of Oxford) and P. Chevillat (IBM Research GmbH, Zurich Research Laboratory)

Authors

Volker Jungnickel, Eduard Jorswiek (Fraunhofer Institute for Telecommunications, Heinrich-Hertz Institut, Germany)

Peter Zillmann, Denis Petrovic, Katja Schwieger, Marcus Windisch (TU Dresden, Germany)

Olivier Seller, Pierre Siohan, Frédéric Lallemand, Guy Salingue, Jean-Benoît Pierrot (France Télécom R&D)

Merouane Debbah, David Gesbert, Raymond Knopp (Institut Eurécom, Sophia Antipolis, France)

Liesbet van der Perre (IMEC Belgium)

Tony Brown, David Zhang (University of Manchester, UK)

Wasim Malik, David Edwards, Christopher Stevens (University of Oxford, UK)

Laurent Ouvry, Dominique Morche (LETI, France)

Ian Oppermann, Ulrico Celentano (University of Oulu, Finland)

Marcos Katz (Samsung, Korea)

Peter Wang, Kari Kalliojarvi (Nokia Research Centre)

Shlomi Arnon (Ben-Gurion University of the Negev, Israel)

Mitsuji Matsumoto (Waseda, Japan)

Roger Green (University of Warwick, UK)

Svetla Jivkova (Bulgarian Academy of Sciences, Bulgaria)

P. Coronel, W. Schott, T. Zasowski (IBM Research GmbH, Zurich Research Laboratory, Switzerland)

Mohammad Ghavami, Ru He (King's College London, UK)

Werner Sorgel, Christiane Kuhnert, Werner Wiesbeck (IHE, Germany)

Marco Hernandez (Yokohama National University, Japan)

Kazimierz Siwiak (Time Derivative Inc, USA)

Ben Manny (Intel Corp, USA)

Lorenzo Mucchi, Simone Morosi (University of Florence, Italy)

M. Ran (Holon Academic Institute of Technology, Israel)

Chapter 8 editors Panagiotis Demestichas (University of Piraeus), George Dimitrakopoulos (University of Piraeus), Klaus Mößner (CCSR, University of Surrey), Terence Dodgson (Samsung Electronics) and Didier Bourse (Motorola Labs)

Authors Vera Stavroulaki, George Dimitrakopoulos Flora Malamateniou, Apostolos Katidiotis, Kostas Tsagkaris (University of Piraeus, Greece)

Karim El Khazen, David Grandblaise, Christophe Beaujean, Soodesh Buljore, Pierre Roux, Guillame Vivier (Motorola Labs, France)

Stephen Hope (France Télécom, UK)

Jörg Brakensiek, Dominik Lenz, Ulf Lucking, Bernd Steinke (Nokia, Germany)

Hamid Aghvami, Nikolas Olaziregi, Oliver Holland, Qi Fan (King's College London, UK)

Nancy Alonistioti, Fotis Foukalas, Kostas Kafounis, Alexandros Kaloxylos, Makis Stamatelatos, Zachos Boufidis (University of Athens, Greece)

Jim Hoffmeyer (Western Telecom Consultants, Inc, USA)

Markus Dillinger, Eiman Mohyeldin, Jijun Luo, Egon Schulz, Rainer Falk, Christian Menzel (Siemens, Germany)

Ramon Agusti, Oriol Sallent (Universitat Politechnica de Catalunya Barcelona, Spain)

Miguel Alvarez, Raquel Garcia – Perez, Luis M. Campoy, Jose Emilio Vila, W. Warzansky (Telefonica, Spain)

Byron Alex Bakaimis (Samsung Electronics, UK)

Lars Berlemann, Jelena Mirkovic (RWTH Aachen University, Germany)

Alexis Bisiaux (Mitsubishi, France)

Alexandre de Baynast, Michael C., Joseph, R. Cavallaro, Predrag Radosavljevic (Rice University, USA)

Antoine Delautre, JE. Goubard (Thales Land & Joint Systems, France)

Tim Farnham, Shi Zhong, Craig Dolwin (Toshiba, UK)

Peter Dornbush, Michael Fahrmair (University of Munich, Germany)

Stoytcho Gultchev (CCSR, University of Surrey, UK)

Mirsad Halimic, Shailen Patel (Panasonic, UK)

Syed Naveen (I2R, Singapore)

Jorg Vogler, Gerald Pfeiffer, Fernando Berzosa, Thomas Wiebke (Panasonic, Germany)

Christian Prehofer, Qing Wei (DoCoMo, Germany)

Jacques Pulou, Peter Stuckmann, Pascal Cordier, Delphine Lugara (FTRD, France)
Gianluca Ravasio (Siemens IT, Italy)

Chapter 9 editors Amardeo Sarma (NEC), Christian Bettstetter (DoCoMo) and Sudhir Dixit (Nokia)

Authors
G. Kunzmann, R. Schollmeier (TU Munich, Germany)
J. Nielsen (Ericsson, Sweden)
P. Santi (IIT/CNR, Italy)
R. Schmitz, M. Stiemerling, D. Westhoff (NEC, Germany)
A. Timm-Giel (University Bremen, Germany)

Foreword by Nim Cheung

Nim CHEUNG
President, IEEE Communications Society

On behalf of the IEEE Communications Society (ComSoc), I would like to congratulate the Wireless World Research Forum (WWRF) for publishing its 3rd book titled 'Technologies for the Wireless Future – Volume 2'. This book contains the up-to-date collection of WWRF's white papers developed during the past 12 to 18 months. The materials cover contributions by hundreds of experts around the world from virtually every sector of the communications and information industry. The book includes an in-depth coverage of technology topics such as ultra wide band, air interfaces, advanced modulation and antenna systems, and multidimensional radio channel measurement and modeling. This is accompanied by an equally rich set of systems chapters on architecture, user interface, network and service management, and emerging areas such as self-organization, cognitive radio, and spectrum management. I am glad to see a chapter on security, trust, malicious code, and identity management. This chapter is particularly timely to address vulnerability issues for the open air interface. I also enjoyed reading the survey of international and regional initiatives in B3G and 4G activities, as well as the projected landscape of the wireless world circa 2005, 2010, and 2017. While it is almost impossible to predict what will happen 12 years from today, I am intrigued by the statement '7 trillion wireless devices serving 7 billion people by 2017' – the designated key technological vision of WWRF. Overall, I believe this book will serve as a useful reference for both experienced researchers and professionals new in the field.

The IEEE Communications Society has been partnering with WWRF in a wide range of publication and conference activities over the past several years. ComSoc's former president Roberto de Marca and WWRF's Rahim Tafazolli and Mikko Uusitalo coedited a special issue in the IEEE Communications Magazine entitled 'WWRF Visions and Research Challenges for Future Wireless World' in September 2004. WWRF was a technical cosponsor of ComSoc's highly successful Symposium on New Frontiers in Dynamic Spectrum Access Networks (IEEE DySPAN 2005) at Baltimore, Maryland in November 2005. I look forward to continuing the collaborations between our two organizations in the dissemination of research and development results in advanced wireless technology. Such collaborations will undoubtedly accelerate the realization of WWRF's goal that 'future systems should be developed mainly from the user perspective with respect to potential services and applications'.

Foreword by Xiao-Hu You

Xiao-Hu YOU
Professor, Southeast University
Chair, China 863 FuTURE Project

It is my great honor to write this foreword for the new 'Book of Visions' of the WWRF.

As this book was being written, it was right at the time that the B3G R&D had arrived at an important stage. In October 2005, the ITU gave 'Beyond 3G' the formal nomination as 'IMT Advanced'. In 2007, WRC'07 will allocate the frequency spectrum for the Beyond 3G. Led by these milestones, countries around the world are working hard toward the goal of creating an even more splendid future for wireless communications.

This book encompasses almost every aspect of the future development of mobile communication technologies, such as the future vision of wireless communications, service infrastructures, user interface and demand, information security and trust, and air interface, ad hoc networks and ultra wide band communications. It may be viewed as a concentration of the research achievements of the WWRF in the Beyond 3G wireless communication area in recent years. I am fully convinced that these achievements will exert active and profound influences on the global development of the B3G technologies.

To my great pleasure, the first chapter of the book introduces some advances in the B3G R&D in China. As part of the global B3G R&D, China launched its B3G research program, the National 863 FuTURE Project, in 2001. While making significant progresses in the B3G R&D, we also established a close cooperative relationship with the 6th Framework Programme of the European Union. A group of Chinese research institutions had entered this program as collaborative companions. The two sides also held two joint workshops in Beijing and Shanghai, respectively. These events enhanced the mutual understanding and promoted further cooperation, between China and the other countries in the world, in the area of future mobile communications.

In October 2005, China Future Mobile Communications Forum was declared to be founded in Shanghai. As the cooperative counterpart unit of the WWRF, the FuTURE Forum is ready for a further clasp with the WWRF to make greater contributions to the development and the standardization of the global B3G wireless communication technologies.

Foreword by Xiao-Hu Yu

Preface

One of the important factors contributing to the success of mobile communications, from the very beginning, has been a good understanding of users' requirements. In its first generation, mobile networks offered users the freedom to communicate irrespective of their locations and the joy of instant access to telephony services. This provided a competitive edge over the fixed wire line communication networks. Continual assessment of users' requirements and the important role that mobile communications play in our daily lives brought about the 2nd and 3rd generations of standards. However, because of the widespread availability of the Internet and high-speed data on mobile networks, the users are becoming more accustomed to the use of such services in carrying out both private and professional aspects of their lives. The users' expectations from mobile networks are becoming more demanding and this trend is expected to intensify in the future. To continue with the successful approach adopted in mobile/wireless communications, a better understanding of the users' needs and requirements is essential to identify appropriate future technologies. However, as it is very difficult to accurately predict how such needs would evolve, the envisaged future systems must possess sufficient flexibility to be able to handle efficiently unforeseen needs, new services, and applications.

This is the very approach taken in the writing of this book, starting with the identification of users' future requirements based on today's observation on the usage of mobile communications and extrapolations of that to the next 10 to 15 years. The book structure also reflects that of the WWRF Working Groups and presents the contents of some of the White Papers produced in the last 18 months. The White Papers are the outcome of a large number of presentations, discussions, and contributions made by international researchers representing both industry and academia who are active in advanced research on future mobile/wireless communications. This is a follow-on book to Volume-1 published in 2004, but with more emphasis on the user requirements and an in-depth analysis of the important technologies.

The book starts with an introduction to worldwide and regional activities on Beyond 3G and 4G and gives a comprehensive overview of the WWRF organizational structure and mode of operation. Chapter 2 presents the WWRF-developed vision and requirements in the future wireless world based on observations made today and their extrapolation to years 2010 and 2017. In Chapter 3, a number of usage scenarios are identified and discussions are presented on how these scenarios should be used in the future development of wireless technologies and in particular, the problems and research issues associated with the advanced user interface in future mobile devices. Chapters 4, 5, 6, 7, 8 and 9 are more technology-oriented chapters addressing various subsystems of a mobile network. In particular, Chapters 4 and 5 cover important areas of future service platforms, generic service elements and enabling technologies, user security and trust requirements, and identify important research issues associated with them as well as challenges in user identity management in future heterogeneous networks. Chapter 6 provides a comprehensive overview of new air-interface technologies and new deployment concepts for wide area networks together with related technologies in achieving high spectral and power efficiencies using advanced technologies such as MIMO, cross-layer optimization schemes, and research issues for resource management. Chapter 7 addresses technologies and research issues for short-range wireless communications and provides a thorough overview of ultra wide band technology and its future perspectives. This chapter also provides a comprehensive overview of requirements and technical challenges encountered in optical wireless communications as well as wireless sensor networking, and identifies a number of interesting research topics. Chapter 8 presents an in-depth discussion on reconfigurability covering aspects such as application scenarios for reconfigurability, essential element management, and Software-defined Radio together with reconfigurable network architecture and support services, and the use of Cognitive Radio for spectrum management. Chapter 9 outlines issues concerning self-organization in communication networks, in particular for Ad hoc and Sensor Networks, and provides an interesting discussion on the potential limitation of self-organization.

Acknowledgements

I would like to thank and acknowledge the contributions of all the researchers and experts from all over the world who passionately and openly shared their ideas in WWRF meetings and influenced the contents of this book. The number is too large to be practically listed here.

The book would not have come to existence without the hard work and continuous encouragement of the WWRF Steering Board members.

I would like to acknowledge the hard work of the Working Groups and Special Interest Groups Chairs in driving and directing the vision within their groups and assisting me in editing the chapters of this book. They are listed below:

Dr Mikko Uusitalo – Nokia Group, Finland – Chair of WWRF
Professor Angela Sasse – University College London, UK – Chair of Working Group 1
Dr Stefan Arbanowski – FOKUS, Germany – Chair of Working Group 2
Professor David Falconer – Carleton University, Canada – Chair of Working Group 4
Professor Gerhard Fettweiss – University of Dresden, Germany – Chair of Working Group 5
Professor Panagiotis Demestichas – University of Piraeus, Greece – Chair of Working Group 6
Dr Mario Hoffmann, Fraunhofer SIT, Germany – Chair of Special Interest Group 2
Mr Amardeo Sarma, NEC – Europe, Germany – Chair of Special Interest Group 3

Professor Rahim Tafazolli
Centre for Communication Systems Research – CCSR
The University of Surrey, UK

1

Introduction

Edited by Mikko Uusitalo (Nokia)

Currently, there are approximately two billion mobile phone users worldwide. A rapidly growing fraction of them is using third generation mobile phones. There were about 50 million 3G subscribers in the beginning of August 2005. Research on what will come beyond the third generation has been active since the change of the millennium.

Wireless World Research Forum (WWRF) [1] was established in 2001 to facilitate the road towards the generation beyond the third. By joining forces early, the community could better understand what is relevant in research and thereby reduce risks in research investments. This should ease future standardization. Interactions in WWRF have also created research cooperation of a magnitude bigger than 100 million euros. This is not formally linked to WWRF. There are also many independent ongoing initiatives. All these activities have contributed to starting a discussion on 3G long-term evolution in 3GPP [2] since the end of 2004 and in 3GPP2 [3] since the beginning of 2005.

However, over the years it has become evident that the future is much more complex than a new cellular radio and the related infrastructure. There has been great focus on user perspective and technologies to make the life of the user simpler with better quality of life. There has been substantial innovation in radio lately. In 2003, WWRF Working Group WG on short-range radio was established to reflect this. There is also a major trend toward the convergence of digital industries. In future, the telecommunications industry will be joined by information technology (IT), consumer electronics, broadcasting and media, and entertainment industries to form a common digital industry. Content will be in digital form and usable across the media and industries. The same generalization will apply to services and applications. The different industries have different modes of operation and their convergence will change these. One example of converging industries is the emergence of the Institute of Electrical and Electronics Engineers (IEEE) 802.XX [4] standards from the communication needs of IT.

On the basis of the experience of the third generation, future systems should be developed mainly from the user perspective with respect to potential services and applications

Technologies for the Wireless Future – Volume 2 Edited by Rahim Tafazolli

including traffic demands. Therefore, the WWRF was launched in 2001 as a global and open initiative of manufacturers, network operators, Small and Medium-size Enterprises (SMEs), R&D centres and academic institutions. The WWRF is focused on the vision of such systems – the Wireless World – and potential key technologies. In the past, it took some 10 years for each generation to come to the market. The future will show whether this pattern will be repeated or whether there will be gradual evolution instead of a clear new generation.

One of the main drivers for the WWRF vision is the introduction of new services and the transformation of the network usage model. I-centric services, adjustable to a vast range of user profiles and needs, along with seamless connectivity anywhere, anytime are considered crucial to the vision of the emerging Wireless World. In addition, the cost/benefit ratio will make such services affordable and less expensive than any alternative or traditional solutions. Flexibility, adaptability, reusability, innovative user interfaces, and attractive business models, will be the key to the success of the systems envisioned for deployment beyond the year 2010. This book outlines this vision of the future considering the environmental, contextual, and technical aspects.

The first version of the Book of Visions [5] was published in 2001 (commonly referred to as 'The Book of Visions 2001') and was an early attempt to document the long-term vision of the Wireless World that each one of us will live in. The second version, Technologies for the Wireless Future, volume 1 [6], is complementary to this book. Since 2001, the WWRF WGs have received a constant flow of contributions. Meetings have been unique opportunities for both industry and academia to harmonize their views on the aroused topics. Some of these topics have mobilized several contributors to write common white papers together. This new Book of Visions is the up-to-date collection of those white papers and added complementary material. Hundreds of researchers around the world from every sector of the information and communications industry have contributed and constructed the Wireless World ahead of us, and this book summarizes their collective wisdom.

The Forum has already played the initiator role in the establishment of the Wireless World Initiative (WWI) [7], a European research project, as an initial step towards the basis of future standardization of the beyond 3G systems. WWI involves some 100 organizations under an umbrella of research activities of the order of 100 million euros. Many other smaller research co-operations have been influenced through interactions with WWRF. Right from the beginning the activities and participation in WWRF have been global, especially as the non-European members in WWRF have advanced similar cooperatives in other regions.

The work of WWRF continues at an ever enlarging volume. It is a unique process harmonizing views from both industry and academia. Through the harmonization of their views the participants can join and focus their resources in research and consequently reduce risk for investments in future system development. The WWRF membership is open to all interested in research on and beyond 3G.

1.1 Goals and Objectives – Shaping the Global Wireless Future

WWRF aims to develop a common global vision for future wireless to drive research and standardization. Major items on the road towards this aim are:

- influencing decision makers' views of the wireless world;

- enabling powerful R&D collaborations;
- advancing wireless frontiers to serve our customers.

The major objectives of the WWRF are:

- to develop and maintain a consistent vision of the Wireless World;
- to generate, identify and promote research areas and technical and society trends for mobile and wireless systems towards the Wireless World;
- to identify and assess the potential of new technologies and trends for the Wireless World;
- to contribute to the definition of international and national research programmes;
- to simplify future standardization processes by harmonization and dissemination of views;
- to inform a wider audience about research activities that are focused on the Wireless World.

The other objectives are:

- to contribute to the development of a common and comprehensive vision for the Wireless World and to concentrate on the definition of research relevant to the future of mobile and wireless communications, including pre-regulatory impact assessments;
- to invite worldwide participation and be open to all actors;
- to disseminate and communicate Wireless World concepts;
- to provide a platform for the presentation of research results.

The WWRF supports the 3GPP, 3GPP2, European Telecommunications Standards Institute (ETSI), Internet Engineering Task Force (IETF), International Telecommunication Union (ITU), Universal Mobile Telecommunications Systems (UMTS) Forum and other relevant bodies relating to commercial and standardization issues derived from the research work. However, the WWRF is not a standardization body. Liaison agreements are established with the UMTS Forum (January 2003), Mobile IT Forum (mITF) (May 2003), IEEE Communication Society (October 2003), SDR Forum (December 2004), and the Next-Generation Mobile Communications (NGMC) Forum (May 2005) [8].The WWRF has also a non-formal liaison via individual WWRF steering board members. Such organizations are 3GPP, 3GPP2, CDMA Development Group (CDG), Defence Advanced Research Projects Agency DARPA, Future Forum of China, and the European Technology Platforms eMobility as well as Advanced Research and Development on Embedded Intelligent Systems (ARTEMIS) [9].

1.2 Structure of WWRF

The WWRF is open to all those sharing the same objectives. There are three kinds of memberships. The normal membership type is a full member. Sponsor members pay higher membership fee and can be members of the Forum Steering Board. Then there are five founding members: Alcatel, Ericsson, Motorola, Nokia, and Siemens. In addition to the founding members, the sponsor members in October 2005 were Bell Mobility, Eurescom, France Telecom, Huawei, IBM, Intel, LG Electronics, Lucent, NEC, Nortel Networks, Raytheon, Samsung, and Vodafone. All members are represented in the General

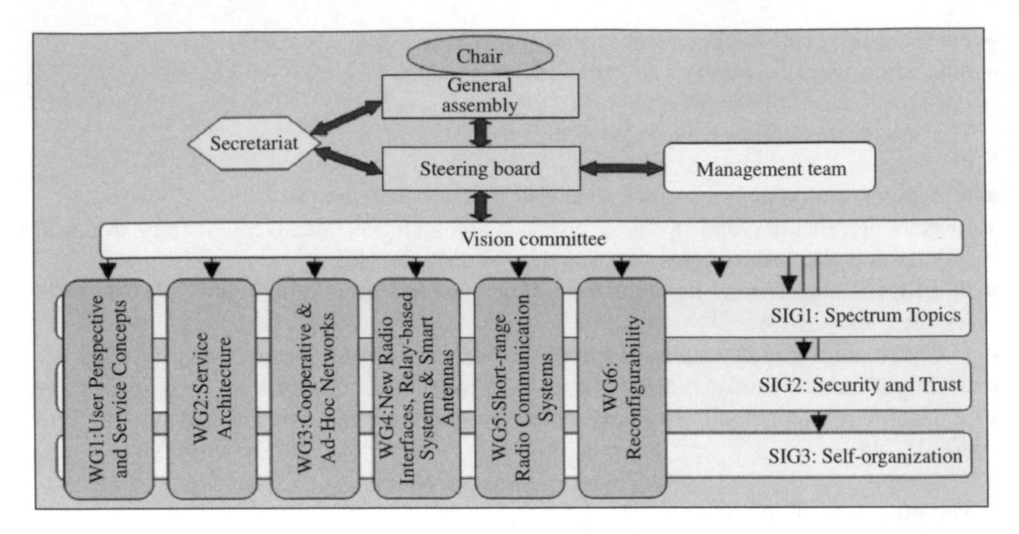

Figure 1.1 Structure of WWRF

Assembly, meeting usually in conjunction with WWRF plenary meetings. The General Assembly is the highest decision making body and it elects WWRF executives: chair, vice chair Americas, Asia Pacific, Europe Middle East, Africa, and the treasurer. The topics are divided between focused WGs. Also, there are special interest groups that cover issues that cross more than one WG subject area, as shown in Figure 1.1. All the groups select their chair and vice chair through voting. The Steering Board operates the forum based on the WWRF objectives and follows decisions made by the General Assembly. The Steering Board consists of WWRF executives, WG and SIG chairs, five representatives of the founding members and 10 representatives from the sponsor members. The Vision Committee coordinates the content work of the WWRF and links the work at WGs and SIGs together. More information on the structure of WWRF can be found from the website [1].

1.3 The International Context and B3G/4G Activities

The WWRF is a consortium of many leading mobile telecommunications companies, research organizations and universities [1]. The goal of WWRF is to ensure momentum, strategic orientation and the impact of the research on wireless communication beyond 3G. WWRF emphasizes that systems in the Wireless World should be the integration of 2G, IMT-2000, their enhancements, and other systems such as wireless local area networks (WLAN) type systems, short-range connectivity, broadcast systems and new radio interface concepts on a common IP-based platform. Continuous connectivity between end-users and a variety of services are considered to play a vital role in enhancing the wireless technology. Other issues, such as short-range spontaneous networking, system coexistence and interworking will also help to bring up the revolution in the wireless technology arena.

There are also many initiatives other than WWRF working on B3G and 4G. The following section will provide further information on these. In addition to the listed major activities and non-mentioned companies involved in these, there are many SMEs and other smaller players trying to bring their disruptive solutions to the markets.

1.3.1 International Initiatives

IEEE

The IEEE [4] is a non-profit, technical, and professional association of more than 365,000 individual members from approximately 150 countries. The full name is the Institute of Electrical and Electronics Engineers, Inc.

Through its members, the IEEE is a leading authority in technical areas ranging from computer engineering, biomedical technology, and telecommunications, to electric power, aerospace engineering, and consumer electronics, amongst others.

Through its technical publishing, conferences and consensus-based standards activities, the IEEE:

- produces 30 % of the world's published literature in electrical and electronics engineering, and computer science areas;
- holds annually more than 300 major conferences;
- has nearly 900 active standards with almost 500 more under development.

The IEEE 802 LAN/MAN Standards Committee (LMSC) develops standards for Local and Metropolitan Area Networks MAN (LAN, MAN) and short-range communications, mainly for the lowest two layers of the Reference Model for Open Systems Interconnection (OSI) [4]. LMSC coordinates with other national and international standards groups. There are several IEEE 802 standards that will play an integral part in the future wireless world.

B3G/4G at the International Telecommunication Union (ITU)

The International Telecommunication Union-Radio (ITU-R) ITU-Radio Sector, Working Party 8F (ITU-R WP8F) and ITU-Telecommunication Sector, Special Study Group (ITU-T SSG) have undertaken high-level vision work to help facilitate global consensus on basic system concepts [10]. These activities are a prerequisite for the World Radio Conferences (WRC) (in particular WRC '07) to discuss and to identify new spectrums for future systems. Its visions and recommendations for technical realization are built on expected user requirements on future mobile telecommunication systems. The ITU-R concentrates on the technical aspect of International Mobile Telecommunications (IMT) 2000 and systems beyond, and is addressing future market and the services aspects, spectrum needs beyond 2010, and high-level goals for the new radio interface(s). The vision stresses seamless service provisioning across a multitude of wireless systems as being an important feature of future generation systems. In general, the views of ITU-R and WWRF are aligned to a large extent, such as the research goals of 100 Mb/s high mobility and 1 Gb/s low mobility. The ITU-R Assembly approved a related recommendation in the beginning of June 2003 on the future framework for IMT-2000 and Systems Beyond (the vision), which serves as an important foundation in developing global consensus and for the forthcoming process of spectrum identification at WRC '07. One of the main activities in the future will be to prepare for the WRC '07 and facilitate global studies on the future market, services, and spectrum needs beyond 2010.

SDR Forum

The Software Define Radio (SDR) Forum is an international, non-profit organization dedicated to promoting the development, deployment and use of SDR technology for advanced wireless systems [11]. The mission of the SDR Forum is to accelerate the development and deployment of SDR technologies in wireless networks to support the need of civil, commercial and military market sectors.

The Technical, Regulatory, and Markets Committees in the Forum, and their WGs:

- develop requirements and/or standards for SDR technologies;
- cooperatively address the global regulatory environment;
- provide a common ground to codify global developments;
- serve as an industry meeting place.

The SDR Forum meets five times a year in major global regions. Typically, meetings involve plenary sessions to address the forum's business aspects, management of work and approval of documents, while industry leaders and key analysts are occasionally invited to speak on SDR topics. Moreover, e-mail reflectors and conference calls are held throughout the year. Work contributions are solicited from both members and non-members. Additionally, the SDR Forum hosts a yearly international technical conference and product exposition on technologies, standards, and business activities related to the forum's interests. Furthermore, the forum periodically schedules seminars and workshops for policy makers, while surveys are also conducted to examine the progress of SDR technologies.

1.3.2 Regional Initiatives

Department of Defence (DoD) Defence Advanced Research Projects Agency (DARPA)

Government institutions, such as the Department of Defence (DoD) Defence Advanced Research Projects Agency (DARPA) in US, have launched research initiatives for advancing the Future Wireless World [12]. DARPA is working to ensure that U.S. forces will have secure, assured, high-data-rate, multi-subscriber, multi-purpose (e.g. manoeuvre, logistics and intelligence) networks for future forces. These networks must be as reliable, available, and survivable as the platforms they connect. They must distribute huge amounts of data quickly and precisely across a wide area. They must form themselves without using or building a fixed infrastructure. They must withstand attempts by adversaries to destroy, disrupt or listen in on them. Towards this end, DARPA is conducting research in areas that include mobile ad hoc self-forming networks; information assurance and security; spectrum management; heterogeneous networks; and anti-jam and low probability of detection/intercept communications. The DARPA programme will develop technology that is useful not only for the military, but also for civilian use.

European Framework Programme for Research (FP6)

European Framework Programmes for research (FP6) [7] are platforms for collaborative research in Europe. The projects in the currently ongoing Framework Programme FP6 are planned to end in 2007. The FP7 has been scheduled to start in late 2006. The part of FPs that is relevant to WWRF is called *IST* (Information Society Technologies). The planned contents for IST in FP7 include:

Information and Communication Technology (ICT) Pillars:

- Nano-electronics, photonics, and integrated micro/nano-systems
- Ubiquitous and unlimited capacity communication networks
- Embedded systems, computing, and control
- Software, grids, security, and dependability
- Knowledge, cognitive, and learning systems
- Simulation, visualization, interaction, and mixed realities.

Integration of Technologies:

- Personal environments
- Home environments
- Robotic systems
- Intelligence infrastructures.

Applications Research:

- ICT meeting societal challenges
- ICT for content, creativity and personal development
- ICT supporting businesses and industry
- ICT for trust and confidence.

European Technology Platforms

European Technology Platforms are collecting together major industrial and other players around important technology areas to formulate common strategic research agendas and then advance those, in preparation for the next Framework Programme for Research, FP7 [7].

From the perspective of the WWRF, the two most important European Technology Platforms are eMobility and ARTEMIS [9].

The Mobile and Wireless Communications and Technology Platform eMobility [13] is reinforcing Europe's leadership in mobile and wireless communications and services and to master the future development of this technology, so that it best serves Europe's citizens and the European economy.

ARTEMIS will pursue a common industrial vision and a strategic agenda to implement its vision on embedded technologies. On the basis of this there is a set-up of a coordination and integration framework where industry, research organizations, public authorities, financial institutions, and other stakeholders across the EU join forces and coordinate their actions for implementing the strategic agenda.

Future Forum of China

The Future Technology for Universal Radio Environment FuTURE Mobile Communication Forum FuTURE Forum of China [14] is a non-government, non-profit organization which was jointly founded by operators, equipment providers, universities and research institutions both at home and abroad. It is dedicated to promoting exchanges and cooperation on B3G research and harmonizing different views on the vision, demands and trends of the next generation of mobile communication systems. Following the principle of openness, cooperation and interaction, the Forum is open to all those involved in the wireless industry.

FuTURE project in the China Communications Standards Association(CCSA)
The FuTURE project in the China Communications Standards Association CCSA [15] is an important part of the wireless communication branch of communication subject of China's 863 Programme. It aims at meeting the trends and needs in the field of wireless telecommunications in the next 10 years. It supports research on key technologies for air interface of beyond 3G/4G mobile communication system, and verification through demonstration of the key technologies. The other objectives of the FuTURE programme are to improve China's overall research capabilities in mobile communications and to enhance its international competitiveness during the standardization process of the future beyond 3G/4G wireless communication systems.

Mobile IT Forum (mITF) of Japan
The Mobile IT Forum mITF [16] was created to realize Future Mobile Communication Systems and Services such as the fourth-generation mobile communications systems and mobile commerce services. Among the forum's activities are; studies on standardization, conducting coordination with other related bodies, collecting information and carrying out promotional and educational activities, thereby contributing to an efficient utilization of the radio spectrum.

National Science Foundation(NSF)
The National Science Foundation (NSF) of the United States of America, [17], is a major funding source for academics. The NSF's Computer & Information Science & Engineering (CISE) Directorate funds research in topics related to B3G systems via three of its units: Computing and Communication Foundations (CCF), Computer and Network Systems (CNS), and Information and Intelligent Systems (IIS). The future wireless world research areas where NSF has shown interest include:

- Control, Networks & Computational Intelligence
- Electronics, Photonics & Device Technologies
- Integrative, Hybrid & Complex Systems
- Computing Research Infrastructure
- Computer Systems Research
- Cyber Trust
- Dynamic Data Driven Applications Systems
- Emerging Models and Technologies for Computation
- IIS
- Next Generation Cybertools

NGMC of Korea
The NGMCs Forum was created to realize the fourth-generation mobile communication systems and services.

The forum has three subcommittees: Market & Service Subcommittee, System & Technology Subcommittee and the Spectrum Ad-hoc Group. Each subcommittee consists of participants from industry, academia, and research institutes. The main activities include executing research and harmonization of opinions through mutual cooperation on NGMC.

IPv6 Forum

A worldwide consortium of leading Internet vendors, Research & Education Networks, make up the IPv6 FORUM [18]. Its mission is to promote IPv6 by improving the market and user awareness of IPv6 benefits and creating a quality and secure Next Generation Internet.

The IETF has sole authority for IPv6 protocol standards. The IPv6 Forum reserves the right to develop IPv6 Deployment Guides to foster the operational use of IPv6. The IPv6 Forum is a non-profit organization registered in Luxembourg since July 17, 1999.

1.3.3 Standardization Initiatives

3GPPx (Third Generation Partnership Project)

Third Generation Partnership Project (3GPP) is a collaboration agreement that was established in December 1998, bringing together a number of telecommunications organization partners; the current organizational Partners are ARIB, CCSA, ETSI, ATIS, TTA, and TTC [2].

The 3GPP system is a combination of the new wide-band Code Division Multiple-Access (CDMA) air interface and related radio access network (RAN) architecture, and evolved Global System for Mobile Communications (GSM) and General Packet Radio Service (GPRS) core networks. 3GPP started long-term evolution (LTE) work in late 2004.

On the other hand, 3GPP2 [3] was established in January 1999 by a joint group of TIA in the U.S., ARIB and TTC in Japan, CWTS in China, and TTA in Korea with a goal to set specific standards of ANSI-41 network and CDMA 2000 technologies. In 2005, 3GPP2 also started the work on LTE.

IETF

The Internet Engineering Task Force (IETF) [19] is a large, open international community of network designers, operators, vendors and researchers concerned with the evolution of the Internet architecture and the smooth operation of the Internet. The work being done at the IETF is relevant to the future wireless world, as Internet is becoming an integral part of mobile and wireless communications. The IETF operates somewhat differently from the standards organizations, with heavy emphasis on implementation aspects, as only the RFCs (Request for Comments) are produced. It is left to the market to adopt the RFCs. If the adoption is on a large scale, they become de facto industry standards.

Open Mobile Alliance (OMA)

The mission of the Open Mobile Alliance (OMA) [20] is to facilitate global user adoption of mobile data services. It specifies market driven mobile service enablers by ensuring service interoperability across devices, geographical boundaries, service providers, network operators, and different networks, while allowing businesses to compete through innovation and differentiation.

The OMA was formed in June 2002 by nearly 200 companies including the world's leading mobile operators, device and network suppliers, IT companies, and content and service providers. The full representation of the entire value chain in OMA signifies a change in the development of future specifications for mobile services. This replaces the traditional approach of organizing activities around 'technology silos', which worked independently in the past.

1.4 Acknowledgement

The following individuals contributed to this chapter: Sudhir Dixit (Nokia), George Dimitrakopoulos and Panagiotis Demestichas (University of Piraeus), Sung Y. Kim and Byung K. Yi (LGE), Larry Swanson (Intel).

References

[1] WWRF–Wireless World Research Forum, http://www.wireless-world-research.org/, (Accessed 2005).
[2] 3GPP, http://www.3GPP.org, (Accessed 2005).
[3] 3GPP2, http://www.3GPP2.org, (Accessed 2005).
[4] IEEE 802, http://www.ieee802.org, (Accessed 2005).
[5] WWRF, *The Book of Visions: Visions of the Wireless World*, Version 1.0, December 2001.
[6] WWRF, *Technologies for the Wireless Future,* Volume 1, Wiley, November 2004.
[7] FP7, http://www.cordis.lu/fp7/home.html, (Accessed 2005).
[8] NDMC Forum, *http://www.ngmcforum.org/ngmc2/eng_ver/eng_1.html*, (Accessed 2005).
[9] ARTEMIS, http://www.cordis.lu/ist/artemis/, (Accessed 2005).
[10] ITU, http://www.itu.int/imt, (Accessed 2005).
[11] SDRF, http://www.sdrforum.org/, (Accessed 2005).
[12] Department of Defense, www.DoD.mil, (Accessed 2005).
[13] eMobility, http://www.emobility.eu.org/, (Accessed 2005).
[14] FuTURE Forum of China, http://www.chinab3g.org/english/futureforum.htm, (Accessed 2005).
[15] FuTURE project, http://www.chinab3g.org/english/futureproject.htm, (Accessed 2005).
[16] mITF Forum, http://www.mitf.org/index_e.html, (Accessed 2005).
[17] National Science Foundation, http://www.nsf.gov/index.jsp. (Accessed 2005).
[18] IPv6 Forum, http://www.ipv6forum.com/, (Accessed 2005).
[19] IETF, http://www.ietf.org, (Accessed 2005).
[20] OMA, http://www.openmobilealliance.org/, (Accessed 2005).

2

Vision and Requirements of the Wireless World

Edited by Mikko Uusitalo (Nokia)

One important lesson learnt from the development of 3G standards was that potential future services, applications, and the expected user requirements should form the bases of future technical specifications and solutions. This approach is essential in ensuring easy market adoption and the economic success of future systems. However, as it is difficult to accurately predict the requirements of 10 years from now, upcoming systems should provide sufficient flexibility in order to incorporate the future requirements of operators and user expectations. In response to this challenge, in this chapter, the vision of the Wireless World Research Forum (WWRF) starts with observations made from the current trends and what will be on offer by 2010 and its projection to 2017 from both user and technology perspectives.

2.1 What we are Observing Today in 2005

Advance of mobile communication: By the third quarter of 2005, there will be more than 2 billion mobile phone users worldwide. This growth rate is expected to be on the rise in the future, considering the constant expanding of markets, such as China and India, as well as the replacement and even high penetration rates of mobile phones in other countries.

Broadband everywhere: Innovation in technology is bringing broadband connection at an increasing pace everywhere, to new fixed locations and mobile environments. The offered transmission rates are increasing while the price per bandwidth is falling steadily.

Advance of the Internet: Internet usage is becoming widespread in homes, offices, and particularly in mobile environments.

Technologies for the Wireless Future – Volume 2 Edited by Rahim Tafazolli

Convergence of digital industries: Currently we are witnessing a convergence among various industries. The best example is the emergence of IEEE 802.XX standards to satisfy the communication needs of information technology. This trend is expected to continue in the future as content will be in digital form and should be usable across various media and networks. This will require the consideration of new business models and a mode of operation incorporating industries such as consumer electronics, broadcasting, media, and entertainment, forming a new common digital industry.

Diversification of technologies for a given application: Owing to the convergence of digital industries, the same content can be provided via different channels. For example, TV is made available via the ADSL, DVB-T, DVB-H or S-DMB systems.

Limited services offer: Current telecommunication services are designed for specific end systems with their own style of presentation, providing weak integration at the service level and no user-initiated service creation support and flexibility.

Deregulation and globalization: Mobile communication is a global business and the ongoing deregulation brings new competitions and opportunities.

Advance of e- and m-commerce: The advances in mobile, Internet, and electronic industries herald new changes to the business processes and commercial transactions.

Heterogeneity and complexity for the user: The number of new technologies is increasing, offering complex features, with not sufficient consideration to the user's techno ability, resulting in user frustration and dissatisfaction.

Services and applications are the key: The end-user is technology agnostic and is mainly concerned with reliability, availability, security, services, and application contents.

2.2 What is on the Way for 2010?

Improved Customer Service

Universal accessibility: There will be a single interface to services and information to manage, for example, personal rights, authentication, and payment. This means physical omnipresence, access and management from anywhere to terminals and equipment.

Augmented reality: Technology will support services enhanced with basic human senses and automated human functions to facilitate more intuitive use of technologies. These services will include other human contexts such as user emotion, environment, location, and gesture. There will be virtual assistants instructed by vocal commands and with learning capabilities.

Proliferation of digital content: Any user could be a potential digital content creator or provider. Part of the content will also be created automatically, based on user context and location.

2.3 Projection for 2017

2.3.1 User Perspectives

- 'I have access to an increasing amount of content and services (artistic creation, purchasing of goods and services, local information, etc.)'.

- 'I have a single point and/or interface for many services/requests for a better customer experience'.
- 'I am known and recognized by different service operators or third parties'.
- 'I have access to and can manage my personal stuff (heating, car, video recorder, etc.)'.
- 'I control my services, my information, and my data (photos, etc.), depending on individual constraints and needs'.
- 'I get into the parking lot, pay for parking, pay for groceries – all with my mobile phone'.
- 'The hardware and services interfaces are customized and adjustable to my needs'.
- 'My terminals can be plugged into any network anywhere, and my services are optimized according to the location, terminal, and network capabilities'.

Human beings have limited communication spaces: 'I do not know everybody', 'I am not interested in everything, etc.'. In general, humans are interested in semantics and not in the kind of presentation of a specific communication system. Humans need telecommunication services to control, communicate and get informed, etc., as a prolongation of their human senses.

Users want access to their personal content, services, and applications anywhere, anytime.

Users want an 'extended communication world', 'always in touch with friends, family and community'.

The vision principles from the user perspective are as follows:

- Users are in control through intuitive interactions with applications, services and devices.
- Services and applications are personalized, ambient-aware and adaptive (I-centric) – ubiquitous from the point of view of the user.
- Seamless services are available to users, groups of users, communities, and machines (autonomously communicating devices) irrespective of place and network and with agreed quality of service.
- Users, application developers, service and content providers, network operators, and manufacturers can create new services and business models efficiently and flexibly, based on the component-based open architecture of the wireless world.
- There is awareness of, and access to, appropriate levels of reliability, security, and trustworthiness in the wireless world.

Users should be able to control and be controlled to the extent they want. Control and usage of services and applications need to be so intuitive that no user manuals are needed. The devices need to guide the user to the degree that is required.

Services and applications need to take into account who is using them, so they need to be personalized. They need to be available everywhere and surround the user ambiently to the degree needed by the user. There needs to be learning based on the behaviour of the user, so there is need for adaptivity. In summary, services and applications need to be I-centric.

In addition to users, groups of users as well as communities and machines need to be served. All need seamless service irrespective of place and network and with agreed quality of service.

All the stakeholders, including users, need to be able to create new services and business models, efficiently and flexibly. Therefore, the underlying architecture needs to be modular and open.

As a great amount of information about a user and its context will be in the common global network with interconnected devices, reliability, security, and trust will become issues of paramount importance to be considered in the design of future mobile systems.

2.3.2 Technological Perspectives

The key technological vision from WWRF is '7 trillion wireless devices serving 7 billion people by 2017'.

All the people should be served with wireless devices. The simplest forms of wireless devices have to be affordable and easily operable.

To fulfil the requirements from the user perspective, the variety of devices serving people will be substantially increased. New cars had more than 10 radios in 2005. Communication between machines will grow faster than communication between humans. Sensors and RFID tags will be added to increase the number of devices, and they will communicate wirelessly. Part of their role is to provide context sensitivity. Sensors are to be embedded in, for example, vehicles, transport systems, weather systems, and building infrastructure (furniture and lights for context sensitivity, doors and windows for security). With a tag and a sensor in a food package, one could read with a handset the origin and history of the food, which would improve food safety, and its ingredients. Sensors and tags will begin to be embedded within many objects, and smart ones will provide the user location and context through capturing the ambient intelligence.

All devices will be part of the mobile Internet seamlessly connected via Internet Protocol, enabling interworking and interoperability between heterogeneous networks with enhanced security and user privacy.

2.4 Acknowledgement

The following individuals contributed to this chapter: Mikko Uusitalo (Nokia).

3

User Requirements and Expectations

Edited by Angela Sasse (University College London, UK)

3.1 Introduction

This chapter reviews the current use of scenarios in the development of future wireless services and technologies. Scenarios are used by a variety of stakeholders involved in the process of designing and envisioning future wireless services and applications, and for a variety of purposes. The reviews find that most current scenarios are devised without empirical foundation or validation. While such scenarios can be a useful tool for the envisionment of, and communication about, future wireless technologies, there is a danger that they morph into specifications. To yield future visions that are technically sound and meet user requirements, they have to be placed in their appropriate role in the development cycle, and used in conjunction with other software engineering and user-centered design methods.

This chapter also provides a thorough review of the current state of User Interface (UI) technology for mobile devices. It describes the handicaps of the devices currently in the market and the emerging technologies to overcome these disadvantages. Additionally, it provides a list of recommendations for industry and research bodies to increase the usability of future mobile devices.

3.2 The Role of Scenarios in The Development of Future Wireless Technologies and Services

The use of scenarios in the conception, design, and implementation of future technologies is currently rather diverse – there are many different types of scenarios, devised by different ways, and serving many different purposes. The fact that scenarios are widely used in research and commercial development projects indicates that they are seen to

Technologies for the Wireless Future – Volume 2 Edited by Rahim Tafazolli

be beneficial. However, it is currently not proven that they actually deliver the perceived benefits. In the context of the WWRF, the expected benefit would be reliable requirements for future technologies, and an improved understanding of how people are likely to use them in the future.

The goals of this section are as follows:

- To review existing scenarios of future wireless technology and services, and assess their efficacy as a tool in the process of developing human-centered and successful wireless technology and services B3G.
- To provide recommendations for devising scenarios for future wireless technologies, and using them at different stages of the user-centered design process.
- To provide recommendations for future research into the development and application of scenarios for the design of future technologies.

This section is the successor of a previous WWRF Working Group (WG) 1 (whose theme is The Human Perspective of the Wireless World) White Paper 'The Use of Scenarios for the Wireless World' [1–3]. Since the original White Paper, many new scenarios have been generated, and some research on the effectiveness of scenarios as a design tool has been conducted. These developments prompted the need to review the scenarios generated, and assess their effectiveness as a tool for identifying, designing, and implementing future wireless technologies and services that are human-centered and commercially successful.

3.2.1 Background

In this section, we review the historical background of scenarios, and how they are currently used in different industries.

3.2.1.1 A Brief History of Scenarios

Scenario planning was originally developed as a strategic management tool in a military context – i.e. the first scenarios were war games. During and after World War II, it moved in to the civil domain. Herman Kahn (of the RAND corporation, later the Hudson Institute) played an important role in the further development of scenario analysis. Early scenario analysis used a predict-and-control approach and did not offer a fundamental advantage over other methods of forecasting. By the 1970s, scenario planning was widely adopted as a tool in the corporate world. Shell was one of the pioneering corporations using scenario planning [4]. Planning is normally based on predictable factors, but scenarios were initially introduced to plan without having to predict factors that were agreed to be unpredictable. The challenge is to separate predictable factors from uncertain ones. Predictable elements of a scenario are known as *pre-determineds*, and will be represented in all version of a scenario in the same way. Uncertain factors are varied to generate multiple, but equally plausible futures, which serve as a test-bed for policies and plans. For instance, the development of the economy is uncertain, and so is the rate at which a certain technology is adopted (Table 3.1).

The aim of a scenario exercise is to be prepared for any possible future, to have a plan for whichever combination of pre-determineds and uncertainties turns out to be the

Table 3.1 Uncertainties in scenario planning

Uncertainties	Assumptions
Economy	– Growth – Recession
Adoption of new technology	– Fast – Slow

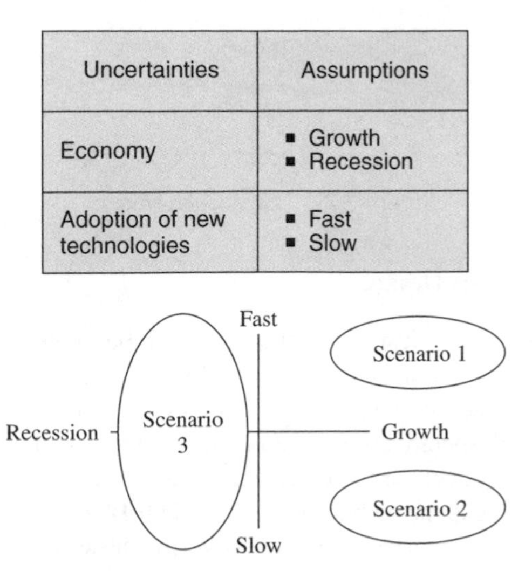

Figure 3.1 Scenarios covering possible futures

future. This is achieved by constructing scenarios that cover all possible assumptions, and considering the implications for the business, and option for response. The participants can then identify and select actions the company can take to be prepared for developments (see Figure 3.1).

The scenarios are the outcome of collective brainstorming and analysis sessions, and are best labelled as strategic planning scenarios.

A key requirement for a successful strategic planning scenario exercise is for participants to escape their existing mindset and to consider all possible futures, and not just the ones most participants consider as being likely (see Figure 3.2). An analysis of past developments helps to identify driving and inhibiting forces for the industry sector.

Some of the driving forces are pre-determineds, i.e. factors that have been important in the past, and are agreed to be likely to continue to be important in the future. Predetermineds form the basis of all scenarios. Other drivers are uncertain, but scenarios explore them systematically in a what-if exercise. The aim is not to generate a better forecast, but to develop plans for any possible future, no matter how unlikely it is deemed to be. As Arie de Geus (Shell) once put it: 'The only relevant questions about the future are those where we succeed in shifting the question from whether something will happen to what we would do if it did happen'.

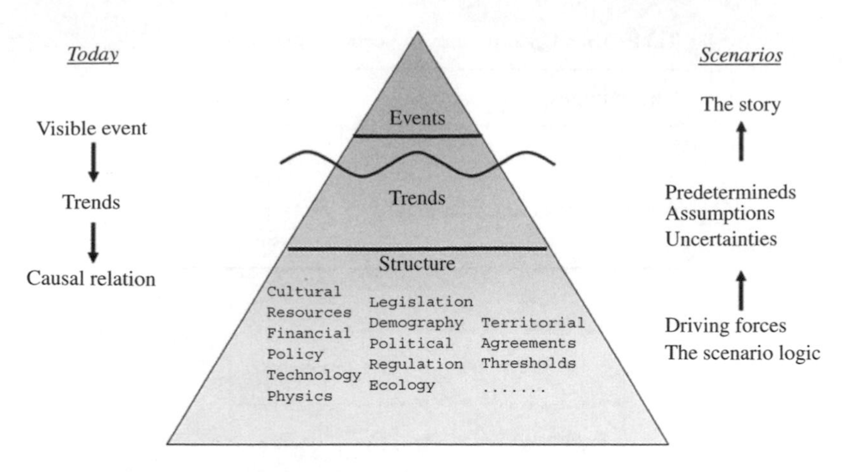

Figure 3.2 The 'iceberg model'

3.2.1.2 Usage Scenarios in Design

The use of scenarios in the design community can be traced back to the early 80s. Due to legal requirements in Scandinavia, employees had to be consulted before new technology was introduced in working environments. But many employees could not understand the technical specifications drawn up by system developers. In response to this problem, researchers developed a participatory design methodology to help employees envision how they would use the technology. The UTOPIA project [5] is one of the earliest documented cases. Usage scenarios – which describe how an employee would interact with the technology to complete tasks – enabled employees to check that the specified functionality was sufficient and correct, and work with system developers to design the screen layout and interactions in a manner that worked best for them. Envisionment can also be presented as storyboards (adapted from film and TV) that 'walk' users through their future interaction with a technology; such storyboards can be presented in paper-and-pencil format, videos, or as multimedia presentations [6] – presents an overview of techniques in use, and case studies of how they have been used to elicit, detail, and validate requirements.

3.2.1.3 Use Case Scenarios in Software Engineering

Use case scenarios in software engineering are used to describe a series of activities that might take place in a certain environment under certain circumstances. Whilst software engineers may use similar techniques for usability engineers, such as storyboards or elicitation workshops, their use case scenarios need to abstract from the views and needs of individual users to a formal notation (usually UML – Unified Modelling Language – see http://www.uml.org/), which models application structure, system behaviour, and architecture, and business process and data structure. The specification systematically de-composes the requirements to specify objects and methods through which the required functionality will be delivered, i.e. it goes beyond a statement of requirements to one that shows in great detail how the resulting technology will work [7]. It provides detailed guidance, case studies and examples on how requirements are elicited and turned into use case scenarios, and from there into designs.

3.2.2 Scenarios for Developing Future Wireless Technologies and Services

Scenarios are currently used to steer the direction of future wireless technology research in two ways. The first are options scenarios, which look at large societal and economic trends to broadly define user needs in terms of gross market movements, and therefore the types of systems, products, and technologies that will be needed in the future. An example of this type of analysis is the Wireless Strategic Initiative (WSI), funded by the European Commission, under the fifth framework program, which looked at the development of the mobile communications market [8]. This use of scenarios is similar to the strategic planning scenarios described in section 3.2.2.1.

The second is the use of scenarios as stories of the use of services, applications, products, and technologies. These scenarios vary from pure technology use cases to scenes of users using services and products in a relatively technology agnostic way. The purpose of these use case scenarios is to link the technologies and services that are being developed or researched to their use by people as a sanity-check that the technological developments are in some way valuable to people and therefore have some sort of market potential, and are not being developed purely out of scientific interest. This usage is now commonplace in the industry, and is well highlighted in many of the European funded sixth framework projects in the strategic objective on Mobile Systems beyond 3G, such as the four Wireless World Initiative (WWI) projects [9].

In order to have a positive effect on the development of technologies in terms of advancements that meet human needs, scenarios should be part of a user-centered design and development approach (which is depicted in Figure 3.3).

Use case scenarios in theory are best driven from some understanding of user needs and values. The WG 1 human-centered reference model [10] has identified a set of general needs and values that will play a role in the use of future wireless technologies, and give designers some areas of requirements to research. However, it is impossible to derive specific requirements for a new technology from such a general model of human needs: most technologies are aimed at people who go about their lives in a particular way, and

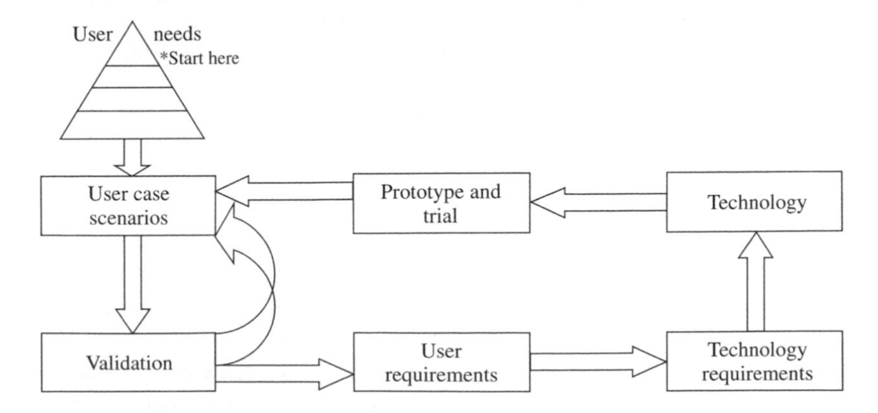

Figure 3.3 Scenarios in the user-centered design process (in theory)

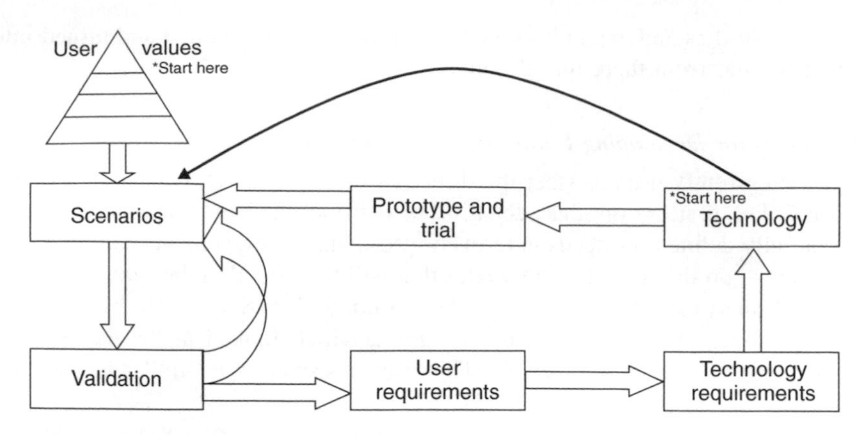

Figure 3.4 Scenarios in the design process (in practice)

are situated in a particular geographic, economic, social, and cultural context, all of which determine their specific, detailed requirements. In addition to the awareness of general requirements, designers thus need data from empirical studies of their target user groups, which make some assumptions on the capability of technologies and the domain of human activity served by the products and services. In practice, however, the nature of current wireless research projects does not allow for extensive empirical studies, and the reality of scenario development is as it is shown in Figure 3.4.

In the majority of cases, scenarios are derived either explicitly or implicitly (by the background of those taking part in the scenario development exercise) from the understanding of the technology and what the technology is capable of doing. In essence, the scenarios are borne from the imagination of the technology aware group in the scenario workshop. Although this seems to be contrary to user-centered principles, it is valid to have the ideas for scenarios come from any source. The key step to allow the model to remain valid is that of validation as shown in Figure 3.4. It is vital that the scenarios are tested against users in some way so that confidence is built before large investments are made in technology development and the building of costly prototype systems. If this step is missed, as shown in Figure 3.5, then the testing of technologies against requirements derived from scenarios becomes a self-fulfilling prophecy.

A scenario validation stage can be carried out by a variety of methods. The WWI carried out a set of stakeholder workshops [11] in focus groups to elicit feedback on a variety of scenarios, where the focus groups had representatives from various parts of the value change (users, operators, manufacturers, service providers). The EU framework 6 MobiLife project [12] arranged focus groups with 17 families in Finland and Italy to test application ideas [13]. Invariably, a study on this scale does not validate the scenarios, in the sense that the ideas are guaranteed value by a mass market, nor do they guarantee the acceptability or usefulness of a technology. They do, however, build enough confidence about the utility of the ideas and technologies to proceed to the next cycle of the user-centered design process [14].

The scenarios then form the basis of a set of user requirements for the project, which in turn are the basis of technology requirements. This step in the process presents some difficulties and the methods chosen depend on the type of technology research targeted by

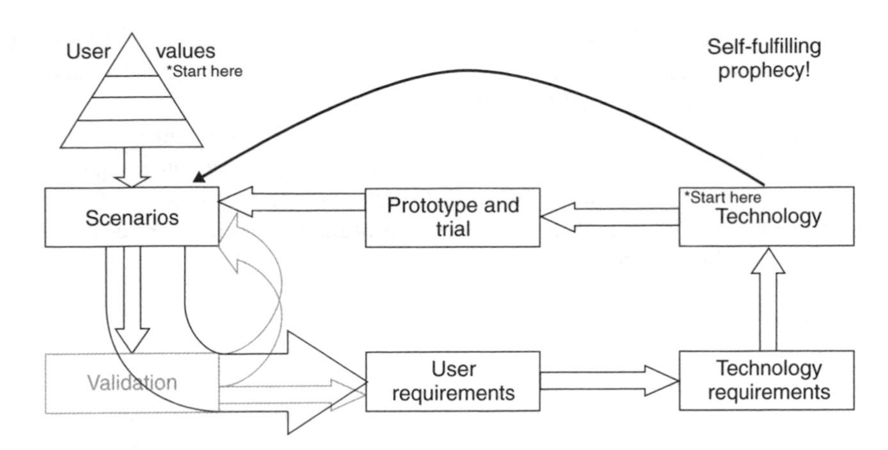

Figure 3.5 Scenarios in the design process (The self-fulfilling prophecy)

a project. In projects developing technologies for the connectivity of future mobile systems, such as the Information Society Technologies (IST) framework 6 projects E2R [15], WINNER [9] and Ambient Networks [16], the scenarios are broken down into technology use cases, and the implications on the technologies being studied identified. From the implications, a technical analysis is made to derive some technology requirements. For example, in a project researching new air interfaces, scenarios may include viewing a video. Assumptions are made about the quality of the video, and data rate requirements are easily calculated. In projects developing service and application technologies, such as the MobiLife project, [12] the user functionalities are identified and classified. These functionalities form a basis for user requirements. Separate technology use cases are developed which can be linked to the project scenarios illustrating the use of the researched technologies. These technology use cases then drive the direction of application development and research, and the linkage to the project scenarios allow feedback of the results of the scenario focus groups. Experience from these projects show that the key to success is two-fold. Firstly, it is important to have usability researchers working with technologists. Secondly, both technologists and usability researchers have to modify their view of requirements. From an engineering perspective, the user requirements from the user studies are not engineering requirements, since from a technical point of view, they are not complete, not sufficiently precise, and may not be consistent. The results of the user studies require in-depth analysis, interpretation and usually a step of application, service or system design before the technology requirements can be derived. From a usability perspective, the results of the user studies are usually a better understanding of how applications are received by potential customers, how they react to the proposed capabilities, and how they fit in their lives and activities. An extra step has to be taken to draw the implications for technologists to derive their requirements.

3.2.2.1 Review of Existing Scenarios for Future Wireless Technologies

Over the past 10 years, there have been many visions of the future of wireless communications using scenarios as a vehicle to communicate the vision. The previous WG 1

White Paper on scenarios [1] catalogued many of the major scenarios and analyzed them. The purpose of this section is to do the same for updated versions of these, and of new scenarios that have been developed in the last few years.

As discussed earlier, scenarios have been used in future mobile and wireless research in two ways. The first is to define a vision, and is usually based on scenarios options method. The second is the use of scenarios to describe stories of the use of new services, applications, products, and technologies as a prelude to technology development or illustrate the capabilities of technologies.

Visionary Scenarios

ISTAG Scenarios

The IST Advisory Group (ISTAG) sponsored the development of a set of scenarios in order to hasten the development of Information and Communication Technologies (ICT), and give European development efforts a heightened focus on future services. The scenarios themselves were developed in 2000 by the IPTS (part of the European Commission's Joint Research Centre), Directorate General (DG) Information Society and with the active involvement of 35 experts from across Europe [17]. The ISTAG scenarios describe life with what they call ambient intelligence. In this world, users are immersed in connectivity and intelligence, having full access to a range of services and applications anywhere, at any time. Four scenarios were developed.

Wireless Foresight

The Wireless Foresight scenarios are a scenario options analysis of the future wireless market [18]. It deviates from the classic four-quadrant approach (see section 3.2.1.1) representing the two most important trends, but identifies 14 trends that it then groups, to define a set of realistic but complementary options of wireless evolution. The report deals with the state of the wireless industry in 2015 and presents four scenarios: Wireless Explosion – Creative Destruction, Slow Motion, Rediscovering Harmony, and Big Moguls and Snoopy Governments. From the scenarios, important areas for technological research are identified. A number of critical challenges facing the industry are identified: the high cost for infrastructure, the slow spectrum release, the stampeding system complexity, radiation, battery capacity, and the threat of a disruptive market change.

Future Mobile Scenarios from Korea

The Korean government has encouraged its industry and universities to take a very visionary approach to next-generation systems.

A Strategic Plan of IT Development in Korea: 'IT839'

The Ministry of Information and Communication (MIC) of Korea has initiated a vision of technology development 'IT 839' [19], which includes project packages for developing leading information technologies in Korea, and released the standardization roadmap of it at the end of 2004. In 'IT 839' strategy, '8' stands for 8 different new information and broadcasting services, '3' refers to 3 advanced IT infrastructures to be built, and '9' represents 9 growth engines in the IT sector. Depending on the degree of standardization maturity and commercialization of 48 essential technologies, the government authorities will classify these technologies into several categories, considering its strategic positioning for international standardization. Based on this roadmap, standardization will be expedited to adopt domestically developed technologies as international standards, while supporting the development of source technologies driving industry innovation.

Software Infrastructure Will Be Added to IT839 Initiative

The MIC is considering the addition of the development of software infrastructures such as embedded software, streaming service and Web service to the IT839 plan, since the importance of software as one of key next-generation IT industries has been recognized. The issue is whether to change the strategy to IT849, adding software infrastructure to it, or to maintain the present arrangement of 'IT 839' by including software infrastructure while excluding IPv6. Because considerably large numbers of infrastructures of IPv6 have already been commercialized, the latter is more likely.

The Ministry is also looking for a new service that will replace VoIP as the eighth essential new service, while seeking ways of including parts and substance industry as one of 9 growth engines. All the revisions are scheduled to be unveiled early in 2006.

IT Mega-trend: IT Vision Project

The Ministry plans to unveil an outline of its information technology trend policies by the end of the year 2005 [20]. The Ministry aims to set the course of IT policy-making through research on technology's influence on future society and analysis on social phenomena. The goal of the future study is to foresee and gird for business trends by considering social, economic, and cultural changes. Given ever-changing environments in technology, business and society, the vision of industrialized society lying on the future technology and business models will be drawn. The MIC plans to unveil the 'IT vision for the following decade' as a part of its post-IT839 strategy. It aims to draw up a draft of the vision by referring to the study of twenty-first century mega trends and listening to leading figures including scientists. The concern is that there is no proved model of ubiquitous society, an IT drive aggressively pursued by the country. Without a future study model that duly understands the present systems, the commitment to research may end up with theories, failing to develop a strategic model tailored for the country.

NGMC 'Vision Book' for The Wireless Future

The Next-generation Mobile Communication (NGMC) Forum in Korea has developed a strategic development plan of wireless technology in 2004, and is now upgrading it as a 'vision book' containing a scenario-based long-term forecast and positioning of future mobile technology. Version 2 of it will be released in October 2005 [21].

Wireless World Research Initiative (WWRI)

The Wireless World Research Initiative (WWRI) project was an accompanying measure to Key Action IV Essential Technologies and Infrastructure, Action Line IST2002 – IV.5.1 'Towards technologies, systems and networks beyond 3G' for the EU funded FP5. The project ran from June 2002, to April 2003. It carried out a 3-quadrant scenario analysis based on a two-axis scenario options development method, with one of the four quadrants not developed [22]. The 'Blue' scenario is a world in 2010 where wireless is the dominant technology in connecting people and machines. Customers have demonstrated an increasing appetite for new wireless technologies and services and use these extensively in all aspects of their lives. The 'Red' scenario is a world in 2010 where customers are highly experimental, intent on finding and trying out the applications and services that meet their needs best. The great success of fixed broadband has resulted in people becoming very familiar with the open nature of the Internet and the ready access to a hugely diverse range of content and applications. Users know that not all the information is of good quality, but they also know that they can lay their hands on lots of content and applications without having to pay much, if any, money. The 'Green' scenario is a world in 2010 where customers primarily want to meet their basic personal communication needs, for example, voice communication. Users do not tend to be experimental and adopt only simple low-cost devices that enable them to keep in touch with their family, friends and colleagues. Financial constraints mean that vendors and operators have not been able to subsidise new devices or services to stimulate take-up while for established products and services, customers, supported by regulators, demand ever lower retail prices.

NTT DoCoMo 'Vision 2010' Scenarios
As part of a far-reaching corporate strategic plan, NTT DoCoMo have assembled an extensive video scenario set entitled Vision 2010 [23]. The scenarios, produced as a video, follow a Japanese family from the 2003 world of 3G systems to a 2010 world of the same sort of ubiquitous computing and communication that appear in the ISTAG [17] and MIT [24] scenarios.

3.2.2.2 Use Case Scenarios

WWI Scenarios
The WWI [9] is a coordinated set of large projects funded by the European Union under the Framework 6 Programme (FP6); E2R [15] develops end-to-end architectures for future mobile systems with an emphasis on adaptive radio systems, MobiLife [12] which develops mobile applications and application enablers, WINNER [25] which is researching future air interface technologies and Ambient Networks which is looking at future network technologies. The E2R, WINNER and Ambient Networks projects have developed essential technology transfer. The E2R scenarios take a bottom up approach to developing the scenarios. Reconfigurable equipment and systems will provide much higher flexibility, scalability, configurability and inter-operability than currently existing mobile communications systems. The end-to-end reconfigurability is a key enabler to support the heterogeneous and generalized wireless access, in a flexible and intelligent manner. In this context, three main system scenarios were elaborated, taking into account the input scenarios coming from the different Work Areas of The Project (WAP). The Ambient Networks project has taken a quasi-top down approach to developing its scenarios. The project started with a number of concepts that are believed will be central to the characteristics of future networks, looking at different types of networks and deployment scenarios. In the WINNER project, a methodology was chosen which makes use of the existing body of scenarios of possible relevance systems beyond 3G. Relevant components are identified from existing scenarios, which can be integrated to create new scenarios targeted for the WINNER project according to a set of motivations identified for users. Within MobiLife, high-level illustrative scenarios were developed by gathering existing scenarios of use from other projects (such as the other WWI projects [3] and foresight initiatives like FMS [26], the Finland TUTTI project [27] and the Wireless@KTH study [28]) as well as a number of scenarios coming from MobiLife technology work packages.

In addition, the WWI cross issue on user requirements gathered all the scenario work from the WWI projects, and analyzed them for the coverage of applications, technologies, and consistency, in preparation to develop a set of user requirements for the WWI. This exercise also resulted in a consolidated set and illustrative single scenario to illustrate the vision of the whole initiative.

Mobile IT Forum Scenarios

The Mobile IT Forum (mITF) is an industry consortium, sponsored by Japan's Association of Radio Industries and Businesses (ARIB). The Forum conducts studies and researches on technologies and standardization. The Forum's goal is to realize an early implementation of Future Mobile Communication Systems including Systems beyond IMT-2000 and mobile commerce. This work is also known as the *Flying Carpet document* [29], and is very well known in the industry. The mITF scenarios are excerpted from the Forum's Information and Communications Council and take the rather unconventional form of cartoons. There are ten scenarios, covering aspects of daily life, work, and medicine.

Project MESA Scenarios

Project MESA is a joint ETSI/TIA effort aimed at the future of public safety communications. MESA stands for Mobile Broadband for Emergency and Safety Applications. Project MESA began as an European Telecommunications Standards Institute (ETSI) and TIA agreement to work collaboratively by providing a forum in which the key players can contribute actively to the elaboration of MESA specifications for an advanced digital mobile broadband standard much beyond the scope of currently known technologies.

The Project MESA scenarios are the least user-centered of the scenarios reviewed here. However, they were included for breadth of coverage, being one of the only future-oriented treatments of public safety communications available.

MIT Project Oxygen Scenarios

Oxygen is a project housed in MIT's Laboratory for Computer Science (LCS). The project's slogan is: 'Bringing abundant computation and communication, as pervasive and free as air, naturally into people's lives'. Oxygen's vision [24] is one of pervasive, human-centered computing, similar to the ISTAG's ambient intelligence [17]. MIT's specific vision is one of ubiquitously available networks, dubbed 'N21's'. Users can avail themselves of services using 'H21's' the handheld devices in this future. In addition, computing devices, called 'E21's' are embedded in everyday objects and communicate across the N21 networks.

3.2.3 How Scenarios Should Be Used in The Development of Future Wireless Technologies

3.2.3.1 A Scenario Framework

As indicated earlier, scenarios are used in a variety of ways by different interest groups: designers, engineers, policy makers, futurologists, strategy makers, and user researchers.

When developing scenarios, it is important to understand who is using the scenario, their needs, and the limitations of particular methods. Carroll [30] points out that 'scenarios have many views' and describes designs 'with respect to many multiple perspectives' and at different levels of resolution. It is therefore important in creating scenarios to use a framework that allows consistence exploration and the extension of scenarios to suit different members of a team.

As an example of framework development, the FLOWS project of FP5 (Flexible Convergence of Wireless Standards and Services) had a remit to create socio-technical scenarios, combining a range of inputs, and suggesting a number of scenarios and use cases that could be used by different groups within the project. [31, 32] There were three primary subgroups: those concerned with local radio environments with multiple users, devices and systems; those with an interest in the communication and services provided to individual users; and those investigating business and market aspects. In addition, there was a need for a common overview and framework to help align the group. The specific technical aim of the project was to explore the potential for the simultaneous use of standards such as WiFi, Universal Mobile Telecommunications System (UMTS) and Global System for Mobile Communications (GSM) from one device, possibly using Multiple Input – Multiple Output (MIMO).

The initial approach conceived three sets of scenarios Figure 3.6: user scenarios, service scenarios (based on descriptive and prescriptive models of network services such as voice communication, streamed video etc.), and technology infrastructure scenarios (based on the descriptions of capabilities of different telecommunications systems deployed in a scenario) [33].

As well as working on updating and modifying the standard descriptions of service characteristics already established in the industry, the main scenario development task was to bring in knowledge and ideas about users to the scenario creation process.

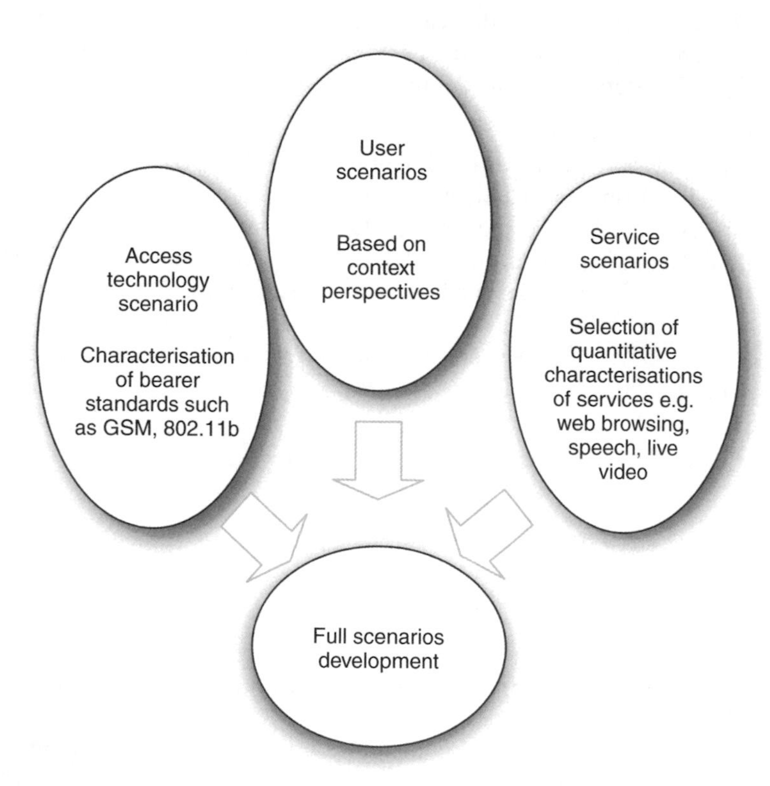

Figure 3.6 Full scenario development

3.2.3.2 Context Perspective Framework

A Context Perspective Framework is a way to structure the bringing together of material for the creation of user scenarios. This approach reflects a number of different analytic perspectives on ICT use, and the needs of this project. Within this framework four perspectives are used: Location, People, Device, and Application Package. For each perspective a number of characteristics are developed, relevant to the characterization of services, i.e. in relation to geographic space, number of users, data applications, demand for services, Quality of Service expected, etc. Quantitative data on these factors is available, although there are still large gaps in the knowledge about what traffic patterns will actually be generated by users. There is also qualitative data, but there are limits on how these can be translated into scales that can be used in scenarios that are useful to engineers. Within each perspective, a number of examples or context scenarios identifying relevant characteristics are developed. This provides a framework not only to understand service requirements, but also to develop further in-depth user analysis.

The context perspectives are not mutually exclusive: users move through particular locations, using particular Application Packages, these Application Packages are used by a range of different users, and locations are inhabited by a range of different users. This overlap strengthens the different scenarios by cross-referencing, and also helps generate a large range of possible user scenarios more easily.

For each perspective, a literature search and a creative exercise were performed to identify a range of different examples, and a set of important dimensions of quantitative and qualitative description created. For example, a key dimension to the use of mobile and wireless services is the location where they are used; the Location Mobile technology opens up many new spaces to the use of ICTs, and this perspective highlights particular spaces that people move through and to in the course of activities, especially those that are by definition related to travel and temporary occupation. The key characteristics of locations are: physical characteristics of the place, the number and type of users occupying or passing through the space, the time and space distribution of users etc., the applications and services that they use, mobility within the location and limitations, and the infrastructure deployment including cell size. For the social and commercial analysis there are characteristics of control of the space, competition and the provision of wireless services, inter-regional and international roaming. The physical descriptions of locations can be obtained from the measurements made in studies of radio propagation. Urban planners and architects, specific studies, and details of current wireless use from mobile operating companies obtain the human use of locations from studies.

Some examples of locations include the airport, the railway station, a tourist city center, industrial estates, a suburban home, an open- plan office, a school, a shopping mall, a high street, a sports center, a stadium, or an exhibition center. It can be extended to rather more restricted spaces such as a car or a train as well. Locations can be quantified in terms of topography, geography, and the type of wireless cell availability, the density and number of users, and the typical use of wireless infrastructure [33]. Each example references original research on these locations, and on the use of existing technologies, such as on the use of mobile phones on trains.

A set of devices, focusing on issues such as form factor, ownership, control over use, technical facilities and integration into network systems, and a range of different people were also selected and descriptions created. The term *Application Package* needs to be described in more detail here. The Application Package relates to the particular activities that a user is engaged in, and the set of data applications and services that are relevant to that activity. The concept of the Application Package links the activities of users to the use of the technology. Application Packages refer to particular activities, such as education, work, leisure, and life management, but in a rather more focused way than application domain. The Application Package scenarios identified include several types of Network Supported Work:

- the mobile office;
- the corporate intranet connection and remote monitoring/control system;
- the 'Traveler's Aid';
- the 'City Survival Kit';
- the 'Media Consumption Portal';
- the 'Personal Networker'.

Characteristics of applications are identified as: information and communication tasks, data applications and usage (time and location). The personal, social, and economic characteristics include:

- immediacy;
- cost;
- competition in provision;
- control over provision and use;
- transferability between devices;
- reliability.

The Context Perspective Framework is now filled with a range of examples that could be drawn upon in creating the working scenarios, and provides a guide to the weakness and gaps in existing knowledge about different factors important in studying users.

3.2.3.3 Creating Working Scenarios

The examples in the Context Perspective Framework (individuals, applications, location, and devices) can now be linked together and developed in a rich and more specific narrative and description e.g. a storyboard (see section 3.2.1.2). This could of course be largely fictional, or based on a particular set of research findings. For example, a scenario developed around a building site would link information about all those who come into the site and the communication needs of key users as they move between the site and other locations. These are then linked to the service scenarios and the technology scenarios mentioned above. However, this scenario is still not sufficient for the requirements of the various WGs, and alternative linked scenarios have to be created, specifically in this example of User-centered Scenarios and Service Provision Scenarios. The User-centered Scenario follows an individual around their daily life, while a Service Provision Scenario describes the use of wireless services by all those in the building site, the physical characteristics of the space, the access infrastructure, and the social relations and control.

3.2.3.4 User-centered Scenarios

These are scenarios based on individuals moving through different places over time, using a range of devices and applications. They are used by engineers and planners working on the design of devices and high-level aspects of communication, e.g. handover between different bearer systems, device power requirements, or requirements of specific end-user markets. Just as the User Scenarios describe more than one particular activity, a complete user-centered scenario is a description of all the different network services across a range of locations that a person or group uses in the user scenario. This could include their home, their car driving around a town or motorways, their office or school, visits to friends, shops etc. It describes what they do in each place, who they communicate with, the technical facilities available, the environment, the social organization of each place, and the position of the person within that organization, and many more factors as are deemed relevant for the particular use of the scenario. These user-centered scenarios are the basis of service product scenarios, developed later.

3.2.3.5 Service Provision Scenarios

Service Provision Scenarios are needed for developers and researchers focusing on locations, developing channel models, or calculating the provision of radio access points that will be needed in a particular place. They require Service Provision Scenarios focused on place, rather than individual users or applications, aggregating all the users in that location. This includes information from the user scenario, such as the number of users (given by per cell or per km^2), the spatial distribution of users, the position and mobility of users, the services utilized by users, the traffic models for those users, and details from the location such as available access technologies, the geographical and topographical descriptions in order to create a propagation scenario. This type of scenario is a richer version of a deployment scenario. Those studying the social aspects of sites will want information on control of the space, social rules, activities and relationships in those sites.

For example, these service provision scenarios could be built on user scenarios such as:

- 50 people in a small airport lounge uploading files from their laptop computers and Personal Digital Assistants (PDAs) within 5 minute periods via a Wireless Local Area Network (WLAN) network, but with UMTS via pico-cells also available; making voice calls from phones, and browsing for travel information on PDAs.

In the FLOWS projects [34], 3 User-centered Scenarios were developed – a businessman on a train, a service engineer visiting sites in a suburban area, and a family visiting a city for the weekend. For the FLOWS research on propagation, 5 locations were identified for testing (office, town square, enclosed corridor, suburban and rural), and the service provision scenarios were based around actual measurements taken in examples of these locations.

However, for some of the technical work, even these scenarios are too broad, so, specific instances of service use need to be described within each scenario and developed to provide an imaginary test environment for the work in hand. In the case of FLOWS, these instances needed to highlight the demands of the simultaneous use of multiple bearer services. Example Scenario Instances include:

- making a multimedia call on a PC while travelling in a train at 120 km/h in an area of macro-cell coverage using UMTS
- switching from a narrowband service to a broadband service during an interactive web browsing session on a web-tablet while stationary in an area with pico-cell coverage (e.g. GSM to WLAN or UMTS).

3.2.3.6 Scenario Connections

Figure 3.7 shows how a set of scenarios are constructed from Context Scenarios and the Service and Access Technology Scenarios. First, a User Scenario is created by combining a number of different context scenarios describing the four key dimensions of end-user use of wireless technology: location, Application Package, user/person, and device. In this case, the user scenarios are centered on a particular user, who moves through several places, using different devices and Application Packages. This User Scenario is used in conjunction with the Service Scenario to describe a User-centered Scenario: all the services and conditions that the user is described as using. Particular instances of the use of convergence technology are highlighted as Convergence Scenarios. Another scenario, this time describing the Service Provision requirements for a particular location is also

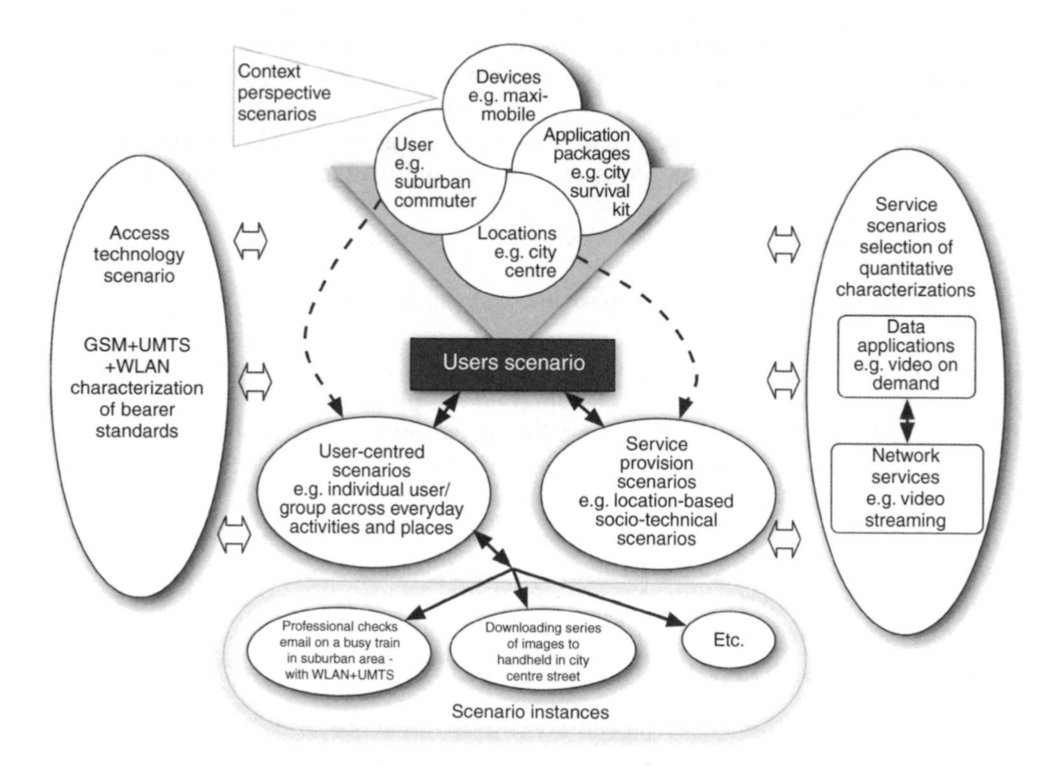

Figure 3.7 Scenario Connections from Stewart *et al.*, 2002 [31]

generated, based on the User-centered Scenario, but this time centered on a particular location within that scenario. It aggregates the users, devices, and applications used within that location.

There are still more scenarios that this technique could be applied to: device-centered and application-package-centered scenarios would be useful in defining and testing the range of conditions under which a particular device or Application Package could be used. In the FLOWS project, the researchers working on market and business aspects needed scenarios Figure 3.8 that addressed these issues. Service Product Scenarios were developed, which described a number of possible products that could be targeted towards particular markets, such as a city dweller's service, or an independent 'traveling business person' service.

This scenario framework provides us with ways to tie a range of user research to the requirements of the engineers, but also opens up an agenda for future social science research. Even though it may appear a little mechanistic, the framework is an important example of the work that needs to be done to systematically draw together qualitative and quantitative work on different people and places, devices and applications and present it in a form that can be quickly turned into scenarios for engineering or service design. It can be used to add depth to the scenarios with reference to supporting research, and build up a library of examples. The user scenarios can also be used as the basis of new user research to investigate people, such as the traveling executive or a family on holiday, and the conditions under which they may turn to technology at different points in their lives.

Engineers conduct much scenario work, with little knowledge of the large amount of research on use of ICTs, and of the methods and approaches developed to understand the relationship between technology and people. Hopefully, this method provides some useful inputs for those within the industry to incorporate into their work.

3.2.4 Summary

This section has outlined how scenarios can be used by a range of stakeholders in the development process for a range of purposes: to further their understanding of implications

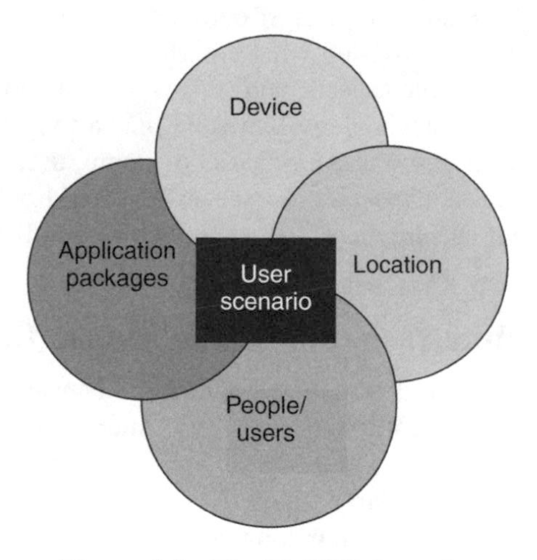

Figure 3.8 The FLOWS Scenarios

for various elements of the technology, to help users envision how they would interact with the technology, to review the existing scenarios. Section 3.2.2 presented the fact that the current practice is to devise a scenario for one particular purpose only, and not conduct the in-depth analysis and development described in section 3.2.3. Thus, the benefits that could be derived to improve technology, usability and market prospects are largely not realized at the moment. There seems to be a competition to produce more visionary scenarios and more polished delivery (e.g. in the form of video stories), and not enough analysis and development work as described in section 3.2.3. A re-distribution of effort from generating scenarios to in-depth analysis and development would help to focus thinking, identify potential issues and problems, and gauge the likelihood that the technology will be accepted by users and generate enough revenue. A further problem is that current visionary scenarios focus on what could be described as glamorous applications for high-end consumers in the developed world. Scenarios rarely address the needs of developing countries – [35] provides a critique of this approach – or less well-off market segments, such as the disabled – see [36] for a study that demonstrates that their user needs are not addressed by current visions.

We would argue that the lack of empirical grounding and in-depth understanding does not only fail to cater to specific user groups, but potentially even the needs and interests of the high-end consumers in developed countries that many scenarios are aiming at. Proponents of scenarios would argue that they are of benefit even if they are used for only one purpose, and without further in-depth analysis: scenarios can provide a vehicle for communication between different stakeholders, inspire investors, motivate technical researchers, or just function as public relations material to promote the image of a forward-looking company (see [37] for a sophisticated example).

However, one danger we see with current practice is that scenarios used in self-fulfiling prophecy modes (see section 3.2.1) may replace proper user needs analysis and requirements engineering. Scenarios derived by technology and market research experts can provide a basis for the discussion of future technologies, but they cannot replace empirical studies on trends, user needs analysis, and iterative prototyping and evaluation that are at the core of the user-centered development approach [14]. At the very least, scenarios must undergo some empirical validation of the sort outlined in section 3.2.1.2: usage scenarios can help potential customers to envision what a particular service would mean for them, identify impractical or undesirable aspects, and suggest how it should be improved. When a technology, such as wireless technology like B3G, has a long lead time, we also need to be aware that users' needs and preferences are not stable: user experience with current technology constantly changes expectations. Scenarios should be constantly revised and updated with insights from current empirical research and trends analysis.

3.3 Advanced User Interfaces for Future Mobile Devices

It was less than two decades ago that the most complex electronic device ordinary people had to operate was the video recorder at home. The interaction with computers, robots, and other more complex machinery was left to a small group of specially educated experts. Nowadays, everyone deals with a large number of high-tech devices almost every day: the mobile phone, the ticket machine, the computerized washing machine, the car stuffed with electronics, the PC on the office desk and the home-entertainment equipment are only a few examples. During the first phase of this development, the manufacturers of mass-market devices, such as Hi-Fi components, for instance, equipped their devices with a large number of switches and dials for the operation of the various built-in features. As the number of features increased, people got overwhelmed with the complexity of operation. The difficulty with programming the video recorder became the synonym for the problems that people had with the increasing complexity of technology. Although a lot of techniques have been introduced to enhance the usability of the devices (e.g. electronic program guides for scheduling video recordings or touch screens for operating ticket machines), the current generation of mobile phones shows that the dilemma is still not solved: for controlling the many dozens of features of modern cell phones each of the keys on the keypad must carry multiple layers of functions. With the continuously shrinking dimensions of the devices, their operation becomes more and more cumbersome. Simultaneously, more and more technologically inexperienced people become users of these complex devices. This leads to the well-known effect that many mobile phone users only use a small portion of the features and functions their device actually offers. This phenomenon is *known* as the *Usability Gap* [38]. Obviously, the Usability Gap has negative consequences not only for the users but also for the device manufacturers and the service providers: it is hard to gain new customers by offering new features or services if the users do not embrace them.

One way out of this dilemma is the improvement of the man-machine-interface so that users can access the device's functions more easily. This section aims at providing recommendations to industry and service providers for improving the user interface for upcoming mobile devices.

To improve the UI of mobile devices, it is necessary to identify the limitations of the current ones. Therefore, this section gives an overview over current man-machine interfaces on mobile devices along with a qualitative discussion of their capabilities.

The main aim is, however, to provide service providers and device manufacturers with recommendations on how the current limitations could be overcome. To this end, we will discuss current research activities on new UIs and we will even present some visions on potential future developments in this area.

Obviously, we cannot hope to cover all potential developments in this domain entirely. But, if we can influence the strategies of device manufacturers and service providers with respect to usability by means of our recommendations, we will have reached our goal.

Man-machine interfaces play a crucial role in all areas of daily life, from infotainment and driver assistance systems in cars to domestic appliances and vending machines. All these areas would profit from an improvement of the UI. Mobile devices put hard constraints on the UI, as the devices have to be cheap, power saving, small, and robust. If a UI can be developed, which works well under these conditions, it will likely influence other application scenarios too.

3.3.1 Description of the Problem

In the following text, a detailed description of the problem that ran under the name Usability Gap is given. The problem with current mobile devices is brought about by a number of concurring developments:

Increasing Number of Features and Functions

Along with the technological evolution, a lot of new functions are built in every new generation of mobile devices. The first generation of mobile phones was limited to voice communication, for instance. Step by step, many more features were built in: text-based communication [Support Mobile Services (SMS)], Internet access (WAP), usage as wireless modem for laptops [InfraRed (IR), BlueTooth (BT)], Personal Information Management (PIM) are typical examples. Current phones transform into multimedia terminals with an HTML browser, an FM radio, a music and video player, a photo and video camera, sound recording, multimedia messaging [Multimedia Message Service (MMS)], 3D-gaming and even built-in TV receivers. Every new feature is built on top of the already existing ones. There are some efforts to produce simple phones with reduced number of features for children and elderly people. However, even functions that seem outdated are hardly ever excluded from mass-market devices. One reason for this can be found in the customers' inertia in changing habits (see below). However, another added technical reason is the compatibility that must be kept within the existing infrastructure. A multimedia UMTS phone must still work in old GSM networks, for instance.

Further, the small form factor and low power constraints lead to a quick integration of new functionality directly into the terminal platform. Even if that functionality becomes outdated after some time, it is sometimes more expensive to remove it from the platform than to leave it in.

Shrinking Form Factor

For a long time, there seemed to be a law that mobile phones always had to be smaller, slimmer and lighter than the previous generation of devices. Currently, the size seems to converge to an optimum size, which makes the best compromise between portability and usability. However, even today, the smallest phones are only the size of matchboxes. Nevertheless, they often contain two displays, a loudspeaker, a microphone, a keypad, a joystick, a camera, a flashlight and several connector sockets (headphone, car antenna, PC interface, IR diode, etc.). In addition, a battery compartment, a SIM-card holder and a slot for storage cards must be built in for the user's access.

Increasing Diversity of User Groups

In many countries, the number of mobile phone users has increased to more than 80 % of the population. This implies that users of all ages and all social, financial, and cultural backgrounds use these devices. In particular, the level of technological knowledge varies largely. Also, the status that a mobile phone occupies in the customers' life is very different: while for business people it may just be a means of communication, for elderly persons it may take the role of a lifesaver in emergency situations. Teenagers express their belonging to a group or their awareness of fashion while parents increase their assurance about the whereabouts of their children by means of the devices (see also [39] for an analysis of user groups).

Increasing Diversity of Usage Scenarios

It was an amusing sensation when in a TV soap opera of the 1980s an actor made a phone call while riding a horse. Today, this sight would not lead to the slightest surprise in any observer. Mainly due to the shrinking size, the long duration of the batteries and the ubiquitous availability of wireless networks, there is hardly any situation in which a mobile phone or a PDA is not used. There even exist waterproof transparent bags to use a PDA under water during diving. But already ordinary scenarios impose strong requirements on the devices: people expect their seamless and robust operation in extreme temperatures (e.g. at the beach or in the snow) and in extreme locations (in a fast train, during a flight, on a mountain top, or in the basement of a steel-concrete building). A good UI must provide optimum usability in all of these situations (e.g. by adapting the size of the controls to the different use cases).

Further, the way people use their devices can be very different: some use it for text-based chatting, some watch movies or listen to music. With modern smartphones and Pocket PCs it is possible to do a lot of the office work (email, text editing, presentations) on the go. Memory cards and miniature hard discs with large capacities allow the transportation of complete photo, audio and video archives and of any personal and business data. Wireless internet access offers the availability of virtually any information anywhere.

This leads to an infinite multitude of different possible usages and there is nothing like a general reference scenario, which contains all possible use cases.

The mutual concurrency between these four developments is obvious: an increasing number of functions and features mean more controls, more sensors and more power consumption. This stands directly against the ever limited space imposed by the shrinking size of the devices. Building in more functions makes the devices less robust, which is opposed to the increasing variety of usage scenarios. In addition, a device with many different features is more complicated to operate especially in extreme situations such as the ones mentioned above. Further, as every user group has its own profiles concerning the desired features, most functions are just cumbersome overhead to many customers. The complexity of operation is even increased by the shrinking size of the devices.

In the light of this analysis it seems impossible to meet all requirements simultaneously with today's devices.

However, there are two possible solutions:

a. Manufacturers build a whole range of devices, each of which is specifically tailored for a certain user group or use case. The aforementioned simple phones are an example: only those functions are built in that are necessary for that specific user.
b. The UI is improved such that, despite the many built-in functions a simplified usage is possible.

The first solution is obviously easier to realize, as it requires only a reduction of technology that already exists. However, it seems that specific characteristics of the mobile phone market hinder the manufacturers to follow this road. Obviously, many of the customers are still in a stage, in which the decision for purchase is mainly driven by the technical capabilities of a device. This is also reflected by the advertising, in which new features are extensively propagated rather than the simple operation of the device.

However, there are also large group of customers who decide on handsets that come with specific price plans (or even for free along with a contract) as long as the basic functionalities (voice, SMS) are supported and the device does not look outdated. For these customers, cheap and simplified devices with a trendy design could be the perfect choice.

From the manufacturer's point of view, however, the development and production of many specialized devices can be much more expensive than the mass production of a single, fully fledged one. However, the idea about one device fitting all possible purposes and user needs holds the danger to being a jack of all trades mastering nothing. Here, a lot of research is required to find the optimum compromise.

Within this section we deal with the second solution, the improvement of the UI. Nevertheless, even for specialized and simplified devices, an advanced UI is beneficial. As we will show, a lot of progress can already be made by improving the UI software without leading to increased costs for the device itself.

3.3.2 UI-related User Needs

All the functions of a device are accessed through the UI. Therefore, the UI defines the user experience in a positive as well as in a negative sense: even the most advanced feature of a device is disrespected if the user's preferences for operating this feature are not well supported by the UI. Any effort to improve a UI must therefore start with an analysis of the user's needs.

These general user needs can be broken down into concrete requirements for the UI. We have selected the following items:

Accessibility: The basis for using any function of a device is its accessibility. Surprisingly often, this trivial sounding requirement is missed in the UIs of mobile devices: for many people the keys of their cell phone are too small to be operated or the displayed text is too small to be read, for instance. In some devices, certain functions are not accessible during a call. Sometimes, the association of keys to functions is illogical (e.g. the red 'end-call' key is also used to switch on the device) or changes depending on the current function. There are also blocks to the device's accessibility which are indirectly caused by the UI: power-hungry displays, for instance, may drain the batteries so fast that the device is low on power just when an important call must be made.

Accessibility needs vary with users, i.e. the handset UIs need to cater to people with special needs, and also with the user environment. For example, the UI needs to be able to provide service accessibility in hands-busy and/or eyes-busy environments.

Finally, some functions may not be accessible for some users just because its operation is too complicated. Good examples are WLAN or BT profiles for which a user must enter complicated network parameters.

Reliability: Closely related with the previous issue is the reliability of the UI. The main concern here is the robustness of the interaction with the device. An example is the accuracy of text-entry mechanisms (e.g. handwriting recognition) and speech input (e.g. voice controlled dialing) in noisy conditions. Another area, in which reliability is important, is the connection of external UI devices such as Bluetooth or IR keyboards. Further, the stability of the underlying UI software is of high importance. The more complex the device becomes, the higher the risk of instabilities. In smartphones and pocketPCs for instance, a reboot of the operating system (a so-called 'soft reset') is sometimes required when the system 'hangs'. This issue is not directly related to the UI. However, if more computationally demanding recognition algorithms will be integrated in the future (e.g. audio-visual speech and mimics recognition), the stability of the software is distinctive.

Efficiency: In particular, business users like devices that allow for fast and seamless work on their tasks. In particular the compromise between portability and ergonomics is important in this respect: fast text processing, for instance, is best done on an ordinary-sized keyboard. The success of the email-push-clients (e.g. the Blackberry) with built-in keyboards (though not full sized) shows this desire.

For reading emails, the screen must be clear and not too small. Further, the main functions must be quickly accessible without scrolling through endless menus. Any gimmicks or unnecessary features should stay in the background.

However, it is not only business users who appreciate an efficient operation of the devices: some teenagers select a specific phone just because it allows for fast text-entry for SMS or on-line chatting. Further, it is also a question of efficiency if the emergency number can be dialed quickly in case of an accident, for instance.

Consistency: Related to efficiency is the topic of consistency. Once users get familiar with a specific handset (layout of the keypad, structure and look of the menus, preferred text-entry method etc.), they may experience frustration when switching to a new device with a different layout of the UI. Therefore, a high degree of standardization between the devices of different manufacturers would be desirable. Even if a new UI feature could be introduced which is obviously superior to an already existing counterpart, it may sometimes be more advisable to stick with the old widely known one in order to prevent user frustration.

Security: Current mobile phones and PDAs allow for large amounts of personal data to be stored within the device. At the same time, sensitive interactions, such as on-line banking can be performed by means of a mobile phone. Therefore, the aspect of data-security is of crucial importance for the customers. For the UI, this means that the device must possess secure interaction channels. Although accurate speech input is important for UIs of the future, every device must provide a silent text-entry facility in addition, for instance.

Another area in which the UI can increase the security of a user is an easy-to-use emergency call mechanism.

With the success of the current camera phones, the topic of confidential company information comes into focus. As some companies already ban such phones, the interesting situation occurs in which an additional feature reduces the ubiquity of a mobile device for the user.

Privacy: Closely connected with the security issue is the demand for privacy. In some respect, this demand is concurrent with the demand for more natural interaction channels such as speech and mimics and with the need for large displays. Particularly when interchanging private information with the device, the accessibility by third parties should be prevented.

Sociality: In contrast to the need for privacy, there is also the users' desire to interact with other people and to build identification within a group. In some situations there is, for instance, a desire to allow public access to a user's device. Especially, the multimedia features of current and future mobile phones call for the capability to show content to other people. Large displays and a built-in loudspeaker can support this need.

The UI can also facilitate the interaction with other people by providing easy-to-use video conferencing and instant messaging functions.

Finally, tuning the outer appearance of a UI (e.g. by so-called 'skins') such that it reflects the belonging to a certain user group or class of users can be important for customers.

Individuality: A device's ease of use can be greatly increased by taking into account the personal preferences of the user. On today's Smartphones and PDAs, the importance of personalizing the UI becomes obvious in the myriad of different so-called 'today screens'. Even on simple mobile phones, a personal ring tone is obligatory to a customer.

The more intelligent the UI becomes, the more it can take personal preferences into account: the sequence in which daily information is presented to the user (appointments, news, weather, etc.), for instance, could be personalized. The same is applicable to menu entries, often used functions, preferred text-entry mode, etc.

However, most users consider a manual adjust of a personal profile as cumbersome. Therefore, an automated personalization of the UI by learning the user's preferences could be advantageous.

Conversely, sometimes this alleged intelligence may also bother the user if the interface varies all the time due to the automatic adaptation. Therefore, this issue is quite sensitive.

A flexible choice, such as which aspects of the UI the user would like to change or personalize, which must stay unchanged and which can be adapted automatically, would be a good compromise.

Despite all personalization, however, the branding of a mobile device is of high value for many users and likewise for the manufacturers. Therefore, a good UI should still allow for a corporate design individual to the specific manufacturer.

Desirability: To many users their mobile phone is much more than just a tool for communication. The outer appearance, the technical specifications and the brand name contribute to the attraction of the device.

The UI is the major contributor to the appearance of a device. Therefore, its design is often the distinctive factor for the purchase. The 'fun factor' and the technological originality should not be underestimated either.

At this point, many more user needs could be mentioned such as ease of use, context dependency, portability, connectivity, ubiquity etc. In most cases, however, these needs can either be subsumed under the more general items mentioned above (ease of use, for instance, shares parts of efficiency, reliability, accessibility, consistency) or additionally, they have the character of a solution to meet a user need (such as context dependency, for instance).

With the selection above we tried to collect a more general set of requirements. The degree of generality can be checked by evaluating UIs for devices other than cell phones with respect to the user needs.

It is the task of researchers, manufacturers, and service providers to analyze a specific device or service with respect to the general user needs.

3.3.3 Current State in UI

The following Table 3.2 lists UI elements versus the user needs from the previous section. The marks between '+' and '−' indicate up to where the UI elements fulfil the corresponding user needs. '+−' means that there are positive and negative aspects and '0' means that the UI element does not address the user need (i.e. 'not applicable').

We have divided the table into three sections. The first two refer to the physical input and output devices and sensors. The third part lists UI solutions in software. These can be implemented based on different physical devices.

Virtual keyboard: This is a keyboard mostly found in smartphones or PDAs, where the keys are displayed on a device's screen. Depending on its size, the user must use either a finger or a stylus to enter text.

Table 3.2 UI-related user needs

UI elements vs. user needs	Accessibility	Reliability	Efficiency	Consistency	Security	Privacy	Sociality	Individuality	Desirability
Input devices									
Virtual keyboard	+−	+−	+−	+	0	+	0	+	0
Built-in keypad	+−	+−	+−	+	0	+	0	−	0
External keyboard	+−	+	+	+	−	+	0	0	+
Touchscreen	+	+	+	0	0	+−	0	0	0
Joystick	+	+	+	+	0	0	+	0	+
Touchpad	+	+−	−	+−	0	0	0	0	0
External mouse	−	+	+	+	0	0	0	+	+
Direction pad	+	+	+	+	0	0	+	−	0
Scrollwheel	+	+	+	+	0	0	0	0	+
Stylus	−	+−	+−	+	0	0	0	+	0
Acceleration sensor	+	−	+−	+−	+	−	+	0	+
Camera	+	+−	+	+	−	−	+	+	+
Output devices									
Built-in display	+−	+−	−	0	−	+−	+−	0	+−
External display	+−	+	+	0	−	−	+	0	+
Loudspeaker	+	+	+	0	−	−	+	0	+
Headphone	+−	+	0	0	+−	+	−	+	+
LED indicators	+−	+	+−	+	0	+	0	0	0
Buzzer	+	+	+	+	0	+	−	−	0
Software									
Handwriting recognition	+−	−	+−	+	+	−	0	+	+
Predictive text input	+	−	+	−	0	0	0	0	0
Multitap	+−	−	−	+	0	0	0	0	0
Keyword speech recognition	+	−	−	−	−	−	0	+	+
Text menus	−	+−	+−	+	0	0	0	0	0
Graphic menu	+	+	+	−	0	0	0	0	0
PIM									
Today screen	+	0	+	−	−	−	+	+	+
Ring tones	+	0	0	−	0	0	+	+	+
Audio- and video-players	+−	+−	0	−	−	+−	+	+	+
Text to speech	+	+	+−	+	−	−	0	0	+

Accessibility and efficiency are good in terms of how many key presses are needed to enter a letter and for entering symbols. However, the lack of tactile feedback may result in poor user experience.

Reliability is good in terms of a one-to-one association between keys and letters (unlike the 12-keypad) but as the keys are usually very small, typos are possible.

Providing the same key layout as typewriters and computers, the consistency of virtual keyboard is very good.

Different 'skins' allow for an individual design of the virtual keyboard.

As with all text-entry methods, privacy is high in comparison with voice input.

Built-in keypad: This is the ordinary keyboard that most mobile phones are equipped with.

In terms of accessibility, reliability and efficiency, the built-in keypad complements the virtual keyboard. Consistency is also high, as all mobile phones have the same keypad layout. Individuality is low as the keypad itself cannot be changed by the user (although there are phones with changeable covers).

External keyboard: Through the IR port, Universal Serial Bus (USB) or BT, most PDAs can communicate with a larger external keyboard. Most external keyboards possess ordinary keys. However, there are also models, which use a laser to project an image of a keyboard on a flat surface and to recognize the user's finger movements on it.

The accessibility of external keyboards is high in terms of the size of the keys. However, in some situations, an external keyboard cannot be connected, as a flat surface is usually required. Reliability, efficiency and consistency are high because of the large keys, the widespread key layout and the tactile feedback (except for the laser keyboard).

Security can be low, as in principle the BT or IR channel can be spied on.

Desirability is high, as the PDA nearly resembles a fully-fledged sub-notebook when an external keyboard is connected.

Touchscreen: All PDAs are equipped with this sort of input device. The user can activate controls and enter text by means of finger presses or using a stylus.

Accessibility, reliability, and efficiency depend on the application that is controlled via a touchscreen. As described before, text entry by means of a virtual keyboard is not always optimum. However, the direct association of icon control and effect in particular (the user can directly press on the control, and see the effect right at that spot) provides ease of use as compared with indirect controls.

Privacy, however, may be low, particularly for large displays with a wide viewing angle as they are built into some PDAs.

Joystick: With the introduction of games in mobile phones, more and more devices were equipped with miniature joysticks. This input method can also be used to scroll through menus.

Particularly for games, a joystick is the optimum control with high accessibility, reliability and efficiency. Although the miniature size of the built-in joysticks requires some dexterity, everyone knows immediately how to operate the game through a joystick when passing the mobile device among different users, so consistency and sociality is also high. As the joystick is the outer sign for the entertainment capabilities of the phone, desirability is high too.

Touchpad: Originally known from notebook computers, a touchpad has recently been introduced in a PDA. Similar to a mouse, the user can move a cursor on the devices screen by means of finger movements. By means of finger taps, controls can be activated.

As the touchpad is built into the device, accessibility is high. The lack of moving parts leads to good reliability, however, the small size of the PDA as compared to a notebook requires that the touchpad is small too. Therefore, accuracy and hence efficiency are low. Known from laptops, touchpads provide high consistency. However, for mobile devices, this input method is rather unfamiliar.

External mouse: Some PDAs can be connected wired or wirelessly to an external mouse by means of a USB host port.

As a flat surface is required, accessibility is rather low. However, a mouse is accurate and efficient to control a cursor on the screen. Consistency is provided as mice are widespread for controlling PCs. The selection of available mice is very large, so individuality is possible. As with an external keyboard, the desirability is high in particular, for business users.

Direction pad: Almost all current mobile phones either possess a joystick or dedicated keys for moving a cursor on the screen. Mostly, these keys are arranged in a square to allow for intuitive cursor control.

The intuitive layout leads to high accessibility, reliability, efficiency and consistency. This is also the reason why Microsoft, for instance, requires the presence of a direction pad for every PocketPC that runs Windows Mobile. This uniformity, however, does not allow for any individual design.

Everyone can control a game through a direction pad, so the devices can be passed between members of a group, improving sociality.

Scrollwheel: Many PDAs possess a wheel-like switch on the devices side, which allow for an up/down movement of the cursor or tune the volume. Sometimes, the wheel can be pressed down to make a selection.

The simplicity and intuitive operation leads to high accessibility, reliability, efficiency and consistency. This makes them a desirable add-on for every mobile device.

Stylus: Every device with a touchscreen is equipped with a separate stylus that can be stored in the device. In particular for high-resolution displays, as they are built into PDAs, a stylus is needed to control small icons and to enter text on a virtual keyboard. Also, a stylus is required for handwriting recognition.

The pen-like design leads to a high degree of consistency. However, reliability and efficiency greatly depend on the application. Accessibility is often low, because no one-handed operation is possible, the stylus is too slim or it gets lost. The possibility to control at least the basic phone applications by finger-input through enlarged icons would be advantageous.

Individuality is provided as many ordinary multi-function ball pens and laser pointers on the market provide a stylus as an add-on.

Acceleration sensor: These are miniature solid-state components (mostly Micro Electrical Mechanical Systems (MEMs)), which can measure acceleration forces in three dimensions. In some devices, this information is used to steer games. However, this input method is currently implemented in a mobile phone for drawing telephone numbers in the air by moving the entire device.

In some sense, an acceleration sensor allows for gesture input, which for humans is a very natural communication channel. Therefore, accessibility and consistency are good. However, this type of man-machine interaction is rather exotic to most people, so a learning phase is probably required to use it efficiently.

For gaming or automatically changing the display orientation, this input method is efficient. However, drawing telephone numbers or letters in the air provides neither privacy nor reliability. Calling an emergency number just by strongly shaking a mobile phone can provide a high degree of security in an emergency situation.

The gaming capabilities and the innovative character lead to a high sociality and desirability.

Camera: Although this is formally an input sensor, a camera is not used as a UI in a strict sense. However, with the introduction of mobile video telephony, the camera image becomes more important for communication purposes.

As the camera is built into the devices and usually works like an ordinary photo camera, accessibility and consistency are high.

The reliability greatly depends on lighting conditions. For collecting picture proofs after an accident, for instance, a camera is very efficient. However, the ubiquitous availability of a camera in a mobile phone causes security and privacy issues.

As camera images can be stored as wallpapers and sent via MMS, they contribute to individuality and sociality. The large success of camera phones proves their desirability.

Built-in display: Except for some exotic devices which are entirely controlled by means of speech, all mobile phones and PDAs possess one or more LCD displays.

The display is probably the device that best shows the discrepancy between usability and size. On the one hand, current technology allows for a presentation of content in high resolution, true colors and wide viewing angle. However, on the other hand, the small size and large power consumption reduces accessibility, reliability, and efficiency.

Only through a display, information can be shown easily to another person, enhancing the sociality. The small size of the built-in displays prevents others from observing the entry of private information. However, a mobile phone left on the desk and displaying private telephone numbers or appointments reduces privacy and security. The desirability depends very much on the form of the display: swivel screens, VGA resolution or Organic Light Emitting Diode (OLED) displays have a high coolness factor.

External display: Either by means of additional hardware (CF or SDIO cards) or by built-in graphics accelerators, some PDAs can access external VGA monitors for presentations or video playback. Mostly, the resolution of the displayed image is larger than the one of the internal screen. With the introduction of TV on the mobile phone, the possibility to easily access an external screen (TV or VGA monitor) will play an important role.

The larger display area enhances the accessibility. However, the connection of a PDA through VGA to a video projector or monitor is cumbersome at the moment. A connection to a public TV set is not yet possible.

Once connected, the reliability and efficiency with which information can be perceived is high. Naturally, security and privacy are low.

Obviously, sharing a video playback or powerpoint presentation among many people enhances the sociality. In particular, for business and multimedia users, the connection to an external display is very desirable.

Loudspeaker: Many mobile phones and all PDAs possess built-in loudspeakers which can be used for freehands telephony or for the playback of music. In some phone-PDAs, this loudspeaker is separate from the one that is used for telephony.

A similar argumentation as the one for external displays with respect to efficiency, reliability, sociality, privacy and security also holds for a loudspeaker. As the loudspeaker is usually built into the device, the accessibility is high.

Headphones: To all mobile phones and PDAs, an external headset for voice telephony or audio playback can be connected either through wiring or through BT.

Accessibility and efficiency are usually high, as background noise is reduced. However, sometimes, the wiring may be cumbersome so that a wireless connection is desirable. The latter, however, has a higher power consumption reducing the uptime of the device.

Privacy and security are naturally high, however, a headphone blocks acoustic information from the environment causing danger in busy surroundings.

Obviously, sociality is reduced. The availability of stylish headphones and BT headsets allows some degree of individuality and causes high desirability.

LED indicators: These indicators are used to provide the user with important information even when the display is switched off. Mostly, network availability, charging status, BT or WiFi activity, alarms and recordings are indicated.

The low power consumption and the visibility in the dark make Light Emitting Diodes (LEDs) accessible. However, the user needs to observe the device, which is unpractical in eyes-busy situations. Reliability is high and consistency is facilitated by intuitive colours (red for alarms, blue for BT, green for charged, etc.). Also, privacy is conveyed, when status messages are indicated optically rather than acoustically.

Buzzer: In noise-sensitive environments, a vibrating device within all phones and phone-PDAs can be activated to announce incoming calls or alerts.

On the go and in loud environments, a buzzer is the ideal alert indicator. It is accessible and reliable even in eyes-busy and ear-busy situations. Consistently, every mobile phone has a buzzer providing a high degree of privacy. In contrast to ring tones, all buzzers sound alike, so individuality and sociality are low.

Handwriting recognition: On most PDAs some sort of natural handwriting input is implemented. A recognition algorithm tries to identify letters, words and symbols based on an internal dictionary. Sometimes, the recognition results can be improved by training the specific handwriting style of the user.

Predictive text input: Based on an internal dictionary all mobile phones can recognize words entered by means of the built-in keypad although each key may carry multiple letters. In particular for writing text on limited keypads, this method can save a lot of keystrokes.

The accessibility of predictive text input mostly depends on the input device (keypad or virtual keyboard). Compared to Multitap, however, the ease of use and efficiency are greatly improved, as less keystrokes are required to enter text.

The reliability, however, depends very much on the built-in dictionary. In particular when mixing languages or when using slang expressions (which is very common in teenagers' SMS texts), predictive text input is error prone. Sometimes, the necessary text corrections can outbalance the advantages.

As different algorithms are on the market [Intelligent Text Prediction (ITAP), T9 and MS-Transcriber] consistency is not yet guaranteed.

Multitap: As the number of keys built into mobile phones is less than the number of possible letters, most keys carry more than one letter or symbol. By pressing the keys multiple times, different layers of letters can be accessed.

In terms of accessibility, multitap has two opposing aspects: the operation principle is quite easy to understand and thus easy to use. However, the high number of necessary keypresses turns out to be cumbersome for entering long texts. This also has a negative influence on the reliability and efficiency.

Consistency, however, is given as the layout of the several layers of letters on the keypad is widely standardized.

Keyword speech recognition: Some advanced mobile phones and PDAs can recognize address book entries or commands from the user's voice. Mostly, the user's voice has to be trained in and/or an internal dictionary is used.

Allowing for hands-free operation, the accessibility of keyword speech recognition is good. Due to the inherent complexity of speech recognition as such, the reliability is still low as many recognition failures particularly occur in noisy environments.

Having to speak slowly and excessively clear, the efficiency is low compared to keyboard input. This may change, however, when the recognition algorithms and the processing power are improved and fluent natural speech can be recognized.

Obviously, security and privacy are low too, as other people can listen to the user's inputs.

Individuality is provided as a personal set of commands can be trained. As spoken text input is still exotic, the desirability is high.

Text menus: The basic form of accessing functions and menu items consists of scrolling through a one-dimensional hierarchically ordered list of entries.

Due to their simplicity, the meaning of the items in text menus is easy to perceive. However, the small font of texts on the internal displays is often hard to read. This also influences the reliability and efficiency.

In contrast to proprietary icons, however, textual menu items provide a high degree of consistency.

Graphic menu: On high-resolution displays the menu entries can be represented by intuitive icons which help the user in memorizing the underlying function.

As known from digital photo cameras, some mobile phones display the menu entries in a circular structure on the screen. Unlike with a simple list, the user can smoothly scroll from the last entry to the first without additional keystrokes.

Graphical menus are complementary to text menus in that they allow for an easy recognition of the icons. However, proprietary icon sets hamper the consistency.

Today screen: A today screen is the information that is displayed right after the device is switched on. In all PDAs, this screen can be customized by the user. Countless so-called *today plugins* display all sorts of information from the current weather forecast to upcoming appointments or the device's battery status.

As the information on the today screen is displayed on the front page and the plugins are optimized for small screens, the accessibility is high. Further, important information can be accessed without going through nested menus, which improves efficiency.

However, most users put their appointments or other private data on the today screen for quick access. This reduces the security and privacy when the device is left unobserved.

The today screen can be personalized in every aspect. Therefore, the consistency between devices of different users is low.

Many users design their today screen with their personal preferences and favorites (animations, wallpapers of celebrities etc.) and show these to their friends. Therefore a high degree of individualism, sociality and desirability is given.

Ring tones: Dedicated sound signals can be associated with different states of the device. Alarms, incoming calls, battery warnings or keystrokes can be indicated by simple beeps, entire melodies or even digitized music. It is also possible to associate individual ring tones to different callers.

In particular, this latter feature can enhance the accessibility of the device as even in hands-busy and eyes-busy situations the identity of the caller is indicated.

As the association of these tones to functions and callers is widely personalized, consistency between devices is not given. However, individuality, sociality and desirability are high as the personal preferences of the user can be expressed in his choice of ring tones.

Audio and video players: Most current cell phones and all PDAs can playback audio and video files either from an on-line stream or from internal memory. The files are either self-recorded by means of the built-in camera, received via MMS or transcoded from DVD or TV. Some mobile phones are even equipped with an FM radio receiver.

For music playback, accessibility is high, as hands-free operation is possible. Video playback, however, suffers from the small screens. The transcoding and streaming of video contents does require some hands on experience.

Also the reliability differs between audio and video: while the mp3, wma and ogg audio formats are widely supported, the restrictions of the mobile platform require highly optimized compression codecs. However, video codecs are still the subject of intense research and no 'standard' format has yet settled. Therefore, videos in a huge variety of codecs can be found and their support by the mobile devices is not guaranteed, while consistency is lacking.

Listening to music blocks out external sounds. This can bring up security problems.

Privacy depends very much on the size and viewing angle of the display and the volume level. As watching videos on a mobile device is still not common practice, curious observers are likely preventing the undisturbed enjoyment of private material.

Conversely, sociality can be conveyed as current advertisements for watching soccer games on mobile phones suggest.

In any case, a very high desirability is given.

Text to speech (TTS): Some devices can generate artificial speech from text. This way, address book entries, ebooks or status information can be read back to users who are either impaired or temporarily unable to observe the display (during driving, for instance).

This hands-free information access improves the accessibility of the device. Also, reliability and consistency are given as clearly spoken text can be perceived accurately.

Sometimes, however, written text can be captured much faster. In emails, for instance, the user can quickly tell the header or signature from the contents and skip these blocks. Ordinary TTS-systems lack this intelligence and read out the complete text, which reduces the efficiency.

At least when switched to loudspeaker output, the security and privacy of the information are not given. However, the 'coolness factor' leads to a high desirability.

Summarizing, current UIs consist of a set of elements each of which has its drawbacks and advantages. It is very difficult to make general statements as the evaluations greatly depend on the application, the usage scenario and the user's preferences.

Nevertheless, the table provides cues on where research activities could provide improvements.

In the following, we will give an overview over these fields of research and the results that are to be expected.

3.3.4 Future Interfaces

This section discusses the evolution of the UI of future mobile devices by considering a number of factors that are influencing it, including:

- the evolution of the mobile terminal form factor and its functionality;
- the focus toward a more natural human-terminal interaction;
- the ubiquity of mobile communications and information access in today's society, which could potentially be provided through device interconnectivity and wearable electronics;
- the need for a more flexible UI that adapts itself and the content to a variety of user environments.

The key driver in these UI considerations is to make the terminal completely intuitive, natural to use and provide blissful convenience. The discussions also highlight a number of technology developments that could be shaping future interfaces and the technical challenges involved for the realization of these novel UI concepts.

3.3.4.1 Evolution of the Device Form Factor

The form factor of the device is evolving and driven by the need to improve the human-terminal interaction for various mobile applications and functionality delivered through size-constrained devices.

Keypad

The keypad is the predominant input channel of current mobile devices. Despite the myriad of mobile data services that require text input, in the majority of current mobile handsets text entry is still based on or around the conventional 12-button numerical keypad, which was initially designed for the specific needs of telephony services. The inability to provide an adequate data input mechanism remains one of the major stumbling blocks in mobile terminal usability and hence a significant amount of research is invested into prototyping and exploring other ways of entering text on a mobile device.

While technologies such as predictive text input have become fairly ubiquitous and go some way towards easing the problem of text entry, it is still a cause of major frustration for expert and novice text users alike. Word ambiguity, switching between predictive text and multitap for non-dictionary words and the customization of the dictionary are major usability barriers. The harmonization of key mapping functionality and easy-to-use command shortcuts (for switching between multitap and predictive text, capitalization, symbol entry, etc.) across different devices and different applications could help increase the usability of such solutions and reduce the learning effort of the user when switching between devices.

Novel keypad designs that enable unambiguous letter entry provide one approach for resolving some of the usability problems pertaining to multitap and predictive text solutions. A typical example is the QWERTY layout, which is integrated in mobile devices and PDAs targeted at business users. This layout has become very popular due to ease of use, speed of entry and accuracy. In the consumer market segment, novel keypad designs with distinct keys for each alphabetical character are also emerging aimed for technologically inexperienced users [40]. However, such designs compromise heavily on key size and are only suitable for clamshell type handsets if terminal size and form factor are not to be adversely affected.

The keypad, however, could soon be a UI element of the past. Future interfaces might utilize handwriting recognition solutions by combining touchscreens and stylus-based input, or speech recognition technologies. The former approach, which is state of the art in PDAs [41], provides a familiar way of entering text as well as maximizing space usage by using the display area for text entry. However, accuracy is a major issue of handwriting recognition solutions, in particular, when the user is not stationary. At the risk of increasing complexity, state-of-the-art PDAs include a language model to guide recognition. This is also state of the art in PDAs already. Other disadvantages include the need for the user to look at the screen, the fact that it is operated with two hands and the lack of tactile feedback. The latter aspect will be discussed in more depth in the Natural Interaction section along with the use of speech recognition for text entry.

Display

Digital video data for mobile consumption is placing increasing demands on the mobile display, in terms of physical screen size, resolution and power requirements. Recent advances in LEDs could hold the key to driving the next generation of mobile displays. This technology is capable of producing high-brightness color displays that are viewable under virtually all lighting conditions and have a 170° viewing angle, while potentially maintaining low power requirements because no separate backlight is necessary.

The user experience of emerging data services, such as mobile TV, is likely to be impacted by the form factor constraints of the mobile device. PDAs and portable media player devices with much larger displays are better suited to providing viewing functionality and are hence more likely to provide a more immersive viewing experience. However, recent advances in the research of flexible displays represent an interesting development as this technology enables the concept of foldable or pull-out displays for the ease of data consumption [42]. Such display technology will overcome limitations posed by the current size of the mobile display and the need for a small form factor device, as well as have positive implications for wearable electronics.

Recent advances in video projector technology suggest an additional solution of the screen size dilemma: there are already battery powered Light Emitting Diodes Digital Light Processing (LED DLP) projectors in the market [43] which can beam a video image onto any flat surface. With further reduction of power consumption or increased battery capacities, it might be possible to integrate such a projector into a mobile device.

Future visual interfaces will also inevitably incorporate 3D displays, providing an immersive user experience for mobile gaming applications, as well as enhanced usability in terms of information visualization. For instance, the additional third dimension could aid the visualization and navigation of complex menu systems.

Head-mounted displays provide an alternative approach to current mobile displays [44]. The novelty of this technology lies in the absence of a screen. Instead high-resolution video images are scanned directly onto the retina creating a large, virtual plasma-size TV screen as viewed from seven feet away, reducing the eye strain often associated with viewing images on a small screen over an extended period of time. The images could be projected in either the left or right eye, increasing user mobility by enabling the wearer to privately watch movies or TV, read text or view pictures on the go. Nevertheless, this is achieved at the expense of a partially obstructed view of the real world. Furthermore, the mechanism used to relay the video data from the handset to the head-mounted display could also impact usability. For instance, the use of cables would limit mobility.

3.3.4.2 Natural Interaction

While people have developed sophisticated skills for sensing and manipulating physical environments, most of these skills are not utilized by the current traditional UIs incorporated in devices. Natural interaction, through speech, haptic, gestures, etc., is currently gaining increasing interest in an attempt to overcome the limitations of existing UIs and improve the human-terminal interaction by leveraging human senses. The focus on natural interaction anticipates an extension, rather than a replacement of traditional UIs.

Speech Interfaces

There is a great deal of ongoing research in the area of speech recognition and Text-to-speech (TTS) technologies for device interaction. This work has been further fueled by the introduction of legislation in some countries that determines the way in which users are allowed to interact with a mobile device while driving. This type of environment is often categorized under hands-busy and eyes-busy and the speech interface is seen as one of the most promising enablers for in-car mobile services and terminal control. Another major driving force behind the use of speech for human–terminal interaction is the recognized necessity to develop mainstream terminals and services that also address the needs of people with special needs, including the elderly and those with visual impairments.

Although researchers have been pursuing this technology for years, it has so far lacked critical capabilities to work efficiently. However, in the last few years, research has yielded new advances in speaker-independent keyword recognition enabling the user to call a contact or dial a number using voice. A number of terminal manufacturers have launched handsets with such capabilities combined with TTS technologies to probe the market's response to such technology.

Speech recognition is also being aimed at addressing usability issues around menu and portal navigation on size-constrained multi-application devices. As the number of commands needed for the vocabulary of the phone menu is relatively small, client-based speech recognition solutions are ideal for cutting through the complexity of menu structures. Speech commands enable the user to directly step to a particular menu item, effectively flattening the menu structure and providing easy and speedier access to services.

Menu navigation effort is also a major hindrance for locating and accessing content on mobile portals. Speech recognition could improve usability by enabling the user to directly access the required information within the portal, such as locating a news article, a weather report or TV listings. Personalization techniques that automatically reorder the menu options to suit each individual user, also improve menu navigation on a user-by-user basis.

Speech commands could additionally enable the management and control of the terminal. 'Zoom', 'next' and 'back' are general commands that could be typically used to navigate through the information presented. Other more application specific commands could also allow the user to control and interact with various applications, such as music and video players on mobile devices. For example, the user can choose to fast forward, rewind or play a music/video clip using voice.

Client-based speech recognition solutions that enable the user to navigate menu structures and control applications using voice commands are slowly appearing in the market [45]. These speaker-independent, language-dependent solutions are limited by the processing power constraints of the terminal and only perform well within a restricted context and for a limited vocabulary. Current solutions are completely software-based but in the near future, the technology will be embedded in silicon chips [46], resulting in greater processing power, capable of supporting larger vocabularies. In comparison to present solutions, this technology is expected to improve performance and robustness to noise by as much as 10 times.

Further research in speech recognition will be vital to overcome the current limitations of poor robustness in terms of speaker-independence and noise tolerance, and a limited vocabulary. The user needs to be able to interact with natural dialogue rather than a number of rigid commands that the recognition engine has been trained to recognize. Hence research is now shifting towards solutions that enable more spontaneous and natural language conversation that will ultimately provide faster and easier text entry on small mobile devices. Such speech-to-text conversion solutions will target applications like form-filling, Instant Messaging and SMS/e-mail dictation. With current dictation solutions, speed is constrained by the need to speak clearly, a limited dictionary of words and manual correction of errors. In order to deliver optimum recognition performance, such solutions might require training prior to use, which represents an initial barrier for usage. Another factor that might hinder the uptake of this technology is the fact that humans are not used to talking the way they write and hence they need to acquire the skill before dictation can become a mainstream application.

An interesting research topic is the attempt to overcome the problems with robustness by integrating visual information in the speech recognition process. One possibility, for instance, is the integration of lip-tracking [47, 48]. In the simplest case, detecting lip-movement could substitute the 'push-to-talk' button, currently needed to activate the speech recognizer and to separate voice commands from background noise. A more advanced version is a fully-fledged audio-visual speech recognition approach in which both visual and acoustical features are fed into the recognition algorithm.

Despite having gathered some momentum, there are several barriers limiting the success of speech technology in mobile handsets. As language interaction is the most natural way of communication, users tend to compare their interactions with the systems to their interactions with humans. Such high expectations for the technology represent a major obstacle.

User acceptance is another major hurdle. Outside of the car environment, where speech technologies might be enforced as the only allowed form of human–terminal interaction, and the special needs area, it is perhaps less clear what value speech recognition adds to the user and whether such an indiscrete mode of input can overcome social taboos, at least for use in public. Perhaps it might succeed as a supplement rather than a direct replacement for more conventional forms of input, such as textual or Graphic User Interface (GUI). When taking a photograph, for example, the user could use speech as a means of annotating the images to provide contextual information for future reference or even to aid its retrieval.

Interpretation of Emotion

Another way of increasing the robustness of speech recognition and providing an enhanced user experience is through the integration of another natural communication channel, namely, the human emotional state. Recent studies in this area are focusing on how the detection of emotion/attitude from the user's voice or hand grip could be exploited to supplement verbal and textual communications. For example, increasing levels of emotion in the tonality of a person's voice or the pressure of the hand grip could be used to prompt a virtual assistant in a customer care application to provide some form of help to ease the difficulty or to route the call to a human operator.

Haptic Interfaces

The integration of a haptic interface in future UIs will be beneficial to the user interaction experience with the handset, as haptic sensations elicit cognitive and emotional responses from users that sight and sound alone cannot accomplish. This interface has positive implications for various applications, such as mobile gaming, messaging, vibrating ring tones and navigational cues [49]. It could also be deployed to overcome the lack of feedback present in current touchscreen technology. Thus, when a soft virtual key is pressed, the screen vibrates and gives the user a more realistic impression that a key press has been registered.

Gesture Input

Various studies demonstrate the potential of gesture input for providing a fast and natural method of interaction [50] that could also supplement conventional input channel methods and improve the usability of the terminal and its applications. The basic technology behind this human–terminal interaction method is the accelerometer, a motion sensor that detects and measures motion. Advances in Micro Electrical Mechanical Systems (MEMS) have resulted in small, low power accelerometers which can be integrated into mobile handsets. Such technology can, for instance, provide an intuitive means of rapidly and easily navigating through multiple pages, scrolling, flipping pages, zooming in and out, etc, simply by tilting the device [51, 52]. Compared to keypad input, gesture input offers the advantage of providing analogue control, which is particularly fitting for gaming applications and volume control. Another advantage of this technology is that it allows the user to operate the terminal with one hand.

An interesting alternative to an accelerometer is the detection of the device movement based on camera images. By tracking an arbitrary object in the image of a mobile phone's camera (e.g. the user's face), it is possible to calculate the relative movement between the object and the device in three dimensions.

Natural movement in 3D space could also play a major role in mobile gaming by providing a more immersive near-reality experience compared to existing navigation keys on mobile phone keypads. Playing ball games is just one example where the user could pretend to be holding a golf stick or a tennis racket instead of the mobile device. The sensor is used to detect the acceleration and direction of a user's hand swing, which are translated and converted into a ball hit. Such 3D games for mobile devices are already hitting the Asian market and are expected to have huge appeal.

In particular in combination with head-mounted 3D displays, the gesture input would be an ideal control for virtual reality: the mobile phone could be used as a pointer to objects or 3D menu items in the virtual world.

Nevertheless, this new and unfamiliar interaction method requires some degree of dexterity and also some practice to get used to. Hence, it remains to be seen whether this innovative way of interacting with the device will gain user acceptance.

Virtual Spatialised 3D Audio

Most of the digital audio data consumed in the mobile space, whether it is speech or music, is either mono or stereo and is heavily constrained by the small spacing of stereo speakers in a mobile device. In the near future, the quality of the audio interface could benefit from virtualized technology whereby multiple sound sources can be virtually positioned around the user [53], allowing the creation of a surround sound affect, thus offering a richer listening experience. Due to the fact that virtual sound sources do not have the restraints of their real counterparts, they can be interactively positioned and moved around the user without the need to physically move sound emitting objects, which makes this technology ideal for mobile applications.

Music and movie playback, mobile gaming, and audio teleconferencing could all benefit from virtualized audio. The impact of advanced graphics technologies in mobile gaming is reduced due to the small screen size inherent to mobile devices and hence audio effects become of primary importance. The aural clues provided by a 3D audio soundtrack could be very useful in 'expanding' the game, making it more immersive and thereby appealing. In the case of audio teleconferencing, speaker intelligibility is improved by generating virtual sound sources of each participant's voice and 'placing' these over a wide 3D area surrounding the user, hence providing increased voice separation and an out of head feel reducing listener fatigue. The combination of this technology with an interface that allows the user to position voices at will and provides visual cues (such as an image or an icon) to indicate who is talking, could provide a powerful and less bitrate alternative to mobile 'video' conferencing.

3.3.4.3 Ubiquitous Communications and Information Access

As mobile terminals are becoming ever more woven into the fabric of today's society, there are increasing expectations for ubiquitous communications and information access. UIs need to develop to facilitate anytime/anywhere communication, make information available for use by services and users independent of the application or device in use and also utilizing the capability of different devices in the environment to assist with rendering content and with the human-machine interaction.

For instance, a portable video player device could alert the user to an incoming call or be used for rendering a picture message. In the car environment, researchers are looking at how mobile devices could leverage the human-machine interaction provided by the car, such as the use of the car display and speakers as assistive rendering systems for content. The aim is to address safety concerns over driver distraction and meet the needs of the user in the car.

Device Interconnectivity

Future interfaces will provide ubiquitous communication and information access through consistent and simple interconnections between the mobile device and other devices (such as consumer electronic devices and PCs) belonging to different environments (home, car, office, etc.) to enable some form of seamless inter-working and provide mobile connectivity. The UI for pure mobile voice communication between two users is already common (for e.g. BT headsets, hands-free kits for in-vehicle mobile device use, etc.). As the trend in mobile communication shifts from pure voice communications to integrated voice and data consumption, easy access to data services becomes more important.

Future interfaces will therefore need to support the exchange of digital media between different rendering and capture devices. Most notable in this arena is the introduction of portable media systems that enable the users to take video, photos and music with them anywhere. These devices will interconnect with TVs for playback, Personal Video Recorders for archiving content and PCs for interchanging content using a variety of ports and removable storage solutions. The mobile industry can leverage its strengths in Digital Rights Management (DRM) and micro-billing to support a ubiquitous content management solution whereby the mobile terminal, by making use of the Subscriber Identity Module (SIM), could be used for controlling the portability and use of digital content between devices.

In the mobile space, the current UI already supports interconnection to the PC, via BT, IR or USB, for synchronization and back-up of personal data as well as the transfer of content, such as music tracks. In future, the UI of mobile terminals could be developed to interface with a domain of interconnected devices such as portable music device, games console, PC, in-car stereo, home Hi-Fi, DVD player, camera, etc. by means of short-range wireless technologies such as BT. The interconnection between devices could also enable the UI of the handset to allow the phone to turn into an impromptu remote control for various consumer electronic devices, such as TVs and Home Theatre systems.

In addition to data transfer and controlling other devices, interconnectivity will also enable the mobile handset to act as a communications module for a range of devices. For instance, while a games console might be used for rendering a football match in a Football management game in the home, the game play could be continued into the mobile space with the user receiving status information on injured players by MMS. Furthermore, the user is able to upload top scores to an on-line tournament competition score table.

Although BT is becoming the de facto technology for interconnecting devices, the interface is not very user friendly and inter-operability caused by different BT implementations is an issue. Future interfaces need to address these problems in order to improve usability.

Near-Field Communications (NFC) technology [54], including Radio Frequency IDentity Module (RFID) [55] and contactless smartcards, is another short-range wireless communications technology that will enable the mobile device to interact with other machines, such as ATMs, vending machines, etc. This technology could also be used for service awareness and adaptation purposes.

Wearable Elements of the UI

It is now possible to incorporate some of the functionality of the mobile phone into clothing/accessories as a means of increasing ease of use [56]. Wearable devices could in fact provide ubiquitous communication and information access through interconnectivity and inter-working. This area could perhaps result in more radical development that might determine how future interfaces will evolve.

This topic has been much discussed in the past, with particular reference to the 'Dick Tracy' watch but which has only recently become a reality due to the modularization of the terminal platform and integration of the chipset into ever smaller die sizes. As a result, technologies such as head-mounted displays for watching high-resolution video data now only weigh a few grams and could be easily attached to ordinary eyeglasses providing comfort and style in a discrete form compared to earlier generation video eyewear products.

Other developments in this area include mounting a camera in a pair of sunglasses, in a baseball cap or other pieces of clothing. The aim is to allow the user to instantly capture moments of everyday lives. The images could either be filtered on the fly or later on through automatic content annotation, classification and metadata generation. Such an innovation could take the concept of mobile blogging to a new dimension. Clearly this sort of innovation raises questions on personal privacy and inappropriate usage, and builds on the current backlash against mobile terminals. Currently, measurements are already being taken to restrict the use of camera phones in social as well as working environments.

Advances in the wearable electronics area are also targeting hands-free phone operation in order to enable movement and communication anywhere and anytime with comfort and convenience. Such devices have appeared in the market in various forms, ranging from sunglasses [57] to belts [58]. The latter example, which is zip-fastened at shoulder height, integrates a speaker, microphone and pouch for the mobile phone. The major drawback with such a device however, is the reduced privacy as anyone within earshot is able to hear both sides of the conversation. Hence it is debatable whether such concepts will catch on, especially now that BT headsets for handling voice calls via speech commands have become commonplace.

However, while such headsets have become popular in Europe and the United States, they have not caught on in Japan, probably because people there are reluctant to be seen as talking to themselves. This has fueled research in another direction resulting in a new kind of wearable mobile phone that utilizes the human hand as part of the receiver [59]. This device is worn on the wrist and converts voice signals into vibration through an actuator, which is then channelled through the bones to the tip of the index finger. By inserting the finger into the ear canal, the vibration can be heard as voice. Since the microphone is located on the inner side of the wrist, the posture of the user's hand, when using the terminal, is the same as when using a mobile phone.

This elegant solution however, presents another dilemma: the number keys on the small mobile phones have reached the smallest dimensions practical for fingertip use. The need for buttons is altogether eliminated by using an accelerometer to detect the tapping action of fingers. Combinations of the finger tapping sequence serve as Morse code-like commands such as 'talk' or 'hang up'. Through a 5-stroke tapping sequence, approximately 30 commands can be issued.

Taking this concept to its logical conclusion might sometime in the future facilitate mobile phone implants whereby a micro-vibration device and wireless receiver are implanted in a tooth during routine dental surgery [60]. Similar to the concept described above, sound is transferred from the tooth into the inner ear by bone resonance. Sound reception would be totally discreet enabling information to be received anywhere and at anytime.

Context-adaptive UI

The ubiquity of mobile communications and information access indicates that users will want to use their mobile devices in different environments, like the home, the office, whilst driving, out in the street, etc. For each of these scenarios, there is a change in the functionality requirements as well as in the interaction method. For instance, a user might be a commuter travelling to work in the morning, a businessman during the day and a father or a friend in the evening. Thus although a QWERTY keypad for simplifying e-mail composition would be ideal for use while traveling to and from meetings, it would perhaps be an overkill for organizing an evening out with family or friends.

Furthermore, until recently, the UI only needed to support one-to-one communications. However, services that enable group-based communications, such as Push-to-Talk and Instant Messaging, are placing new requirements on the UI to provide an easy way to set up and manage one-to-many communications.

The need to SMS in different contexts places heavy constraints on the interface and is a major influencing factor in the design of future UIs. As a result, we may see a shift towards flexible UIs that adapt content and services to the context in which they are going to be used and through the most appropriate interaction channel. The adaptation is based on information elements that characterize the user situation, such as location, environment and activity.

Context-adaptive UIs will also need to support the dynamic transfer of a particular task between interconnected devices in a seamless manner. The idea is to enable the user to continue with an activity while being able to move freely between different environments populated by different devices that support different interaction methods. Such a concept would need to rely on various sensors and intelligent mechanisms to detect changes in the user context and availability of devices.

The introduction of different sensor capabilities in the mobile device is likely to be a major area for the development of context-adaptive UIs. Activity sensors, such as accelerometers, are just one example. This type of sensor can be used to determine if the user is stationary or involved in some kind of physical activity. The UI can then select the most appropriate interaction modes for data entry and presentation of information. For example, if the user is jogging, voice commands and TTS would be more fitting. On the other hand, information, such as call and messaging alerts, could be relayed to a watch instead.

Mobile phone cameras can also be considered as an input sensor in the mobile environment. When combined with location and position-based sensors, the resulting information could be used for automatic tracking of the user environment and making mobile devices more perceptive. For example, by taking a picture of a building, the user is able to obtain information about the building and its location. This kind of technology is aimed for 'Where am I' applications. This so-called *image-based localization* could also be used to offer rich augmented reality services, where the view of the real world is enriched text, graphics, 3D maps, animations and audible output superimposed on the traditional GUI.

In addition to sensors integrated in the device, future UIs could also garner information on the user context through ambient sensors and machine-to-machine dialogue. For example, the user initiates a route-finding application through the mobile device to get directions to a meeting. A Global Positioning Systems (GPS) receiver incorporated in the terminal returns the user's current location rather than relying on the user inputting this data manually. Directions to the meeting venue are given to the user via TTS through the in-car audio system. On route, the user is alerted of a major accident and obtains an alternative route through the mobile navigation application. During the journey, the user is alerted that the car gas tank needs refilling and is provided with spoken directions to the nearest gas station.

Virtual Personal Assistants
A vision which reaches further out into the future is the integration of an intelligent personal assistant into the mobile device [61]. The idea is to implement an anthropomorphic avatar in the mobile device that acts as a personal secretary for the user. Communicating with the owner via natural channels such as speech, gesture and mimics, the virtual assistant can fulfil routine tasks: reading emails, searching the web for information and audio-visually answering calls on behalf of the user. Equipped with a high degree of artificial intelligence, the avatar can interpret the intentions of the user and adapts to his personal preferences just like a good human secretary.

3.3.5 Recommendations

In the following section, we will summarize the contents of this section in a couple of recommendations for research and industry. We group these recommendations according to a rough timeline starting out from improving already existing technology and research areas up to visionary future developments.

3.3.5.1 Improving Mobile Devices Using Current Technology

As we have already described with currently existing technology, the usability of mobile devices' interfaces can be improved. In addition to the standard equipment of current handsets, the following features can improve the usability:

Input: Standardized intuitive keypad layout, context aware graphical menu structure, touchscreen, handwriting recognition, the possibility of attaching an external keyboard and mouse, predictive text input, and speech recognition

Output: Configurable sound signals for incoming calls (e.g. caller dependent) and system events, text-to-speech output, the possibility of connecting an external screen or projector, context aware graphical menu structure, user-definable start-screen

Although these features are relatively easy to integrate based on current technology, there is still no device in the market that includes them all. In particular, the interoperability with external input/output devices is missing in nearly all devices.

Strong standardization efforts should always accompany the integration of new features, software, and interfaces.

3.3.5.2 Short Term Research Issues

Many of the more advanced features we listed in this section are on the cusp of being integrated in the next generation of mobile devices. However, there is still some research required to find the optimum implementation. In particular, usability studies and real-world tests are necessary. In the following text, we list these research issues.

Speech recognition:
- increased the robustness of the recognition by incorporating more language knowledge and visual features;
- speaker-independent recognition;
- natural language recognition.

Gesture recognition:
• research on which functions profit from gesture control;
• increase robustness of recognition to unintended operation.
Speech output:
• natural language output;
• seamless integration of voice output with phone functions.
Context aware menus:
• personalized menus;
• self-learning menu preferences;
• context dependent help functions.

All research efforts should always be accompanied by quantitative acceptance and usability tests to provide potential industry partners with reliable figures on the market chances of the new technologies.

3.3.5.3 Promising Future Directions

Based on the contents of the chapter on future UIs we list some promising areas for joint academic and industrial long-term research.

Advanced input devices:
• haptic interfaces;
• gesture-controlled virtual reality;
• localization based on object recognition.
Advanced output devices:
• foldable displays;
• built-in video projectors;
• 3D head-mounted displays;
• 3D audio menus.
Intelligent assistants:
• anthropomorphic avatar;
• virtual personal secretary;
• natural human-device dialog.

Naturally, the predictions in this field have a more speculative character. Therefore, virtual or physical concept demonstrators are necessary.

3.3.6 Summary

In this section, we have analyzed the advantages and drawbacks of current UI elements for mobile devices. Based on this analysis, we described ways to overcome these drawbacks in the future, making use of already existing technology.

Further, we have described promising research efforts aiming at an enrichment of man-machine interaction.

Finally, a list of recommendations aims at providing a rough guideline for industrial and academic research in the field of UI technology for mobile devices.

3.4 Acknowledgment

The following individuals contributed to this chapter: James Stewart (University of Edinburgh, UK), Andy Aftelak (Motorola, UK), Hans Nelissen (Vodafone, NL), Jae-Young Ahn (ETRI, Korea), Axel Steinhage (Infineon Technologies, Germany), Maria Farrugia (Vodafone Group R&D, UK), David Pollington (Vodafone Group R&D, UK).

References

[1] L. Marturano, T. Turner, P. Pulli, P. Excell & M. Visciola: *The Use of Scenarios for the Wireless World*. WWRF WG 1 White Paper, 2003.
[2] R. Tafazolli [Ed]: *Technologies for the Wireless Future: Wireless World Research Forum (WWRF)*. Wiley 2004.
[3] K. Crisler, T. Turner, A. Aftelak, M. Visciola, A. Steinhage, M. Anneroth, M. Rantzer, B. von Niman, M. Sasse, M. Tscheligi, S. Kalliokulju, E. Dainesi, A. Zucchella, (2004). Considering the User in the Wireless World, *IEEE Communications Magazine*, vol. 42 no. 9, pp. 56–62.
[4] K. van der Heijden: *Scenarios – The Art of Strategic Conversation*. Wiley 1997.
[5] P. Ehn: *Work-oriented Design of Computer Artifacts*, Second edition. Erlbaum 1989.
[6] M. B. Rosson, J. M. Carroll, *Usability Engineering Scenario-based Development of Human Computer Interaction*. Morgan-Kaufmann 2001.
[7] I. Alexander & N. Maiden [Eds]: *Scenarios, Stories, Use Cases Through the Systems Development Life-Cycle*. Wiley 2004.
[8] Wireless Strategic Initiative, http://icadc.cordis.lu/fep-cgi/srchidadb?CALLER=PROJ_IST&ACTION=D&RCN=53660&DOC=1&CAT=PROJ&QUERY=1.
[9] Wireless World Initiative, http://www.wireless-world-initiative.org/.
[10] K. Crisler, M. A. Sasse, M. Anneroth & A. Steinhage: *A Human-centered Reference Model for the Wireless World*. WWRF WG 1 White Paper, 2005.
[11] A. Aftelak: *Consolidation of the Scenarios from the Wireless World Initiative (WWI)*. Contribution to WG 1 at WWRF12, Oslo, June 2004.
[12] The Mobilife Project, https://www.ist-mobilife.org/.
[13] E. Kurvinen, M. Lähteenmäki, D. Melpignano, O. Pitkänen, H. Virola, O. Vuorio & I. Zanazzo: *User Responses to Scenarios of Future Mobile Services in Mobilife*. Contribution to WG 1 at WWRF13, Jeju, March 2005.
[14] S. Kalliokulju, T. Liukkonen-Olmiala, J. Matero, M. Asikainen, H. Mansikkamäki, A. Aftelak & M. Tscheligi: *A User-centered Design Process for Wireless World Research*. WWRF WG 1 White Paper, 2003.
[15] End-to-End Reconfigurability E2R, http://e2r.motlabs.com/.
[16] Ambient Networks, http://www.ambient-networks.org/.
[17] K. Ducatel, M. Bogdanowicz, F. Scapolo, J. Leijten & J-C Burgelman: *STAG, Scenarios for Ambient Intelligence in 2010*. IPTS-ISTAG, EC: Luxembourg, 2001.
[18] B. Karlson, A. Bria, J. Lind, P. Lönnqvist & C. Norlin: *Wireless Foresight: Scenarios of the Mobile World in 2015*. Wiley 2003.
[19] IT 839 Strategy: *Ministry of Information and Communication*. Korea, http://www.kora.or.kr/eng/index.jsp.
[20] *Socio-cultural Impacts of IT on Korea: Mega-trends in the 21st Century*. http://www.kisdi.re.kr/kisdi/wwbs/eng/news/eNewsView.jsp?kind=21&seq=4733&PAGE=4&searchKey=.
[21] http://www.ngmcforum.org/.
[22] NTT Docomo Vision 2010, http://www.nttdocomo.com/vision2010/.
[23] Wireless World Research Initiative, www.ist-wwri.org.
[24] *MIT Project Oxygen: Pervasive Human-Centered Computing*. http://www.oxygen.lcs.mit.edu/.
[25] Wireless World Initiative New Radio WINNER, https://www.ist-winner.org/main.htm.
[26] Future Mobile Communications Markets and Services, http://fms.jrc.es/pages/about.htm.
[27] Systematic Product Concept Generation Initiative TUTTI, http://www.machina.hut.fi/project/tutti/.
[28] Wireless@KTH, http://www.itsweden.com/main.aspx?newsid=381&id=13&type=news.

[29] Mobile IF Forum: Flying Carpet II: Towards the 4th Generation Mobile Communications System, http://www.ttc.or.jp/e/external_relations/cjk/cjk_5th/CJK5_008.pdf.

[30] J. Carroll: *Making Use: Scenario-Based Design of Human Computer Interactions*. MIT Press 2000.

[31] J. Stewart, L. Pitt, M. Winskel, R. Williams, I. Graham, J. Aguiar, L. M. Correia, B. Hunt, T. Moulsley, F. Paint, S. Svaet, B. Michael, A. Burr, T. G. Eskedal, V. Yin & C. Stimming: *FLOWS Scenarios and Definition of Services*. FLOWS-IST Project Deliverable, D06, University of Edinburgh/FLOWS/European Commission IST Programme: Edinburgh, 2002.

[32] J. Stewart: Conxt Perspectives for Scenarios and Research Development in Mobile Systems, In: L. Hamill & A. Lasen [Eds.] *Wireless World: Mobiles – Past, Present and Future*. Springer 2005.

[33] J. Aguiar, L. Correia, J. Gil, J. Noll, M. E. Karlsen & S. Svaet: *Definition of Scenarios (D1). FLOWS-IST Deliverable, D1, IST-TUL/FLOWS Project/EC IST Programme*. Lisbon 2002.

[34] *Flexible Convergence of Wireless Standards and Services- FLOWS Project*. http://www.elec.york.ac.uk/comms/FLOWS.html.

[35] J. Sherry, T. Salvador & H. Ilhaiane: *Beyond the Digital Divide: Thinking through Networks*. Contribution to WG 1 at WWRF 11, New York, Oct. 2003.

[36] E. Price, J. Gandy, J. Mueller: *User Needs Analysis of Persons with Disabilities using Cellular Phones*. Contribution to WG 1 at WWRF 11, New York, Oct. 2003.

[37] Vodafone Futurevision, www.vodafone.com/futurevision.

[38] http://www.etsi.org/pressroom/Previous/2003/2003_10_UI_stf231.htm.

[39] A. Steinhage *et al.*, User-Interface-Technology and – Techniques, In: Rahim Tafazolli [Ed.] *Technologies for the Wireless Future*. John Wiley & Sons, pp. 60–73, Oct. 2004.

[40] http://www.digitwireless.com/.

[41] http://www.cic.com/Apps/PRDetails.aspx?id=237.

[42] http://www.research.philips.com/technologies/display/ov_polyled.html.

[43] http://www.mitsubishi-presentations.com/projectors.asp.

[44] http://www.microoptical.net/Applications/mobile.html.

[45] http://www.voicesignal.com/shared/vstdemo/index.html.

[46] http://www.theregister.co.uk/2004/09/14/speech_silicon/.

[47] http://www.newscientist.com/article.ns?id=dn6228.

[48] M. Sanchez & J. P. de la Cruz: AudioVisual Speech Recognition Using Motion Based Lipreading, *Proceedings of the 8th International Conference on Spoken Language Processing ICSLP (Interspeech)*. Sunjin Printing Company Oct. 2004.

[49] http://www.immersion.com/mobility/.

[50] http://www.dcs.gla.ac.uk/people/personal/rod/publications/CrosMur04.PDF.

[51] http://www.ecertech.com/.

[52] P. Eslambolchilar, R. Murray-Smith: Tilt-based Automatic Zooming and Scaling in Mobile Devices – a state-space implementation, *Proceedings of Mobile Human-Computer Interaction – MobileHCI 2004: 6th International Symposium*. Springer-Verlag, pp. 120–131, 2004.

[53] http://www.sonicspot.com/guide/3daudio.html.

[54] http://www.nfc-forum.org/.

[55] http://www.rfidjournal.com/.

[56] D. Marculescu, R. Marculescu, N. H. Zamora, P. Stanley-Marbell, P. K. Khosla, S. Park, S. Jayaraman, S. Jung, C. Lauterbach, W. Weber, T. Kirstein, D. Cottet, J. Grzyb, G. Tröster, M. Jones, T. Martin, Z. Nakad, Electronic Textiles: A Platform for Pervasive Computing, *Proceedings of the IEEE*, vol. 91, no. 12, Dec. 2003, pp. 1995–2018.

[57] http://oakley.com/o/c771s/.

[58] http://www.networkitweek.co.uk/computeractive/hardware/2013186/orange-wearaphone.

[59] http://www.nttdocomo.com/corebiz/rd/fingerwhisper.html.

[60] http://amo.net/NT/06-19-02Tooth.html.

[61] A. Steinhage: A Virtual Personal Assistant as Natural User Interface. *Proceedings of the 11th Wireless World Research Forum*, Oslo June 2004.

4

Service Infrastructures

Edited by Stefan Arbanowski (Fraunhofer FOKUS, Germany)
and Wolfgang Kellerer (DoCoMo Euro-Labs, Germany)

4.1 Introduction

Industry is pushing new standards that allow high data rate multimedia applications as well as seamless communication across heterogeneous radio and network technologies. The success of the next generation mobile telecommunication systems will, however, depend on the services and applications that can be provided. These future systems are expected to integrate the paradigms of traditional mobile telecommunication systems with the Internet protocols (IPs) and state-of-the art software engineering methods. In addition, new paradigms will emerge. For example, customer acceptance is considered widely increased by tailoring services and applications to actual user needs, their preferences and the context a user is in. A well-engineered next generation service platform should provide all capabilities to allow innovative services to be created and deployed in a short time, addressing user needs. Third party interfaces allow a chaining of expertise in service provisioning. Semantic technologies may help to structure knowledge about the user environment.

On the basis of this consideration, rich Internet services enhanced by personal context aware mobility will be the ultimate application. This requires a convergence of Internet technologies and mobile telecommunication technologies but also means a convergence of current providers' respectively stakeholders' roles.

Standardization might be one answer to achieve a globally accepted mobile service architecture comprising emerging capabilities. Whereas before standardization, the industry was mainly concentrating on radio and networking aspects, the variety of service capabilities and the need for their global inter-working among several providers requires common interfaces also on the service platform level. Discussion beyond 3G service architectures is just starting. This chapter summarizes first steps such as motivation, scenarios, and requirements for service architectures and generic service element for B3G mobile communications.

Technologies for the Wireless Future – Volume 2 Edited by Rahim Tafazolli

4.2 Requirements for Future Service Platform Architectures

A next generation mobile service network has to establish its own value proposition between the stakeholders of next generation mobile service provisioning. It has to place its unique selling points into the overall service-centric environment dealing with a number of different access systems. Mobile networks have the potential to act in a central role within this service environment and is therefore required to be capable of acting as service control environment. It is not the intention to copy all successful Internet services but to support them in an efficient and trustworthy way.

The Service Platform Architecture is interfacing to the service and applications at the upper layer of a communication system [1]. The Service Platform Architecture covers service support components (such as Generic Service Elements), their relationship and their internal and external interfaces. It usually targets the upper system layers as depicted in Figure 4.1. However, some functions or parts of them are residing in lower system layers, for example, location or mobility. When considering ubiquitous communications environments, such as sensor networks, the layering is diminishing anyway.

Service Architecture for future mobile systems must be able to satisfy requirements from a broad range of services and different kinds of service usage. This can be reflected in the consideration and analysis of various future life scenarios. With this approach a scenario-centered design process can be applied.

During requirement analysis, a number of different scenarios have been considered. They have been obtained from a number of different sources, including project results, dedicated studies and published literature. The following list provides their references [2–11].

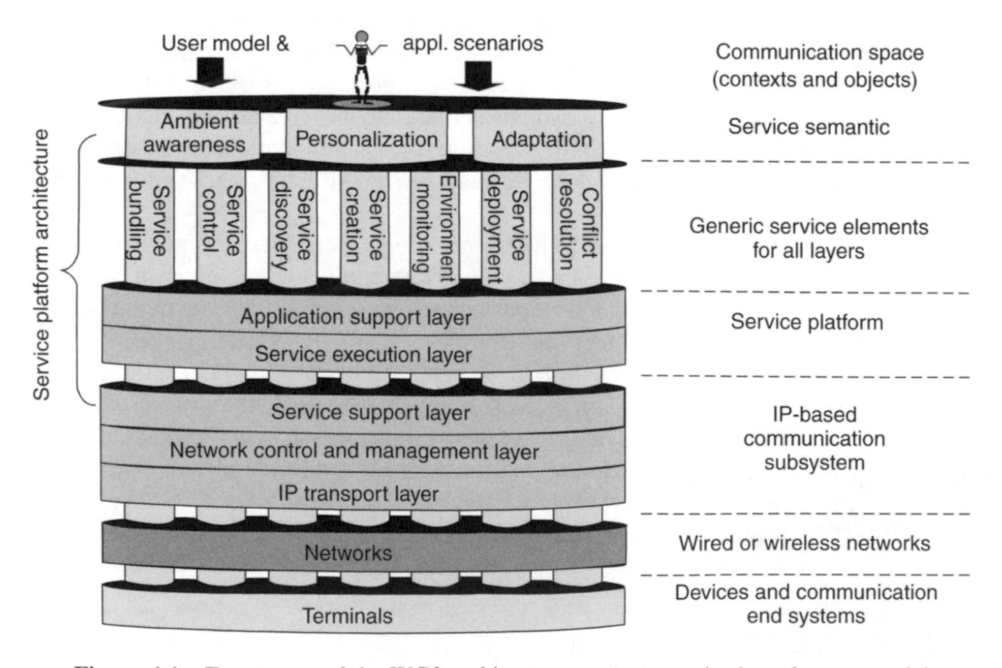

Figure 4.1 Target area of the WG2 architecture work shown in the reference model

Although there are a lot of scenarios available, their impact on service architectures is rarely obvious. The reason might be that service architectures have not been in the focus of the scenario creator at all. Alternatively, the scenarios might be tailored to specific service architectures, that is, those activities do not comply with the scenario-centered design process. In cases where scenarios were in fact considered for the design process, these are often describing very specific aspects and the resulting service architectures are very much tailored to them.

To gain the desired demands from future life scenarios on service architectures, it is required to evaluate a selection of scenarios to derive commonalities and their specific features. A first analysis resulted in a number of aspects, which can be taken as initial input for the service architecture design: Future Services will be context-sensitive, adaptive and personalized. They will be available in different networks, with different bandwidth/QoS and for different devices or multimodal User Interfaces (UIs) respectively. Service architectures need to support sophisticated charging and billing, security and privacy, identity management, Digital Rights Management (DRM) and trust. The complete service lifecycle is to be reflected: from service creation and composition to service discovery and delivery.

4.2.1 Challenges in Future Service Provisioning and Interaction

Personalization is one of the most important characteristics of future telecommunications. Personalization means the tailoring of services and applications to the very specific needs of a user to a ubiquitous comfortable service environment together with a single bill service independent of the partners involved in the end-to-end value chain. Personalization also means to allow fully manual service configuration, carrier access without any automatic vertical handover procedures, and individual invoicing of the individual service partners, in case this is required by any user for any reason.

Market participants are faced with the following trends:

- Content and services are distributed across domains including fixed line and mobile networks as well as open (e.g. World Wide Web) and closed service environments. Business relations might be distributed.
- Internet services are mainly used by fixed line access, whereas hotspots have just started.
- Technology evolution separates call control and transport. The directions of centralized or spread out of the service control are regarded as a chance and a threat for mobile operators.
- Transport charges are dropping tremendously due to excessive IP transport capacities in backbone networks and due to increased competition in access networks (flat rates).
- There is a strong impact of IT technologies on telecommunications protocols, system design and interfaces.
- A growing importance of peer-to-peer services integrates the single user into the group of service and application providers.
- An increasing importance of ubiquitous computing technologies (sensor networks, Radio Frequency Identification (RFID), Near Field Communication (NFC)) leads to pervasive interaction with service systems and provides new possibilities for services based on rich information.
- Trust in context aware services still suffers from the privacy concerns of the users. In addition, legal aspects also restrict providers from offering rich-context aware services.

These aspects lead to the following conclusions:

- Users will maintain several business relations in different service domains.
- Access procedures including authentication and authorization will become more and more complex and will happen more often.
- IP-based services will dominate.
- Service control is not per definition in the domain of the telecommunication operator.
- Telecommunication network operators can hardly survive by providing bit pipes only. Value-added services are a prerequisite for the future business of telecommunication operators. Mobile network operators have to identify their unique selling points and to force them into the market.
- A trustworthy position in the market is important for acceptance by the users.

Figure 4.2 illustrates our vision of a multidomain network and service environment. Mobile services platforms of telecommunication providers and mobile operators in parallel to third-party service environments provide service through application servers to the customers. Additional functionalities, for example, identity manager or billing, are depicted as separate functions that may also be provided by third parties.

4.2.1.1 Primary Service Features

In the future, features will, to a great extent, be considered with regard to the value they provide to the user. But what is the value for a broad range of different users? There are basically two groups of users:

- Professionals
  For business users the world is quite simple. The value of a service or a feature is considered as the ratio of cost savings in the business processes to the expenses for the service/feature.

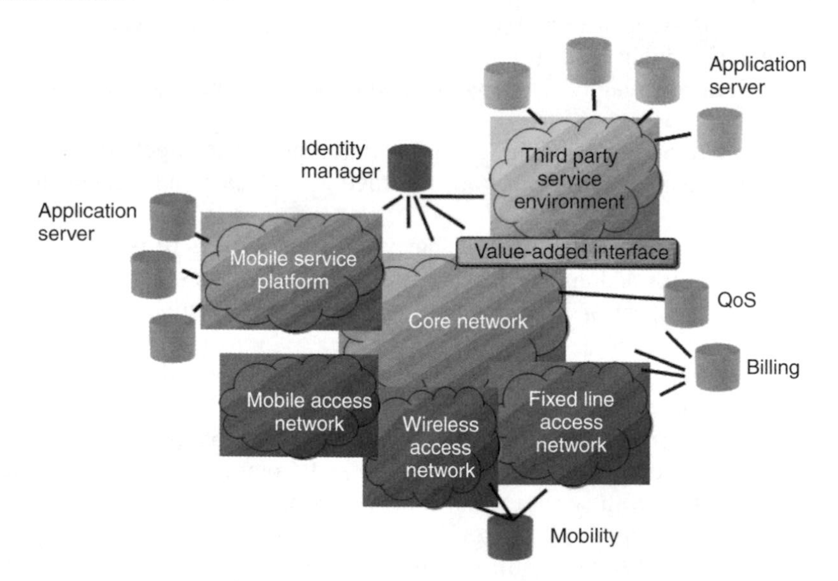

Figure 4.2 Platforms, networks and roles in architectural view

- Consumers
  In case of private users, it is more complicated due to the impact of status and fun in consumer goods (the consumer market is very segmented). Nevertheless, there are also clear constraints given by the economic situation of private households.

  The following value-added services have been identified in the discussions of WG2:

- Personalization
  Mobile services are very personal since they accompany the user together with a personal device. This is in contrast to desktop equipment, where multiple users might gain access to services via the same device. Therefore, personalization as defined in Section 4.2.2.3 makes special sense in the mobile context.
- Context awareness
  In addition to personalization the context of the user, in contrast to fixed line access, is less predictable in mobile usage scenarios, which gives a special value to context aware controlling of communication processes. Nevertheless, context awareness is one of the most complex features since there are various potential context parameters to be taken into account.
- Mobility and seamless service access are native properties of mobile networks which are meeting a basic requirement of business users.
- Adaptation
  In order to support a wide adoption of services in a diverse environment (heterogeneous devices, networks, usage patterns), services adapt to the context of a user.
- Usability
  Ease of use is a requirement for mobile services other than voice telephony to become accepted for the mass market. It not only includes UI, but also includes information to the user about the status and costs of active services and reasoning for performed adaptations (e.g. degradation in video quality).
- Identity management/authentication/authorization/service access control
  A trusted ID management as well as reliable authentication and authorization procedures are a valuable legacy of GSM networks. On the basis of this well-established customer relationship, a service access control system can be established serving not only within the telecommunication network but offering this service to other application domains.
- Security, privacy and trust
  The success of the future systems will depend in totality on the consumers' trust. The current Internet clearly demonstrates that 'add it later' security does not work. Security, trust and privacy must be addressed from the very beginning of system design and on all levels: hardware, operating system (OS), protocols, middleware, and applications.
- Accounting/Charging/Billing
  A trusted and secure charging and accounting system has been established for existing 2G/3G communication systems. Offering this environment to the applications in different service domains would be a huge benefit for all participants.
- Service discovery and composition
  The service architecture of the future should be able to support distributed and composition views that consider a service as just a component that can be used to compose additional value-added services. In other words, a service is just a building block that is used along with others to create 'more' value-added services.

4.2.1.2 Potential Roles of Stakeholders

For future mobile service platforms, the traditional roles of the stakeholders are not fixed any longer. Furthermore, additional stakeholders enter the market place. In general, we can distinguish the following stakeholders:

- End-users
  The end-user is the stakeholder that uses services. Furthermore, it is useful to observe that an end-user can have different kinds of roles depending, for example, on whether the business relationship with the provider is through a private or business-related contract.
- Service platform operator
  The service platform operator provides the service environment such as service functionalities and the execution environment. The actual environments can be distributed among the core network, edge devices, and user terminals. Therefore, the management of the distributed service environment has to be done partially by the end-user and partially by the service platform operator.
- Service providers
  The service provider operates services offering them to the end-users. The service provider may be the operator itself or may be a third party. It can be distinguished between elementary service providers (providing own services) and service aggregators (providing bundled services that consist of services offered by elementary service providers).
- Mobile network operator
  A mobile network operator provides mobile network infrastructure like core network, radio access network, and interconnectivity network to service providers that want to offer mobile services. Typically, mobile network operators also act as service providers, offering mobile services to end-users. Nevertheless, mobile network operators resell their network infrastructure to third parties, which do not have their own network.
- Content providers
  The content provider offers content such as video or audio to the users via a service provider. Content providers can be distinguished as elementary content provider (offering their own content) and content aggregators (offering content aggregated from different sources or players).
- Service developers
  The service developer develops service-related functionalities either by implementing functionality components or by linking components together. Usually, the result of the service development process is then offered to end-users via a service provider.
- Third parties
  A third party is a stakeholder taking over particular responsibility from any other stakeholder. This helps to open market segments, but requires well-defined interfaces between all stakeholders and the third party. In particular, security and authorization aspects have to be enforced by such interfaces.

New roles are emerging that have not existed or have not existed as a role that could be provided by a separate entity. We distinguish the following roles:

- Network provider
  The network provider operates the connectivity between the stakeholders and ensures adequate service quality in the core network.

- Access/mobility provider
  The access/mobility provider maintains the access network connecting the end-user terminals, for example, with service functionalities or other users. There are several different kinds of access providers from small to big players (from hotspot provider up to global coverage).
- Service provider
  Similar to the content provider, the stakeholder service provider adds the dynamic aspect that every stakeholder might act in the role of a service provider to the traditional stakeholder concept. For example, a terminal provider, which preinstalls games or portal access on handsets he sells, will then become a service provider.
- Service broker
  A service broker announces services to end-users by providing service discovery facilities. Service brokering can be proactive (end-users are contacted by the broker) or reactive (end-users do contact the broker).
- Content provider
  The role content provider adds some dynamic aspects to the traditional content provider stakeholder. Since even the end-users can become content providers, exchanging content in an infrastructure or peer-to-peer–based manner, the monolithic structure of content providers will vanish.
- Reseller/Retailer
  Resellers or retailers buy services, functions or network capacity from other players and resell them typically bundled with other offerings they might have. For example, a new agency might bundle the information they have together with messaging capabilities they buy from a mobile operator to offer them a mobile news service.
- Portal provider (ease of use)
  A portal provider provides a specialized interface (portal) that suits the requirements of (mobile) clients. This portal acts as a single service access point to help end-users access services. Typically, the portal can be customized to reflect user preferences. The mobile device or the portal itself might manage these preferences.
- Terminal provider
  A terminal provider sells or leases mobile terminal equipment either to end-users or third parties, which then also become a terminal provider.
- Identity provider
  The identity provider manages end-user–related information such as customer relationship data, presence information, and context information. When interacting with any kind of service, the identity provider can provide this information to improve the service experience.
- Adaptation provider
  Since adaptation is not an easy task to do, there is a big opportunity for third parties offering specialized adaptation functions to service providers or end-users. Adaptation providers realize a special kind of service enabling the adaptability feature that was introduced earlier.
- Billing provider
  The billing provider offers a single point of contact between the end-user and other stakeholders acting in different roles. It provides one bill for several services. The

end-user has only a business relationship with the billing provider, whereas the billing provider is maintaining the financial relationship to any other role.

- Trusted third party
  One important aspect that has been introduced with third-party stakeholders is contract enforcement. This should ensure that the third party behaves as agreed between cooperating roles. A trusted third party has prearranged policies that ensure trustful cooperation. It can also serve as a trusted mediator for any other cooperation between stakeholders.
- License provider
  A license provider owns licenses dedicated to specific regulatory issues (i.e. spectrum usage license, service provisioning license etc.). If regulation allows, the license provider can give these licenses to other parties. Special license providers are the governmental authorities that issue the licenses to be used by any stakeholder.

This observation of new additional stakeholders and especially the emergence of new roles in service provisioning show that the traditional roles in service provisioning are blurring. As discussed earlier, this is a result of the convergence of the telecommunications world and the IT world. In future, many different cooperation models of stakeholders and their respective roles are imaginable.

One extreme case would be, for example, a service provider that is providing services independently of any own network infrastructure relying on mobile network operators for connectivity. In addition, content and services may come from content providers and service developers as external stakeholders. The service provider has the role of an orchestrator independent of traditional service and telecommunications infrastructure.

On the other hand, based on its traditional position in service provisioning, the mobile operator could easily provide most of the listed roles that are usually not all provided today or are only provided for its own domain original services. Such roles include billing providers, portal providers, retailers, brokers, trusted third parties, and identity providers. This would empower the mobile operator to control the whole service provisioning process, which is of benefit to the user, for example, in terms of personalized services.

Another extreme case is a totally distributed situation, where all roles are fulfilled by independent stakeholders described as third-party providers who collaborate in providing billing, adaptation, brokerage, portals, identity, service platforms, terminals, and so on.

Another essential aspect of these emerging roles is the empowerment, emerging from the IT world, of the user to take over additional roles such as those of service provider or content provider. Considering the widespread use of private open wireless local area networks (WLAN)/DSL access points, the user might even become an access provider himself.

Service provisioning is expected to change with coming generations. New capabilities such as personalization and context awareness have to be considered. In addition, the roles of stakeholders are blurring. In this motivation, we have highlighted the most important challenges, capabilities and roles identified within WG2. This motivation will be detailed and set in more concrete terms by the following discussion of scenarios and requirements.

4.2.2 Functional Requirements

Published visions, applicable to the needs of enterprise, personal, and embedded computing, of the wireless future can be summarized as follows. Future applications and platforms will be context aware, adaptive and personalized. In addition, they need to be run, in a reasonable and secure manner, in a variety of execution environments: anywhere, anyhow, anytime, and by anyone. These properties have been examined by the wireless world research forum (WWRF) [2] in detail.

In this section, we briefly elaborate the functional requirements for systems beyond 3G. We have divided the requirements into nine groups: 1) High-level user and stakeholder requirements, 2) General system requirements, 3) Business model requirements, 4) Regulative perspective, 5) Applications and services perspective, 6) Technology perspective, 7) Multidevice, personalization and multimodal user interaction requirements, 8) Privacy and trust requirements, and 9) Context awareness requirements.

4.2.2.1 High-level User and Stakeholder Requirements

These eight high-level requirements are derived from the issues discussed in the WWRF WG1. They reflect the value plane in the WWRF Reference Models and fundamental system properties.

Best possible user experience. Users have expressed their traditional requirements that service offering has to be fast, simple and useful, and that charging mechanisms have to be flexible. The user should know what the cost is of accessing/using the service, as it is the situation today with the plain old telephone services (POTS). They are not interested in technologies, for example, access mechanisms, which should be hidden from them. Complexity has to be avoided on all levels. The UI should contain the minimal set of 'actions' and 'buttons'. Multimodality shall be supported in the UIs (several senses to be leveraged). It is to be noted that some interfaces need to be supported for long periods like 10 years (e.g. car applications).

User in control. Instead of automatically managing, for example, scheduler entries, the system should help the user in managing them. Also, the user needs to be in control:

- of whether they do or do not receive 'pushed' information (proactive suggestions made by the system, for example, location and/or context-based alerts or profile-related offers).
- adjustments made to the interface of his/her device.
- who gets to see and access their content.

Provisioning of new services and applications supported. It is quite a clear requirement that the technologies and business models should allow anyone at anytime and anywhere to take the role of the content/service provider. They should also support the short and cost-efficient time-to-market requirement of a new service development. That requires standardized interfaces (UI, systems) and service deployment/delivery/operability practices. Additionally, the distributed content provisioning should be supported. The short lifecycle of content requires fast development/delivery models.

The ownership relations of data are complicated and existing models are not sufficient in the future; the DRM is to be developed.

Seamless access. As has already been stated above (User Experience), the end-users are not interested in technologies, which is also valid for w.r.t. access types; users are access (or – in general – technology) agnostic, they want to see services, not technology or bandwidth. Anything, that is comparable with the complexities of, for example, the general packet radio service (GPRS) access configuration, will not be acceptable. Also, the handovers between different access types should be invisible to users (if their reaction/feedback is not needed).

The look and feel of the services should not depend on the system or country.

Mobile access and terminal should mainly be seen as a temporary substitute to access some services, for example, TV programs, e-mail or WWW pages, while the 'primary' means are not available.

QoS. Users should not be bothered, mainly, with the questions of QoS, because the majority of them do not even have a knowledge of the related concepts; that is, QoS should be just right for the service in question – with an affordable and predictable price. If the quality is too low or the price too high, users will simply not use the service. Would any service provider be willing to offer a service of low quality? If yes – why? However, there may be some user groups, who would be willing to increase the level of quality explicitly, and would like to pay an extra fee for it; how many are there?

Security, privacy and trust. For simplicity (from the end-user's point of view), there might only be a few trusted entities. Users want to have control on information created from them; they also want to know who has information on them, and what information. Also, the level of privacy should be changeable/adjustable by a user according to the ongoing task. If the required level of privacy and trust is threatened, the user should be informed of it.

Owing to mobility it will be important to harmonize the related national and international legislations. New technologies, for example, based on biotechnologies, will emerge in the area of security. Standardization is needed to homogenize the systems of security, and their accesses.

Finally, seamless sign-on is needed to avoid unnecessary complexity.

Deployment and operation. There is a requirement for decentralized service management or management by third parties; this should not take place at the cost of reliability. However, the relation between the personal area network (PAN) and public network (PN) management should be defined.

Users want to have 'some' access to the management systems to control their own private communication environment. This seems too complicated for the masses (4 billion people) – how will it be realized in practice? The interoperability of heterogeneous systems is needed. Operation Expenses (OPEX) is usually underestimated; could self-organization be a resolution to this problem?

Migration. Both revolution and evolution is foreseen: revolution in service provisioning and evolution in systems (especially network systems) development. The changing work and enterprise patterns may contribute to the revolution. Also, there should be a paradigm shift from managing networks/systems to managing services.

The excessive complexity of systems and interoperability should be avoided; it would make the terminals and systems too complicated to implement. Extendability of the subsystems and components would support evolution.

4.2.2.2 General System Requirements

These 13 requirements are essential system properties needed to realize context aware, adaptive, and personalized applications usable everywhere and by everybody.

The architecture shall support the user-centric approach. The end-user is the main stakeholder in the system and uses the services. The User Centric Design (UCD) framework recognizes that users are a central part of the software development process, and that their needs should be considered from very early in the development process.

The architecture shall support mobility. Support for users with mobile devices and mobility are important requirements for the software infrastructure.

The architecture shall support context aware systems. The system uses information about the state of the context of the user to adapt the service behavior to the situation the user is in. Contextual information includes low-level context data such as location, time, temperature, noise, as well as higher-level context data such as user situation (in a meeting, with friends).

The architecture shall support context management. Context data management is an essential part of context aware systems. Managing context information efficiently is a challenge. Different types of context data originate from various distributed sources. This data must be gathered and made available to those components that need them. Context reasoning mechanisms must be created to elaborate and infer information from raw data.

The architecture shall support adaptation. The basic principle of adaptation is simple: When the circumstances change, then the behavior of an application changes according to the desires of a user or more precisely according to principles ascribed to her. Services and context information should be dynamically adapted in the future to the context by the use of an automated learning functionality.

Learning the wishes and desires of a user is a crucial part of adaptation. Adaptation must also be proactive, which, in turn, requires predictability of the near future.

The architecture shall support service personalization. Personalization includes the ability of the framework to acquire and manage personal information about the user, including user preferences, and the ability to use this information to adapt an application's behavior to specific user needs.

End-users may affect device characteristics by configuring its parameters. Additionally, users may have global preferences affecting all services and specific service settings, for example, their favorite language.

The architecture shall support privacy and trust. Privacy and trust are among the most important features to increase user acceptance of services. Privacy is needed to guarantee that personal user data is only revealed according to the user preferences (expressed as policies). The trustworthiness of components, identities and information need to be guaranteed to establish a basic user – application trust level.

The architecture shall support group communities. The user centricity approach also includes an approach that considers a group of users. The architecture must support the creation of groups of users and functionalities to manage the users in the group for example, enter a group, leave a group, privacy and trust issues, and so on.

The architecture shall support both managed services and spontaneous services. The framework supports services that are controlled and maintained by a service provider for example, through a service portal; and services that are provided in an ad hoc manner directly between users without the control of a third party, for example, pure peer-to-peer services.

The architecture shall be open to emerging value networks. The framework is not restricted to current providers – consumer value chains, but shall be flexible regarding new ways of service provisioning such as new ways of incorporating third party service providers.

The architecture shall support seamless service access via multiple-access technologies. The framework shall support service provisioning within the framework independently of the access technology. While the transfer between different access networks should in general be transparent for the application, in some cases an adaptation of the application is needed to ensure an optimal service experience.

The architecture shall support service lifecycle management. In addition to enhanced service features supporting user services, the framework also includes support for the service lifecycle including development, deployment, operation and removal phases for services and their constituent components.

The architecture shall support backward compatibility. Backward compatibility plays an important role in the successful deployment of service architectures. It enables an easy migration for incumbent service providers. The service architecture(s) proposed in the past and which were not backward compatible did not go far. However, there are several levels of backward compatibility and full backward compatibility may be an impediment to progress. The service architecture should be backward compatible from the following points of view:

- It should be possible to deploy it in existing 3G systems. This will enable easy migration for service providers. They can start deploying it in their 3G systems if they wish. They can then migrate gradually to post-3G.
- End-users can roam while still having access to their services with the same look and feel. This roaming includes roaming from an existing 3G system to a post–3G system. However, it excludes roaming from a post 3G system to a 3G system

4.2.2.3 Business Model Requirements

The seven requirements in this group elaborate ways in which B3G systems could become commercially viable.

The increasing role of communities should be empowered. There is a need to provide the user with the possibility to control the communities that she is a part of. Usually, some kind of negotiation with current members is needed before she can be seen as part of the community. In concert with other members, users should be able to form, control and manage the rules related to the communities and groups they are part of. The group policy determines the rules. Many communities cut across the work and home domains calling for mobile applications and services able to support the user in her multiple roles.

Service architecture and service creation should support segmentation. Segmentation in this sense means user clusters and mass customization in order to create services that satisfy the different needs of different people.

The users' personal wishes and needs point out the need for personalization functions. The customer selects the services and what kind of features they will include.

The services are more like collections of service components than single services and are thus possible to use according to the customer's expectations and decisions.

In mass customization, the customer defines the content of the service by herself and this means the definitions of dynamic services.

User should be able to access and modify her personal information, profile and preferences regardless of place, time and devices she is using. Users are using not one but many types of devices. Ideally, they should be able to access and change their personal data, preferences and profiles and so on with one device, so that the information is consistently updated everywhere.

It should be easy for the consumer to discover services suitable in the current context. Many consumers are not even conscious of many existing services. Easy discovery is especially essential in services that are to be used in mobile context.

Support value–based pricing mechanisms. Previously accepted pricing structures for mobile data services, based on charging per megabyte, are insufficient for the introduction of wireless broadband (3G) services. Charging per megabyte would make some services too inexpensive to generate sufficient revenue, whereas others would be too expensive to create any demand. Therefore, pricing needs to be value based, for example, measured from demand or related to the context of use. The pricing of services, therefore, needs to be taken at a service-specific level to generate satisfactory profits in 3G. The basis on which each service is charged is likely to vary over services. The system should support the gathering of relevant pricing information and the pricing process.

Support revenue sharing mechanisms. In the network of actors, the services are bundled from components created by different actors. Some actors collect the revenues and share them according to some 'rules' with other actors. The sharing of investments, costs and revenues is complex, and should be such that each actor is properly compensated. The system should support the gathering of information and performance monitoring on which revenue sharing may be based.

Support cost effective service provisioning and short service lifecycles. In order to provision viable and sustainable services, the architecture should support cost-effective provisioning. Provisioning should be affordable even for short or very limited periods.

4.2.2.4 Regulative Perspective

The Directive on Privacy and Electronic Communications (2002/58/EC) gives the regulatory requirements for future systems. The essence of the directive is captured in the following three requirements.

To increase trustworthiness, the system should emphasize its compliance with policies. Privacy and data protection is a good example: Even though end-users themselves are not necessarily aware of privacy issues or at least they do not behave accordingly, they appreciate a company that has a reasonable privacy policy. Likewise, it is easier for customers to buy services and products from companies that respect consumer protection law and consequently seem trustworthy.

The system should enable digital rights management (DRM). Digital rights management should be supported.

User should have simple and free-of-charge means to control the processing of their personal information. The user must be able to separately accept the use by each service of their personal information, and the system must continually provide the possibility of using a simple and free means of temporarily refusing the processing of personal information for each connection to the network or for each transmission of a communication.

4.2.2.5 Applications and Services Perspective

The three requirements in this group discuss very high-level issues related to applications and services. These requirements are further elaborated in later subsections.

The system should be able to scale content across platforms, devices and modalities, and the user should be able to control and adjust how the content is provided for her. To meet the multimodality needs means that the user can use the voice and data, for example, e-mail services, at the same time. Here it is a question of real multimodality meaning that the content of the services are delivered by using different applications and the contents have to be defined accordingly. In the usage situation, the content or messages from different channels are combined. This means that the identification as well as trust and privacy issues are of huge importance.

The users' content production, elaboration and delivery should be empowered. Supporting multidevice and multichannel content self-production is a challenge in multiple dimensions, with special regard to storage, delivery, control mechanism and possibilities for the moderation of the content. The UI, identification, payment systems and registration topics are important. The user should also be able to easily transfer her data from one storage provider to another.

Support media sharing, integrating automation of regulative mechanisms. Media – especially music – sharing, despite several legal battles and drawbacks has gained robust tractions with users. Portability is a great enhancer of this phenomenon, as it can guarantee even more freedom and agility in sharing and delivering content, bypassing fixed and wireless network operators, infrastructures with local sharing, either by using increasingly cheap storage portable devices or wirelessly over personal networking communication channels.

4.2.2.6 Technology Perspective

These nine requirements summarize necessary functionalities to be enabled by technologies when the vision of the wireless future has materialized.

Support convergence in different areas: fixed-wireless, mobile media, home technology. As a result of convergence (fixed-wireless convergence, mobile media convergence, home technology convergence) technologies that were applied in one area of applications can now be applied to another area. This creates new competition and can create competing business models. Players that are strong in one area can enter new areas. New entrants are more likely than before. Business models that apply in the fixed networks can be applicable for wireless networks, too, and vice versa.

Support using unlicensed frequencies. Unlicensed frequencies, like using WiFi for providing broadband Internet access in hotspots, make wireless access provisioning possible for new entrants like corporations or groups of users. On the other hand, telco operators can extend their current cellular service to broadband access.

Support peer-to-peer models for communication and sharing of data. P2P environments utilize resources contributed by the participating nodes. Control and coordination are at the edge of the network. P2P systems enable new types of applications and business models.

Enable proximity communication. Proximity technologies (RFID, proximity transactions) can enable the use of mobile devices for industries that are currently not utilizing mobile terminals. This may be an opportunity for players controlling the mobile area to provide value for players like retail merchants.

Enable easy identification of users. Easy identification becomes important when consumers are accessing a multitude of systems and services. Identity management can be a valuable asset and control point. Players that can manage consumer identity effectively and are trusted to maintain user privacy can sell this service to other players.

Enable open interfaces. Standardization (adoption of formats, protocols etc.) can happen through standards organizations or by industry (de facto standard). By definition of open interfaces, standardization can make the entrance to markets easier. However, strong players can affect (de facto) standardization.

Support diversity of terminals. Diversity of terminals (different size, performance and features) makes deployment challenging. The adaptation to terminal characteristics may become a part of the value chain. On the other hand, dominant suppliers/distributors of terminals will guide application and services development. Operators can develop proprietary UIs and service platforms to achieve vertical control. Lack of interoperability, the ability of software and hardware from different vendors to work together, hinders technology adoption. Industry, in order to be successful, needs to drive interoperability. Still, dominant players can guide application and services development.

Support promotion and discovery of services. Difficulties in service discovery, finding and initiation of services, prevent services from becoming widely used. There should be methods for service providers to promote the services and for users to discover them.

Support openness in a managed way and provide protection from spam and viruses. As part of convergence, the mobile space is exposed to the negative aspects of openness evidenced in the current Internet: spam and viruses. Successful management of the openness and protection from the negative aspects should be enabled.

4.2.2.7 Multidevice, Personalization and Multimodal User Interaction Requirements

These eight requirements are based on the needs in multimodal user interactions in dynamic execution environments.

Resource discovery. The r esource discovery function should be able to provide information about the availability of devices nearby, and their capacities. It needs to receive notifications from underlying OS/drivers to discover new devices or services upon notifications from them. It should also be able to detect that a previously available device/service is not reachable anymore.

It should be easy to access the same data with different devices and in different locations, even within the same house:

- The user wants to move easily across different rooms and still have consistent access and functionalities;
- The user wants to have seamless access to the same service as when moving across different devices.

Resource handling. The resource handling should be able to hide the heterogeneity of the different devices and networks.

Modality fission and fusion. The UI must be able to integrate various input modalities originating from the user; these inputs shall be intelligently combined to a consistent input data to the application. Also, in a wider sense, sensory inputs can be a part of the integration with intentional user inputs.

When providing information from the application to the user, the output manager shall be able to transform multimodal information to unimodal parts in a synchronized fashion. Otherwise, for example, showing directions on a map would not be in synchrony with audio guidance.

The system shall be able to decide which devices, modalities and content transformation should be used to present content to the user (in the best usable way).

Context gathering and interpretation. Context gathering and interpretation should provide robust (against e.g. noise) and grounded (based on observations) lower and higher-order situation interpretations of an individual user's multimodal application or service usage, physiological parameters, for example, heart rate and body movement, and environmental parameters, for example, global positioning system (GPS) location, altitude or temperature, all in terms of instances of an open upper ontology.

Context reasoning. Context reasoning is needed to personalize applications and services for the user. Context reasoning is the component, which infers the appropriate actions to be performed to personalize applications and services.

Presentation adaptation. Presentation adaptation should be able to select among different views of the same content to match user preferences or device capabilities. It should also be able to convert text/image information into a printable format. When the user requests output to more than one modality, the presentation adaptation has to go through the list of modalities and do adaptations for each modality. The presentation adaptation has to adapt both content and user interaction controls (e.g. whether verbal commands are allowed or not), to device-specific representations.

Content handling. Content handling must be able to decide which devices, modalities and content transformation should be used to present the content to the user (in the best usable way). Content received by content handling should have some associated metadata to support decision (e.g. 'private information'), thus it should take into account when deciding how and where to present the content, and it must have reasoning capabilities.

Content routing. Redirect data flow from sensors and other input devices: Data gathered from input devices is sent to the content routing to be dispatched to the correct 'target'.

4.2.2.8 Privacy and Trust Requirements

The five requirements in this group are the basis for developing applications that can be executed in a secure manner on various execution environments.

The user has to be able to trust the system. Communications must be secured against eavesdropping. Personal profile (PP) information is protected against unauthorized access. Location information is not revealed and distributed unnecessarily to unauthorized parties.

Access control, timeouts, and authentication methods have to be adequate. The communication involves profile information and a possible wide range of multimodal devices. Access control must take the restrictions of these devices into account, like cryptographic computing capability, the ability to verify a presented security token, communication delays and timeouts.

The user should easily switch her role, but the privacy of each role must be protected. Typical user roles are employee, family, hobby club member, virtual community member and so on. Each role has its own data set that relates to the user activities in this role for example, pictures, documents, calendar entries, and so on. The user should be able to synchronize or to separate data pieces from each other. The data usually resides in different administrative domains for example, company intranet, hobby club web server, home-PC and so on.

Creation of new trust relations involves user interaction. Users have existing trust relationships. If a user wants to create a new trust relationship or extend an existing one, he needs to get involved in the creation process. This requirement creates user awareness for his trust relationships and also prevents malware from creating 'unwanted' trust relationships.

Personalization is subject to privacy and trust constraints. Personalization of content increases the usability, but it might also be used for spamming purposes, customized advertising and hence should be in-line with the user preferences on her personal data.

4.2.2.9 Context Awareness Requirements

The eight requirements in this group identify the functionality in key enablers to develop context aware applications.

Default behavior of applications and basic set of functionalities guaranteed on all platforms. End-user applications must guarantee default behavior because the context may not always be available. The platform may limit and restrict the availability of different functionalities, but the access to the data should be guaranteed.

Ability to handle different approaches to context reasoning. The context management framework shall support the discovery and management of different reasoning components; combined usage should also be supported. The framework enabling context awareness should be able to deal with different approaches to context awareness.

Exchange of context information should be supported. The context representation format shall at least support the exchange of context information with hypertext transport protocol (HTTP) as transport protocol and eXtensible markup language (XML) as syntax and the most common data types (e.g. integer, double, string, date, time, etc.). This requirement enables loose coupling between context providers and context consumers, based on for example, the web services.

Context representation must be extendible. It must be possible to add new types of context information. Domain managers must be able to specify their own domain-specific context information elements.

Context representation shall provide a mechanism to link context information to context ontologies. Upper context ontology can act as the central knowledge base that describes the (qualitative) context parameters that are used within the system, and the (logical) relation between those parameters. When actual context information is exchanged, an explicit link to the ontology can be established.

Ontologies should be simple, consistent, extendible and have practical access. Ontologies should be concise and simple enough to be easily utilized by context providers and should enable practical, meaningful, intuitive, and simple queries and subscriptions to the represented context information. If a well-founded ontology representation is chosen, established methods for ontology management, querying and reasoning shall be applied (e.g. subsumptional reasoning in the case of OWL-DL). Context ontologies shall be defined using a well-defined representation model. If the chosen model is sound, complete and decidable (e.g. OWL-DL), context ontologies shall be consistent.

Context management framework shall support privacy and security. Context management framework must support privacy on all levels and must include mechanisms to restricted access to data.

Context management framework shall be scalable and reliable. Context management framework must include the ability to scale large numbers of sources of context information, reasoning mechanisms and applications and services that use the context information. It must always be available, since many different applications and services rely on high- as well as low-level context information, while reasoning mechanisms, in return, rely on lower-level context information in order to provide higher-level context information. Even in the complete absence of suitable sources of context information, the system must be able to provide information accordingly.

4.2.3 Summary

In this section, we elaborated the functional requirements for systems, applications, and platforms that will meet the challenges of the wireless future. One way of summarizing the 64 identified requirements is to specify generic functions (or functional blocks) that would realize an infrastructure, which enables an environment for mobile computing, communication, and seamless service delivery. With this infrastructure, users are able to access services anytime and anywhere that best tailor their preferences and environment. Context aware systems aim at providing relevant information about users in order to provide services and applications that tailor their preferences and environment. These systems reflect the societal trends and concerns in terms of communications between individuals. This envisioned environment will be open, distributed and scalable, and integrates heterogeneous components.

On the basis of the requirements, we have identified the following seven functional blocks that can capture most of the requirements identified in this section:

User Interface Adaptation Function: Today, a great variety of mobile terminal devices exist that users can employ to access services. The devices are heterogeneous with respect to their capabilities to handle input, present UIs or the media they support. Additionally, network capabilities are also changing when using mobile devices to access a service. In consequence, the function is needed to properly handle these discrepancies and to provide the best user experience. The UI Adaptation Function provides functionalities to allow service developers to make services available through multiple devices, using multiple modalities.

Context Awareness Function: The function takes care of raw, interpreted and aggregated context data. This function handles context data related to individual users and to groups of users. It supports the service developer by providing users' and groups' current context information through well-defined interfaces. New context information and changed context information can be notified to interested components and application services, and context information can be requested from the Context Awareness Function.

Privacy and Trust Function: Services and applications often deal with data related to the user, which raises the issue of trust and privacy of the personal user data. The user with the service does not only share personal user data as it is the case with traditional single user-application interaction but groups of users can be involved. The trust and privacy management of personal user data needs to be supported by a suitable policy system.

Personalization Function: The function provides profiles and preferences of users and groups to the services, applications and components, and supports the learning of user and group interests as well as user and group preferences. To support user feedback for profile learning, it is directly related to the services and applications.

Operational Management Function: The function supports the management of the whole lifecycle of services. To perform this, it needs access to information about services stored in the SPF, and access to applications and services, and to the user agents running on users' devices. To manage the privacy aspects, it also needs access to the Privacy and Trust Function.

Service Usage Function: The function covers all aspects related to service usage; in particular it covers every step in the 'timeframe' between service discovery and service offering.

Service Provisioning Function: The function holds a repository of services known to the system, their descriptions and properties and offers functionalities for service discovery, proactive service provisioning and service composition. In practice, the Service Usage Function and the Service Provisioning Function have strong relationships with each other and they are almost always discussed together.

4.3 Generic Service Elements and Enabling Technologies

The next generation (beyond 3G) of mobile services and applications must offer ever-increasing levels of value and differentiation (i.e. they have to be attractive, intuitive and easy-to-use, with personalization and ubiquitous access.). These services must also be developed easily, deployed quickly and, if necessary, altered efficiently.

Recent developments in mobile communication and small computing devices have had a tremendous impact on our societies. They have brought the vision of ubiquitous or invisible computing and communication closer to reality. Already, in the near future, communication and computing devices will be in a state that technology enables mass-market scale ubiquitous services and applications. Therefore, the main challenge will be software that fulfills the needs of personalized, ambient aware services and solutions. The ISTAG Visions for 2010 imply the use of computing and communication services and applications anywhere, anyhow, anytime. A key enabler for this vision is personal networking supporting communication in different kinds of environments: in personal domains, in ad hoc communities, in digital homes and other smart places, and in infrastructure-supported networks. Sensors contribute to this vision the necessary information of local context and actuators the means to affect the real-world context situation.

The current trend in developing communication software and services is to utilize Internet protocols; not only IP, but also Internet solutions both above and below the IP. Another significant trend is the requirement of ever-faster service and application development and deployment. The immediate implication has been the introduction of various service/application frameworks/platforms, usually referred as middleware.

But nevertheless it becomes more and more time consuming to write user-centric applications as the needed required mechanisms themselves become more and more sophisticated, requiring high expertise in many different fields. The Reference Model, which has been sketched by the Working Group 2 'Service Architecture for the Wireless World' of the WWRF, introduces three areas of research that are considered as being of the utmost importance for supporting the concept of I-centric applications, those applications devoted entirely to the user, that put him/her in the center of everything. The previous white papers released by this WG2 have then identified a number of mechanisms in the fields of Ambient/Context Awareness, Personalization – Customization – Profiling and Adaptability that needs more research to reach their maturity. These three high level concepts, but also more generally the writing of applications, needs some more generic supporting functions or components both at design and runtime.

The purpose of this chapter is to identify those generic functions and components that are needed for writing the next generation of user-centric applications. Some of them are very basic and will serve to register, to discover or to combine applications, while some of them will be more advanced providing higher-level functions to support personalization or adaptability (e.g. classification engine, knowledge bases, ontology servers, inference engine etc.).

It is worth noting that the Generic Service Elements (GSEs) as they are represented in the WG2 Reference Model, are spread over the whole set of layers from the physical one to the service one. It simply means that for some generic functional aspects, support is needed at all layers. For instance, discovery mechanisms need to be available almost at all layers of the model with tight cooperation between the different pieces of solution that have been implemented in the different layers (physical, network, service).

Furthermore, this chapter highlights the different enabling technologies that are needed or available to realize the mentioned generic services and outlines several needed fields of research.

4.3.1 Generic Service Elements

4.3.1.1 How to Build Abstract Services

The technical approach for a service execution environment within WG2 follows the idea of reusable service building blocks called *GSEs* (see Figure 4.3 below). A GSE is a functional software component that can be used (invoked) by other GSEs, services or applications and it is hosted by the service execution environment. In contrast to traditional approaches, the envisioned Service Execution Environment (SEE) is fully distributed and composed of a set of coexisting and loosely coupled GSEs. This allows for maximum flexibility regarding business model support and maximum stability regarding the needs of the application developer.

The SEE supports the development (creation), operation and management of applications and services by means of an execution environment for GSEs, support for (re-)deployment (hot-plugging) of GSEs, (re-) binding (configuration) of GSEs, inter-working of applications/services and GSEs, resolving inter-and intra-GSE dependencies and the mapping of GSE-specific protocols and technologies.

GSEs should be usable by widespread Internet development tools and dependencies on programming languages or execution platforms should be minimized or, when possible, eliminated. Consequently, GSEs facilitate a more simplified development of new wireless-enabled, user-centric applications. Corresponding GSE service discovery mechanisms, inter-working protocols, invocation methods, APIs, data coding schemas and dynamic configuration and orchestration methodologies are required.

A GSE operates as a black box exposing common, well-defined interfaces (Open Service API) hiding the specifics of all the underlying technologies. The functionality provided by

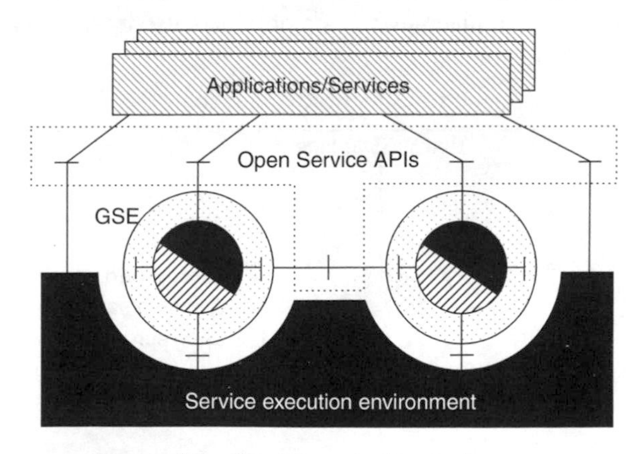

Figure 4.3 Service execution environment

a GSE can be described, discovered, monitored, configured, and controlled through these interfaces in an open and clearly defined manner. A GSE is generic in the way that it:

- can be easily plugged into the service execution environment;
- can be discovered, monitored, configured and controlled using for all GSE's common interfaces;
- can be managed by the service execution environment;
- ensures QoS and security issues by cooperating with the service execution environment;
- can communicate with other GSEs;
- can be dynamically grouped with other GSEs for a certain task, and
- it provides all common functions through well-defined Open Service APIs hiding all the underlying technologies that are used to provide a dedicated functionality.

Besides the provision of these common functions, a GSE consists of a specific part that implements the 'real' functionality provided by the GSE towards end-user applications or services. Figure 4.4 shows how GSEs are embedded in the SEE.

Services and applications can be built just by programming the logic leaving everything that is not directly related to the service logic up to the GSEs by calling functions of the specific part. Therefore, a set of needed GSEs has to be specified to ensure that a sufficient set is available for application developers to build on a solid base of functionalities. Following the user-centric idea, at least four GSEs have been already identified. These basic building blocks provide ambient information, profile information, adaptation functions, and content delivery functions.

Figure 4.4 illustrates the envisaged structure of a GSE in more detailed way. Each GSE provides three different interfaces (specific interface, common interface, and common execution environment interface) compliant to an Open Service API. According to their names, these interfaces provide the functions of the specific part and the common part of a GSE towards services and applications or other GSEs. The common execution environment interface established a well-defined interaction point between a GSE and the service execution environment to provide all the plug' n play and management features. The functions needed to serve this interface are provided also by the common part because each GSE has to support them.

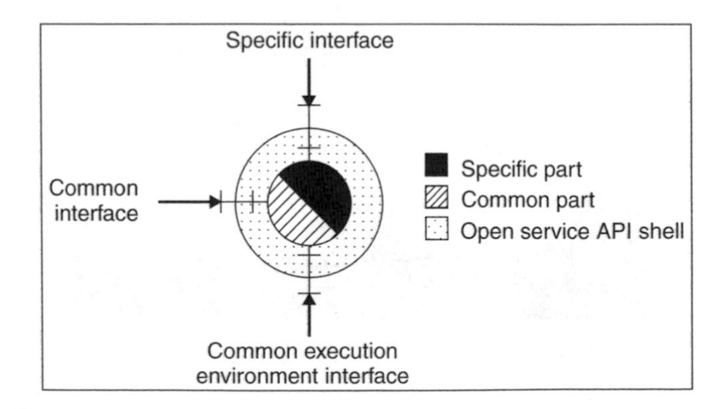

Figure 4.4 GSE interfaces: Open Service API vs. internal interfaces

The 'Open service API shell' implements the Open Service API toward other components and provides a mapping to the GSE internal interfaces of the specific and common part. That enables more flexibility in developing and implementing GSEs.

This approach is flexible in terms of reusing once realized GSEs and the fast realization of new GSEs. Application programmers can use well-defined APIs to find, analyze and use existing GSEs as well as being able to plug new GSEs into the platform. Therefore, a GSE framework specifying formalisms for the development and runtime behavior of GSEs and their interrelations and guidelines for the creation of Open Service APIs should be developed. Owing to the fact that GSEs can be easily integrated in any kind of new application, changing business models can be addressed by just selecting/introducing the right set of GSEs that fulfill the required functionalities. A GSE Framework simplifies the work of service implementers by providing a set of common functionalities that no longer have to be developed repeatedly. Moreover, GSEs provide additional flexibility in terms of reusing once realized GSEs and the fast realization of new GSEs by allowing for dependency on specific programming languages or execution platforms.

GSEs form a highly distributed infrastructure without any inherent need for centralized, technical or business control. GSEs cut across the whole 'Service platform' and even go beyond that, spreading across many physical and logical components like particular wireline and wireless networks, mobile communication devices, moving vehicles or independent business entities (see Figure 4.5). These autonomous GSEs may pool for solving

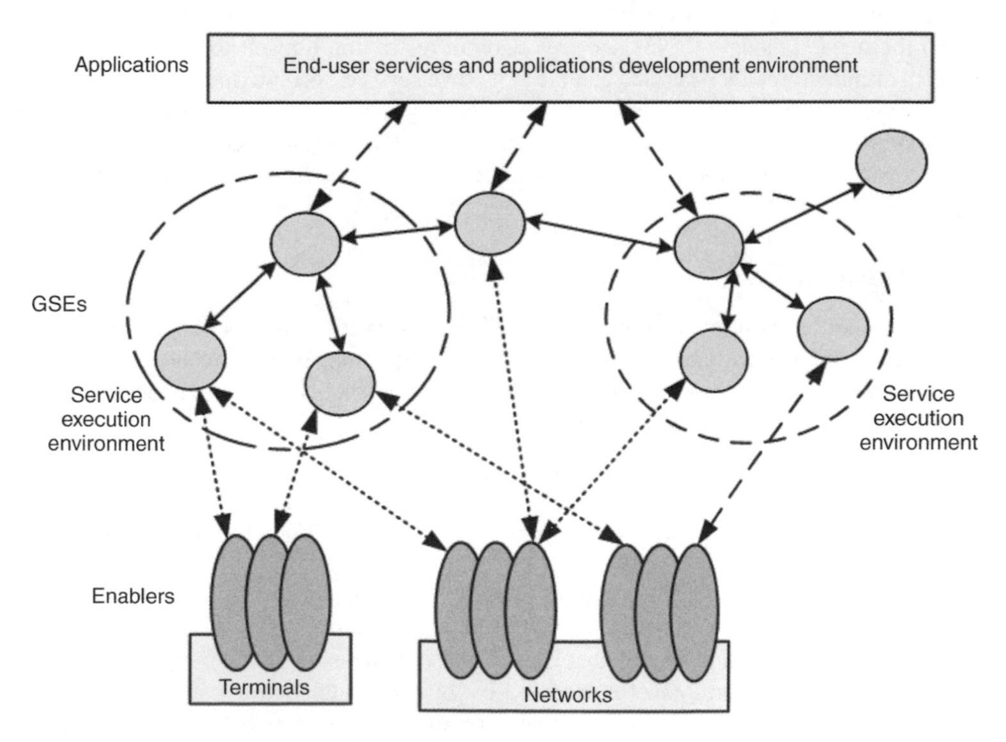

Figure 4.5 Distributed GSEs

problems, for composing applications, and for sharing contents. Each GSE performs its task either autonomously or cooperatively with other GSEs.

Modifying a GSE does not involve modifying the application invoking the enabler. The linkage of GSEs, network and terminal enablers and external services is demand driven and dynamic. The GSE composition is a distributed process and is independent of compile-time bindings. Sub-sets of GSEs may federate to provide more complex functionality, to reflect changing business roles or to adapt to new system and service demands. Sub-sets of GSEs may form hierarchies, trust and/or business domains. GSEs can act as coordination entities for such groups. Replicated instances of GSEs may realize load-balancing networks capable of fulfilling the users' QoS demands and to reflect the geographically distributed mobile networks architectures.

Developing New Services Based on the GSE-concept

Service creation is one of the most crucial processes for the future success of 3 GB systems and environments because they are mainly driven by new I-centric services. In order to create new services, the developer is traditionally mostly strongly related to the underlying platform and its necessities.

Incorporating the concept of GSEs, a new methodology for service design and development can be used. New services can then be easily created without having any details about the underlying service execution environment in mind. The whole service logic can be specified and implemented in a fully independent way. After finishing the service logic implementation, this service can easily be integrated in a GSE in order to fit it into the desired service platform and execution environment.

That leaves the service developer the whole freedom to design a service architecture, and all the needed models for his services without restricting himself to a specific service platform architecture. Each service can build its own service-specific interfaces to organize these services in a distributed environment. To plug any kind of services into existing service platforms, these services must only be encapsulated within a GSE. This GSE shell will provide all needed and necessary interfaces specific for the service platform, where these GSEs should fit in.

4.3.1.2 Event Modeling

Event models consist of event sources, event listeners, notification services, filtering services, and event storage and buffering services. In addition, there may be one or more authentication schemes to enforce security and access control. This section focuses on the general definition of events and event models.

Events

An event represents any discrete state transition that has occurred and is signaled from one entity to a number of other entities. For example, a successful login to a service, the firing of detection or monitoring (DOM) hardware and the detection of a missile in a tactical system are all events. The firing of each event is either deterministic or probabilistic. A source can generate a signal every second, making it deterministic. A stochastic source follows some probabilistic model that can be described using, for example, a Markov chain. Both event qualities can be modeled by building statistical or stochastic models of the firing behavior of the event source. For example, a correlation analysis can be made between a series of event occurrences in time or between two event sources. Such an analysis would measure how strongly one event implies the other or how two event source firings are related.

Events may be categorized by their attributes, such as which physical property they are related to. For instance, spatial events and temporal events note physical activity. Moreover, an event may be a combination of these, for example, an event that contains both temporal and spatial information.

Events can be categorized into taxonomies based on their type and complexity. More complex events, called *compound events*, can be built out of more specific simple events. Compound events are important in many applications. For example, a compound event may be fired:

- in a hospital, when the reading of a sensor attached to a patient exceeds a given threshold and a new drug has been administered in a given time interval;
- in a location tracking service, where a set of users are in the same room or near the same location at the same time, or
- in an office building, where a motion detector fires and there has been a certain interval of time after the last security round.

Event-based interaction can be:

- discrete or
- continuous, as event streams.

Events can also have different prioritizations. Event aging assigns an expiry time to each event notification. Event expiration prevents the spreading of obsolete information.

Event Model

The standard client/server communication models in distributed object computing are based on synchronous method invocations. For example, COM+, Java RMI, and Common Object Request Broker Architecture (CORBA) use synchronous calls (CORBA 3.0 supports asynchronous invocations). This approach has several limitations [12]:

- *Tight coupling of client and server lifetimes*: The server must be available to process a request. If a request fails, the client receives an exception.
- *Synchronous communication*: A client must wait until the server finishes processing and returns the results. The client must be connected for the duration of the invocation.
- *Point-to-point communication*: Invocation is typically targeted at a single object on a particular server.

Mobile clients and large distributed systems motivate the use of asynchronous and anonymous one-to-many distributed computing models. Event-based models address the limitations of the standard client/server paradigm by introducing two roles: consumers and producers. Since event models employ differing technical terms, in this chapter we consider event consumers, listeners, sinks, and event producers, sources, and suppliers to be synonymous.

Event Routing

The event model consists of event listeners and event sources. A listener expresses interest in an event supported by an event source and registers to receive notifications of that event

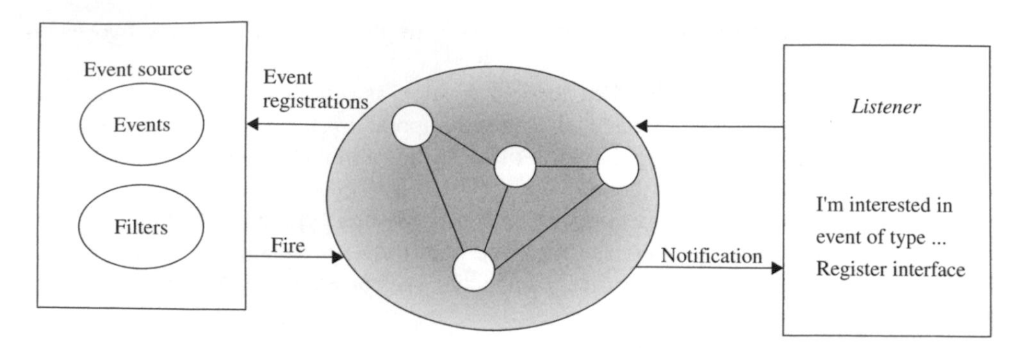

Figure 4.6 General model of the event source and event listener

based on a set of parameters. Figure 4.6 presents a general model of the listener–source paradigm, where the actual filtering and notification are treated as a black box, which can reside either on the source or on the network. Ideally, the event source does not have knowledge of all the parties that are interested in a particular event.

Event source fires events, and the listener is notified using some mechanism on the network or in the client. The event system is a logically centralized component that may be a single server or a number of federated servers. In a distributed system consisting of many servers, there are two approaches for connecting sources and listeners:

- The event service supports the subscription of events, and it routes registration messages to appropriate servers (for example, using a minimum spanning tree). One optimization to this approach is to use advertisements, messages that indicate the intention of an event source to offer a certain type of event, to optimize event routing.
- Some other means of binding the components is used, for example, a lookup service.

In this context, by event listener, we mean an external entity that is located on a physically different node on the network. However, events are also a powerful method to enable inter-thread and local communication, and there may be a number of local event listeners that wait for local events.

Event routing requires that store-and-forward type of event communication be supported within the network on the access nodes (or servers). This calls for intermediate components called *event routers*. Each event source is connected to at least one router. Each router needs to know a suitable subset of other routers in the domain. In this approach, the request, in the worst case, is introduced at every router to get a full coverage of all message listeners. This is not scalable, and the routing needs to be constrained by locality or by hop count. Effective strategies to limit event propagation are zones used in the ECO architecture, the tree topology used in JEDI, or the four server configurations addressed in the Siena architecture. Siena broadcasts advertisements throughout the event system; subscriptions are routed using the reverse-path of advertisements, and notifications are routed on the reverse-path of subscriptions. IP Multicast is also a frequently used network-level technology for disseminating information and works well in closed networks; however, in large PNs multicast or broadcast may not be practical. In these environments, universally adopted standards such as TCP/IP and HTTP may be better choices for all communication [13].

4.3.1.3 Basic/Low level GSE

Registration, Repository and Discovery facilities

Under this section are grouped all mechanisms (languages, protocols, descriptions) that allow to register or to discover the availability of a resource, a device or a service given search criteria. This includes:

- Formalism used to describe the entity
- Descriptions management and storage/retrieval
- Formalisms used to advertise the available entities
- Formalism used to discover available entities including search criteria
- Formalism used to capture the semantic associated with the entity

Component Orchestration and Choreography

Given a set of (web-) services that appear as the basic bricks that are necessary to reach a goal such as the run of a business logics, orchestration and choreography consists of the mechanisms used to interconnect these bricks together, to decide which one has to communicate with another one and in final how these different bricks will be executed (in sequence, in parallel etc.) and following which sequencing or order (workflow).

While standards such as (XML, Simple Object Access Protocol (SOAP) and web services description language (WSDL) [14]) provide languages and protocols to describe information, access and invoke, another set of standards provide the means to define how the elementary web service calls have to be sequenced and coordinated in order to implement a whole business process. Three main orchestration specifications exist today:

- WSCI
- BPML
- BPEL4WS

Brokering/Aggregation

The broker acts as a reference point between a calling party and a group of supplying actors in the sense that it will negotiate (see next section) on the caller's behalf access to information or services provided by the suppliers based on given criteria, as in Figure 4.7.

If only a part of the requested information or a subset of the needed services have been obtained at different service provider places (meaning then that the broker has not been able to find out the entire information at one location), the broker might have to aggregate

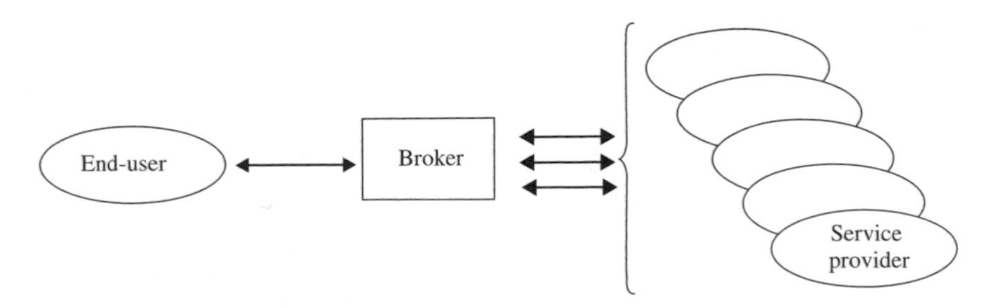

Figure 4.7 Basic concept of a broker

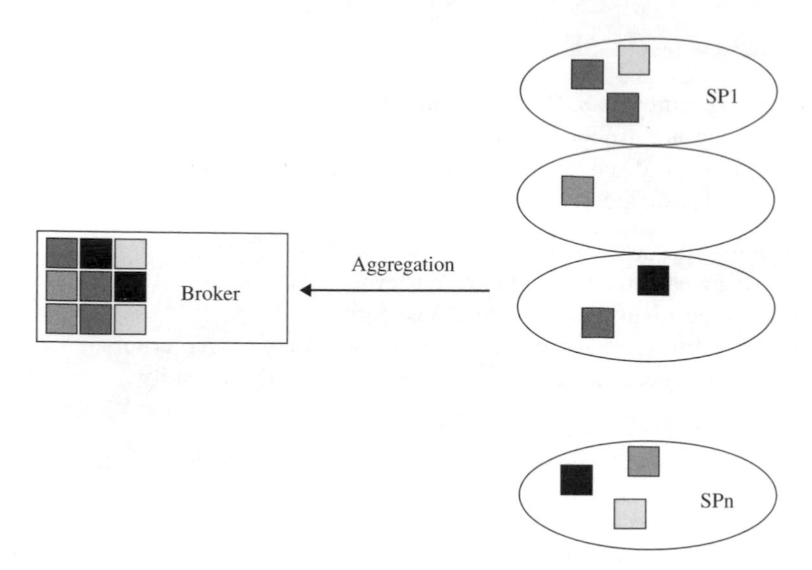

Figure 4.8 Service aggregation by a broker

these different pieces of information before presenting the whole result to the caller, see Figure 4.8. Aggregation may also imply some post-computation and presentation work.

Clearly, in order to accomplish such a task (recombining information fragments), the broker needs 'to understand' both the caller request and the domain of discourse accurately. When a caller requests a particular service (the service class of which is present in the reference service taxonomy), the broker basically has two possibilities offered.

To find out one or more instances of this service type offered by one or many service providers. In this case its main task is to select the most appropriate one, that is, the one that offers the best fit with the caller's expectations (captured in the search_criterion). Case (a) in Figure 4.9, to be able to understand how this service can be recombined from an orchestration of more basic (sub-) services, is the ability to find out over the Net. Once again, it means that the service semantic is well captured and that the process of recombination can be deduced, case b) in Figure 4.9.

4.3.1.4 User-centric GSE

GSE for Personalization/Profiling

In order to support personalization, specific GSEs to handle the user's preferences and his behavior are needed. Details about profiling technologies and related approaches can be found in the personalization. In order to capture the user's preferences and behavior, at least the following functionalities should be realized as GSEs:

- *Click stream analyzer*: These are mechanisms to capture and analyze the user's behavior on the UI.
- *Statistic-based user profiler*: Based on the statistics of common user behaviors, such a GSE could group users automatically by identifying common interests. This is basically done by applying collaborative filtering mechanisms.

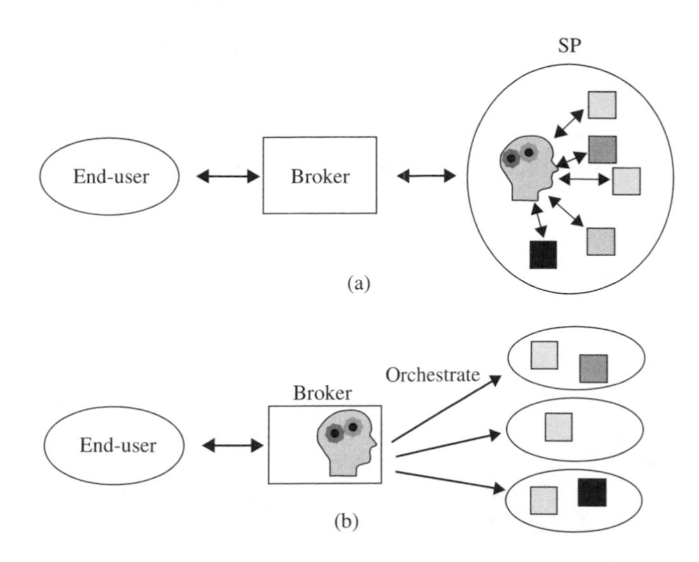

Figure 4.9 Dynamic combination and aggregation of services by a broker

- *Profile management*: Explicit profile management is needed to enable the user to personally modify and enter his preferences. A profile management system can be quite complex, especially taking into account the aspects of distributed profiles.

Details on these functionalities can be found in the [2].

GSE for Ambient Awareness

Typical GSEs that provide ambient awareness support are described in this section as abstract components. Together they provide for sensing, exchanging, and interpreting ambient information in a multidomain distributed context environment and delivering the relevant ambience parameter set – at the right time – to applications, application and network providers, and other (subscribed) stakeholders. The ambience parameters focus on the user's ambience in the sense of the user's situation and environment. Three main parts are identified: spatial information (geographical data like location, orientation, speed and acceleration), physiological information (depicting life conditions like blood pressure, heart rate, respiration rate, muscle activity, and tone of voice) and environmental information including temperature, air quality, and light or noise level. The following GSEs (sketched as abstract components with their core functionality) can be distinguished:

- *Ambience Monitor*: allows for monitoring the ambience and taking appropriate actions to ensure that the software representation of the ambience information is correct and updated. It uses a Discovery Provider for finding other sources of ambient information that can either come from sensors or various processing components that interpret multiple ambient information objects. The Ambience Sensor allows the sensing and acquiring of information about the situation of an actor, by making use of concrete elements. The Ambience Sensor delivers raw ambience information. The interpretation components are generalized by the Context-Based Reasoner. These Ambience Interpreters allow for inferring (new) relevant ambience information either by aggregating multiple information objects, extrapolating information over time or space by anticipating current contextual information, and so on. Where superfluous ambience information is present, for example, network location coordinates, GPS coordinates, schedule meeting location data, the relevant location information object is specified via the Reasoner-Interface, that also allows for specifying rules (if... then...) which data prevails. Situational models (like topology information) are present in knowledge bases (both domain-specific ontologies and heuristic knowledge are stored there) and are interpreted by the Inference_Engine in order to retrieve higher-level concepts (like he is in a particular room). In this way, integrating sensed information with contextual models enriches ambience information. The upgrading of contextual information into higher levels is, for example, reflected in the five-layer model for structuring context aware applications in [15]. Typically, reasoning techniques are rule-based, model-based, or case-based [16] and inference rules can for example, make use of the forward chaining, backward chaining or behavior description techniques. Reasoning can deal with partial context information and implicit information can be inferred. Upper context ontology typically uses standard vocabulary like web ontology language (OWL) [17] and is supported by off-the-shelf reasoning components like RACER [18]. The following special types of Context-Based Reasoners can be distinguished:

- *AmbienceStorage*: allows for (rule-based) storage, synchronizing and retrieval of (new deduced or user-specified) situational information. Part of it can be a history of ambience parameters.
- *AmbienceInfoAggregator*: allows for (rule-based) collecting of heterogeneous ambience information, and combining, digitizing and filtering this data, hence delivering 'smart' ambience information. The smart ambience information is the first layer of context abstraction and this information can be made available through a repository interface.
- The Merino architecture [19] also deals with context (aggregation and inference) of both multidomain distributed context and multi–interpreted level (from raw sensor data to higher semantics). However, a specific implementation is prescribed: Context information is stored as fields of objects that represent particular context entities. A context repository is implemented as a type of distributed object database. Objects are stored in servers that are distributed across several administrative domains and are arranged in a tree. Within a domain, a tree search is carried out in a similar way to Gloss systems [20]); beyond a domain, a distributed hash table approach is followed. Communication at a local server is done with multicast message-based protocols; between servers it is unicast message-based. Agents can subscribe and register with context repository servers. Two classes of agents are distinguished in the – second layer of – context abstraction: (1) rich-context providers that infer higher semantics context at different granularity levels by using raw context data, user model information and historical data and (2) performance enhancers that also interact with environmental knowledge over the smart sensor layer (using e.g. learning and reasoning algorithms).
- Shielding of low-level sensing implementations from the high-level applications can be done by using context widgets or a library of procedures that forms a middleware abstraction for context acquisition, see for example, the work of Dey [21]. In addition,

abstractions of real world entities (such as mobile phones, home appliances, and also various software components) are required, which can be done by so-called *super distributed objects* (*SDOs*). The concept of SDO is developed by the object management group (OMG) Domain-SIG for Super Distributed Objects (OMG DSIG SDO). An SDO can be seen as the counterpart of real-world entities, representing physical objects and individuals as abstract objects. In [22] a (Java-based) implementation with SDOs are described, specifying both the SDO usage interface and non functional interfaces needed for management and platform integration issues, allowing for the discovery, reservation, configuration and monitoring of SDOs.

- *AmbiencePredictor*: allows for extrapolating ambience information over time and/or space by anticipating current ambience information and can request for complementary ambience information. For example, use the current location of a mobile user, combine it with additional information on speed, direction and the (railway-) road map to predict the most likely next user positions as a function of time, enabling intelligent usage of context information. When combined with network cells, intelligent handovers of running sessions to other networks can be triggered [23] instead of depending on coincidental network solicitation order. When combined with user preferences, intelligent pushing of potential interesting services can be realized (upcoming train station with a certain residence time), and so on.
- *ContextServiceAdapter*: allows for the adjustment of generic context information to special-purpose needs of application and network services. It uses a DiscoveryCrossInfo abstract component that uses interfaces to get service-specific information. (The csa functionality is offered by the infrastructure as support to, for example, applications. It need not necessarily be used; applications can also choose to adapt the context to their specific needs themselves.) AmbienceSources in different (operator or network) domains can have bi lateral agreements about the type of context that can be exchanged between them, the format used to exchange it, and the allowed receiving parties (via the DiscoveryPolicies functionality).
- In [23] such an approach – that allows for the separation of application-independent and application-specific context exchange-is advocated and demonstrated. The implementation of these application-specific modules in network nodes and mobile devices can be done by programmable platforms [24, 25] that also enable the interaction among different protocol layers through the use of dynamic cross-layer interfaces. Since different sets of server-client CSA modules can be required at different times and places, a Service Deployment Framework (SDF) can be included that is capable of controlling the provision of application-specific services in a network. The configuration of the exchange protocol depends on the location of Ambience Source, for example, retrieving context information from other networks depends upon the policies that rule the protocol among operators.

Furthermore, specific components for resolving ambiguities that might arise due to contradicting and superfluous contextual information can be included [26]. Such a component explicitly enables certain contextual information to prevail over the other by using rule-expressed preferences, or weight factors, and so on. It can, for example, find the contextual information that is best suited given all the available contextual data or it can extend the complexity of inference rules to resolve the ambiguities (in where a hierarchy of trials or rules is applied before the optimal context is found.)

GSE for Adaptation

This section gives an overview of typical GSEs needed to ensure Service Adaptation. By Service Adaptation, we mean the ability of services and applications to change their behavior when the circumstances change in the environment during service execution. The environment comprises of a network environment, an execution environment and the user's environment, as we will see in the following paragraphs.

According to adaptation, the following GSEs can be identified:

Environment Monitoring: allows to observe the environment and to detect changes in it. It is likely that not only one GSE for environment monitoring is needed, but also somewhat different ones specialized with regard to the nature of the considered environment, that is, radio, network, personal, service execution and so on. The environment monitoring GSE must include a language to define the criteria leading to the firing of events based on raw observations: definition of threshold, values, conjunction and disjunction of events and so on. Correlation capabilities might also be useful to code complex behaviors made of different perceptions appearing at different points in time. These filters will then be helpful to identify among the enormous quantity of various information the one that is relevant to a given service or application candidate to adaptation. Figure 4.10 shows an example of what can be environment monitoring in the field of medical monitoring.

Event Notification: allows notifying changes to the services and applications that have to adapt; Event Notification is responsible for transporting an event that has been previously observed toward the components that have subscribed to those kinds of events. It then needs to solve the correct set of recipients for a given event and to deliver the event to the recipients using a proper representation format and medium.

Mobile Distributed Information/Knowledge Base: for accessing information like (raw) perceptions, user preferences and profiles, device capabilities, network characteristics and so on from everywhere at any time with a high level of availability, consistency and efficiency.

Modeling Service: The purpose of the Modeling Service is to build a model of the environment or a part of it (a particular phenomenon) based on available collected information as well as learning mechanisms and rules dictated by the modeling service client. A click stream analyzer is an example of modeling service focusing on profiling the customers of a web site, for instance.

Ontology Service: See Section 4.3.3.1.

Semantic Matching Engine: The Semantic Matching Engine is responsible for using the ontological information provided by the Ontology Service to match and reason over instances of ontological knowledge such as profiles. It needs to be able to determine whether a given document of semantic information complies with a related ontology.

In addition to these GSEs, the Service Adaptation assumes a surrounding **Distributed Application Framework** including high-level service registration and discovery and auto-reconfiguration. The distribution application framework will allow the application to seamlessly replace some of its components by new ones, potentially running at another location hosted by another service platform, according to the detected environment change and the replacement strategy implemented. Another possibility for adaptation being, of

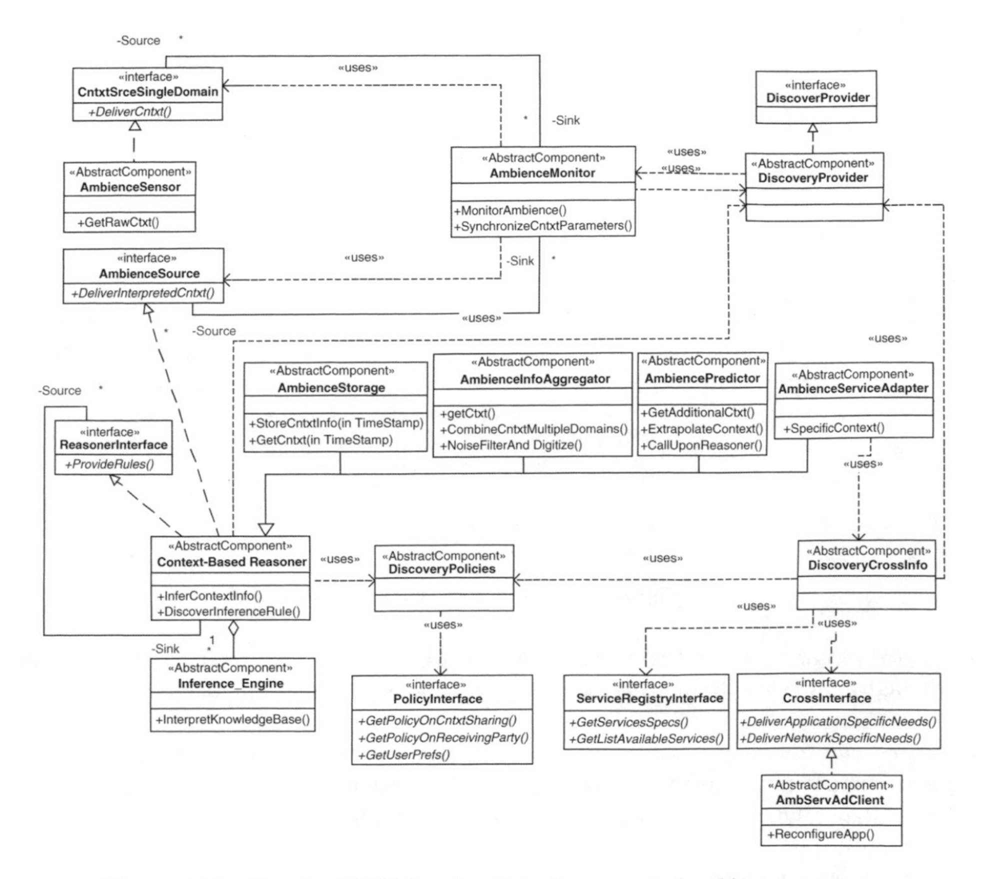

Figure 4.10 Sketch of GSE functionalities in support of ambient awareness

course, to internally modify to the service logic the service/application behavior according to the detected change, assuming that all kinds of changes have been foreseen at the service/application design.

More focus is put on the adaptation decision process in [27]. In Figure 4.11, the authors suggest the following Service adaptation loop:

In this loop three processes take place:

1. Analysis of the environment and service (environment to be interpreted in the broad sense i.e. network, user, service, etc.)
2. Decision process about adaptation when relevant
3. The adaptation itself.

While the first process refers to already identified GSEs, the second and third processes put forward the following requirements for new GSEs:

- *Adaptation policy coding (2)*: a number of techniques can be used ranging from hard-coded sequences of if-then-else constructs to expert-and rule-based systems that reason over events resulting from the observation process (1);

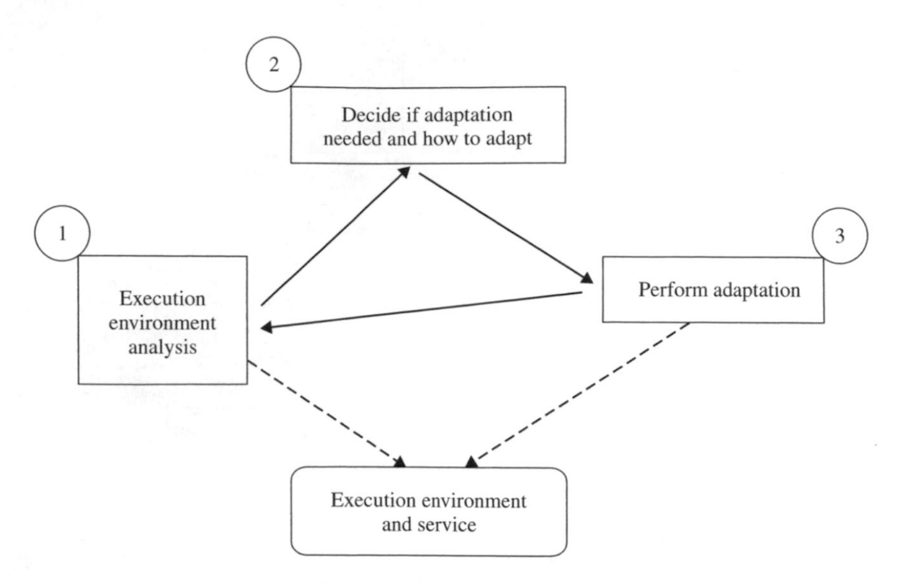

Figure 4.11 Service adaptation loop

- *Service replacement and component change (3)*: already identified as part of a distributed component framework such as service description, composition/orchestration and discovery;
- *Service migration (3)*: have a look at mobile agent technology (not a GSE but another programming paradigm) where the code itself is capable of migrating to new servers (for instance when the old server is too loaded or when a local resource is no longer available);
- Media stream adaptation (3);
- Data filtering (3);
- Transcoding (3).

4.3.2 Enabling Middleware Technologies for the GSE-concept

Middleware is a widely used term to denote generic infrastructure services above operating system and protocol stack. Although the term is popular, there is no consensus of a definition, but a good discussion can be found in Internet engineering task force (IETFs) RFC 2768 (Feb 2000). Nevertheless, the role of middleware is to ease the task of designing, programming and managing distributed applications by providing a simple, consistent and integrated distributed programming environment. The programming models supported can be used to classify the middleware platforms: Object-Oriented Middleware (OOM), Message-Oriented Middleware (MOM), and Event-Based Middleware. Recently, some have started to speak about Reflected Middleware as a new class of middleware. Nowadays, Wide-Area Distributed Systems are usually considered to be a part of Web Systems.

4.3.2.1 Object-oriented Middleware

The OOM is the most widely used platform. The prime examples are OMG's CORBA, Sun's Java 2 Enterprise/Standard/Micro Edition, and Microsoft's DCOM. OMG's CORBA is a mature specification that is supported by many commercial products and is also available for small embedded devices (e*object request broker (ORB) by Prism Technology, Orbix/E by Iona, ORBExpress by OIS, VisiBroker by Borland). Recent OMG specifications relevant for the SmartWare project include Telecom's Wireless CORBA (OMG document formal/2003-03-64), SDOs (dtc/2003-04-02), and Smart Transducers (formal/2003-01-01). The Wireless CORBA provides the basic support of terminal mobility and the minimal means to discover CORBA-services. The SDO specification provides general mechanisms to model and to control real-world entities in a unique manner by providing mandatory interfaces, for example, discovery, monitoring, configuration, and reservation. The Smart Transducers specification adds CORBA-based management support of tiny devices like actuators and censors.

Java Community Process has produced several Java Specification Requests (JSRs) enlarging the functionality of Java 2 Micro Edition (J2ME). The J2ME architecture builds on configurations, profiles, and optional packages. Currently, there are two J2ME configurations: the Connected Limited Device Configuration (CLDC), and the Connected Device Configuration (CDC). The available J2ME technologies for CLDC include Mobile Information Device Profile version 2 (MIDPv2), Wireless Toolkit and Oracle9i JDeveloper J2ME Extension. The available J2ME technologies for CDC include Foundation Profile (FP), Personal Basis Profile (PBP) and PP. Currently available optional packages are shown in Table 4.1

Despite wide and rapid developments in J2ME specifications, the server functionality and ad hoc communities are almost completely overlooked.

4.3.2.2 Message-oriented Middleware

The MOM can be seen as a natural extension of packet-based (connectionless) communications. Unlike object-orientation, which is based on method invocation or remote procedure call (RPC), MOM utilizes asynchronous communication. The underlying programming model is publish-and-subscribe. MOM systems are used for database access

Table 4.1 J2ME optional packages

		CLDC	CDC
J2ME RMI	JSR-66		×
JDBC	JSR-169		×
Information module profile	JSR-195	×	
Java Technology for Wireless Industry	JSR-185	×	
Mobile media API	JSR-135	×	
Location API for J2ME	JSR-179	×	×
Session Initiation Protocol (SIP) API for J2ME	JSR-180	×	
Wireless messaging API	JSR-120	×	×
Security and trust services API for J2ME	JSR-177	×	×
J2ME Web services	JSR-172	×	×
Mobile 3D graphics	JSR-184	×	
Bluetooth API	JSR-82	×	×

in large business applications. Today, the difference between MOM and OOM is not so obvious. CORBA has one-way operations, Event Service, and Notification Service. In the Java world, Java Message Service (JMS) and Sum™ ONE Middleware can be used for asynchronous messaging. The major MOM productions include IBM's WebSphere™ MQ, Microsoft Message Queue Server (MSMQ), Bea Systems' MessageQ, and TIBCO's ActiveEnterprise™ (including TIBCO Rendezvous™. Open Source alternatives include ObjectWeb's JORAM and xmlBlaster.

4.3.2.3 Event-based Middleware/Systems

Event-based Middleware is not always considered to be a class of its own since its close resemblance to MOM. They are both based on asynchronous messaging, but their origins are different. The first MOM systems were designed to free the clients from waiting for replies from database queries. The event-based systems have the target in unpredictable events to which the applications must react.

Java Delegation Event Model

The Java Delegation Event Model was introduced in the Java 1.1 Abstract Windowing Toolkit (AWT) and serves as the standard event processing method in Java. The model is also used in the Java Beans architecture and supported in the Personal Java and Embedded Java environments. In essence, the model is centralized and a listener can register with an event source to receive events. An event source is typically a Graphical User Interface (GUI) element and fires events of certain types, which are propagated to the listeners. Event delivery is synchronous, so the event source actually executes code in the listener's event handler. No guarantees are made on the delivery order of events [28]. The event source and event listener are not anonymous, however, the model provides an abstraction called an adapter, which acts as a mediator between these two actors. The adapter decouples the source from the listener and supports the definition of additional behavior in event processing. The adapter may implement filters, queuing, and QoS controlling.

Java Distributed Event Model

The Distributed Event Model of Java is based on Java Remote Method Invocation (RMI) that enables the invocation of methods in remote objects. The architecture of the Distributed Event Model is similar to the architecture of the Delegation Model with some differences. The model is based on the Remote Event Listener, which is an event consumer that registers to receive certain types of events in other objects. The specification provides an example of an interest registration interface, but does not specify such.

The Remote Event is the event object that is returned from an event source (generator) to a remote listener. Remote events contain information about the occurred event, a reference to the event generator, a handback object that was supplied by the listener, and a unique sequence number to distinguish the event globally. The model supports temporal event registrations with the notion of a lease (Distributed Leasing Specification). The event generators inform the listeners by calling the listeners' notify method.

The specification supports Distributed Event Adaptors that may be used to implement various QoS policies and filtering. The handback object is the only attribute of the Remote Event that may grow to unbounded size. It is a serialized object that the caller provides to the event source; the programmer may set the field to null. Since the handback object carries both state and behavior, it can be used in many ways, for example, to implement an event filter at a more powerful host than the event source.

A mediator component can register to receive events and give a filter object to the source. Upon event notification, the filter is handed back and the mediator can use it to filter the event before handing it to the original event listener. The specification supports recovery from listener failures by the notion of leasing. A lease imposes a timeout for event registrations. This is used to ease the implementation of distributed garbage collection. Since this model relies on RMI, it is inherently synchronous. Each notification contains a sequence number that is guaranteed to be strictly increasing.

Java Message Service

The JMS [29] defines a generic and standard API for the implementation of MOM. The JMS API is an integral part of the Java Enterprise Edition (J2EE) version 1.3. The J2EE supports the message-driven bean, a new kind of bean that enables the consumption of messages. However, JMS is an interface and the specification does not provide any concrete implementation of a messaging engine. The fact that JMS does not define the messaging engine or the message transport gives rise to many possible implementations and ways to configure JMS.

JMS supports a point-to-point (queue) model and a publisher/subscriber (topic) model. In the point-to-point model, only one receiver is selected to receive a message, and in the publisher/subscriber model many can receive the same message. The JMS API can ensure that a message is delivered only once. At lower levels of reliability, an application may miss messages or receive duplicate messages. A standalone JMS provider (implementation) has to support either point-to-point or the publish/subscribe approach, or both. Normally, JMS queues and topics are maintained and created by the administration rather than application programs. Therefore, the destinations are seen as long lasting. The JMS API also allows the creation of temporary destinations that last only for the duration of the connection.

The point-to-point communication model consists of receivers, senders, and message queues. Each message queue is addressed to a particular queue, and receivers extract messages from the queues. Each message has only one consumer and the client acknowledges the successful delivery of a message to the component that manages the queue. In this model, there are no timing dependencies between a sender and a receiver; it is enough that the queue exists. In addition, the JMS API allows the grouping of outgoing messages and incoming messages and their acknowledgements to transactions. If a transaction fails, it can be rolled back. In the publish/subscribe model the clients address messages to a topic. Publishers and subscribers are anonymous, and messaging is usually one-to-many. This model has a timing dependency between consumers and producers. Consumers receive messages after their subscription has been processed. Moreover, the consumer must be active in order to receive messages.

The JMS API provides an improvement on this timing dependency by allowing clients to create durable subscriptions. Durable subscriptions introduce the buffering capability of the point-to-point model to the publish/subscribe model. Durable subscriptions can accept messages sent to clients that are not active at the time. A durable subscription can have only one active subscriber at a time. Messages are delivered to clients either synchronously or asynchronously. Synchronous messages are delivered using the receive method, which blocks until a message arrives or a timeout occurs. In order to receive asynchronous messages, the client creates a message listener, which is similar to an event listener. When a message arrives, the JMS provider calls the listener's onMessage method to deliver the message. JMS clients use Java Naming and Directory Interface (JNDI) to look up configured JMS objects. JMS administrators configure these components using facilities specific to a provider (implementation).

The CORBA Event Service

The CORBA Event Service specification defines a communication model that allows an object to accept registrations and send events to a number of receiver objects [30]. The Event Service supplements the standard CORBA client–server communication model and is a part of the CORBA-Services that provide system level services for object-based systems. In the client-server model, the client makes a synchronous interface definition language (IDL) operation on a specified object at the server. The event communication is unidirectional (using CORBA one-way operations) [31]. The Event Service extends the basic call model by providing support for a communication model where client applications can send messages to arbitrary objects in other applications. The Event Service addresses the limitations of synchronous and asynchronous invocation in CORBA.

The specification defines the concept of events in CORBA: An event is created by the event supplier and is transferred to all relevant event consumers. The set of suppliers is decoupled from the set of consumers, and the supplier has no knowledge of the number or identity of the consumers. The consumers have no knowledge of which supplier generated the event. The Event Service defines a new element, the event channel, which asynchronously transfers events between suppliers and consumers. Suppliers and consumers connect to the event channel using the interfaces supported by the channel. An event is a successful completion of a sequence of operation calls made on consumers, suppliers, and the event channel.

W3C DOM Events

W3C's Document Object Model Level 2 Events is a platform and language neutral interface that defines a generic event system [32]. The event system builds on the DOM Model Level 2 Core and on DOM Level 2 Views. The system supports the registration of event handlers, describes event flow through a tree structure, and provides contextual information for each event. The specification provides a common subset of the current event systems in DOM Level 0 browsers. For example, the model is typically used by browsers to propagate and capture different document events, such as component activation, mouse over, and clicks. The two propagation approaches supported are capturing and bubbling. Capturing means that an event can be handled by one of the event's target's ancestors before being handled by the event's target. Bubbling is the process by which an event can be handled by one of the event's target's ancestors after being handled by the event's target. The specification does not support event filtering or distributed operation.

The specification 'An Event Syntax for XML' is a W3C Recommendation and defines a module that provides XML languages with the ability to integrate event listeners and handlers with DOM Level 2 event interfaces [33]. The specification provides an XML representation of the DOM event interfaces. The ability to process external event handlers is not required.

Web Services Eventing (WS-Eventing)

The Web Services Eventing (WS-Eventing) specification describes a protocol that allows Web Services to subscribe or to accept subscriptions for event notifications [34]. An interest registration mechanism is specified using XML-Schema and WSDL. The specification supports both SOAP 1.1 and SOAP 1.2 Envelopes. The key aims of the specification are to specify the means to create and delete event subscriptions, to define expiration for subscriptions, and to allow them to be renewed. The specification relies on other specifications for secure, reliable, and/or transacted messaging. The specification supports filters by specifying an abstract filter element that supports different filtering languages and mechanisms through the Dialect attribute. The filter is specified in the Filter element [35, 36].

COM+ and. NET

Standard COM and OLE support asynchronous communication and the passing of events using callbacks; however, these approaches have their problems. Standard COM publishers and subscribers are tightly coupled. The subscriber knows the mechanism for connecting to the publisher (interfaces exposed by the container). This approach does not work very well beyond a single desktop. Now, the components need to be active at the same time in order to communicate with events. Moreover, the subscriber needs to know the exact mechanism the publisher requires. This interface may vary from publisher to publisher making this difficult to do dynamically (ActiveX and COM use the IconnectionPoint mechanism for creating the callback circuit; an OLE server uses the method Advise on the IOleObject interface). Furthermore, this classic approach does not allow the filtering or interception of events [37–39].

4.3.2.4 Reflective Middleware

Reflective middleware is a newcomer incorporating self-awareness of its own state and surrounding into middleware in order to achieve flexibility and adaptation. It augments event-based middleware by conceptual models that describe the runtime system and the surrounding. The applicability of event-based and reflective middleware in mobile, dynamically changing environments is still primarily a research issue, experimental systems include Siena, Cambridge Event Architecture, Scribe, Elvin, JEDI, ECho, Gryphon, and Fuego Event System.

4.3.3 Semantic Support

In order to enable the meaningful communication between different GSEs developed by different vendors, a common understanding of used terms and definitions need to be achieved. But as there is no global common understanding and use of terms, this understanding needs to be established between communicating parties on the fly. In order to enable this mechanism, semantic descriptions using taxonomies and ontologies need to be used. The following section will introduce the state of the art in respect to ontologies and taxonomies, describing GSEs, and will identify further research topics in these fields.

4.3.3.1 Ontologies

Ontologies are for **knowledge sharing and reuse**, while languages such as XML are perfect to **express information** in a **structured and efficient way**. To be able to discuss with one another, communicating parties need to share a common terminology and meaning of the terms used. Otherwise, profitable communication is infeasible because of a lack of shared understanding. With software systems, this is especially true – two applications cannot interact with each other without common understanding of terms used in the communication. Until now, this common understanding has been achieved awkwardly by hard-coding this information into applications. This is where ontologies come into the picture. Ontologies describe the concepts and their relationships – with different levels of formality – in a domain of discourse. An ontology is more than just a taxonomy (classification of terms) since it can include richer relationships between defined terms. For some applications, a taxonomy can be enough, but without rich relationships between terms it is not possible to express domain-specific knowledge except by defining new terms.

Ontologies have been an active research area for a long time. The hardest issue in developing ontologies is the actual conceptualization of the domain. Additionally, to be shared, the ontologies need a representation language. Languages like XML that define the structure of a document, but lack semantic model, are not enough for describing ontologies – intuitively an XML document may be clear, but computers lack the intuition. In recent years, ontology languages based on Web technologies have been introduced. DARPA Agent Markup Language DAML+Ontology Interference Layer (OIL) [27], which is based on Resource Description Framework-Schema RDF-S-Schema [40], is one such language. It provides a basic infrastructure that allows machines to make simple inferences. Recently, DAML+OIL language was adopted by W3C, which is developing an OWL [41] based on DAML+OIL. Like DAML+OIL, OWL is based on RDF-Schema [40], but both of these languages provide additional vocabulary – for example, relations between classes, cardinality, equality, richer typing of properties, characteristics of properties, and enumerated classes – along with a formal semantic to facilitate greater machine readability. The OWL language has strong industry support, and is therefore expected to become a dominant ontology language for Semantic Web. In this section, we use OWL language when giving examples of ontology encoding.

The following sections tackle different aspects associated with ontologies. They will provide, in particular, a short description of the OWL language, the OWL-S ontology and will then point out critical issues to be dealt with as far as ontologies construction, usage and management and maintenance are concerned.

Ontology Construction

This section gives an introduction to OWL and OWL-S and an overview of the Protégé tool, which provides ontology editors and tools to access information available within the ontology.

OWL – The Web Ontology Language

OWL – the Web Ontology Language – is intended to provide the explicit meaning (semantics) of information while languages such as XML mainly focus on the information structuring.

OWL then facilitates the automatic handling and understanding of information by computer program (for instance, reasoning about this information) and then enhances the level of cooperation between machines. OWL goes far beyond previous initiatives such as RDF, RDF-S (RDF-schema) providing extended capabilities and expressive power. OWL has been built on top of W3C recommendation stack XML → XML-Schema → RDF →

RDF-Schema. With respect to these languages, OWL adds more vocabulary for describing the classes and the properties that hold between classes and individuals. This enhanced vocabulary includes set-theoretic constructs such as cardinality, intersection, characteristics of properties and enumerated classes. As far as expressive power is concerned, OWL is divided into three dialects (from the less expressive but most decidable and complete) to the most expressive (but less decidable and complete) that are respectively OWL-lite, OWL-DL and OWL-FULL.

Very quickly and in addition to the RDF-S construct, OWL-lite provides equivalence between classes and properties, equality and inequality between individuals (member of a class), inverse of a property. Properties can be stated as transitive and symmetrical and restrictions can be applied to cardinality and to values on properties associated to classes. OWL-DL and OWL-FULL use exactly the same vocabulary (extending OWL-LITE) but some restrictions apply to OWL-DL (making the stuff complete and decidable). They provide enumerated classes, classes disjunction, union, complement and intersection, and finally, cardinalities are no longer restricted to 1 or 0. Please refer to the OWL Web Ontology Language guide for more details about the specifications of these three dialects and full details about their capabilities.

OWL-S – Semantic Mark-up Language for Web Services

OWL-S has been especially designed to allow the definition of (web-) service ontologies. It is based on the OWL language briefly described above. OWL-S is expected to allow computer software or agents to make intelligent use of web services and will, unsurprisingly focus on:

- *service formal descriptions (aka service profile)*: or in other words: 'What the service does'. Useful for advertising, registering and discovering services;
- *process model*: providing a detailed description of the service operations;
- *grounding*: which provides the necessary details to communicate with the service via exchange of messages.

These three aspects can be summarized in Figure 4.12.

This Web Service ontology will then propose a main class `service`. With this class are attached three properties (as described above) the name of which are `presents`,

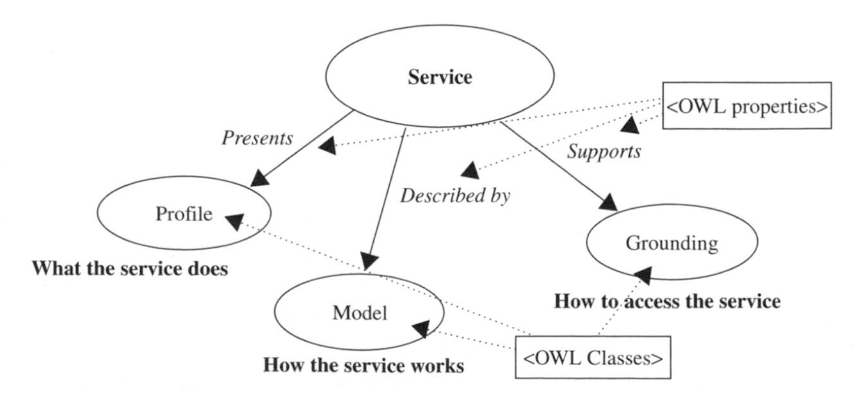

Figure 4.12 The three concepts of OWL-S

`describedBy` and `supports`. Their respective ranges (as described in OWL) are respectively the following classes: `serviceProfile`, `serviceModel` and `service-Grounding`.

Obviously, the ontology only defines the framework – that is, the vocabulary – needed to describe the three aspects of a (web-) service. The exact content of these classes will vary from a given service to another. The three following sections introduce in more detail the purpose of the three 'aspects' of a service description, which can also be used for describing GSE functionalities:

serviceProfile: The *serviceProfile* provides three kinds of information that characterizes what the service does:

- *Contact information*: about the entity that provides or operates the service
- *Functional description*: in a nutshell it specifies what the needed inputs to run the service are, what the outputs resulting from the service execution are, what the precondition that has to be valid before the service executes is and finally what the effect of the service execution (effect not captured by the output parameters) is
- Properties such as category within a classification system, QoS information, rating, service level agreement information.

The purpose of the *serviceProfile* is essentially service selection. It is, therefore, well suited to service descriptions located at a repository or registry side. Once the service has been sought and selected, the profile is no longer of use, as all information that is to be used for service interaction is found in the *serviceModel*. Obviously, maximum consistency has to be ensured between the formal description of the service as a set of pre-post conditions associated with inputs and outputs and the implementation of the service itself, which, of course, has to comply with the logical description.

processModel: This class model services as process and builds on a variety of fields such as Artificial Intelligence (AI), agent communication languages, work on programming languages and distributed systems and so on. The class Process provides information about how a given 'kind' of service is to be constructed. These pieces of information relate to inputs, outputs, pre-conditions, conditional post-conditions (effect) and also how a process can be split into sub-processes and which execution model applies them to the process components (in sequence, in parallel etc.). A party that volunteers in providing a real implementation of a given kind of service (characterized then by an ontology) will have to make sure that its implementation complies to the process model described within the service ontology.

Grounding: The grounding provides all the details required to invoke the service. It gives an account of protocols and message format. It represents the mapping between a formal specification as described in the ontology to the concrete specification as implemented by the service provider. It then allows a calling party, a consumer, to build the right messages allowing the consumer to invoke and make use of the service offered by the service provider, assuming this service complies with the declared service characterization.

Ontology Integration

Ontology integration means reusing available ontologies in order to build a new ontology for a given application. Available ontologies may only be used partly, and further extended, if needed, to better suit the application. Pieces of the reused ontologies are usually about different domains and subjects in the domains.

Figure 4.13 depicts the ontology integration process at a conceptual level. Existing ontologies $O_1, O_2, \ldots, O_n$ about different domains $d_1, d_2, \ldots, d_k$, respectively, are integrated to build a new ontology O about domain d. To illustrate the integration, let us consider three ontologies. The location ontology has been used in GSM positioning application, where the location – constituted by longitude and latitude – is used in providing location-based applications for mobile phone users. Network ontology is a fragment of the one we presented in [42]. The device ontology models devices and their relationships with networks. It has been used in content-adaptation application, where content is formatted based on the characteristics of a device.

Now, let us assume that there is a need for an application that combines the features of those previous applications, and we need to integrate the three ontologies. As we can see in Figure 4.13, there are both syntactically and semantically overlapping concepts, and one could also find relationships between the concepts of the three ontologies. The concepts that we are highlighting are painted with white background color. The syntactically and semantically similar concepts are easier cases; only the relation stating that the concepts are equivalent is needed. This is the case with *Location, Network*, and *Phone* concepts. The next thing is to find the concepts that are semantically similar, but have different names. There are two pairs of these kinds of concepts, *WirelessNetwork/Wireless* and *WirelineNetwork/Wireline.* As with the previous case, it is safe

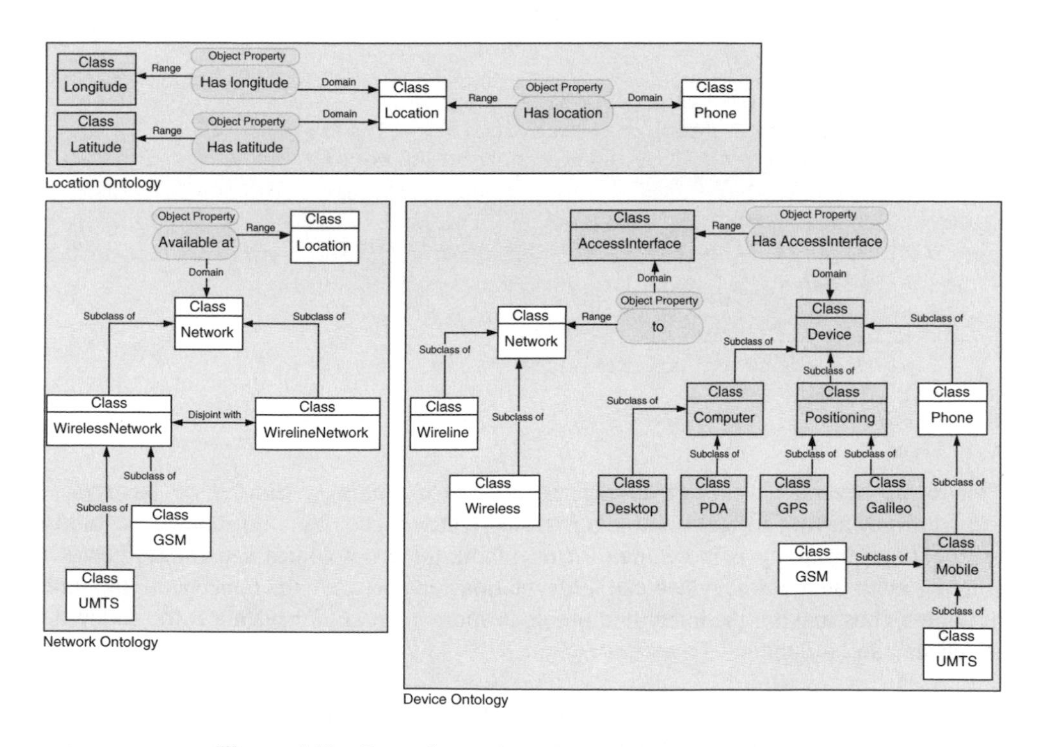

Figure 4.13 Example ontology fragments for the integration

to define that in the integrated ontology these are equivalent. The syntactically similar but semantically different concepts are more problematic. In our example, the *Universal Mobile Telecommunications System* (*UMTS*) and *GSM* are such concepts. In the network ontology, they refer to network types, whereas in the device ontology they represent devices capable of operating in the networks. Therefore, in the integrated ontology, we have to state that although these concepts have the same name, they do not mean the same thing.

In the following a complete OWL [41] description for these relationships is given:

```
<?xml version="1.0" encoding="UTF-8"?>
<rdf:RDF
  xmlns:rdf="http://www.w3.org/1999/02/22-rdf-syntax-ns#"
  xmlns:rdfs="http://www.w3.org/2000/01/rdf-schema#"
  xmlns:owl="http://www.w3.org/2002/07/owl#"
  xmlns:net="http://helluli.com/network#"
  xmlns:location="http://helluli.com/location#"
  xmlns:device="http://helluli.com/device#"

  <owl:Ontology rdf:about="">
    <rdfs:comment>An sample integrated ontology</rdfs:comment>
  </owl:Ontology>

  <owl:Class rdf:ID="&net;WirelessNetwork">
    <owl:sameAs rdf:resource="&device;Wireless"/>
  </owl:Class>

  <owl:Class rdf:ID="&net;WirelineNetwork">
    <owl:sameAs rdf:resource="&device;Wireline"/>
  </owl:Class>

  <owl:Class rdf:about="&net;GSM">
    <owl:disjointWith rdf:resource="&device#GSM"/>
  </owl:Class>

  <owl:Class rdf:about="&net;UMTS">
    <owl:disjointWith rdf:resource="&device#UMTS"/>
  </owl:Class>
</rdf:RDF>
```

There are several benefits of having the integrated ontology. Firstly, we have saved time from not having to build ontologies from scratch. Secondly, and more importantly, the integrated ontology is more than a sum of the three ontologies separately, because, using the integrated ontology, we can infer relationships between the concepts in the three ontologies, thus making the integrated ontology more general. For instance, the following inferences can be done:

- Because Location is composed of Latitude and Longitude, the Network is available at certain {latitude, longitude}-coordinates.

- Because Phone has a Location, all the Mobile phones (e.g. GSM and UMTS in our ontology) also have a location.
- Because all the phones have a location, and the networks are available at certain locations, in the application we can easily have a functionality that maps the phones to a certain network based on the location.

In addition to these, we can define new relationships between the concepts between the integrated ontology, if we have sufficient domain knowledge. For instance, in our example we can define a relationship *provides* between the *Positioning* and the *Location*, which means that all the positioning devices are able to provide location. The fragment of OWL description to be added would be as follows:

```
<owl:ObjectProperty rdf:ID="provides">
  <rdfs:domain rdf:resource="&device;Positioning"/>
  <rdfs:range rdf:resource="&location;Location"/>
</owl:ObjectProperty>
```

Merge of Existing Ontologies

Merging ontologies differ from the integration of ontologies in such a way that, in merging, existing ontologies are about the same domain and subjects. Merging is about unifying ontologies: removing duplicate subjects, and finding similar subjects (and perhaps treating them as the same). As in the case of integration, the output of the ontology merge operation is a new unified ontology.

In the ontology merge process, existing ontologies $O_1, O_2, \ldots, O_n$ about a domain s are merged together to build a harmonized ontology O about the same domain. As an example of the ontology merge process, let us consider a situation where a mobile operator operating GSM, GPRS, and UMTS network wants to extend its operation by providing Internet hotspots in a certain locations, such as hotels and airports. The mobile operator has been using a wireless ontology for mobile networks, but now seeks an ontology for Internet service providers as well. Figure 4.14 shows the two ontologies: Internet service provider ontology on the left and mobile operator on the right.

To merge these two ontologies, (1) the identical concepts are unified, (2) the identical properties are unified, and (3) the duplicate concepts are removed. Also, should there be concepts with the same names but denoting different things in the domain, this must be expressed in the new ontology. In our example, however, there are no such concepts. The identical concepts are painted with white background. These concepts are *Network, WirelessNetwork, CircuitSwitched*, and *PacketSwitched.* These can be unified using the following OWL description:

```
<?xml version="1.0" encoding="UTF-8"?>
<rdf:RDF
  xmlns:rdf="http://www.w3.org/1999/02/22-rdf-syntax-ns#"
  xmlns:rdfs="http://www.w3.org/2000/01/rdf-schema#"
  xmlns:owl="http://www.w3.org/2002/07/owl#"
  xmlns:mobile="http://helluli.com/mobileoperator#"
  xmlns:internet="http://helluli.com/internetoperator#"
```

```
  <owl:Ontology rdf:about="">
    <rdfs:comment>A merged ontology</rdfs:comment>
  </owl:Ontology>

  <owl:Class rdf:ID="&mobile;Network">
    <owl:sameAs rdf:resource="&internet;Network"/>
  </owl:Class>

  <owl:Class rdf:ID="&mobile;WirelessNetwork">
    <owl:sameAs rdf:resource="&internet;WirelessNetwork"/>
  </owl:Class>

  <owl:Class rdf:ID="&mobile;CircuitSwitched">
    <owl:sameAs rdf:resource="&internet;CircuitSwitched"/>
  </owl:Class>

  <owl:Class rdf:ID="&mobile;PacketSwitched">
    <owl:sameAs rdf:resource="&internet;PacketSwitched"/>
  </owl:Class>
</rdf:RDF>
```

There are also two identical properties: *operatedBy*. Although the domain – *Network* – is the same, the range is not. In mobile operator's ontology the range is *MobileOperator*, whereas in the Internet service provider's ontology it is *InternetOperator*. However, using the domain knowledge while merging the ontologies, we can remove the *InternetOperator* concept and the *operatedBy* property associated with it, and keep the *MobileOperator* concept and the corresponding property. After this, we have a merged ontology for a mobile operator, which is also depicted in Figure 4.15.

Now, the mobile network operator could benefit from this new ontology in several ways. One clear benefit, as with the integration presented earlier, is that there is no need to design the Internet service provider's part of the ontology from scratch. The other thing is that the existing concepts within the mobile operator can be included in the Internet service provider ontology as well. For instance, the network identifier, which is used as a unique identifier of a mobile network, can be mapped to WLAN hotspots as the operator-specific Service Set Identifier (SSID), and to phone numbers for modem dial-up lines.

4.3.3.2 Taxonomy for Event-systems

Event models can be grouped into taxonomies clustered by their properties. As contrasted with the client–server paradigm, event models involve one-to-many communication. Other important aspects for event model classification are [28]:

- Does the model support distributed operation, local operation, or both? In a centralized event model, the event sources and listeners are located on the same host, whereas in the distributed model they can be located on different hosts.
- Support for detecting composite events (compound events). Compound events require more complicated filtering and history mechanisms.

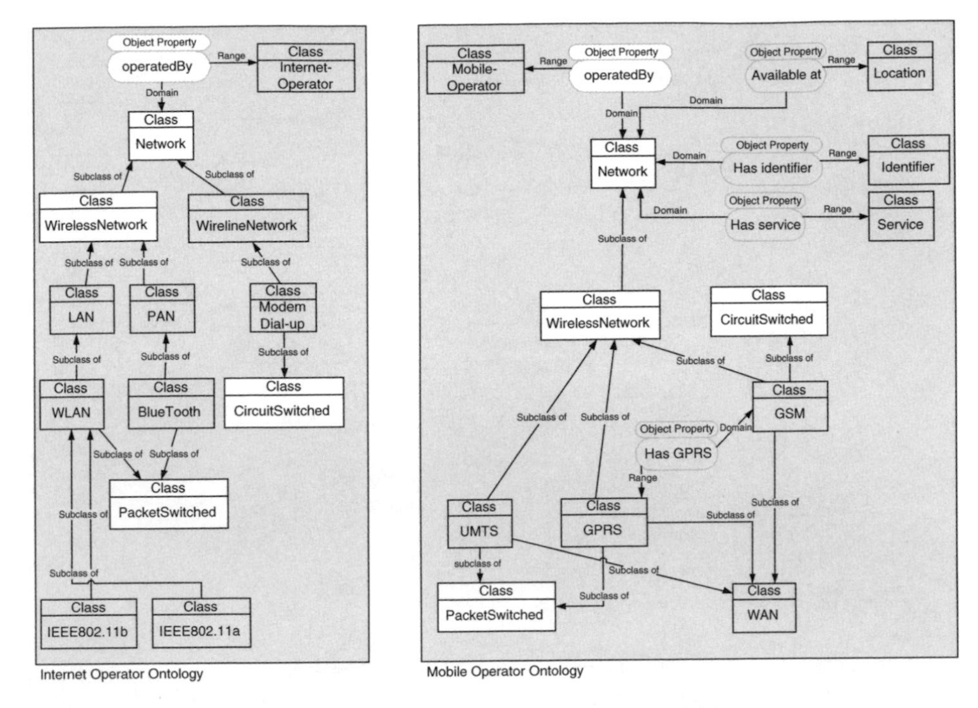

Figure 4.14 Example ontology fragments to be merged

- Support for QoS requirements, for example, delivery semantics (best-effort, at-most-once, etc.).
- Support for typed events, generic events, or both. Typed events have a well-defined structure, for example, a set of ordered strings, and generic events do not have an expressive structure (datatype any).
- How decoupled are the event listeners from the event sources?
- Is the model subscription-based or advertisement-based?
- Support for channel-based, subject-based, or content-based routing.

Additional aspects are:

- Support for wireless systems and disconnected operation.
- Does the model support event routing, direct notification, and so on?
- How are interests defined and discovered? Not all models include discovery functionality.

Figure 4.16 presents an example taxonomy based on the event architectures explored in this review.

4.3.3.3 Reasoning and Rule-based Systems

Reasoning is the general process of inferring new knowledge from a base of existing knowledge. The production of this new knowledge is made possible by applying 'rules' over the set of existing knowledge, also called classically 'base of facts'.

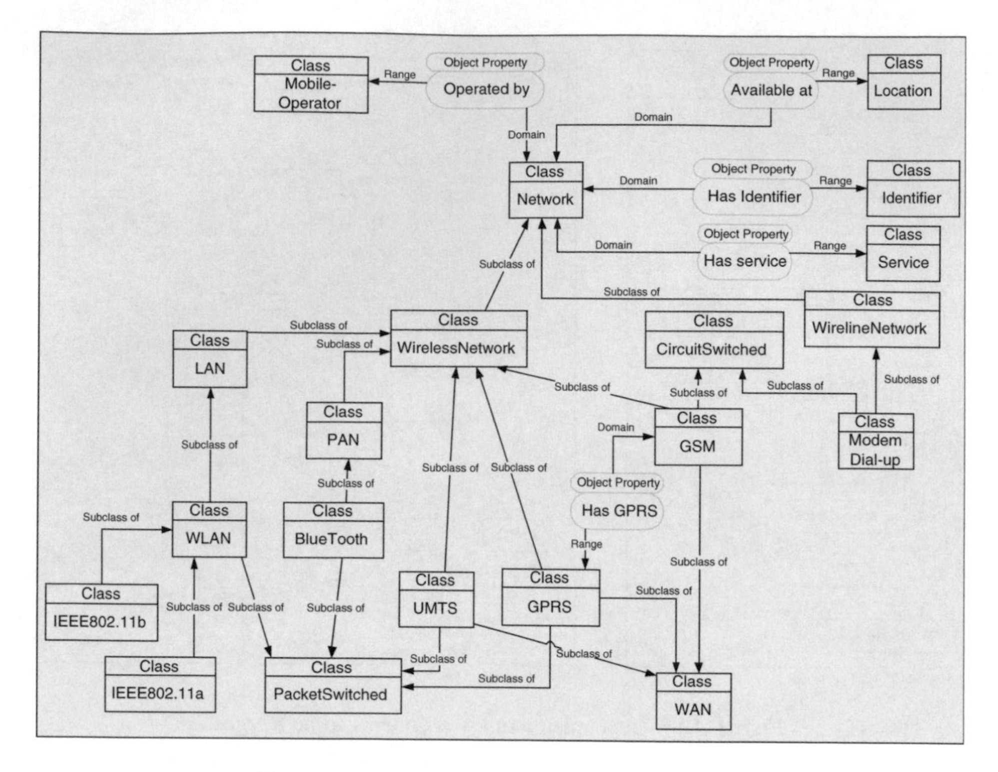

Figure 4.15 Example ontology after the merge

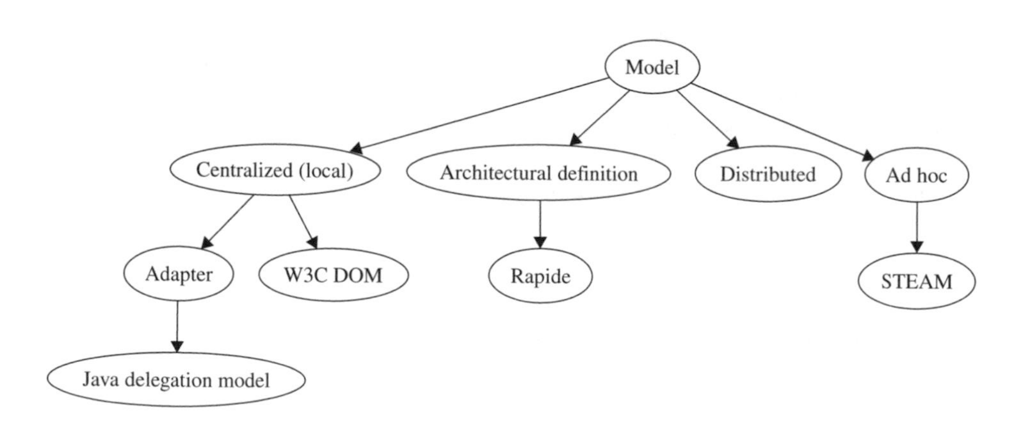

Figure 4.16 Example event model taxonomy

Applying a rule can result in both creating new knowledge (or new facts) and modifying/deleting existing knowledge (or existing facts). Owing to the inherent complexity of (mathematical) deduction, it is not expected to demonstrate logical formulae using a commercial inference engine (except in really rare cases). Inferencing is more to be used in the spirit of expert systems, where the efficiency of the system directly depends on how the rules are written. Rule-based systems (the name explains it all) consist of an algorithm that continuously applies the existing rules to a base of facts and a language (classically built around C, C++ or Java) that allows the implementation of the rules. The rules are built generally in the spirit of an if-then construct, where the condition (the left hand side of the rule) is made of a set (conjunction) of patterns and the then part (the right hand side) is made of actions. Basically, the algorithm selects a rule among the set of rules that is applicable and applies it. Applicable means that there exists a match between the pattern, which is expressed in the lhs of the rule and the facts that are present in the current base of facts. Applying the rule means that the substitution, which is obtained after pattern-matching in the previous step (between the fact and the pattern), is first propagated to the right hand side of the rule, and that the actions expressed in the rule are executed. These actions generally result in deleting/modifying existing facts from/in the base or creating new facts in the base. The action (or part of it) can also result in any method call to an object existing in the base. It is worth noting that the rule language sometimes proposes temporal modalities that allow extending the reasoning to temporal reasoning, that is, reasoning about how and when events appear as the time runs.

The inference engine can be easily embedded in a GSE in such a way that it provides API to create new rules, to manage facts. It can then be used to filter or correlate events such as ambient information, or to infer higher-order facts from the raw events (see for instance the Environment Monitoring). There have been experiments and prototypes that embed an inference engine within a mobile agent in order to implement mobile expert systems. The low weight of the engine is then really compatible with mobile code.

Figure 4.17 shows a rule-based system and Figure 4.18 how the rule-based system can be used in the field of medicine to monitor persons, for example, elderly people or a person with heart deficiencies.

4.3.4 Future Research and Development

For realizing future service platforms, current middleware solutions need to be augmented. In particular, conceptual models for enabling reflection are needed. The reflection, or self-awareness, is needed for reasonable initial configuration and dynamic reconfiguration after essential changes in the execution environment.

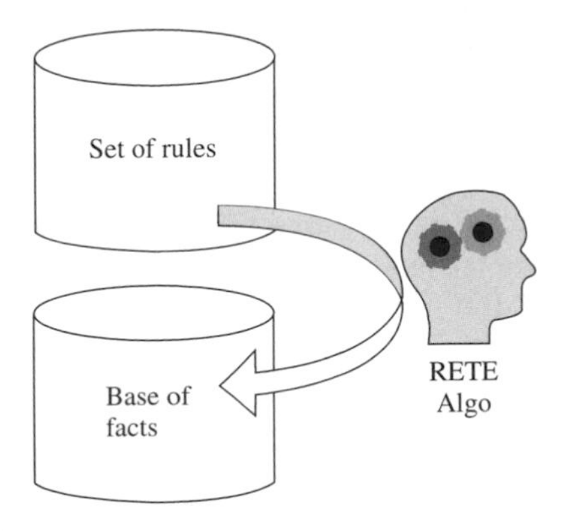

Figure 4.17 Basic concepts of rule-based systems

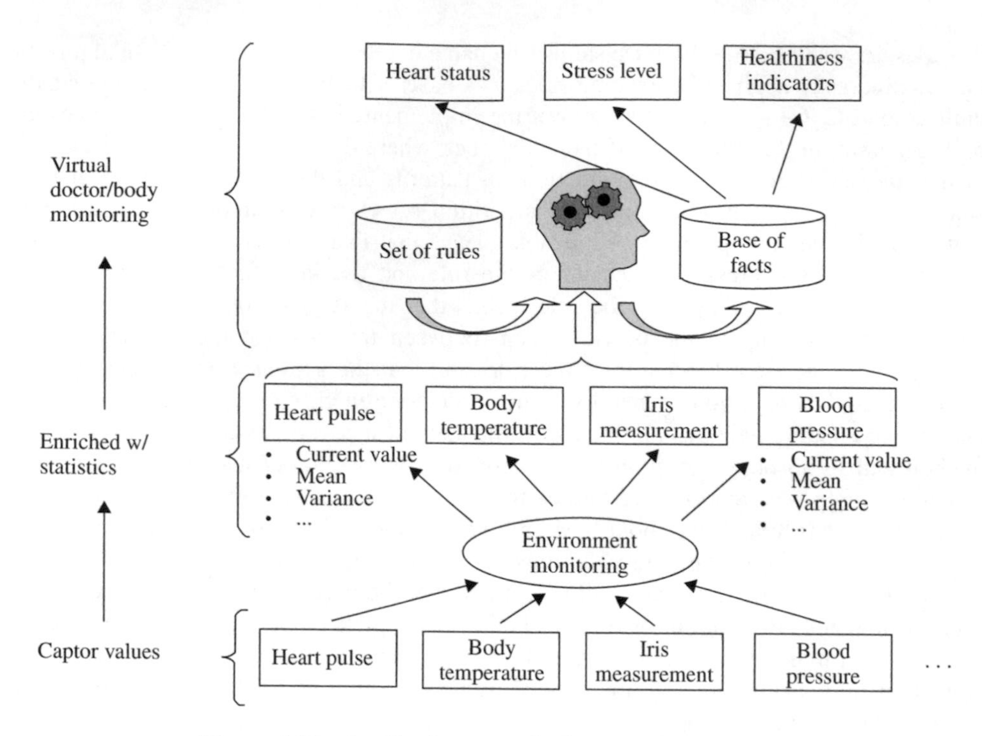

Figure 4.18 Application example for a rule-based system

Another important area for R&D is seamless roaming between networks with and without infrastructure support. In ad hoc communities, we also need a new approach to fault-tolerance. The traditional solution based on replication needs to be replaced by a novel distribution of state providing robustness against disappearing members during the processing of a distributed task.

In respect to generic service elements, it needs to be investigated which functionalities should be commonly available to deploy them as GSEs.

Based on these decisions the available middleware and service execution environment approaches should be enhanced to cope with this kind of flexibility and to open the market for new services and applications.

4.3.5 Summary

Openness is one of the major issues for platforms that should serve the I-centric communication space. Therefore, technologies and software approaches should be used to support standardized open APIs so that, on the one hand, new services can be developed and deployed easily, while on the other hand, existing technologies should be used in such a way that each application and service can make use of the available resources.

With this chapter we propose to start an in-depth research activity for generic service elements within the I-centric communication contexts and the needed processes and technologies to make the I-centric vision happen.

4.4 Acknowledgment

The following individuals contributed to this chapter: Stefan Gessler (NEC Europe, Germany), Mika Kemettinen (Nokia, Finland), Michael Lipka (Siemens, Germany), Kimmo Raatikainen (High Intensity Interval Training (HIIT)/University of Helsinki/Nokia, Finland), Olaf Droegehorn (University of Kassel, Germany), Klaus David (University of Kassel, Germany), François Carrez (Alcatel CIT, France), Heikki Helin (TeliaSonera, Finland), Sasu Tarkoma (HIIT/University Helsinki, Finland), Herma Van Kranenburg (Telematica Instituut, The Netherlands), Roch Glitho (Ericsson Canada/Concordia University, Canada), Seppo Heikkinen (Elisa, Finland), Miquel Martin (NEC Europe, Germany), Nicola Blefari Melazzi (Università degli Studi di Roma – Tor Vegata, Italy), Jukka Salo, Vilho Räisänen (Nokia, Finland), Stephan Steglich (TU Berlin, Germany), Harold Teunissen (Lucent Technologies, Netherlands).

References

[1] S. Arbanowski, P. Ballon, K. David, O. Droegehorn, H. Eertink, W. Kellerer, H. van Kranenburg, K. Raatikainen, and R. Popescu-Zeletin, "I-centric communications: personalization, ambient awareness, and adaptability for future mobile services". *IEEE Communications Magazine*, 42(9), pp. 63–69, 2004.

[2] R., Tafazolli [Ed.,] *Technologies for the Wireless Future: Wireless World Research Forum (WWRF)*, ISBN 0-470-01235-8, Wiley 2004.

[3] ISTAG; Scenarios for Ambient Intelligence in 2010; Final Report, February 2001, EC 2001. Available at http://www.cordis.lu/ist/istag.htm

[4] The Activities of 4th Generation Mobile Communications Committee (FLYING CARPET Ver.2.00), March 2004, available at http://www.mitf.org/public_e/archives/contents.html.

[5] ETSI/TIA Project mesa scenarios, Fall 2003. Available at http://www.projectmesa.org.

[6] MIT Oxygen project scenarios: Pervasive Human-centred computing. Available at http://oxygen.lcs.mit.edu/Overview.html.

[7] NTT DoCoMo Vision 2010 scenarios. Multimedia content available at http://www.nttdocomo.com/vision2010/.

[8] CyPhone Mediaphone Project – Taxi Trip Scenario, 1999. Available at http://paula.oulu.fi/Publications/Submited/ICAT99.pdf.

[9] IST-SIMPLICITY project scenarios. Available at http://www.ist-simplicity.org/.

[10] The Vision of ITEA (SOFTEC Project); Technology Roadmap on Software Intensive Systems. Available at http://www.itea-office.org/newsroom/publications/itearoadmap.htm.

[11] Bo Karlson *et al.*, *Wireless Foresight: Scenarios of the Mobile World in 2015*, Wiley 2003.

[12] Pradeep Gore, Ron Cytron, Douglas Schmidt, and Carlos O'Ryan, "Designing and optimizing a scalable CORBA notification service", *ACM SIGPLAN Notices*, 36(8), pp. 196–204, 2001.

[13] IBM, Gryphon: Publish/subscribe over public networks, December 2002. (White paper) http://www.research.ibm.com/gryphon/Gryphon/Gryphon-Overview.pdf..

[14] E. Christensen, *et al.*, "*Web Services Description Language (WSDL) 1.1*", W3C Note, 15 March 2001, http://www.w3.org/TR/wsdl.

[15] H. Ailisto, P. Alahuhta, V. Haataja, V. Kyllönen, and M. Lindholm, "*Structuring Context Aware Applications: Five-Layer Model and Example Case*", Workshop in Ubicomp, Gothenburg, Sweden, 2002.

[16] G. F., Luger, and W. A. Stubblefield, *Artificial Intelligence. Structures and Strategies for Complex Problem Solving*, Addison Wesley Longman 1998.

[17] World Wide Web Consortium, *Web Ontology Language (OWL) Guide Version 1.0*, Available at http://www.w3.org/TR/2002/WD-owl-guide-20021104/, (Accessed 2004).

[18] http://www.sts.tu-harburg.de/~r.f.moeller/racer/, (Accessed 2005).

[19] B. Kummerfeld, A. Quigley, C. Johnson, and R. Hexel, "Merino: Towards an intelligent environment architecture for multi-granularity context description", workshop on *User Modelling for Ubiquitous Computing*, Pittsburgh, USA, June 2003.

[20] A. Dearle, G. Kirby, R. Morrison, A. McCarthy, K. Mullen, Y. Yang, R. Connor, P. Welen, and A. Wilson, "Architectural support for global smart spaces", *Proceedings of the MDM'2003*, Springer-Verlag, Melbourne, Australia, pp. 153–164, 2003.

[21] A. K. Dey, D. Salber, and G. D. Abowd, "A conceptual framework and a toolkit for supporting the rapid prototyping of context aware applications". *Human-ComputerInteraction*, vol. 16, pp. 97–166, 2001.

[22] St. Arbanowski, St. Steglich, and R. Popescu-Zeletin, "Super Distributed Objects – an execution environment for I-centric Services", In *Proceedings of the 9-th IEEE International Workshop on Object-oriented Real-time Dependable Systems (WORDS 2003F)*, Capri Island, Italy, October 2003.

[23] Paulo Mendse, Christian Prehofer, and Qing Wei, "Context management with programmable mobile networks", In *IEEE Computer Communication Workshop (CCW 2003)*, San Diego, USA, October 20–21, 2003.

[24] C. Prehofer, L. Yao, N. Kawamura, and B. Souville, "Middleware and network support for re-configurable terminals", In *Software DefinedRadio Technical Conference and Product Exposition*, San Diego, USA, November 2002.

[25] C. Prehofer, W. Kellerer, R. Hirschfeld, H. Berndt, and K. Kawamura, "An architecture supporting adaptation and evolution in fourth generation mobile communications systems". *Journal of Communications and Networks*, 4(4), pp. 336–343, 2002.

[26] H. van Kranenburg, and E. H. Eertink, "Processing heterogeneous context information", submitted to SAINT workshop *Next Generation IP-Based Service Platforms for Future Mobile Systems*, http://www.saint2005.org/workshops/wk-cfp-6.txt, 2004.

[27] http://www.aosd.net, (Accessed 2006).

[28] René Meier. *State of the Art Review of Distributed Event Models*, Technical Report TCD-CS-2000-15, Department of Computer Science, Trinity College, Dublin, Ireland, March 2000.

[29] Sun Microsystems, *Java Message Service Specification*, June 2001.

[30] Jon Siegel. "An overview of CORBA3", In *Proceedings of the Second International Working Conference on Distributed Applications and Interoperable Systems*, July 1999.

[31] Object Management Group, *CORBA Event Service Specification v.1.1*, March 2001.

[32] World Wide Web Consortium, *Document Object Model (DOM) Level 2 Events Specification, Version 1.0*, November 2000. [Recommendation] http://www.w3.org/TR/DOM-Level-2-Events/

[33] WorldWideWeb Consortium, XML Events – An Events Syntax for XML, February 2003. [Candidate Recommendation] http://www.w3.org/TR/2003/CR-xml-events-20030207.

[34] BEA, Microsoft, TIBCO, Web Services Eventing (WSEventing), January 2004. http://xml.coverpages.org/WSEventing200401.

[35] IBM Web Services Architecture Team *Web Services Architecture Overview – The Next Stage of Evolution for e-business*, http://www-106.ibm.com/developerworks/webservices/library/w-ovr/?dwzone=webservices, (Accessed 2004).

[36] D. Box, D. Ehnebuske, G. Kakivaya, A. Layman, N. Mendelsohn, H. F. Nielsen, S. Thatte, D. Winer, *Simple Object Access Protocol (SOAP) 1.1*, W3C Note, 08 May 2000, http://www.w3.org/TR/SOAP/.

[37] David Platt, The COM+ event service eases the pain of publishing and subscribing to data. Microsoft Systems Journal, September 1999. http://www.microsoft.com/msj/0999/comevent/comevent.aspx

[38] Paddy Srinivas, Introduction to COM+ events, March 2001. http://www.idevresource.com/com/library/articles/com+eventsintro.asp.

[39] Microsoft. Message Queuing in Windows XP: New Features, 2002. (White paper) http://www.microsoft.com/msmq/MSMQ3.0_whitepaper_draft.doc.

[40] http://www.aspectj.org, (Accessed 2005).

[41] M. N. Huhns, Multiagent Systems. Tutorial at the European Agent Systems Summer School (EASSS'99), *Proceedings of the First European Agent Systems Summer School (EASSS'99)*, Utrecht University/Agent Link, Utrecht, The Netherlands, 1999.

[42] A. K. Caglayan, and C. G. Harrison, *Intelligente Software-Agenten*, Carl Hanser Verlag München Wien 1998.

5

Security and Trust

Edited by Mario Hoffmann (Fraunhofer SIT), Christos Xenakis, Stauraleni Kontopoulou (University of Athens), Markus Eisenhauer (Fraunhofer FIT), Seppo Heikkinen (Elisa R&D), Antonio Pescape (University of Naples) and Hu Wang (Huawei)

5.1 Introduction

During the past 30 years, computing has moved from highly centralised stationary computers to increasingly distributed and mobile environments. This evolution has profound implications for the security models, policies and mechanisms needed to protect user's information and resources in an increasingly globally interconnected computing infrastructure. The vision of 'Ambient Intelligence (AmI)' focuses on the user as an individual, in the centre of future developments for an inclusive knowledge-based society for all. However, this vision also challenges existing paradigms for software engineering, because the availability, location and configuration of important system components and resources cannot be assumed *a priori*. Security, privacy and trust are important areas where existing software engineering paradigms are predicted to fail.

The aim is to improve our understanding of dependability, trust and security in a dynamically changing context, and make this understanding operational through the development of a trust management. This requires a deep understanding of trust, dependability, security, resilience, privacy, and the specification and enforcement of policies in a vast network of collaborating autonomous entities.

Identity management (IdM) is about the management of attributes and the provision of them when appropriate. In some systems, it is enough to limit oneself to the identifier part of the identity and the additional semantic data thereof. This is true in the case of authentication, where an entity presents information as an assurance of its linkage to a certain identity through the used identifier. The level of assurance is dependant on the semantic interpretation of this information, that is credentials. A simple example of this

Technologies for the Wireless Future – Volume 2 Edited by Rahim Tafazolli

is a login as an identifier and a password as credentials, where the semantic interpretation relates to the comparison of bit strings.

Currently research projects – especially EU funded integrated projects-have started fundamental research on security and trust for B3G mobile systems.

5.2 Trust Management in Ubiquitous Computing

Trust is a multilateral relationship amongst users, actors and stakeholders and is an estimation of the level of reliability between these players. Trust is certainly a continuous quality and can be expressed as the degree of reliance.

A traditional, identity-based, security mechanism cannot authorise an operation without authenticating the user. This means that no interaction can take place unless both users are known to each other's authentication framework. In the current environment, a user who wishes to partake in collaboration with another party has the choice between enabling security and hence disabling spontaneous collaboration or disabling security and thereby having spontaneous collaboration. This is clearly unsatisfactory, as mobile users and devices should have the ability to autonomously, quickly and reliably authenticate and authorise each other without having to rely on a common authentication infrastructure. A common security infrastructure will restrict the provision of demanding applications and services in the new emerging heterogeneous mobile network.

Two fundamentally different approaches to trust management have emerged from the computer security communities.

In computer security, all entities are assumed to be trusted to a certain degree and the role of trust management is to determine whether a particular principal is trusted sufficiently to allow a given operation. This implies that trust is an inherently static property of all principals, which can be objectively quantified and used in pre-defined security policies at the time of evaluation. It is therefore often the task of the requesting principal to prove that (s)he is sufficiently trustworthy for the security policy evaluator to allow access to the requested resource, for example through disclosure of credentials in the form of certificates.

In open distributed computing, all entities can be assumed to be initially unknown, but potentially everyone is trustworthy, so the role of trust management is to determine whether a sufficient level of trust can be established in order to allow a certain operation. This means that the roles are reversed, because the security policy evaluator must autonomously decide whether the requesting principal is sufficiently trustworthy to be allowed access to the requested resource. Credentials may be used in this evaluation, but past experience, recommendations from trusted third parties and input from reputation systems are equally important. Recently, dynamic approaches to trust management, the notion of reputation, and the associated mechanisms have attracted much attention from the research community. The currently employed reputation mechanism in commercial applications such as Peer-to-Peer (P2P) systems and web sites like e-Bay and Amazon, represent only the initial steps in what appears to be a major new area for network security research. This dynamic form of explicit trust management supports secure collaboration with newly encountered parties, because the decision to trust is based on all the evidence (e.g. observations, recommendations, and reputation) available at the time when the decision is made.

Explicit trust management has additional advantages, for example it allows a principal to select the service provider that is most likely to provide the required service whenever it is faced with a number of previously unknown service providers. We therefore believe that a general trust management middleware will provide the enabling technology for secure and reliable collaboration in highly dynamic (mobile) computing environments.

5.2.1 Trust Requirements

To ensure high usability and acceptance, trusted actions must be both transparent and trustable.

The requirement of trustability is that users must be confident that only the appropriate part of their identity will be revealed for any given trust transaction or relationship, and that the system will not reveal information allowing the user to be impersonated or defrauded, even if physical devices holding the data are stolen or tampered with.

Transparency implies discovery and categorisation of elements joining into trust management. This includes subjects (entities, devices, security agents ...), objects (documents, resources ...), and contexts (situations, temporal and spatial variables ...).

The importance of considering the user perspective is twofold: On the one hand, security and trust issues directly involve the users' privacy and therefore, the development of such concepts is impossible without taking user requirements into account. On the other hand, a precondition for the success of mobile applications is acceptance by the users.

Usability for example requires that users should not have to continually configure or interact directly.

Trust management should include autonomic capabilities:

self-configuration (for example, to automatically select the most appropriate trust policy based on context information)

self-adaptation (for example, to recognise parties that have been dishonourable in past relationships and to make policy adjustments accordingly)

self-protection (for example, to detect 'abnormal' behaviours or requests)

Devices with different capabilities (from a simple Global System for Mobile Communications (GSM) phone to a complex Personal Digital Assistant (PDA)) must be handled, attending to their restrictions and possibilities, in order to ensure backwards compatibility and to address open issues on actual environments.

5.2.2 Trust Life Cycle

The four elements of the trust life cycle are illustrated in Figure 5.1.

Establishing an initial trustworthiness between collaborators is referred to as trust formation. Trustworthiness thus deals with the evaluation of the past interactions. Evidence relevant to the current context should carry the most weight. The evidence is derived from interaction monitoring and leads to the allocation of privileges. Initially, new entities have no evidence of past behaviour to establish a base for interaction. To form an opinion of trustworthiness in this case requires the presence of some optimistic entities willing to take risks in unknown situations, allocating privileges judiciously until experience shows that it was unwise.

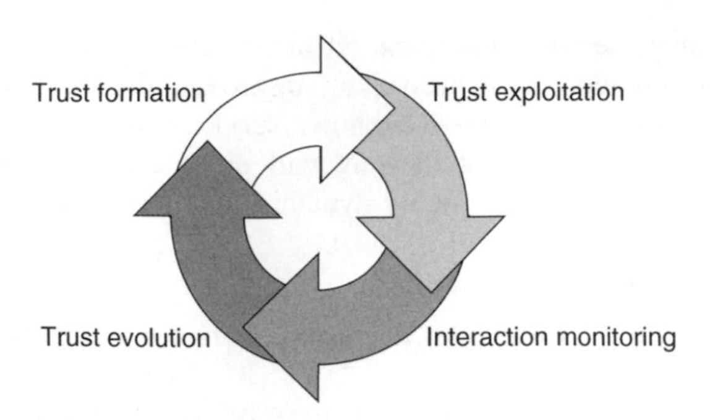

Figure 5.1 The trust life cycle

Trust evolution recurs to the iteration of trust formation as additional evidence becomes available. Accumulation of evidence with experience of new interactions must modify the level of trust to be placed in an entity, incrementing the summary information to maintain accuracy. The risk assessment for an entity performing an action in a particular context will change depending on how much is known about positively or negatively perceived actions in the past.

Trust exploitation refers to the problem to determine behaviour on the basis of trust, balancing appropriately risk and benefit within the context. The security policy for access control is expressed in terms of trust and specifies the level of positive experiences required to allow access to a specific resource. The policy will determine whether an entity is optimistic or pessimistic about an interaction depending on the risks involved.

A trust and security framework will incorporate models, policies and mechanisms that allow an initiator to assess the risk of a potential collaboration and to explicitly reason about the trustworthiness of the other party in order to determine if the other party is trustworthy enough to mitigate the risk of collaboration. Confidence can be built gradually based on previous relations according to a reputation-based trust management model.

5.2.3 Trust Management

Network and information security can be understood as the ability of a network or an information system to handle accidental events and malicious actions. The layered nature of most communication systems permits the introduction of such protection mechanisms at multiple layers (physical link, network, or application layer). A major limitation of contemporary security technologies is that the application of security functionality at multiple layers, and the usual lack of a cross-layer design (that is the usual price that is paid for achieving a modular design), lead to excessive resource consumption and/or unnecessary complexity due to the unnecessary duplication of similar security functionality across multiple layers. Notice that an uncoordinated deployment (duplication) of similar security functionalities does not necessarily increase the overall resilience of the system to attacks; on the contrary, the addition of excessive mechanisms and code is opening new possibilities for exploitation and security compromises. Even if this is not the case, duplication is clearly not desired because of the (unnecessary) cost that it entails for deploying the various mechanisms at different levels.

One solution for the above problems would be to implement general trust management in a middleware, which will provide the enabling technology for secure and reliable collaboration in highly dynamic (mobile) computing environments.

A middleware aims at facilitating communication and coordination of distributed components, by concealing the complexity raised by mobility from application developers. Typically, application designers building distributed applications have to guarantee the following nonfunctional requirements: scalability, openness, heterogeneity, fault tolerance, and resource sharing. In standard middleware, the complexity introduced through distribution is handled by means of abstraction. Implementation details are hidden from both users and application designers and encapsulated inside the middleware itself. The fundamental problem is that by hiding implementation details, the middleware has to take decisions on behalf of the application.

Ubiquitous computing introduces major new challenges for middleware to overcome. Research into middleware for ubiquitous computing has concentrated on overcoming problems related to the characteristics and technologies of wireless networks, such as the following:

- Mobile devices have scarce resources, low battery power, slow Central Processing Units (CPUs), and little memory.
- Communication faces poor network quality of service, variability of network bandwidth, and temporary and unannounced loss of network connectivity when users move.
- Connection sessions are usually short, and need to discover other hosts in an *ad hoc* manner.
- The system is required to react to frequent changes in the environment, such as change of location, availability of interaction partners (human–human, human–machine, and machine–machine interaction), and context conditions.

In order to cope with these limitations, many research efforts have focused on designing new middleware systems capable of supporting the technological requirements imposed by mobility. As a result of these efforts, extensions to well-established platforms for fixed networks have been proposed.

5.2.4 Research Issues

To establish a trust and security framework in an open environment that comprises heterogeneous mobile networks is one of the major challenges of the future. A trust and security framework should take into account all the involved factors (e.g. network technologies, network architecture, the employed security mechanisms, users' requirements, application characteristics, etc.). The analysis of specific network technologies (Internet Protocol (IP), Global System for Mobile Communications (GSM), General Packet Radio Service (GPRS), Universal Mobile Telecommunications System (UMTS), Wireless Local Area Network (WLAN), etc.), security mechanisms (link layer security, network layer security, application layer security, WLAN security, UMTS security, etc.) and their application and operation modes will guide the design of abstract trust and security models for trust and security principles (transparency, responsibility, traceability, etc.), security objectives (integrity, availability, confidentiality, etc.), security policies (protection, deterrence, vigilance, etc.), and security functions (identification, authentication, access control, management of secret elements, privacy, etc.). The development of such an abstraction can be quite beneficial by:

- allowing for a common description of different security mechanisms;
- allowing for their comparison under the same setting;
- laying the ground for identifying problems, open issues, and designing solutions.

The incorporation of such abstract security models in a trust and security framework will enable it to adapt incompatible and heterogeneous security mechanisms and it will allow the optimised configuration and cross-layer collaboration of the employed security mechanisms providing more efficient and effective security services.

Another trust-related problem is to allow higher flexibility and autonomy in collaboration between communicating peers. The support of secure, trusted and personalised access to services for communities of mobile users is a future challenge: enabling of content distribution and service provision between sources in a decentralised way while taking all precautions to protect privacy and filter out un-requested materials and unreliable information sources.

5.3 Identity Management

Generally speaking, an identity can be considered to be a collection of attributes, characteristics, and traits and these attributes can have further semantic information attached to them. Some of these attributes can be seen as weak and some strong depending on how defining and binding they are. For example, the attributes that have the property of uniquely naming the identity can be considered to be strong. The collection of attributes can also change over time, although in a philosophical sense one can question whether we are then still discussing about the same identity. One further interesting point with regard to the attributes is the concept of identifier. These are generally strong attributes that are used to provide a handle to the identity in some namespace, and sometimes in speech they are used as synonymous concepts. Identifiers can also have credentials attached to them. In other words, some extra information that is used to signal the ownership of the said identifier and thus, promoting the uniqueness of the identifier in a certain environment. The relationship may not be as straightforward in all the cases: the actual identity may be anonymous and only exists in the logical sense, even though presenting just the credentials may be deemed sufficient to gain access to some resource, for example.

Another question that can be asked is that who or what has an identity. A simple answer is 'every entity'. Another thing is, whether this is viewed from the conceptual or practical viewpoint. In practice, one needs to bind some attributes to the entity in order to be able to define the boundaries of its existence and to point to it. These attributes define a subset of the identity that is meaningful to act upon and in different contexts the subset can be different even though logically the actions point to the same identity. This has a strong linkage with the privacy aspects, that is the entity can decide which facet of its identity it is revealing (Figure 5.2).

A more complex case is, for example, the use of public keys: a public key can be used as an identifier, whereas credentials relate to the ability of solving a mathematical problem, which is generally seen as a proof of possessing a private key. A private key in this case as an attribute belongs to one facet of the identity that is not to be released. One

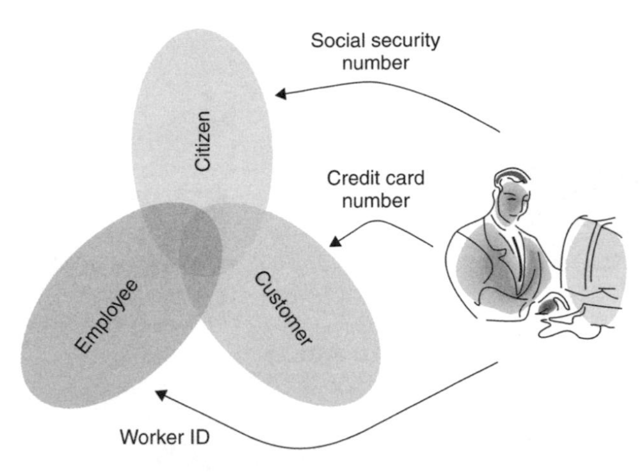

Figure 5.2 Some facets of identity and potential handles

further example is a shared secret. Depending on the context, it can either be interpreted as an identifier or credentials or both, even though for an external observer being out of the scope of the namespace the distinction is hard to make. The same is true also for the case where the amount of identities sharing the secret grows beyond mutual relationship.

The above-mentioned procedures can be executed locally, or it is possible to outsource them to some external, trusted party that is in charge of obtaining and validating the identity claims. The term *identity provider* (IdP) is generally used for such a party. The IdP is responsible for providing the authentication information either directly to the entity on whose behalf it is working, or it can also work through the subject of the authentication event by providing its own assertion be it in the form of security token or some other. In the similar sense, authorisation to some resource or action can be provided, that is some subject is deemed to be entitled to the access. In this respect it may not be necessary to tell who the actual subject is to the entity hosting the resource, but just the access decision. Additionally, it is equally possible to address the possession of certain attributes of the subject, that is give guarantees that the said attributes are part of identity of the subject. Here the IdP acts as an attribute provider. In order to be able to make such access decisions on behalf of some entity, the IdP has to be aware of the policies related to the resource and therefore act as a policy decision point, even though the actual policy enforcement is still done by the resource owner. As the information connecting an identity and a resource access may not be suitable to be disclosed to the external, even though trusted, parties, it is sometimes ruled out of IdM. If it is desired that the distinction is made, some propose the use of the term *identity and access management* (IAM) instead. With regard to access and authorisation it is good to keep in mind that the identity attributes are also subject to the authorisation decision, that is a certain identity may want to control which of the attributes are disclosed to which parties. In this respect it is a question of whose identity and attributes the IdP is managing, that is which entity has the role of service provider and which the user. The IdP is also in the potential position to account all these security events (Figure 5.3).

One additional aspect of identity provisioning is the concept of federation. Federation refers to the process of linking different identities or rather, different subsets of the same

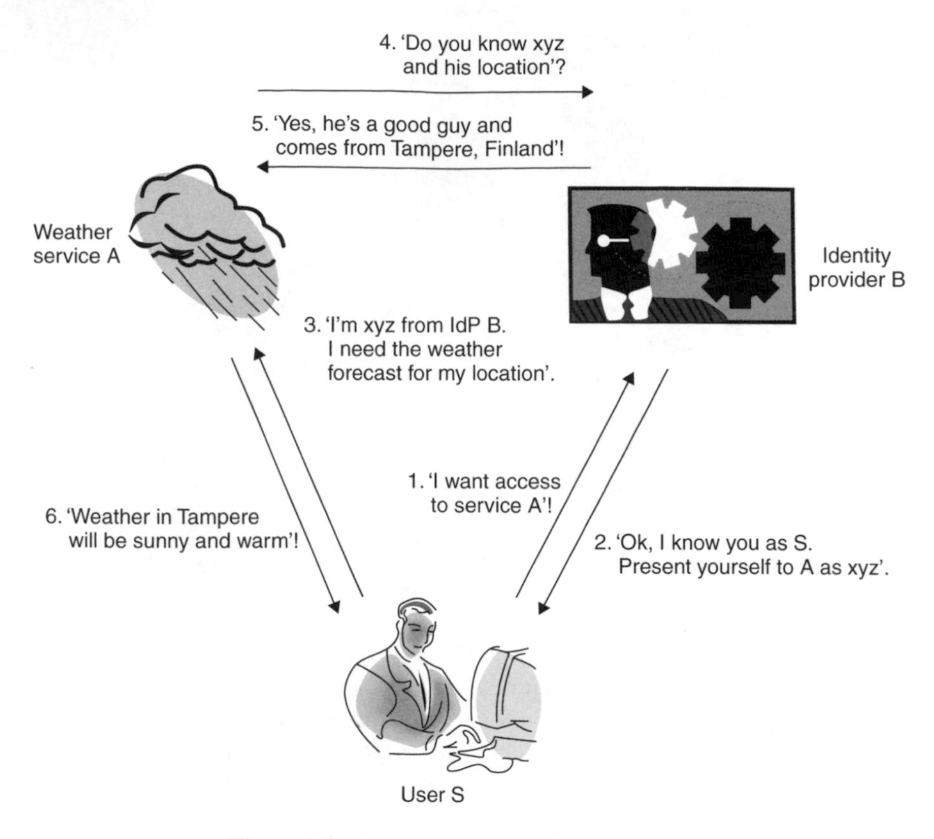

Figure 5.3 User case for identity management

identity. These subsets typically are identified by different identifiers in different contexts. With the help of IdP it is possible to do the linking without making the other parties aware of it. Often this type of linking is used to promote single sign on (SSO), so that the authentication of one identity is enough, from the point of view of the IdP, to provide authentication of the other identity, hence mitigating the burden of remembering several credentials and increasing usability. However, this may require attaching some additional information to the authentication event, like the strength of the authentication method employed, which may dictate applicability of the authentication in the different context. Another way of seeing federation is as a simple trust statement that a trusted identity makes for some other identity. In this respect a trusted identity could devise additional, possibly ephemeral, identifiers for its own use without involving external IdPs.

To summarise, the IdM can be seen containing issues involving provision of authenticated identities (or subsets of them) and identity attributes. Also, adding the access component results in the inclusion of resource authorisation for the particular identity. If viewed from the technical perspective, most of the management actions relate to the management of credentials and their provision in different forms in order to facilitate authentication and authorisation needs. This can also include the management of security tokens and derived key material. As mentioned earlier, the technical binding between an identifier and credentials may not exist in all the cases or the binding might be to something ephemeral.

5.3.1 Benefits of Identity Management

One clear benefit that the IdM brings to the user is the increased security especially against the identity theft that has been increasing at an alarming rate. One reason for this is that the users are burdened with several different types of credentials of varying quality, which they have to use in different services. Especially when the users have to remember several passwords or something very abstract like Personal Identification Number (PIN) numbers they are likely to choose very easy ones or write them down somewhere. This increases the danger of disclosure. With the help of IdM it is possible to provide the user just one set of credentials with the controlled level of security. The benefit becomes even clearer when one considers the heterogeneous nature of the access networks, which the user ought to be able to use seamlessly.

The previous benefit also enables the enhanced user experience as the amount of authentication events that involve user interaction is reduced. Additionally, it is also possible to provide identity-based information to the services, which are able to personalise themselves according to the user needs and context. This can also take into account the various requirements on user privacy as there is no need to disclose the actual user identity but just the minimum set of relevant information.

With standardised solutions, it is also possible to offer better and more secure services as it is easy for the service building blocks to include authentication and authorisation measures into the service compositions. Such solutions thus enable crossing of administrative boundaries and promoting greater flexibility and interoperability. Additionally, the management costs of systems can be reduced, even though this may be more prominent within organisations offering a multitude of different services to their users.

5.3.2 Examples of Identity Management

Many IdM products have resorted to proprietary means to control and provision identity information, but lately the trend has been to support more standardised solutions that enable the interoperability across different administrative domains.

One such example is Liberty Alliance (LA) that aims at providing specifications and frameworks for holistic IdM in the web oriented world [1]. LA builds on the concept of network identity that can be shared between several different partners provided that they have negotiated the relevant agreements and have trust relationships. The specifications pay special attention to the privacy requirements of the user and let the user be in control of what is shared and with whom. The specifications are meant to enhance the user experience by introducing the support for SSO and the means to convey identity-based data that can be used to personalise the different web services according to the preferences of the user. The intention of LA is to take advantage of the existing technology as much as possible in order to guarantee easier acceptance and wider support. Some of these technological choices include Security Assertion Markup Language (SAML) and HyperText Transport Protocol (HTTP). Also, the emerging paradigm of Web Services is taken into account and a framework building on those concepts is introduced. So in the technological sense LA relies much on eXtensible Markup Language (XML) and hence the relevant multitude of tools and security specification can be used. In a similar sense, the Web Services Enabler specifications that Open Mobile Alliance (OMA) has been designing base their IdM parts on the work of LA [2].

SAML is not an IdM system itself, but it is an important enabler that makes it possible to make assertions about the authenticity or authorisation state of some entity [3]. In other words, a trusted party can ascertain some facet of a certain identity. This could be something simple as stating that the identity contains a certain attribute or it could be something more complex like granting an anonymous authorisation to a resource hosted within a service. Even though LA is based heavily on SAML, LA has also contributed to the latest version of SAML.

When talking about IdM systems, Microsoft's Passport is also often mentioned [4]. It is, however, a rather centralised system that is under the control of Microsoft and even though it can contain some profile information it is rather perceived as an authentication system for providing SSO. The service does not enjoy very widespread fame outside Microsoft services and it is conceivable that it will remain so as Microsoft seems more interested in investing on Web Services–based solutions.

Microsoft has been active with IBM in making several specifications for the Web Services world that define methods and procedures for establishing trust and exchanging identity-based information. One such specification is the Web Services (WS)-Federation that aims at federating identity, attribute, authentication, and authorisation information [5]. WS-Federation does not provide a complete framework on its own, but also relies on other similar specifications like WS-Security, WS-Policy and WS-Trust. It should be noted that only WS-Security has an official endorsement of a standards body, that is OASIS [6]. It might happen, though, that there will be convergence with LA. Lately, Microsoft has also brought forward a new term for its vision of an interoperable identity meta-system that collects the various WS-specifications into a unified architecture [7].

Shibboleth is an IdM system that aims at providing a secure framework for exchanging information about the users [8]. Much like in LA the user privacy and federated administration has high priority. Additionally, it considers the usage and enforcement of policies imposed on different kinds of resources. It also uses the same base technologies and similar kinds of information flows, so in principle it would be possible to make the different systems interoperate even though it might require some sort of gatewaying functionality. It is worthwhile to note that Shibboleth has its roots in the academic world and its emphasis and use cases have been mainly there, especially within the Internet2 project.

5.3.3 Principles and Requirements

One central requirement that the future systems have to fulfil is the protection of privacy. It is essential that the IdM functions are able to provide means to refer to users without revealing their true identity. This usually means that the IdM system is able to assign to the user pseudonyms or ephemeral identifiers, which may have very limited lifetime and are only valid in a certain context, like just between two different systems. This is also reflected in the question, whether a user has a global name or only within some specified context. Generally, a global name may not be a very feasible solution as there will be political questions to consider as well.

Another prominent requirement is the concept of federation. In the future systems it is likely that there are several administrative domains that have control over the different facets of the identity. In order to promote the enhanced service experience there is the need to be able to exchange information about the user and the level of authentication required and performed. The ability to extend an authentication done on one identifier to other identifiers also has the potential of increasing the security. Of course it is required that the different systems are able to negotiate and agree on the terms and trust relationships concerning these identifiers. Also, the user has to be able to be in control of this process and make the decision when and with whom the federation is done.

Authorisation is often associated with authentication, but it should not be required that both of these have to always take place. It should be possible to get authorisation to some resource, be it access to network or some other, without having to authenticate first, even though in many modern systems the authorisation to access network comes implicitly from the authentication. The issue of decoupling authentication and authorisation has also privacy preserving qualities.

5.3.4 Research Issues

Generally, the federation concept is seen as an enabler for the SSO kind of functionality. Usually this, however, relates to the web applications that are used through browser environment, therefore one challenge is to study the applicability of these ideas in a wider scope. It would be interesting to learn, whether it would be possible to cross different application domains or even different layers. Even though it might be possible to technically implement it, it still needs to be studied whether it would be feasible from a holistic security point of view, that is does it increase the security of the overall system or does it just create some new vulnerabilities or decrease the effectiveness of the security measures.

It is also good to remember that even though IdM can bring more flexibility and increase security, some aspects of it may in fact increase the risks. If one, for example, considers the concept of federation, then the user is relying quite heavily on the identity that is used in authentication. If this were compromised, the potential attackers could thus get access to all the other linked identities as well. Therefore, this one identity would seem a more attractive target of potential attack.

As IdM suggests the role of IdP to some actor it actually also suggests the role of a trusted party. Implicitly, this actor also has the possibility to function as a trusted party between two different service providers. Another thing is, whether this is acceptable in a different context between these same providers. The parties may just want to limit themselves to certain usage scenarios, rather than making all-inclusive agreements. The implications and problematics of this and the required negotiations, possibly happening on the fly, may provide an interesting research problem, not just from the technical, but also from the economical perspective as well.

One further issue with IdM comes from the legislation perspective. As there is heavy involvement with privacy and data protection issues there is also the need to take into account the different laws and regulations that dictate the applicability of certain technological solutions. The relevant research should therefore remember to consider these aspects as well.

5.4 Malicious Code

Modern nomadic society depends heavily nowadays on convergent mobile and wireless Information and Communication Technology (ICT). ICT has pervaded in all traditional infrastructures, rendering them more intelligent but more vulnerable at the same time. Our new economy is highly dependent on such trustworthy and secure information infrastructure services – they are to be considered as critical information infrastructures.

Examples can be found in the European funded project *Mitra* [9], taking advantage of such infrastructures monitoring transports of hazardous materials. Other examples residing in the area of civil protection and emergency services count on confidential and easy to establish and maintain wireless *ad hoc* networks.

5.4.1 What is Malicious Code?

On questioning more than 200 experts in the field of the mobile and wireless world, 70 % expect the introduction and a substantial impact of malicious mobile code, that is viruses, worms, trojans, spam, and spy-ware, within the next two years. The questionnaire was held at the Wireless World Research Forum (WWRF) in June 2004 [10] – just a few days before the first proof of concept of mobile worms, Cabir, appeared.

Remarkably, the expected motivation of potential attackers does not indicate a winner and seems to be fuzzy. Except 'Trial and Fun', which got more than 60 % for a high and very high probability, all other possible motivations asked for are in the range from 35 % to 50 %: 'Criminal/Terroristic', 'Making/Saving Money', 'Spy out Personal Data', and 'Proof of Concept'. Thus, people obviously know that they have to take into account the same hype of malware we already have to face on the Internet. At the same time the opponent and his motivation have not been clearly identified and prioritised.

5.4.2 Background

In winter 2002, mobile network operators (MNOs) introduced first mobile service platforms providing information and Location-Based Services (LBS). Past as well as recent studies promise a valuable plus in multimedia enabled services over the next five years, however, nowadays most end-users have only adopted Short Message Service (SMS) and ring tones as reasonable services to pay for.

Following a study published in August 2004 by Strategy Analytics [11], the global sales volume of mobile services will grow from 61 Billion US Dollars to 189 Billion in 2009. The main reason is supposed to be the introduction of multimedia enabled mobile devices accessing 3G networks. Athur D. Little expects a similar prospect [12] in the UMTS 2005 study performed together with Exane BNP Paribas for Europe.

At the same time, new threats well known from the Internet pass through proof of concept implementations targeting mobile systems. F-Secure [13] reported several examples of phone malware during recent months. Many of those were Cabir (a virus) and Skulls (a worm) variants, which affect Symbian [14] Series 60 phones. More precisely, since June 2004 to March 2005 the following malware has been discovered and classified [15]: *Cabir* (Bluetooth-Worm, June 2004), *Mosquit* (Trojan, August 2004), *Skuller* (Trojan, November 2004), *Lasco* (Bluetooth-Worm/Virus, January 2005), *Locknut* (Trojan, February 2005), *Comwar* (MMS-Worm, March 2005), *Dampig* (Trojan, March 2005), *Drever* (Trojan, March 2005).

Moreover, in November and December 2004 the ITU (International Telecommunication Union), the University of St. Gallen and bmd wireless have performed the world's first collaborative empirical study about the effects of mobile spam on consumer behaviour and MNO's actions. The study 'Insight into Mobile Spam' [16] introduced at 3GSM World Congress [17] in Cannes on February, 2nd states that spam on mobile phones is an increasing phenomenon worldwide.

In order to assess security issues in 3G networks and devices an internal security study conducted by Fraunhofer SIT in November 2004 has shown principal vulnerabilities of all German mobile service platforms. Not one platform – neither Vodafone Live, T-Zones, E-Plus nor O2 – is reasonably prepared to protect itself against URL-attacks, WML-Script, and phishing.

Integrated research projects in the Sixth European Research Programme such as Ambient Networks of the Wireless World Initiative address security requirements and privacy constraints in mobile networks beyond 3G (B3G). The high-level architecture approach, however, is designed to suit mobile systems in a medium term. Currently, there is no publicly known project that specifically deals with malware for mobile communications.

5.4.3 Requirements and Research Issues

Principle security requirements and privacy constraints of the parties involved in the mobile business can be easily identified.

- For mobile end-users, ease of use and simple accounting are the crucial objectives. Moreover, in terms of security, confidentiality and authenticity are the main conditions to be met.
- For manufacturers of mobile devices such as Smart phones and PDAs, worms, viruses, and trojans attacking and undermining their operating systems are the most threatening challenges.
- For providers of mobile services availability, accounting and nonrepudiation are the critical points to provide and to address.
- For network operators, availability and integrity of the network are the requirements to meet.

Security goals such as confidentiality, availability, integrity, authenticity, and nonrepudiation can only be reached and maintained, if it is an inherent part of both the design and the implementation of mobile platforms and systems. Considering security as a module or as a function that can be easily added afterwards has proven to be deceptive and expensive. Therefore, research and development have to be based on taking security concerns seriously into account from the very beginning as part of a holistic approach.

Basically, this can be achieved in two phases in order to fulfil the major requirements sketched above. The first phase basically contains the detailed analysis of current mobile service platforms and convergent networks with respect to malware and its corresponding attacker models (Figure 5.4). This phase comprises the following:

- the description of the target of evaluation and its interaction with currently existing mobile and wireless environments;
- a detailed threat analysis that includes the execution of attacker models in specific application scenarios, an analysis of network and service communications, and penetrations tests;

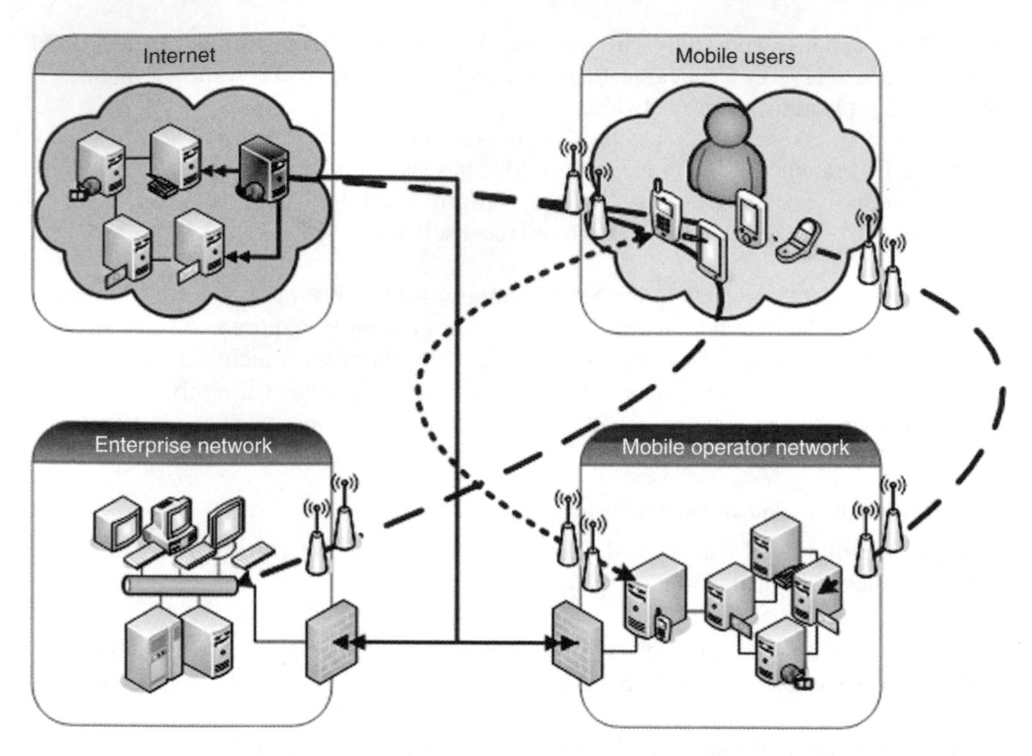

Figure 5.4 Attack paths for malicious code in convergent networks

- identification and prioritisation of threats and its potential impact to business models at the network and service level as well as the reputation of the provider in a risk analysis;
- indication of requirements and recommendations for a trustworthy and secure operation.

The second phase is driven by the vision of realising a user-friendly, trustworthy, robust, and open platform for mobile services. Taking into account the results and lessons learnt from the first phase, the second phase follows a well-defined and diligent software engineering process instead of time-to-market constraints. This phase emphasises the following:

- strict separation of the Internet and the platform;
- transparent SSO, client authorisation, and mutual authentication;
- end-to-end security including protection of personal information and data;
- provision of seamless access to generic/enabling services of network operators;
- support of approved service providers certified by common standards;
- negotiation of security policies to integrate third party service providers on runtime.

These characteristics are the most innovative aspects. Designing and implementing a service platform for mobile services based on a holistic security approach assures continuous trust relationships between mobile end-users, service providers, and network operators. This, finally, leads to robust, secure, and trustworthy information infrastructure services upon convergent networks 3G and beyond. Our focus is to deter potential attackers, to protect and control mobile and wireless information infrastructure services and networks, and to detect and respond to attacks based on malicious mobile code.

5.4.3.1 Relevant Standards

In order to identify a proper and definite methodology, the knowledge of international standards as well as results from scientific and research works are required. As for the standards in the security field, the following are the most prominent:

- Guidelines for the Management of Information Technology Security (GMITS), also known as ISO/IEC TR 13335 1–4: it gives comprehensive guidance on information security management.
- ISO/IEC 21827 (System Security Engineering Capability Maturity Model): it gives standard metrics for measuring IT security.
- ISO/IEC 17799-2000: it defines security controls and best practices. It is composed of two main parts. First, a code of practices [called *ISO 17799*]. Second, a specification for an information security management system [called *BS7799-2*].
- ISO/IEC 9797, Information technology and Security techniques: it defines data integrity mechanism using a cryptographic check function employing a block cipher algorithm.
- Internet Engineering Task Force (IETF) opsec Working Group, Operational Security Capabilities for IP Network Infrastructure: this working group will list capabilities appropriate for device use in both ISP and Enterprise networks codifying knowledge gained through operational experience. Wireless devices, small office–home office (SO-HO) devices, firewalls, Intrusion Detection System (IDS), Authentication Servers, and Hosts are excluded from the working group activities.

As for standards in the field of 3G security the following list contains the main reference documents:

- 3G TS 33.120 Security Principles and Objectives [18]: it contains 3G security principles as compared to the security of second generation systems. These principles can secure the new services and new service environments offered by 3G systems.
- 3G TS 33.120 Security Threats and Requirements [19]: it contains an evaluation of perceived threats to 3rd Generation Partnership Project (3GPP) and it produces a list of security requirements to address these threats.
- 'On the Security of 3GPP Networks' [20]: it gives an overview on 3GPP Security architectures.
- 3G TR 33.900 A Guide to 3rd Generation Security [21]: it contains a general description of the security architecture and features of 3G security. It also provides the purpose of identifying the potential risks and threats during the implementation of a 3G mobile system.
- 3G TS 33.102 Security Architecture [22]: it defines 3G security procedures performed within 3G capable networks (R99+), that is intra-UMTS and UMTS-GSM.
- 3G TS 33.105 Cryptographic Algorithm Requirements [23]: it constitutes a requirements specification for the security functions which may be used to provide the network access security features defined in 3G TS 33.102.

5.4.3.2 Current Studies

The 3G Wireless Security project of the Lucent Mobile Networking Research Department [24] is aimed to identify vulnerabilities of current 3G wireless networks for better protection against a variety of threats such as wireless specific Denial of Service (DoS). One of the main results of this project is that 3G networks not only share all kinds of wireline network vulnerabilities but also have their specific vulnerabilities. Moreover, the 3G specific vulnerabilities cannot be protected using existing defence approaches. Such result confirms that in order to obtain a reliable network infrastructure several enhancements in the 3G network architecture, 3G network protocols, and 3G standards are needed.

In [25] authors propose a security mechanism aiming to enhance the security abilities of mobile devices in 3G/4G networks. The proposed architecture adopts the SESAME [26] systems to support authentication, call set-up and to provide security services for multimedia applications. Vergados *et al.* [27] discuss how 3G/UMTS networks may be used in Tactical Communication Networks showing that GPRS and UMTS can provide the military world with access to many different demanding tactical services. Nevertheless, [27] also highlights that in order to use 3G networks in such scenarios several issues must be addressed, including security mechanisms and fraud detection.

Recently the Australian service provider Hutchison has deployed Cisco Secure IDSs to protect its 3G network from viruses and other interference. The Cisco equipment was installed in Hutchison's data centre in central Sydney [28].

In [29] the authors show how to enlarge the 2G End-to-End Security Solutions to 3G Networks. In [30] the authors present their experience with the development of mobile devices and solutions. They stated that there is the need of products with continuing demand for high assurance and end-to-end secure information services. Also, the idea of inter-working between mobile systems and WLANs holds great promise when security aspects will be addressed. In [31] the authors focus on security aspects of 3GPP-WLAN inter-working. The main issue that they suggest to investigate is the need for even better identity protection.

5.5 Future Steps

Future research programmes such as the Seventh EU Framework Programme (starting 2007) will focus on technologies enabling AmI. The concept of AmI is a vision where humans are surrounded by computing and networking technology unobtrusively embedded in their surroundings providing and maintaining context information. Currently, specific support activities such as the SWAMI project (Safeguards in a World of AmI) identify major application scenarios as well as common security requirements and privacy concerns. A workshop in 2005, for example, has identified three major challenges – so-called *dark scenarios*: (1) Loss of privacy, (2) increased possibility of surveillance, and (3) profiling.

Nevertheless, especially context-aware technologies can better serve users according to the environmental and technical context and the user's preferences; at the same time user privacy and context information have to be protected against any type of misuse, security attack and unnecessary disclosure. Particular research topics comprise protection of location privacy, secure context information of user behaviour, usage patterns, and so on. In addition, one of the latest research areas is dedicated to user-centric mechanisms allowing

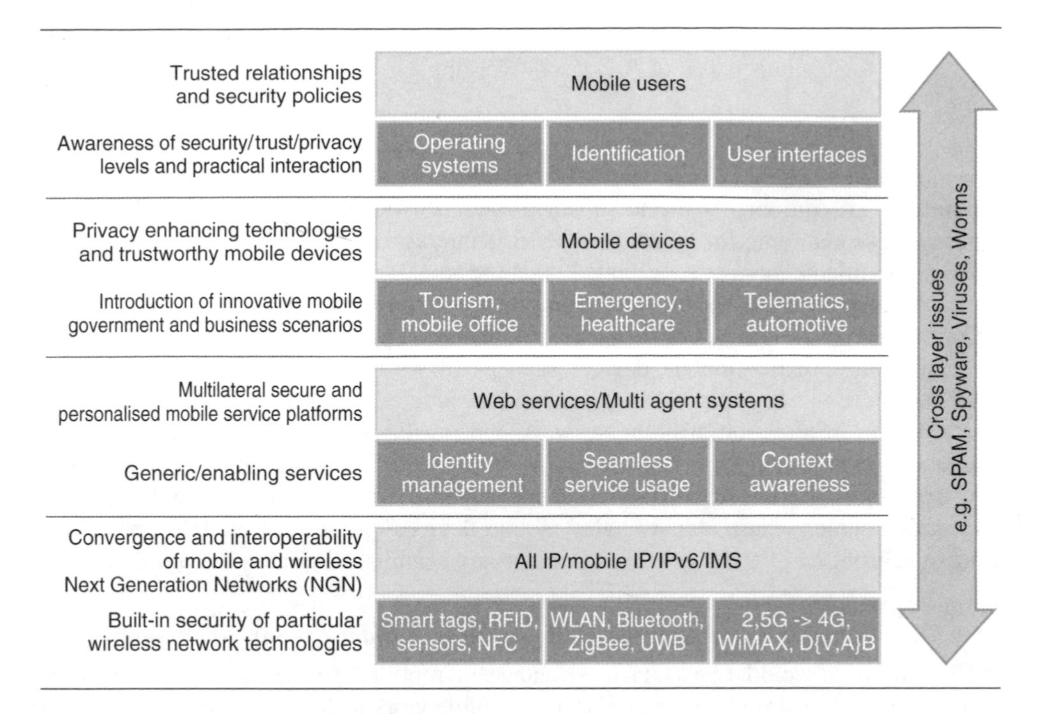

Figure 5.5 Top 5 security challenges

controlled release of personal information. Similar to Digital Rights Management (DRM) content in terms of user's profile data and preferences may be sent to services assigned to a specific purpose of use and protected by an explicit date of expiry.

Moreover, an online questionnaire addressing experts of WWRF – especially SIG2 – in September 2005 asked for the major research challenges in the field of mobile and wireless security within the next two to five years. Thus, the next paragraphs emphasise the results of this questionnaire: The top five security challenges. According to Figure 5.5 the topics are arranged from the user's point of view down to the network layer.

5.5.1 Usable Security

Nowadays, mobile technologies not only burden end-users with maintaining access names and corresponding passwords or phrases, average users also have to remember dozens of account information. Unfortunately, most of them use almost the same combination for several or even all accounts or write them down – mostly on post-its at the computer screen or underneath their keyboards. For the user it is even getting worse when thinking of installing and maintaining personal firewalls and anti-virus software even on mobile devices such as PDAs and Smart phones considering that the end-user is also an administrator.

In general, users should not be confused by the configuration of security systems or in choosing secure parameters. However, there are also still users who like to have a clear control of the security of the system's terminal, communication and service, and so on. Thus, it is important to study how to make the proposed security service practically usable to users and how to avoid security issues in user interaction with the system. The next year's joint white paper of SIG2 and Working Group 1 will address this topic in detail.

5.5.2 Trusted Computing Platforms in Mobile Devices

Mobile devices equipped with the so-called *Trusted Computing Platform* (TPM) are an important research topic for DRM. Besides, it is interesting to research on the security for software modifiable devices (not limited to end terminals) that can change their behaviour and capabilities, and the secure reconfiguration of mobile devices.

Current TPMs are only integrated into Personal Computers and Laptops. However, in the near future smaller mobile devices such as PDAs and Smart phones will become more and more the linchpin (1) of receiving video, television, and entertainment, (2) of synchronising information at home, at work and on the way, and (3) of organising and maintaining everyday lives taking advantage of smart software and service agents. So the requirements of keeping the hardware under some control and authentic will rise – not only in companies where already today mobile devices circumvent modern firewalls by being synchronised at the user's desk without any established quarantine procedures.

5.5.3 Security for Fast Intra/Inter-technology and Intra/Inter-domain Handover

The mobile user would like to take advantage of mobile services no matter where, no matter when, and definitely no matter what kind of network technology is being used – just context aware, user-friendly, and secure. However, security mechanisms and Quality of Service may differ from one network technology to another. In order to offer a seamless experience to the mobile user, intelligent and transparent handover mechanisms have to be established between each network technology. A video stream started at home using Wireless Fidelity (WiFi), for example, might be continued on one's way taking advantage of UMTS and finally making use of Worldwide Interoperability for Microwave Access (WiMAX) near the train station. Ideally, the mobile user does not have to authenticate himself many times to network operators or service providers. The individual session and the corresponding Quality of Service and security parameters will just be mapped to the next network technology – the process is transparent and seamless.

5.5.4 Trust Development and Management in Dynamically Changing Networks

The number of future access network operators, service and application providers, and users will increase to a level very difficult to manage by using present technologies due to lack of trust relationships arranged in advance. For example: Current roaming agreement frameworks cannot work well due to a large number of network and service providers of various types. On the other hand, to manage trust for a large number of devices can be a problem for end users to access the ubiquitous communication environment surrounding them. Thus, trust development and management in dynamically changing networks is an important research item to make systems Beyond 3G work in a secure way. Specific topics are the following:

- flexible/dynamic establishment of security associations (e.g. opportunistic TLS with DNSsec);

- role of authorisation and service compensation in trust development and related security issues; In the changing ecosystem there is the need to guarantee that service is delivered and relevant compensation is received as the dynamic relationships between multiple actors cannot be treated feasibly in a similar way as, for example, current roaming agreements are handled.

5.5.5 Security for Ambient Communication Networks

The raising complexity on the network level necessitates security mechanisms in the whole bandwidth of all kinds of networks and topologies. From well-organised infrastructures such as cellular networks of MNOs to self-organising P2P networks of, for example, fire sensors (cf. 'smart dust') in the woodland, major security goals remain the same: availability and integrity. Whereas other security goals such as authenticity, confidentiality and nonrepudiation are addressed in the middleware or application layer, availability and integrity of the underlying network infrastructure are basic requirements.

Moreover interoperability and interconnectivity have to be addressed by defining a standardised network access layer. Providers of generic, enabling, and value-added services rely on an abstract and common network access layer in order to develop and roll-out a new generation of mobile context-aware services spending less time on considering specific network facilities and more time on software quality and built-in security. MNOs and their component suppliers, for example, promote the IP multimedia subsystem (IMS) based on All-IP.

Concluding Remarks: In terms of security and privacy from the mobile user down to network level, major efforts have to be made taking into account not only current trends such as network and media convergence but also future visions such as AmI. A balanced, holistic and sustainable approach considering security requirements and privacy concerns of all involved parties is necessary in order to raise user acceptance, establish multilateral trust relationships, and realise promising business models. Dedicated WWRF conference meetings and particular White Papers of Special Interest Group 2 'Security and Trust' will underpin those efforts.

5.6 Acknowledgement

The following individuals contributed to this chapter: Mario Hoffmann (Fraunhofer SIT), Christos Xenakis, Stauraleni Kontopoulou (University of Athens), Markus Eisenhauer (Fraunhofer FIT), Seppo Heikkinen (Elisa R&D), Antonio Pescape (University of Naples) and Hu Wang (Huawei).

References

[1] Liberty Alliance Project, http://www.project-liberty.org.
[2] Open Mobile Alliance, OMA Web Services Enabler (OWSER), *Network Identity Specifications*, 2004.
[3] OASIS, *Assertions and Protocols for the OASIS Security Assertion Markup Language (SAML)*, V2.0, 2005.
[4] Microsoft, Microsoft, NET Passport Review Guide, 2003, http://www.microsoft.com/net/services/passport/review_guide.asp (as of Jan. 2004).
[5] IBM, Microsoft, *et al.*, *Web Services Federation Language (WS-Federation)*, 2003.
[6] Web Services Security, *SOAP Message Security*, 2003.
[7] Microsoft Corporation, *Microsoft's Vision for an Identity Metasystem*, 2005.

[8] Shibboleth Project, http://shibboleth.internet2.edu/.
[9] Monitoring and Intervention for the TRAnsportation of Dangerous Goods (MITRA), European Project web site, http://www.mitraproject.info/ (as of Sep. 2005).
[10] M. Dillinger, E. Mohyeldin, J. Luo, M. Fahrmair, P. Dornbusch, E. Schulz, 11th Wireless World Research Forum Meeting, *Services and Applications Roadmaps – Invigorating the Visions*, 10–11 June 2004, Oslo (Norway).
[11] Strategic Analytics Web Page, http://www.strategyanalytics.com/ (as of Sep. 2005).
[12] D Arthur, Little Central Europe web page, http://www.adlittle.de/default_eng.asp (as of Sep. 2005).
[13] F-Secure web site, http://www.f-secure.com/ (as of Sep. 2005).
[14] Symbian Operating System web site, http://www.symbian.com/ (as of Sep. 2005).
[15] Kasper sky lab web site, http://www.kaspersky.com/ (as of Sep. 2005).
[16] ITU, University of St. Gallen, and bmd wireless, Mobile Spam Project, http://www.mobilespam.org/ (as of Sep. 2005).
[17] 3GSM World Association, *The 3GSM World Congress 2005*, http://www.3gsmworldcongress.com/2005/ (as of Jul. 2005).
[18] 3rd Generation Partnership Project, *Technical Specification Group Services and System Aspects, 3G Security: Security Principles and Objectives*, http://www.3gpp.org/ftp/tsg_sa/WG3_Security/_Specs/33120-300.pdf.
[19] 3rd Generation Partnership Project, *Technical Specification Group Services and System Aspects, 3G Security: Security Threats and Requirements*, http://www.3gpp.org/ftp/tsg_sa/WG3_Security/_Specs/21133-320.zip (as of 2001).
[20] Michael Walker, On the Security of 3GPP Networks?, http://www.esat.kuleuven.ac.be/cosic/eurocrypt2000/mike_walker.pdf.
[21] 3rd Generation Partnership Project, *Technical Specification Group SA WG3, A Guide to 3rd Generation Security*, http://www.3gpp.org/ftp/tsg_sa/WG3_Security/_Specs/33900-120.pdf.
[22] 3rd Generation Partnership Project, Technical Specification Group Services and System Aspects, 3G Security: Security Architecture (Release 1999), ftp://ftp.3gpp.org/Specs/2000-12/R1999/33_series/33102-370.zip.
[23] 3rd Generation Partnership Project, *Technical Specification Group Services and System Aspects, 3G Security: Cryptographic Algorithm Requirements (Release 1999)*, ftp://ftp.3gpp.org/Specs/2000-12/R1999/33_series/33105-360.zip.
[24] Lucent Technology Mobile Networking Research Department, 3G Wireless Security Project, http://www.bell-labs.com/org/113440/#projects (as of Sep. 2005).
[25] J. Al-Muhtadi, D. Mickunas, R. Campbell, A lightweight reconfigurable security mechanism for 3G/4G mobile devices, *IEEE Wireless Communications*, vol. 9, no. 2, Apr. 2002, pp. 60–65.
[26] The SESAME project web page https://www.cosic.esat.kuleuven.ac.be/sesame (as of Sep. 2005).
[27] D. D. Vergados, C. Gizelis, D. J. Vergados, The 3G wireless technology in tactical communication networks, *Proceedings of the 60th IEEE Fall Vehicular Technology Conference*, VTC2004 2004, 26–29 Sep. 2004, vol. 7, pp. 4883–4887.
[28] Uk 3G web site, Intrusion Detection Capabilities Help Protect 3G Network, http://www.3g.co.uk/PR/April2005/1367.htm (as of Sep. 2005).
[29] T. Sandberg, P. Kennedy, *Extending 2G end-to end Security Solutions to 3G Networks, Secure GSM and Beyond: End to End Security for Mobile Communications*, IEE Seminar on (Digest No. 2003/10059), 11 Feb. 2003, pp. 4/1–4/7.
[30] O. Johnson, *The Challenging Road Ahead to 3G security, Secure Mobile Communications Forum: Exploring the Technical Challenges in Secure GSM and WLAN*, 2004, 2nd IEE (Ref. No. 2004/10660) 23 Sep. 2004, pp. 4/1–4/6.
[31] G. M. Koien, T. Haslestad, Security aspects of 3G-WLAN interworking, *IEEE Communications Magazine*, vol. 41, no. 11, pp. 82–88, Nov. 2003.

6

New Air-interface Technologies and Deployment Concepts

Edited by David Falconer (Carleton University), Angeliki Alexiou (Lucent Technologies), Stefan Kaiser (DoCoMo Euro-Labs), Martin Haardt (Ilmenau University of Technology) and Tommi Jämsä (Elektrobit Testing Ltd)

6.1 Introduction

Future-generation wireless access systems will be mostly Internet Protocol (IP) packet-based and must provide satisfactory quality of service (QoS) with respect to bit and packet error rate (PER), end-to-end and link delays, outage probability, packet loss and packet delay variation for different traffic classes.

Multimedia services with peak user bit rates ranging from a few Kbps up to at least several tens of Mbps are foreseen. Aggregate data rates will be on the order of 50 to 100 Mbps, or even higher in hot spots [1–3]. Transmission will occur in varied radio propagation environments, with moving terminals, and with sharing of the radio spectrum by many user terminals. Future systems must use spectrum-efficient air-interface technologies that can compensate for the severe frequency selectivity and other radio impairments, while keeping signal processing complexity within reasonable limits.

Spectrum availability for next generation broadband wireless services is presently uncertain. All wireless user terminals should be as efficient as possible in spectrum utilization. Efficient spectrum utilization may require cognitive radio operation, in which the transmitted signal bandwidth and center frequency or frequencies must be adaptively and dynamically adjusted in response to the current interference environment, shared by many user terminals and perhaps multiple service providers. Intelligent spectrum-sharing capability and context-awareness requirements will affect the air-interface design, as well as dynamic resource allocation, scheduling and medium access control (MAC) design. Reconfigurability in response to varying network and user requirements will be made possible by Software- (SW) defined radio technologies.

Technologies for the Wireless Future – Volume 2 Edited by Rahim Tafazolli

An important additional requirement is low cost: low terminal cost and power consumption for subscribers, high spectral efficiency and coverage, and ease of deployment for service providers. Concentrating as much signal processing as possible in the base station (BS) or access point, instead of the user terminal, is a step in this direction. The choice of modulation and coding schemes can significantly influence performance and cost for a given range of bit rates.

The adoption of Smart Antenna techniques, that is, of multiple antenna processing schemes at the transmitting and/or the receiving side of the communication link, is indeed expected to have a significant impact on the efficient use of the spectrum, the minimization of the cost of establishing new wireless networks, the enhancement of the quality of service, and the realization of reconfigurable, robust, and transparent operation across multitechnology wireless networks.

6.2 Broadband Frequency-domain–based Air-interfaces

Future-generation cellular wireless systems, transporting high bit rates in nonideal radio propagation environments, must be robust to severe frequency-selective multipath. Further requirements include moderate terminal and BS hardware costs, high spectral efficiency, and scalability of the cost of terminals with respect to their maximum bit-rate capabilities. Reconfigurable air-interfaces, based on *frequency-domain transmission and reception methods*, best meet these requirements, by adaptively selecting the Uplink (UL) and downlink modulation and multiple-access scheme that is most appropriate for the channel, interference, traffic, and cost constraints. This approach also leads to a general unified framework for possible air-interfaces.

Terminal specifications should be scalable with respect to user bandwidth requirements. Simple user terminals may be limited to low bit rates, while more sophisticated user terminals will be able to employ the full range of bit rates on demand. The cost and complexity of terminals restricted to low bit rates should not be driven by the requirements and system architecture of high bit-rate terminals. Systems should also be flexible with respect to duplex traffic symmetry or asymmetry.

In this section, we outline a *generalized multicarrier* (*GMC*) (Earlier versions used the term *generalized frequency domain* (*GFD*) instead of GMC. The GMC term is felt to be more descriptive. It has been applied previously to code division multiple-access (CDMA) systems [4], and more recently in [5].) approach to link transmission, reception, and multiple-access for future-generation systems in Sections 6.2.2 and 6.2.3. This class of signals includes parallel modulation (such as orthogonal frequency division multiplexing (OFDM), orthogonal frequency division multiple-access (OFDMA), multiple carrier CDMA (MC-CDMA), and spread spectrum multicarrier multiple-access(SS-MC-MA)) and serial modulation (such as conventional single carrier, single carrier Direct Sequence Code Division Multiple-Access (DS-CDMA), and frequency-domain orthogonal signature sequences (FDOSS)).

Common signal processing operations within this general class facilitate reconfigurability of the air interface in response to user requirements and the radio environment. Section 6.2.4 discusses bit-error rate (BER) performance characteristics of parallel- and serial-modulated signals. Section 6.2.5 discusses spread spectrum (CDMA) signaling within these two main classes. In Sections 6.2.6 and 6.2.7, we consider the impairments of phase noise, frequency offset, and power amplifier nonlinearities and their effects on these types of signals. Spectrum flexibility, including transmission over several disjoint sub-bands in a crowded spectrum, is readily achieved with frequency-domain signaling, and is discussed in Section 6.2.8. Sections 6.2.9 and 6.2.10 conclude with brief discussions of further research issues and a summary, respectively.

6.2.1 Frequency-domain–based Systems

In 100 Mbps systems, with delay spreads of several μs, the application of traditional time domain equalization methods would require prohibitively complex and high-speed signal processing, since the inter-symbol interference (ISI) could span hundreds of data bits. A better equalization performance/complexity trade-off is obtained by doing transmission and reception operations on a block by block basis in the frequency domain, using discrete Fourier transform (DFT) processing, for which the signal processing complexity per bit increases only logarithmically with ISI span, when the DFT is implemented as a fast Fourier transform (FFT). Modulation and equalization schemes in which signal processing is done in the frequency domain include 'parallel' schemes such as OFDM [6, 7] and its variants [8], and 'serial modulation' schemes such as single carrier with frequency-domain equalization (SC-FDE) [9, 10].

6.2.1.1 Parallel Modulation Methods

In 'parallel' modulation methods, coded data symbols are transmitted in parallel on narrowband subcarriers. Prime examples include OFDM, OFDMA [7, 11, 12], MC-CDMA [8], and >SS-MC-MA [13, 14]. The transmitter's modulation process is carried out on blocks of data symbols (typically (QAM) quadrature amplitude modulation), using a computationally efficient inverse FFT (IFFT) operation. At the receiver, an FFT operation decomposes each received signal block into the narrowband subcarrier components, which are independently equalized and passed on to the decoder. A cyclic prefix (CP) precedes each transmitted block to act as a guard interval to avoid inter-block ISI, and to produce the appearance of cyclic convolution by the channel impulse response (CIR), thus facilitating the FFT operation. The CP is a copy of the end of the transmitted block, and its length should be at least equal to the maximum expected channel delay-spread.

The CP (and maximum expected delay-spread) should be a small fraction (e.g. <1/8) of the FFT block length for two main reasons: (1) to minimize the percentage of overhead caused by the CP; (2) to ensure that the inter-subcarrier frequency spacing is much less than the channel's coherence bandwidth; then each transmitted subcarrier passes through an essentially nonfrequency selective channel, and therefore inter-subcarrier interference is minimized.

Since time domain cyclic convolution is equivalent to frequency-domain multiplication of FFTs, equalization is carried out by a simple complex gain multiplication on each received subcarrier. It can be incorporated within the decoding process in coded OFDM. Coding and inter- or intra-block interleaving are essential for OFDM systems used on frequency-selective channels; the decoder sees the equivalent of a fast-fading channel, since each coded data symbol emerging from the equalizer has a different instantaneous signal-to-noise ratio (SNR), depending on the channel gain at its subcarrier frequency. Channel state information (CSI) about the gain of each OFDM subcarrier is generally used in decoding, since it significantly enhances BER performance. The coding/decoding process provides frequency diversity, which is essential for OFDM or OFDMA systems used on frequency-selective channels.

Sequences of zeroes of the same length as the CP can replace the CP's separating successive blocks. This approach, called *zero-padding* [15] requires slightly more complex equalization, but has the advantage that nulls in the channel's frequency response do not hinder equalization. A further variation, called *pseudorandom postfix* (*PRP*) [16–19] replaces each zero-sequence with a fixed sequence of training symbols multiplied by a pseudo-random rotation factor to avoid spectral peaks. The interspersed training sequences combine the functions of periodic channel estimation and prevention of inter-block interference. This combination reduces overhead. Equalization is similar to methods employed for zero-padding.

The CP overhead is eliminated completely in a scheme called *OFDM/IOTA* [20, 21]. In it the data symbols are transmitted using Offset QAM (OQAM) modulation, with filtering by a Gaussian-like waveform following the IFFT operation.

6.2.1.2 Serial Modulation Methods

Equalization in the frequency domain, using computationally efficient FFT and IFFT operations, can also be applied to traditional ***serial modulation***, or 'single carrier' (SC) modulation schemes, in which data symbols are transmitted sequentially, with a high symbol rate, on a single carrier. In serial modulation systems with frequency-domain equalization, the FFT operations are all done at the receiver: an FFT followed by simple independent equalization of each frequency component, followed by an IFFT to restore the equalized serial data stream [9, 10, 22]. As with parallel modulation, transmission is in blocks, separated by cyclic prefixes. In the conventional way of generating these signals, the blocks are sequences of low-pass-filtered coded data symbols, and the cyclic prefixes are formed by appending a copy of the final data symbols at the beginning of the block before filtering. The block and CP lengths should obey the same constraints as for parallel modulation. Unlike OFDM, the decoder sees the equivalent of a fixed additive noise channel during a block. Frequency diversity is exploited by the averaging over the frequency band that is inherent in the equalization process.

As for OFDM, zero-padding can replace the CPs [23]. The need for the CP can be eliminated in serial-modulated systems by transmitting data symbols sequentially, not in blocks, and using overlap-save frequency-domain processing at the receiver [24].

For both parallel and serial modulation, the received CP at the beginning of each received block is discarded, and FFT processing is done on each received block. Because of the influence of the transmitted CP, the received block appears to the receiver as if it were one period of a periodic waveform. This fictitious periodic waveform is equivalent to a multicarrier waveform consisting of subcarrier at frequencies that are at multiples of the reciprocal of the period. The multicarrier characteristic is obvious for OFDM and its variants. For serial modulation, the complex-valued samples modulating the virtual multiple subcarrier are revealed by the DFT operation at the receiver.

6.2.2 Generalized Multicarrier Signals

Figure 6.1 shows a more general transmitter architecture that includes the aforementioned parallel and serial types of block transmission with CPs. The IFFT block is

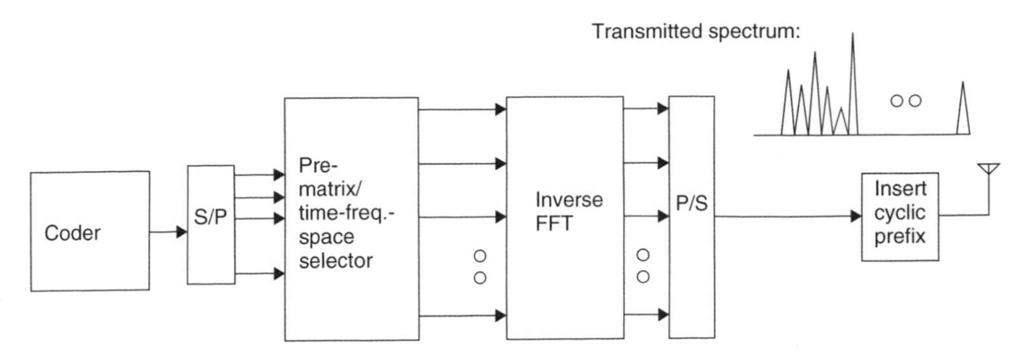

Figure 6.1 Generalized multicarrier transmitter (FFT–fast Fourier transform operation)

preceded by a general matrix operation, which may also include spreading, a selection mechanism and/or an allocation to multiple transmitting antennas in a MIMO (multiple-input, multiple-output) or space-time code. This structure, called *GMC CDMA* in [4], can be specialized to OFDM, SC-FDE, and other frequency-domain–based modulation and multiple-access schemes. Recognition of this generalized structure can also be found in [23, 25, 26] and [27]. We will call the initial matrix block the *pre-matrix*, and the corresponding class of signals *GMC* signals. The GMC architecture provides a general framework for air-interface technologies that are applicable to next generation wireless systems. It also suggests a modular signal processing architecture that can be exploited to allow modulation and multiple-access schemes to be changed dynamically in response to user requirements and the radio environment.

The structure envisages data being transmitted in blocks, separated by cyclic prefixes, zero-padding, or PRP. The organization of data into blocks is carried out by the serial-to-parallel converter (S/P). Following the parallel-to-serial converter (P/S) after the IFFT, time-windowing may be applied to reduce the sidelobes of the transmitted spectrum.

Figure 6.2 shows the details of the block labeled 'matrix time-frequency selector' for several options to be described next. The space-frequency selector matrix maps M_N input samples onto a desired set of frequencies, and also, if desired onto different antennas in a MIMO system. It can also include the functions of bit- and power-loading of subcarriers. The FFT matrix is omitted for OFDMA, but is used for serial modulation.

Conventional OFDM results from making the pre-matrix operation a simple identity matrix. A *spreading* pre-matrix (with M input symbols and MK output chips, where K is the spreading factor) produces MC-CDMA. If the pre-matrix functions simply as a selector, distributing data symbols onto a number of subcarriers frequencies within a band, the result is an OFDMA signal. In this case, signals to or from different user terminals occupy different segments of the same DFT block, and thus do not interfere with one another. A block of M_N coded data symbols $\{A_\ell, \ell = 0, 1, \ldots M_N - 1\}$ to or from a particular user is transmitted as the waveform consisting of N_c samples

$$s(k) = \frac{1}{N_c} \sum_{\ell=0}^{M_N-1} A_\ell \exp\left(j\frac{2\pi k f(\ell)}{N_c}\right), \text{ for } k = 0, 1, \ldots N_c - 1, \tag{1}$$

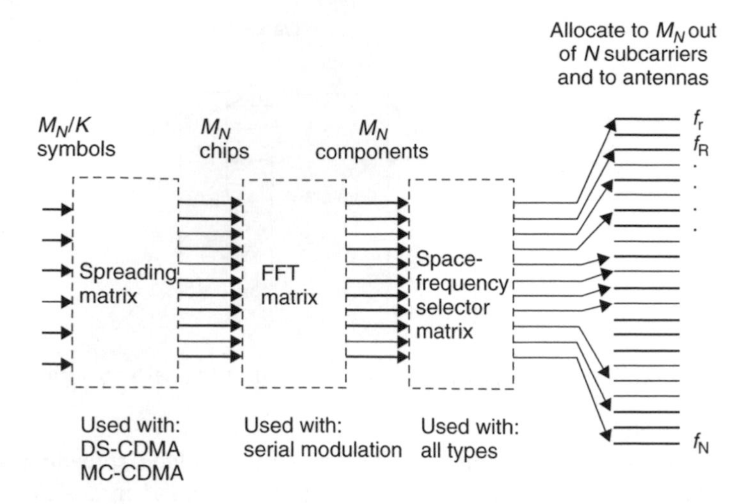

Figure 6.2 Details of the pre-matrix time-frequency selector in Figure 6.1 (dashed boxes are optional)

where $\{f(\ell), \ell = 0, 1, \dots M_N - 1\}$ represents a set of M_N frequencies to which the data symbols are mapped for that user. In the case of MC-CDMA with a spreading factor of K, the $\{A_\ell\}$ are chips representing M_N/K data symbols, with $\ell = 0, 1, \dots M_N - 1$, where M_N is the number of chips per block.

Low and high bit-rate user signals can be multiplexed onto a common sequence of FFT blocks by allocating different numbers of time and frequency slots. FFT block lengths would be long – on the order of thousands of symbols – to accommodate both low and high bit-rate users in the same frame.

If the pre-matrix includes an FFT matrix, the resulting transmitted signal is a block *serial-modulated* single carrier signal. In this case, the coded data symbols are denoted $\{a(m)\}$, and

$$A(\ell) = \sum_{m=0}^{M_N-1} a(m) \exp\left(-j\frac{2\pi m\ell}{M_N}\right). \tag{2}$$

Substituting (2) into (1), with $f(\ell) = \ell$ results in the traditional form of a serial-modulated waveform

$$s(k) = \sum_{m=0}^{M_N-1} a(m) g\left(k - m\frac{N_c}{M_N}\right), \tag{3}$$

where $g(n) = \frac{1}{N_c} \exp\left(j\frac{\pi}{N_c}(M_N - 1)n\right) \frac{\sin\left(\frac{\pi M_N}{N_c}n\right)}{\sin\left(\frac{\pi}{N_c}n\right)}$ is a bandlimited pulse waveform.

Traditionally, serial-modulated signals are generated by multiplying a serial train of pulses by data symbols and filtering the result. The potential usefulness of the alternative structure of Figure 6.1 and Equations (1), (2), and (3) to serial modulation comes through additional operations that may be done between the FFT and IFFT; examples are frequency-domain filtering for spectrum-shaping, and distribution of FFT coefficients to a noncontiguous set of frequencies. Spectrum-shaping of both parallel- and serial-modulated signals can be effected by time-windowing the waveform $s\ (k)$, or by filtering it.

The subcarrier frequencies $\{f(\ell)\}$ used by any given user may or may not be contiguous. A serial modulation and multiple-access scheme sometimes called *interleaved frequency division multiple-access* (*IFDMA*) [28] or FDOSS can be generated from Figure 6.1, by mapping M FFT outputs from each of up to K users onto $N_c = MK$ frequencies as follows: user p $(p = 0, 1, \ldots K - 1)$ is assigned to the frequency set $\{p, p + K, p + 2K, \ldots p + (M - 1)K\}$. Each user thus occupies a unique set of M frequencies in an interleaved fashion; there is no inter-user interference, even for frequency-selective channels. Furthermore it can be shown [29, 30] that the resulting waveform from a given user is equivalent to a serial-modulated waveform, which results from rearranging the order of the chips of a DS-CDMA waveform with a spreading factor of K. It therefore has the low peak to average power ratio (PAPR) properties of serial-modulated waveforms. This type of signal has been rediscovered and named a number of times. For example: IFDMA [28], FDOSS [31, 32], Orthogonal Frequency Division Multiplexing-FDMA (OFDM-FDMA) [33], UP-OFDMA (unitary-precoded OFDMA) [34], VSCRF-CDMA (variable spreading and chip repetition factor CDMA) [35, 36] and carrier interferometry OFDM (CI-OFDM) [37].

Reiterating, it is clear that 'serial modulation' and 'parallel modulation' signals, transmitted in block format, preceded by an adequate CP, can be considered as special cases of the GMC signal format. The transmitted signal is equivalent to a sum of discrete frequency components, each modulated by data that have passed through the pre-matrix. The signal is, furthermore, organized in successive blocks, or frames, each preceded by a CP. The main difference between the serial modulation and parallel modulation forms of GMC is in the presence or absence, respectively of the FFT operation at the transmitter in Figure 6.1. Their transmitted waveforms differ in that their data symbols are transmitted serially and in parallel, respectively. Wang and Giannakis [38] point out that more general complex-field coding operations are possible prior to the transmitter's IFFT operation, to create multicarrier signal classes with greater diversity gains than are possible with pure OFDM. The FFT operation is a special case of such a complex-field code. As we point out later, the FFT operation significantly reduces the transmitted signal's instantaneous power dynamic range, and therefore the back off required for a nonlinear power amplifier.

Figure 6.3 shows the general receiver structure for GMC signals. The first element is an FFT operation, followed by a selector (or sampler) if necessary, and an equalizing frequency-domain filter. For coded OFDM, the function of the equalizing filter can be subsumed within the soft decision decoding effort. The job of the equalizing filter is to scale each frequency component in amplitude and phase, either to completely compensate for the channel's frequency selectivity (as in the case of linear equalization), or to partially compensate for it (as for decision-feedback equalization [9, 39, 40]). The linear operation after channel equalization depends on the transmitter's pre-matrix operation: detection and decoding for OFDM or OFDMA, a correlation operation for MC-CDMA, IFFT for serial modulation, and so on. All the basic forms of GMC systems offer an overall (transmitter plus receiver) signal processing complexity that is approximately proportional to the logarithm of the maximum CIR length. Pure time domain equalizer signal processing complexity typically would grow at least linearly or quadratically with impulse response

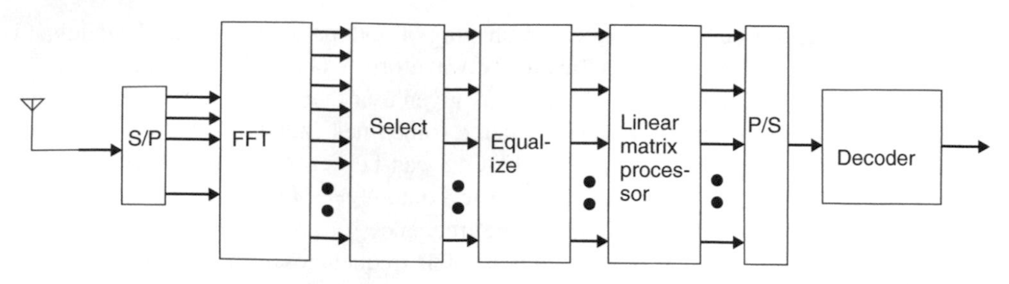

Figure 6.3 Generalized multicarrier receiver

length. For impulse responses spanning more than about 10 or 15 symbols, frequency-domain–based systems offer a significant complexity advantage over their time domain counterparts.

6.2.3 BER Performance of Parallel- and Serial-modulated Systems

OFDM and OFDMA typically make use of powerful (BICM) bit-interleaved coded modulation schemes [41] that exploit estimated knowledge of the SNR at each frequency component. OFDM and OFDMA transmitters can also 'adaptively load' subcarriers; that is, to optimize the allocation of coded bits and power to each subcarrier, according to its channel gain or Signal-to-Interference and Noise Ratio (SINR). OFDM or OFDMA systems with adaptive loading approximate the optimum 'water-filling' prescription for frequency-selective channels arrived at from information-theoretic considerations [26, 42–44]. Adaptive loading requires feedback from the receiving terminal to the transmitter, and accurate tracking of time-varying channel frequency responses. For time division duplexing (TDD) systems with slow channel variations, the downlink frequency response is available at the BS from frequency response estimation of the received UL signal. Given such channel information at the transmitter, adaptive bit loading combined with adaptive coding and power allocation to the subcarrier give OFDM and OFDMA unrivalled flexibility and BER performance on frequency-selective channels [45–47].

In serial-modulated systems, each data symbol is transmitted over the entire signal bandwidth. After equalization, each data symbol in a block thus sees essentially the same SNR; the equalized channel appears to the decoder as an additive noise channel with a constant SNR, which is determined by the channel's fading and frequency selectivity characteristics and the equalizer's compensation of them. A coded modulation scheme tailored to additive white Gaussian noise channels is appropriate, for example pragmatic trellis-coded modulation or turbo coding with symbol interleaving, which is confined to one block.

Comparisons of coded linear SC-FDE systems with nonadaptively loaded coded OFDM systems have shown that the two systems offer similar, but not identical, BER performance [9, 10, 48], for the same average received SNR. Figure 6.4, from [48], shows the BER performance of coded OFDM and coded single carrier systems with frequency-domain linear equalization (SC-FDE) signals over indoor frequency-selective Rayleigh-fading channels with a maximum delay-spread of 0.66 μs, corresponding to over 40 coded data symbols, and bit rates from 60 to 300 Mbps (depending on the code rate). The SC-FDE systems used trellis-coded modulation, while the OFDM systems used BICM. For

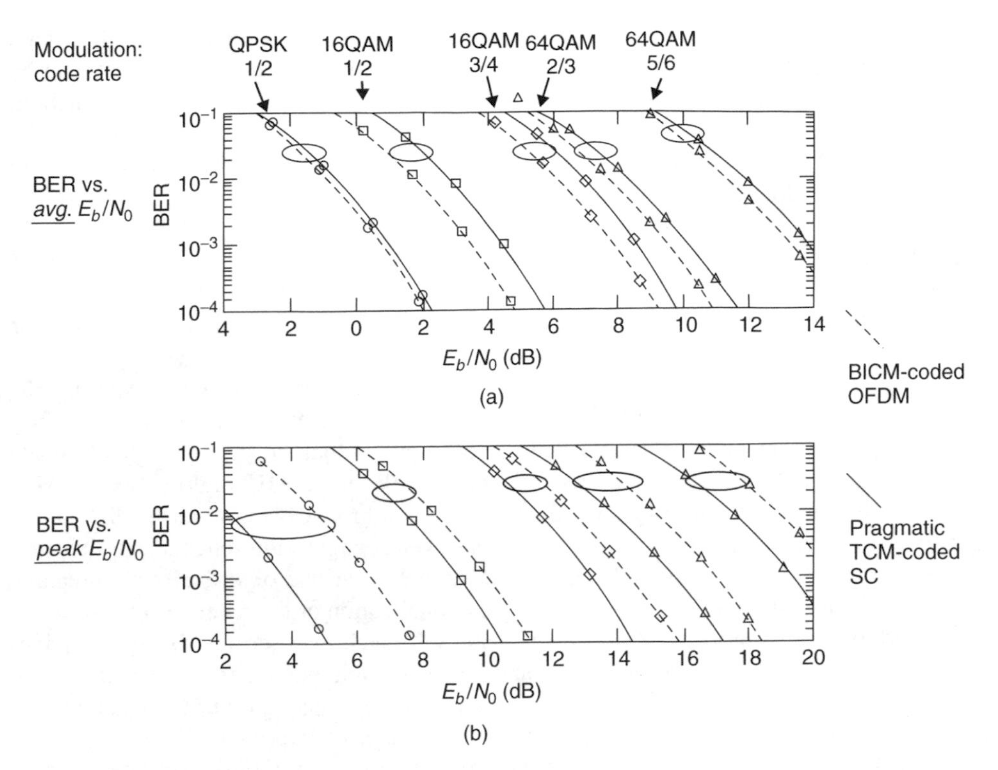

Figure 6.4 (from [48]) BER performance for pragmatic TCM-coded SC-FDE (solid lines) and BICM-coded OFDM; ○: QPSK, rate 1/2; □: 16QAM, rate 1/2; ◇: 16QAM, rate 3/4; △: 64QAM, rate 2/3; ▷: 64QAM, rate 5/6. Part A shows BER versus average E_b/N_0; part B shows BER versus *peak* E_b/N_0

code rates of about 1/2 or less, nonadaptively loaded coded OFDM shows a 0.5 to 1 dB average SNR gain over coded linearly equalized SC-FDE. Higher code rates tend to favor SC-FDE. Similar conclusions were reached in [49]. Both systems' performance can be further enhanced – for example, SC-FDE by the addition of full-length or sparse [9] time domain decision-feedback equalization (DFE), iterative equalization [50–52] or turbo equalization [53, 54], and OFDM by iterative detection/decoding [14, 55].

Figure 6.4 also shows that when bit error performance is measured against *peak* SNR (where the peak amplitude per block is averaged over many blocks), SC-FDE is consistently superior to OFDM. In these performance results, the transmitted OFDM signal undergoes clipping and filtering to reduce its peak power. In effect, this comparison is between OFDM and SC systems, which are using the same power amplifier with a fixed peak transmitted power output. The worse performance of OFDM is a consequence of the higher PAPR of the OFDM signal waveform.

6.2.4 Single- and Multicarrier CDMA

A *spreading* pre-matrix (with M input symbols and MK output chips, where K is the spreading factor) produces the multicarrier CDMA (MC-CDMA) form of parallel modulated signals [8, 20, 43]. All the chips from a given block of data can be transmitted in one IFFT block, in which case spreading is only in frequency, or they can be distributed among several successive IFFT blocks, in which case spreading is in both time and frequency. The latter case corresponds to signals with both serial and parallel transmission. Different users' data can be multiplexed onto different spreading codes, in which case spreading code chips can occupy both frequency and time dimensions. The serial-modulated version of CDMA is direct sequence CDMA (DS-CDMA). It can be generated using the GMC system of Figure 6.1 and Figure 6.2 by including the FFT operation as well as the matrix spreading operation. Figure 6.5 illustrates MU OFDMA, OFDM with Time Division Multiple-Access (TDMA), MC-CDMA and FDOSS signals in time, frequency, and code dimensions.

MC-CDMA signals from multiple users with different spreading codes sharing the same frequency and time chip allocations suffer from multiuser (MU) interference. They use a frequency-domain correlation receiver to suppress but not completely eliminate interference, or a MU detector receiver, which can eliminate MU interference. If MC-CDMA is used in the downlink path (BS or access point broadcasting to terminals), all the intra-cell interference arrives at a given terminal over a single radio channel; in this case MU interference can be removed relatively simply by the use of orthogonal spreading codes and the use of adaptive frequency-domain equalization at the receiver, to restore the codes' orthogonality properties. However in the UL, each user terminal's signal arrives at the BS over a different radio channel, and hence a single equalizer is not valid for all users. Individual frequency-domain MU detectors can be employed, but their complexity increases significantly with spreading gain or the number of interferers.

MC-CDMA systems can be an attractive alternative to pure OFDM systems because they can exploit antenna sectorization and spreading code orthogonality to achieve more efficient frequency reuse in cellular systems, and also because their resistance to multipath is enhanced by their spectrum-spreading [13, 14, 56]. Moreover, MC-CDMA offers efficient MU detectors with reasonable complexity compared to DS-CDMA, including iterative Turbo detectors with soft interference cancellation [45]. The MC-CDMA signal design allows the spreading factor to be much smaller than the number of subcarrier, which in combination with frequency interleaving gives high diversity gain at reduced receiver complexity. This flexibility is known as *M&Q* modification [8]. However in situations where MU intra-cell interference is problematic, systems such as OFDM, OFDMA,

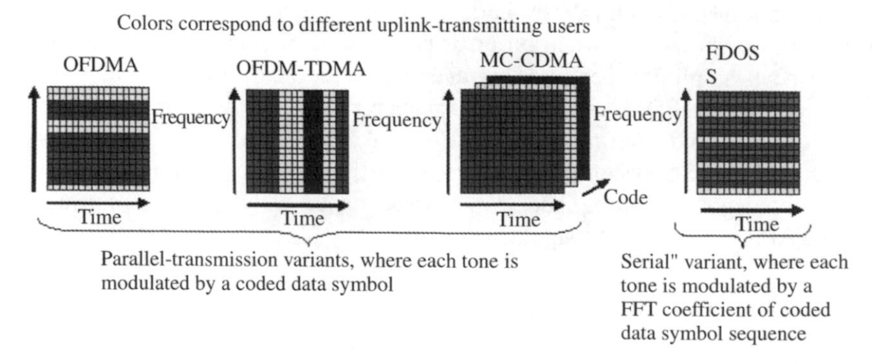

Figure 6.5 Illustrations of multiuser GMC signals: OFDMA, OFDM-TDMA, MC-CDMA, FDOSS

SS-MC-MA, or FDOSS may be preferable since they avoid MU interference [36], at the expense of somewhat reduced multipath immunity. MC-CDMA used in the downlink is more robust to multipath than is serial-modulated DS-CDMA with RAKE reception [8, 57]. For the UL of cellular systems, parallel-modulated SS-MC-MA or serial-modulated DS-CDMA is found to yield better performance than MC-CDMA because of more efficient channel estimation from transmitted pilots [8, 57]. For a fixed bandwidth, each user's spreading factor varies inversely with its bit rate. The DS-CDMA BS receiver could carry out conventional RAKE or correlation processing in the frequency domain to combat multipath and separate different user signals. Higher capacity and spectral efficiency would be achieved by minimum mean squared error (MMSE) MU detection using frequency-domain processing of the received UL DS-CDMA signals. Unfortunately, the complexity of the MMSE receiver's adaptation processing increases sharply with the spreading factor or number of interfering signals [25, 58]. For low bit-rate signals, with large spreading factors and numbers of interferers, the MMSE MU detection adaptation and complexity would be prohibitive.

Two promising alternatives to DS-CDMA in the UL exist:

1. Use of the FDOSS, or IFDMA signals mentioned earlier. They have the serial modulation, low PAPR and DS-CDMA properties, and have the advantage that MU interference is eliminated simply by appropriate sampling of the DFT of the received signal blocks at the receiver. However they have two disadvantages relative to conventional serial DS-CDMA:
   - They have a lower diversity order, since an FDOSS signal occupies only the M subcarriers in a block, whereas a DS-CDMA signal occupies up to MK, where M is the number of data symbols per block, and K is the spreading factor. FDOSS diversity can be improved by inter-block interleaving and FH, at the expense of extra delay.
   - FDOSS signals have no spreading code protection from FDOSS signals in adjacent cells.

   In an interference environment, the MU interference robustness of FDOSS signals more than compensates for their diversity loss. The problem of interference from adjacent cell UL FDOSS signals can be solved by optimal spatial combining at the BS and/or by applying direct sequence spreading to the data symbols before FDOSS transmission [35].
2. SS-MC-MA combines the advantages of OFDM and spectrum-spreading. As with FDOSS, MU interference is avoided and channel estimation and detection are simplified. The used subcarriers (subcarriers) per user are reduced compared to MC-CDMA, which reduces the diversity gain. However, due to frequency interleaving, the achievable diversity gain is still the spreading code length. The PAPR is higher in the UL than with FDOSS but less than with MC-CDMA since fewer subcarriers are used. The resistance to inter-cell interference with SS-MC-MA is comparable to that of MC-CDMA.

6.2.5 Zero-padded OFDM (ZP-OFDM) and Pseudorandom-postfix OFDM (PRP-OFDM)

The Zero-Padded OFDM (ZP-OFDM) modulation scheme has been studied in [59, 60]. The idea is to replace the guard interval sequence of CP-OFDM by a zero-padding sequence of identical length. With the same transmission scheme, several different equalization approaches are possible in the receiver [60] ranging from low-complexity/medium performance (Overlap-Add based) to high-complexity/high performance (MMSE-based equalization). In particular, [60] demonstrates that (in contrast to CP-OFDM) symbol equalization is possible even if frequency-domain channel nulls the fall onto data carriers. However, ZP-OFDM still requires channel estimates, which are typically obtained by learning symbols and/or pilot tones.

The Pseudo-random-Postfix OFDM (PRP-OFDM) modulation scheme has been studied in [16] for the single-antenna context and in [19] for the multiple-antennas context. The idea is to replace the zero-sequence of ZP-OFDM by a pseudo-randomly weighted deterministic sequence [19]. These weighting factors prevent any signal stationarity and thus spectral peaks. This is desirable, since in the presence of frequency-selective fading the spectrum should be as flat as possible – otherwise, the system performance would depend on the band affected by the fading.

Muck *et al.* [16] demonstrates that PRP-OFDM keeps all advantages of ZP-OFDM: different equalization approaches are possible in the receiver ranging from low-complexity/medium performance (Overlap-Add based) to high-complexity/high performance (MMSE-based equalization). Also, symbol equalization is possible even if the frequency-domain channel nulls the fall onto data carriers. Additional to these features, PRP-OFDM allows simple channel estimation and tracking based on the deterministic postfix sequences: A first idea consists in exploiting that the OFDM data symbols are zero mean; a simple mean-value calculation is sufficient to extract the postfix sequence convolved by the channel [16]. The channel itself is extracted by de-convolution. Muck *et al.* [18] demonstrate that such an approach must be refined in practice if higher order constellations are used (QAM64 and higher). Since the postfix sequence and the CP-OFDM guard interval are of the same power and duration as the guard interval of CP-OFDM, a higher spectral efficiency can be obtained; in particular, the typical overhead in terms of learning symbols and pilot tones for CP-OFDM is avoided.

PRP-OFDM typically is a suitable choice if the target application requires: (i) a minimum pilot overhead; (ii) low-complexity channel tracking (i.e. high mobility context) and (iii) adjustable receiver complexity/performance trade-offs (available due to the similarities to ZP-OFDM) without requiring any feedback loop to the transmitter.

6.2.6 OFDM/OffsetQAM (OFDM/OQAM) and IOTA-OFDM

OFDM/OffsetQAM is an alternative to conventional OFDM modulation. Contrary to it, OFDM/OQAM modulation does not require the use of a guard interval, which leads to a gain in spectral efficiency. Although a guard interval is a simple and efficient way to combat multipath effects, better performance can be reached by modulating each subcarrier by a prototype function. To obtain the same robustness to the multipath effects as OFDM with a guard interval, this prototype function must be very well localized in both the time and frequency domains. The localization in time aims at limiting the inter-symbol interference and the localization in frequency aims at limiting the inter-carrier interference (*e.g.* because of Doppler effects).

The orthogonality between the subcarriers must also be maintained after the modulation. The optimally localized functions having these properties exist but only guarantee orthogonality on *real* values. An OFDM modulation using these functions is denoted as

OFDM/OffsetQAM. We can note that in OFDM/OQAM, each subcarrier carries a *real* valued symbol but the density of the subcarriers in the time-frequency plane is two times greater in OFDM/OQAM than in conventional OFDM, also called *OFDM/QAM*, with no guard interval.

A particular prototype function called *IOTA* (Isotropic Orthogonal Transform Algorithm) is discussed in [61]. By construction, the IOTA function has the same shape in both the time and frequency domains. We typically call IOTA-OFDM an OFDM/OQAM system using the IOTA function. As stated before, with IOTA-OFDM *real* valued symbols are transmitted at twice the rate of conventional OFDM in the case of no guard interval inserted between the symbols [62].

6.2.7 Effect of Phase Noise and Frequency Offsets

Frequency offset and phase noise are impairments to the received signal that result in inter-carrier interference for parallel-modulated systems (OFDM, OFDMA, and MC-CDMA) [63–68]. These impairments are especially significant for UL signals from multiple users with different local oscillators. For OFDM, OFDMA and MC-CDMA, the degradation from a frequency offset Δf is approximately proportional to $(N_c \Delta f)^2$, where N_c is the number of subcarriers [63]. Clearly the sensitivity of parallel-modulated systems to phase noise and frequency offset increases with the FFT block length N_c. There exists receiver signal processing techniques for OFDM systems to counter frequency offset and phase noise, but they are relatively complicated, since they involve mitigation of frequency-domain inter-symbol interference [69, 70].

For serial-modulated systems, relatively less degradation results from frequency offset and phase noise (proportional to Δf^2) [63]. In these systems, a constant frequency offset just produces a slowly linearly increasing or decreasing phase shift over the received sequential data symbols. There is negligible additional inter-symbol interference. The slow phase shift variation is easily estimated by simple decision-directed techniques, and removed at the receiver prior to detection.

6.2.8 Power Amplifier Efficiency

Power amplifier efficiency is important since the power amplifier usually constitutes a substantial portion of the cost of a wireless user terminal, and power amplifier cost rises very sharply with peak power rating. Peak output instantaneous power exceeds a power amplifier's linear range, causes significant spectral regrowth, and BER degradation resulting from nonlinear distortion. Since a transmitted parallel-modulated signal consists of many parallel-modulated subcarriers, it has a significantly higher PAPR than a comparable serial-modulated signal. As a result, parallel-modulated systems require several dB more power back off and more expensive power amplifiers than do comparable serial-modulated systems for the same average power output [71, 72]. For a given power amplifier the larger the power back off, the lower the cell coverage. This power back-off penalty is especially important for subscribers near the edge of a cell, with large path loss, where lower-level modulation such as Binary Phase-shift Keying (BPSK) or Quadrature Phase-shift Keying (QPSK) modulation must be used. Recall from Figure 6.4 that BER performance is evaluated as a function of the *peak* bit energy to noise ratio, averaged over an ensemble of channels, coded SC-FDE is consistently superior to coded OFDM over a wide range of code rates and QAM constellation sizes [48]. In effect, this comparison is between parallel- and serial-modulated systems, which are using the same power amplifier with a fixed peak transmitted power output.

In practice, it is usually found that the power amplifier nonlinearity has a more critical effect on the regrowth of power spectrum sidelobes than on BER performance. In other words, at levels of nonlinearity that significantly affect spectrum sidelobes, the effect on received BER performance is negligible. This is especially relevant when spectral masks permit only very low out of band sidelobes.

Figure 6.6 shows examples of the instantaneous to average power ratio (IAPR) distributions for single-user OFDMA and serial-modulated signals transmitted in the UL. N_c is 2048, M_N is 256. The waveforms were generated using Expression (1). No explicit PAPR-reduction schemes were applied (unless the extra FFT operation in serial modulation is thought of as a PAPR-reduction scheme for OFDMA). Spectral sidelobes were reduced by applying a 6 % raised cosine time window to each waveform. It is clear that the serial-modulated waveform has a significantly lower dynamic range than the corresponding OFDMA waveform. This implies that the serial-modulated waveform will require a lower power back off at the input of a nonlinear transmitter power amplifier. For example, Figure 6.7 shows the power spectra of the two waveforms at the output of an amplifier whose nonlinear characteristic is modeled by a AM/AM conversion Rapp model [73], with parameter $p = 2$. The power backoffs of the two types of waveforms are adjusted to just conform to an European Telecommunications Standards Institute (ETSI) 3GPP power

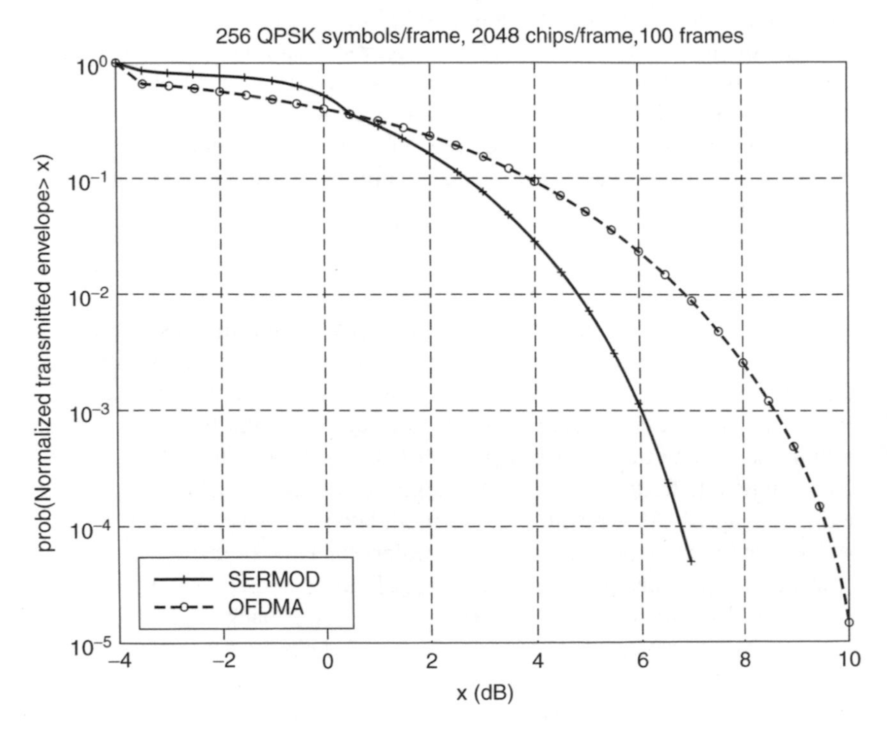

Figure 6.6 Distribution functions of instantaneous average power ratio (IAPR) of $S = 1$ sub-band OFDMA and serial-modulated signals, $N_c = 2048$, $M_N = 256$, $S = 1$

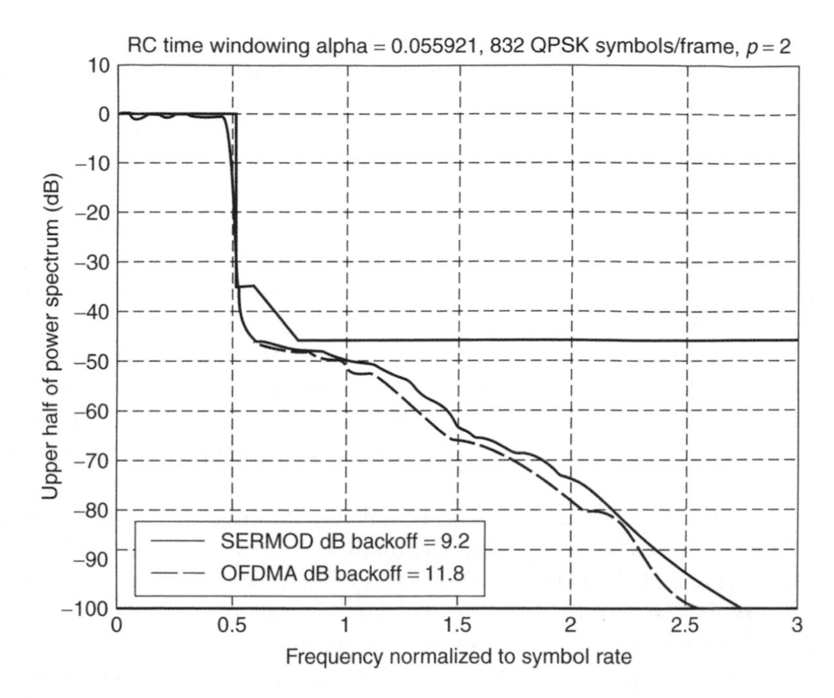

Figure 6.7 Output power spectra of signals passed through a Rapp model $p = 2$ nonlinearity; Backoffs of OFDMA and serial-modulated signals are adjusted to produce approximately – 48 dB relative spectral sidelobe levels. Scaled ETSI 3GPP mask is also shown

spectrum mask. In this example, the serial-modulated waveform requires 2.6 dB less back off than the OFDMA signal.

Neighboring user spectra may be received with large power variability due to differing path losses. Avoidance of adjacent channel interference then requires rather stringent spectral masks. For example, allowable interference to adjacent-frequency receivers is usually specified in terms of maximum interference power at a certain difference and at a certain frequency offset from the interferer's carrier [74]. Under typical transmitted power and path loss conditions, this may imply spectral masks with as much as 40 to 60 dB of attenuation of transmitted spectrum sidelobes at a normalized frequency offset of 1.

There are signal processing or clipping techniques for reducing the PAPR and spectral regrowth of parallel-modulated systems. These include selective mapping and partial Transmit (TX) sequences [6, 75–77], reference signal subtraction [78], coding [62, 79, 80], and clipping and filtering [81–83]. All of these approaches entail extra signal processing complexity, especially at the transmitter; they may also require extra overhead signaling or redundant symbols, and can also degrade performance and bandwidth efficiency [84, 85]. It can be shown that for the GMC transmitter of Figure 6.1, with symbol rate sampling, minimum PAPR is achieved if the pre-matrix and Inverse Discrete Fourier Transform (IDFT) matrix combine to form an identity matrix. That is, the pre-matrix should be a DFT, and the transmitted signal should be serial-modulated [33].

The transmitted serial-modulated signal is still in the form of Equation (1), and hence can also be considered a 'multicarrier' signal. However, as we have seen, using the DFT as in (2) reduces the signal's dynamic range. Thus the DFT operation, which creates a serial-modulated signal can also be considered as a linear PAPR-reduction operation on an OFDMA signal. Furthermore, it is a very simple operation, in comparison to alternative linear operations such as selective mapping and partial TX sequences, and requires no side information to be transmitted.

The back off and power amplifier efficiency are less important issues for the downlink since the cost of the BS power amplifiers is shared among many terminals. Also, since the BS usually has to TX many signals simultaneously through a common power amplifier, the resulting PAPR will be high regardless of whether individual downlink signals are serial- or parallel-modulated. However, the power amplification issues can significantly influence the cost of the wireless subscriber unit. Thus a very sound air-interface approach in cellular systems is to use serial modulation in the uplink (with or without CDMA) and parallel modulation in the downlink [9, 22, 30]. The main virtue of serial modulation for the terminal to base link is the lower peak power of its transmitted waveform and lower required power back off.

The peak transmitter power requirement of a mobile terminal (MT) is also influenced by the multiple-access method. Possible UL multiple-access methods, with comparable spectral efficiencies, include TDMA, CDMA, and FDMA, or combinations of them. Pure TDMA requires that all terminals TX at the same high aggregate bit rate and with the same high peak power. This places a severe cost penalty on terminals that only need to TX uplink at very low bit rates. CDMA and FDMA are preferable to pure TDMA since they allow each terminal's peak power, as well as its average power, to be proportional to its bit rate.

6.2.9 Spectrum Flexibility

A useful feature of GMC signals generated as in Figure 6.1 and Figure 6.2 is their flexibility to shape the transmitted spectrum to occupy any segment or segments of the frequency band, by means of the selector matrix shown in Figure 6.2. For example, if the transmitted spectrum is required to fit in a single band of width M_N, starting at frequency f_1, the selector distributes the $\{A(\ell), \ell = 0, 1, \dots M_N - 1\}$ into the set of frequencies

$$\Im = \{f_1 + \ell; \ell = 0, 1, ..M_N - 1;\}. \tag{4}$$

Zeroes are inserted into the remaining $N_c - M_N$ frequencies constituting the block.

If instead, the only vacant spectrum available for transmission of $\{A(\ell), \ell = 0, 1, \dots M_N - 1\}$ is in two disjoint parts ('sub-bands'), denoted respectively $\Im_1 = \{f_1, f_1 + 1, f_1 + 2, \dots f_1 + M_1 - 1\}$ and $\Im_2 = \{f_2, f_2 + 1, f_2 + 2, \dots f_2 + M_2 - 1\}$, with bandwidths M_1 and M_2 respectively, such that $M_1 + M_2 = M_N$, the set of data is partitioned into two sets $\mathfrak{a}_1 = \{A(\ell)\}_{\ell=0}^{M_1-1}$ and $\mathfrak{a}_2 = \{A(\ell)\}_{\ell=M_1}^{M_N}$. $\mathfrak{a}_1$ is mapped onto frequency set $\mathfrak{F}_1$, and $\mathfrak{a}_2$ is mapped onto frequency set $\mathfrak{F}_2$. The signal waveform is then generated from the N_c-point inverse DFT of the resulting block of data symbols and zeros, where we have defined $M_0 = 0$. If more than one user's signal is to be frequency-multiplexed in the downlink, each user's signal would occupy a disjoint set of frequency sub-bands and be generated in this way. This generalizes in an obvious way to more than two sub-bands whose aggregate bandwidth exceeds M_N.

Thus a cognitive radio system [86] can characterize potential interference in time and frequency, and dynamically configure its transmitted spectrum by appropriate selection of

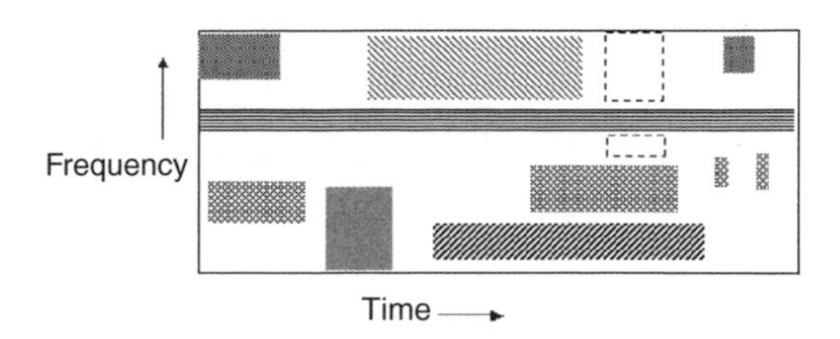

Figure 6.8 Illustration of spectrum-sharing. Each user is represented by a different shading pattern. A new user's spectrum is tailored to fit into the temporarily unoccupied 'white space' denoted by the dashed lines. Such a segmented spectrum can be generated by a suitable time-frequency selector matrix

frequencies to carry the coded data symbols for a given user [87]. Figure 6.8 illustrates this spectrum-sharing concept.

If the signal, with CP, is considered as being periodic, with period N_c, it will have a line spectrum

$$S(f) = \begin{cases} A(f - f_i) \text{ for } f \in \Im_i \\ 0 \qquad\qquad \text{otherwise} \end{cases} \tag{5}$$

The actual transmitted spectrum will be this line spectrum convolved with the $\frac{\sin(\pi f)}{\sin\left(\frac{\pi f}{N_c}\right)}$ spectrum of the rectangular time window of N_c samples. The resulting relatively high sidelobes can be suppressed by using a nonrectangular time window on the block plus its CP, such as a raised cosine window. Figure 6.9 shows an example of a multiband signal's line spectrum, with S sub-bands.

For serial-modulated signals transmitted over S disjoint sub-bands, it is straightforward to show from Equations (1) and (2) with the above partitioning of $\{A(\ell), \ell = 0, 1, \ldots M_N - 1\}$ and $\Im$, that the counterpart to Equation (3) is

$$s(k) = \frac{1}{N_c} \sum_{i=1}^{S} \exp\left(j\frac{2\pi k f_i}{N_c}\right) \sum_{m=0}^{M_N-1} d_i(m) g_i\left(k - m\frac{N_c}{M_N}\right), \tag{6}$$

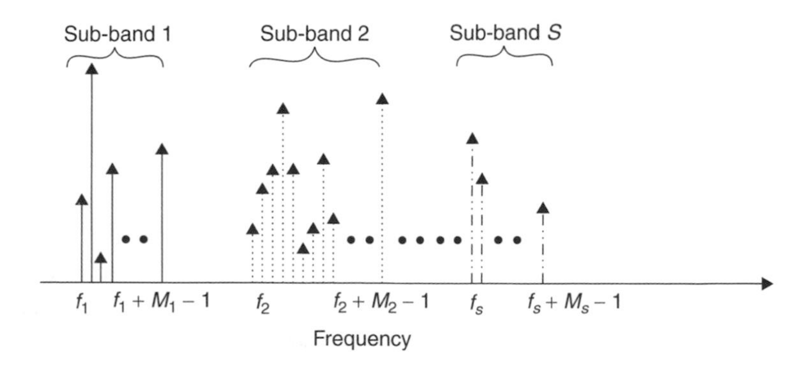

Figure 6.9 Illustration of a multiband line spectrum

where

$$d_i(m) = a(m) \exp\left(-j\frac{2\pi m}{M_N}\sum_{i'=0}^{i-1} M_{i'}\right) \text{ for } m = 0, 1, 2, \ldots M_N - 1, \tag{7}$$

and

$$\begin{aligned} g_i(k) &= \sum_{n=0}^{M_i-1} \exp\left(j\frac{2\pi nk}{N_c}\right) \\ &= \exp\left(j\frac{\pi(M_i - 1)k}{N_c}\right) \frac{\sin\left(\frac{\pi M_i k}{N_c}\right)}{\sin\left(\frac{\pi k}{N_c}\right)} \text{ for } k = 0, 1, 2, \ldots N_c - 1. \end{aligned} \tag{8}$$

Expression (7) represents a rotated complex data symbol. Expression (8) represents a sampled pulse waveform, similar to a modulated sinc pulse. Thus the waveform Expression (6) represents the sum of S complex single carrier waveforms, up-converted by the frequencies $\{f_i\}$. For relatively small values of S, like 2 or 3, its PAPR will be significantly less than that of a corresponding multiband OFDMA signal, which is the sum of M_N complex exponential waveforms.

For both OFDMA and serial modulation, essentially the same multiband average power spectra will be produced. Figure 6.10 illustrates the average power spectra for serial

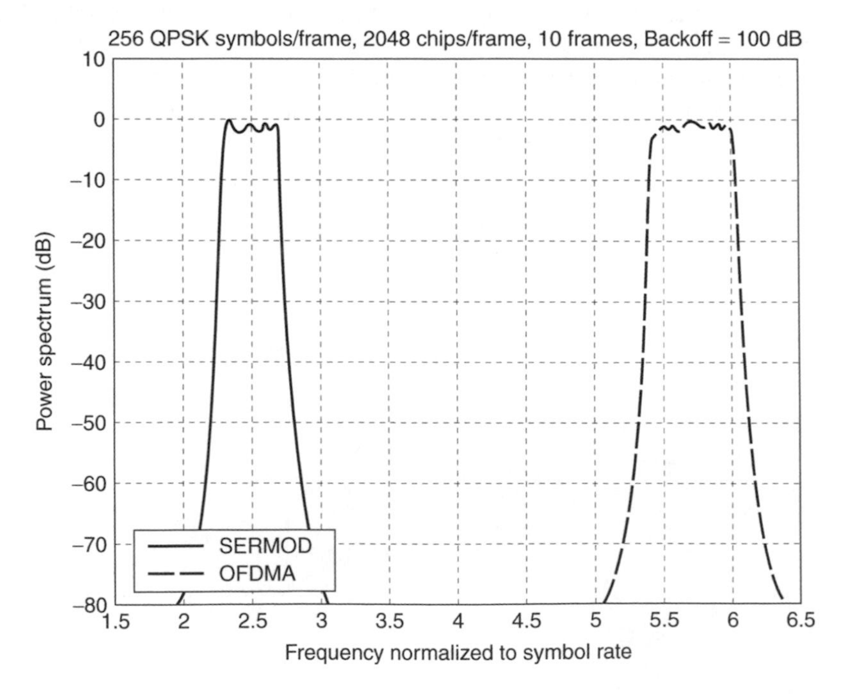

Figure 6.10 Power spectra for $S = 2$, $N_c = 2048$, $M_N = 256$ for OFDMA and serial modulation

modulation and OFDMA for two sub-bands, with $M_N = 256$ data symbols within a total bandwidth of $N_c = 2048$. The other parameters for this figure are $f_1 = 589$, $M_1 = 103$, $f_2 = 1390$, and $M_2 = M_N - M_1 = 153$. Note that the OFDMA and serial modulation spectra are almost identical.

The spectrum flexibility feature of GMC signals also allows a BS to transmit to low cost, low data rate user terminals that have limited receiver bandwidths. Such a terminal would use the same DFT block duration as that of the transmitted signal, but extracts from it, equalizes and decodes only the portion of the frequencies in the block that are relevant to it.

6.2.10 Some Issues for Further Research

The impact of the proposed mixed mode on the MAC layer and resource allocation needs to be investigated. For example, feedback information must be exchanged for adaptive modulation, power control, and synchronization, parameter control for coding, modulation, and antennas. Mobile systems' MAC should be modified to take into account the higher expected mobility and rate of channel change in future mobile broadband systems. Also a more detailed study should be made of means to accommodate, within a common allocated frequency band, user terminals with vastly different bit rates; for example, many small embedded devices sporadically transmitting and receiving say a few hundred bits per second, coexisting with 1, 10 and 100 Mbps terminals.

Efficient frequency-domain parameter estimation methods using training words and/or pilot tones are known for OFDM [88–90] and SC-FDE systems [9]. However, high bit rates and multipath delay spreads call for large FFT block lengths; there is then a question of the ability of receivers to estimate and track the resulting large number of frequency-domain parameters and Doppler shifts over rapidly time-varying channels. Investigations should include adaptation using training and blind adaptation as well as the use of unique words (UW) for channel tracking. Since the UW is known at the receiver it can effectively be used for tasks like synchronization (carrier frequency, clock frequency, etc.), channel estimation, and preventing the error propagation in the DFE [23, 40]. The CP can be replaced by a short time domain training sequence, which is processed and extrapolated in the frequency domain to estimate the required equalizer frequency-domain parameters [9, 23].

In this document we have focused on block transmission schemes using frequency-domain processing at the transmitter or receiver, and with cyclic prefixes at the beginning of each block, to avoid inter-block interference and to simplify frequency-domain filtering with DFTs. Other frequency-domain processing methods can be used for systems that do not organize data in blocks separated by cyclic prefixes. The motivation is to eliminate the CP overhead or to achieve compatibility with legacy systems that did not use CPs. For serial-modulated SC-FDE systems, overlap-save processing can be used [24]. OFDM systems without the CP have also been proposed [91].

6.2.11 Summary and Recommendations

We have proposed GMC signaling as a common framework for future-generation wireless systems which must accommodate a very wide range of user bit rates with low cost and high reliability in a difficult, spectrally crowded radio environment. Within this framework, adaptive, reconfigurable transmission techniques are desirable, including not only power and bit rate, but also spectrum occupancy and the choice of the modulation and multiple-access/multiplexing scheme itself. For example, downlink parallel modulation together with UL serial modulation is a good configuration when user terminal power efficiency is a major issue. Flexibility for multiband operation is also relatively easy to achieve for GMC signaling.

The proposed GMC modes are quite compatible with MIMO and space diversity, and also with spectrum-spreading. A number of variations of OFDM and SC-FDE are useful, such as OFDMA, MC-CDMA and FDOSS.

6.3 Smart Antennas, MIMO Systems and Related Technologies

The adoption of smart antenna techniques and MIMO systems in future wireless systems is expected to have a significant impact on the efficient use of the spectrum, the minimization of the cost of establishing new wireless networks, on the optimization of the service quality provided by wireless networks, and the realization of transparent operation across multitechnology wireless networks. Nevertheless, the design of future-generation smart antenna and MIMO systems involves a number of challenges, such as efficient MU downlink precoding, the scheduling of users in space, time and frequency, reconfigurability to varying scenarios in terms of propagation conditions, traffic models, mobility, transceiver architectures, MT resources (i.e. battery lifetime), QoS requirements for different services and interference conditions. The design of robust solutions matched to the reliability of the available channel information is required in order to account for the impact of channel estimation errors and feedback quantization and delay. The techniques addressing these challenges need to be assessed by adequate performance evaluation based on realistic modeling assumptions both at the link and the system level. Moreover, in the design of smart antenna techniques, system architecture, implementation, and complexity limitations need to be taken into account. Finally, the requirement for future-generation systems operating in multitechnology networks introduces a further set of challenges associated with the design of Smart Antenna systems, which enhance the performance and facilitate the interoperability across different wireless technologies.

In this section, we provide an overview of the most recent advances on smart antenna and MU MIMO algorithms, system-level performance optimization strategies, channel and interference modeling, realistic performance evaluation and implementation issues for the deployment of smart antennas in future systems.

To this end current research effort in the area is focusing on the following:

- the design and development of advanced and innovative MIMO processing algorithms that allow for increased network capacity, adaptivity to varying propagation conditions, and robustness against network impairments [92–96];
- the design and development of innovative Smart Antenna strategies for optimization of performance at system level and transparent operation across different wireless systems and platforms;
- the realistic performance evaluation of the proposed MIMO algorithms and Smart Antenna strategies, based on the formulation of accurate channel and interference models; introduction of suitable performance metrics and simulation methodologies;

- the analysis of the implementation, complexity, and cost efficiency issues involved in the realization of the proposed Smart Antenna techniques for future-generation wireless systems.

An important research topic is the study of MU MIMO systems. Such systems have the potential to combine the high throughput achievable with MIMO processing with the benefits of space division multiple-access (SDMA). In the downlink scenario, a BS is equipped with multiple antennas and it is simultaneously transmitting to a group of users. Each of these users is also equipped with multiple antennas. In this case, the BS has the ability to coordinate the transmission from all of its antennas. The receiving antennas are associated with different users that are typically unable to coordinate with each other. The BS exploits the CSI available at the transmitter to allow these users to share the same channel and mitigate or ideally completely eliminate multiuser interference (MUI) by beamforming (linear precoding) or by the use of 'dirty-paper' codes. It is essential to have CSI at the BS since it allows joint processing of all users' signals, which results in a significant performance improvement and increased data rates. All precoding techniques can be classified considering whether they allow MUI (as zero or nonzero MUI techniques) and by their linearity (as linear and nonlinear techniques). Linear precoding techniques require no overhead to provide the mobile with the demodulation information and are less computationally expensive than their nonlinear counterparts. However, nonlinear techniques provide a much higher capacity.

In this section, the latest trends and future directions in the area of smart antennas and MIMO systems are analyzed. In Section 6.3.2, an overview is presented of how smart antennas can improve the performance of mobile communication systems. Section 6.3.3 addresses MIMO transceivers that support reconfigurability and robustness to varying propagation and traffic conditions. Smart antenna system-level strategies are discussed in Section 6.3.4 by investigating novel radio resource management techniques based on reconfigurable scheduling schemes and cross-layer optimization designs as well as network optimization exploiting contextual information. MU MIMO downlink precoding schemes are summarized in Section 6.3.5, before cross-layer optimization strategies of smart antenna systems are provided in Section 6.3.6. Performance evaluation strategies for the deployment of smart antennas are treated in Section 6.3.7, and the deployment of smart antennas in future systems is discussed in Section 6.3.8. Finally, Section 6.3.9 summarizes the main conclusions.

6.3.1 Benefits of Smart Antennas

Smart antennas are generally used at the BS for UL reception and downlink transmission with multiple antennas. They allow spatial access to the radio channel by means of different approaches, for example, based on directional parameters or by exploiting the second-order spatial statistics of the radio channel. Thus, space-time processing reduces interference and enhances the desired signal. Moreover, adaptive antennas can exploit long-term and/or short-term properties of the mobile radio channel to achieve improved channel estimation accuracy and reduced computational complexity [97]. Some antenna array-processing techniques are classified in Figure 6.11. This classification is based on the propagation channel properties, that is, on the structure of the spatial correlation matrix at the antenna array.

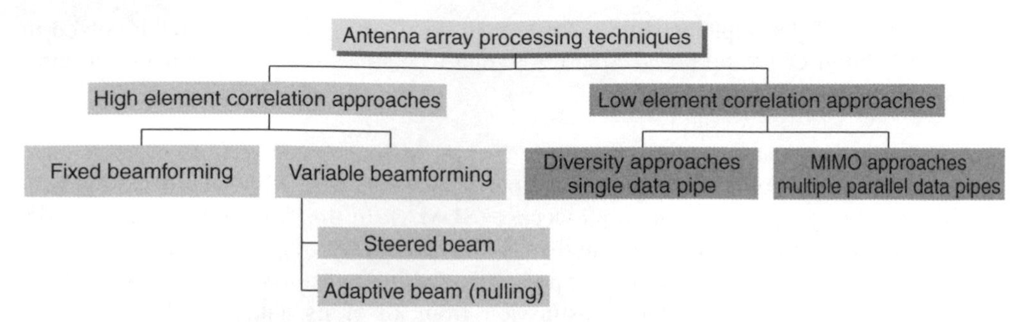

Figure 6.11 Taxonomy of smart antenna techniques

Spatial antenna processing is seen as a key technology for future high performance 4G wireless systems. Future system designs are likely to embody a combination of beamforming, MIMO, interference cancellation, and other related technologies in order to maximize system capacity and high data rate coverage across the network. To this end, we therefore propose the study of the intelligent use of multiple antennas at the BS and the user terminal. Antenna array processing will provide the basis for improved system operation and performance, and result in a substantially lower cost per user within the network. The multiple antenna concepts should be targeted at scalable, tiered cell deployments, with the emphasis on providing high capacity, high data rate wide-area coverage to mobile and nomadic users. Optimized spatial processing concepts will be identified for both uplink and downlink aspects of the system design.

The gains achievable with multiple antenna systems on the transmit side as well as the Receive (RX) side can be classified as follows [98]:

- The array (or beamforming) gain is the average increase in signal power at the receiver due to a coherent combination of the signals received at all antenna elements. It is proportional to the number of RX antennas. If channel knowledge is available at the transmitter, the array gain can also be exploited in systems with multiple antennas at the transmitter.
- The achievable diversity gain depends on the number of TX and RX antennas as well as the propagation channel characteristics, that is, the number of independently fading branches (diversity order). The maximum diversity order of a flat-fading MIMO channel is equal to the product of the number of RX and TX antennas. TX diversity with multiple TX antennas can, for instance, be exploited via space-time coding and does not require any channel knowledge at the transmitter.
- The interference reduction (or avoidance) gain can be achieved at the receiver and the transmitter by (spatially) suppressing other co-channel interferers. It requires an estimate of the channel of the desired user.
- The spatial multiplexing gain can be obtained by sending multiple data streams to a single user in a MIMO system or to multiple co-channel users in an SDMA system. These techniques take advantage of several independent spatial channels through which different data stream can be transmitted.

Multiple antennas at the BS and also at high performance terminals achieve significantly higher data rates, better link quality, and increased spectral efficiency [99]. Hence, more users can be accommodated by the system and a corresponding capacity increase achieved. To obtain very high data rates, multiple-input multiple-output (MIMO) processing techniques such as spatial multiplexing and space-time coding can be used [100]. Alternatively, MIMO techniques can be used in order to reduce the total transmitted power, while preserving the data rate. This in turn reduces the overall system interference. Overall, the adoption of MIMO processing techniques in future wireless systems is expected to have a significant impact on the efficient use of the spectrum, the minimization of the cost of establishing new wireless networks, on the optimization of the service quality provided by those networks, and to facilitate the realization of transparent operation across multitechnology wireless networks, see Figure 6.15.

Performance enhancements achieved by the use of multiple TX and/or RX antennas in wireless telecommunications can be summarized as follows:

- Increase of channel (and hence system) capacity. Investigations show that in case of uncorrelated Rayleigh-fading, the channel capacity limit (Figure 6.12) grows logarithmically (Figure 6.13), when adding an antenna array at the receiver side (Single-Input Multiple-Output system – SIMO), and in the MIMO case as much as linearly with min (MR, MT}, where MR and MT denote the number of antennas at the receiver and the transmitter, respectively (Figure 6.14).
- Decrease of the BER without any bandwidth expansion or TX power increases when RX diversity and space-time linear processing or coding, that is, TX diversity, are used jointly, or alternatively with cell range expansion if performance is traded-off for coverage.
- Decrease of the impact of fading effects, this potentially leading to enhanced robustness and reliability in detected data.
- Decrease of total transmitted power/system interference for the same data rates.
- Increase of the packet call and cell throughput at the system level.
- Improvement of coverage for high data rates.

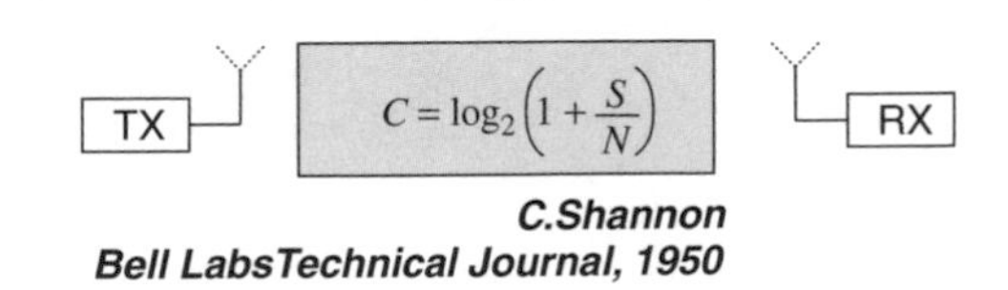

Figure 6.12 Single-Input Single-Output (SISO) capacity limit label

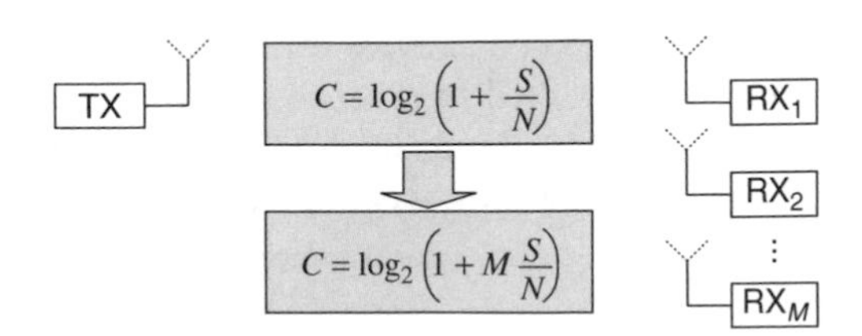

Figure 6.13 Adding an antenna array at the receiver (Single-Input Multiple-Output – SIMO) provides logarithmic growth of the bandwidth efficiency limit

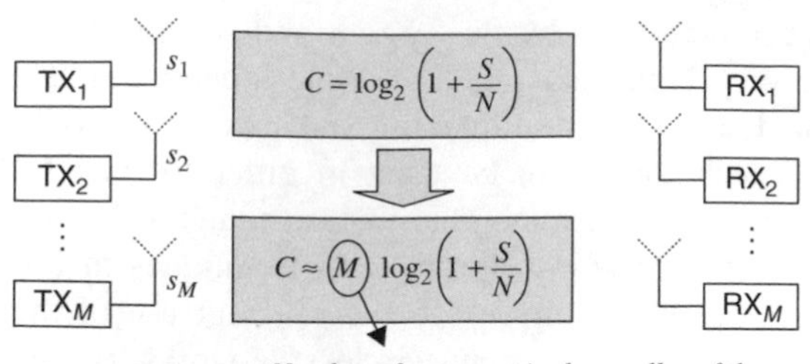

Figure 6.14 A Multiple-Input Multiple-Output (MIMO) provides linear growth of the bandwidth efficiency limit

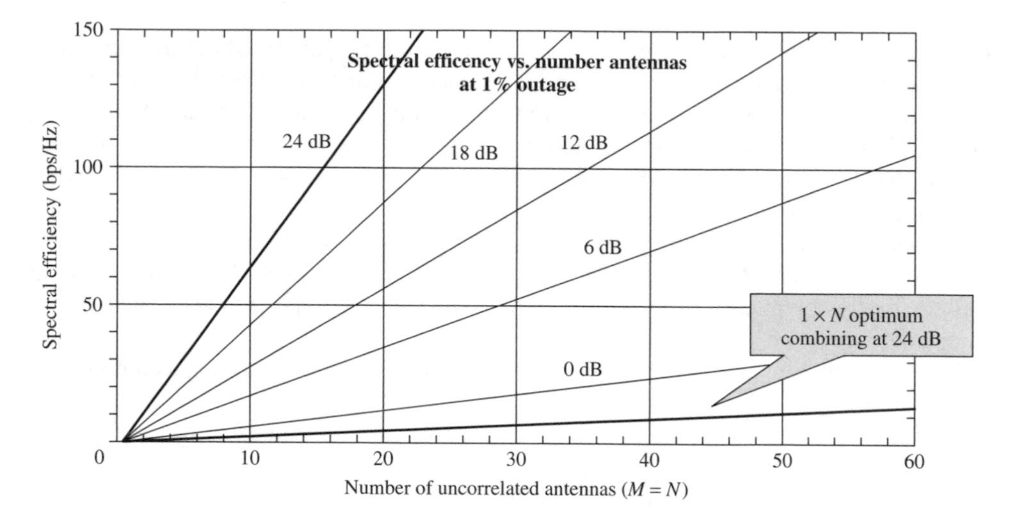

Figure 6.15 Spectral efficiency versus number of antennas

6.3.2 MIMO Transceivers

6.3.2.1 MIMO Transceivers Design

In a multiple TX multiple RX antenna system as the one illustrated in Figure 6.16, the data block to be transmitted is encoded and modulated to symbols of a complex constellation. Each symbol is then mapped to one of the TX antennas (spatial multiplexing) after spatial weighting of the antenna elements or a linear space-time precoding technique is applied. After transmission through the wireless channel, at the receiver demultiplexing, weighting, demodulation, and decoding is performed in order to recover the transmitted data.

The transmission schemes over MIMO channels, designed to maximize spectral efficiency, typically fall into three categories, corresponding to the maximization of the following criteria:

a. Diversity,
b. Data rate, and
c. SINR.

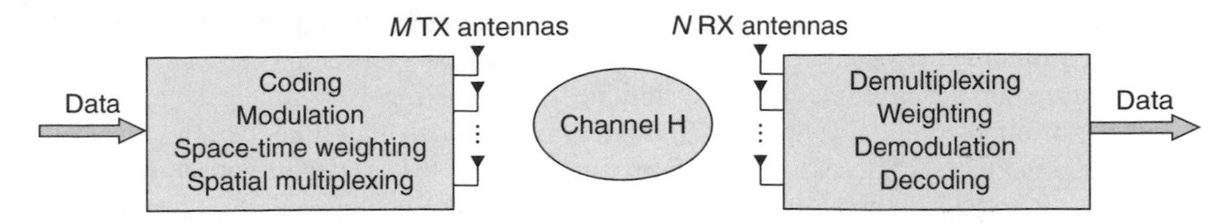

Figure 6.16 MIMO transceiver

- In the case of maximization of diversity, joint encoding – space-time coding – is applied and thereby the level of redundancy between TX antennas is increased as each antenna transmits a differently encoded fully redundant version of the same signal. Space-time codes were originally developed in the form of Space-Time Trellis Codes (STTC) [101], which required a multidimensional Viterbi algorithm for decoding at the receiver. These codes can provide diversity equal to the number of TX antennas as well as coding gain depending on the complexity of the code without loss in bandwidth efficiency. Space-Time Block Codes (STBC) [102, 103] offer the same diversity as the STTC but do not provide coding gain. However, STBC are often the preferred solution against STTC, as their decoding only requires linear processing.
- The maximization of data rate is achieved by performing spatial multiplexing, that is, by sending independent data streams over different TX antennas. BLAST (Bell labs Layered Space-Time) technology takes advantage of several independent spatial channels through which different data stream can be transmitted [100, 104–107]. The receiver must demultiplex the spatial channels in order to detect the transmitted symbols. Various techniques have been used for this purpose, such as Zero-Forcing, which uses simple matrix inversion but results in poor results when the channel matrix is ill conditioned; MMSE, which is more robust in that sense but provides limited enhancement if knowledge of the noise/interference is not used; and Maximum Likelihood (ML), which is optimal in the sense that it compares all possible combinations of symbols but can be too complex, especially for high-order modulation.
- The maximization of SINR is achieved through focusing energy into the desired directions and minimizing energy toward all other directions. Beamforming [108] allows spatial access to the radio channel by means of different approaches, for example, based on directional parameters or by exploiting the second-order spatial statistics of the radio channel. Thus, space-time processing reduces interference and enhances the intended signal. Moreover, adaptive antennas can exploit long-term and/or short-term properties of the mobile radio channel to achieve improved channel estimation accuracy and reduced computational complexity.

These transmission strategies require efficient techniques to separate the signals of multiple users sharing the spectrum resources at the receiver and to cancel interference, under various interference scenarios [109, 110]. Some of them use pilot signals known by both the transmitter and the receiver (nonblind techniques), whereas others only employ a priori knowledge of the received signals (blind/semi-blind techniques). Another class of techniques assumes that the transmission channel is known at the transmitter, this knowledge being obtained either by feedback or by the TDD channel reciprocity assumption.

Depending on the multiple-access technique, different receiver strategies have been proposed, from theoretically optimal strategies to practical ones like Parallel Interface Cancellation (PIC) or Successive Interface Cancellation (SIC), decorrelation, and Joint Detection. Reduced complexity schemes may use Turbo techniques for performance enhancement [111]. The channel estimation process can also be incorporated within the Turbo mechanism, which can enhance the channel throughput [112].

Technologies using a priori channel knowledge at the TX have been recently proposed, which allow for independent data streams to be sent and received without interference, relying on the computation of the channel matrix spaced vertical dipoles (SVD) [113, 114].

The key parameters to be taken into account when designing MIMO transceivers are the following:

- *Channel knowledge*: what are the assumptions (narrowband, frequency-selective, long-term properties, short-term properties, open loop or close loop systems, feedback data rate in close loop systems), how reliable is CSI and how robust algorithms are to potential CSI mismatches.
- *Deployment environment*: picocell, macrocell, and so on.
- Fading correlation seen by the antenna elements on both TX and RX antennas.
- Dispersion in the channel.
- *Characteristics of the transmitted signal*: modulation format, multiplexing of traffic and training information, and so on.
- Sample size in the number of observations with respect to the dimension of the array.
- Level of desired computational complexity, and its partitioning between TX and RX.

6.3.3 Reconfigurable MIMO Transceivers

Wireless systems beyond 3G require signal processing techniques that would be capable of operating in a wide variety of scenarios [115–118] with respect to the following:

- propagation environment (indoor/outdoor, rich scattering/specular, etc.);
- traffic environment (hot-spot/spotty, uniform/directional, dense/sparse, etc.);
- interference environment (intra-cell/inter-cell, same system/other system, etc.);
- user mobility (static/mobile users/users in bullet train, speed of interference, etc.);
- antenna configuration (number of antennas at the terminal/base, antennas correlation/ bandwidth, antenna characteristics, etc.);
- radio access technology (single or multiple parallel technologies, frequency, modulation/waveform, etc.);
- reliability, availability or unavailability of information on the channel prior to transmission (feedback channel, TDD vs. Frequency Division Duplex (FDD) mode)

There are two main approaches to allow wireless communication transceivers to operate in a multi-parametric, continuously changing environment:

- **reconfigurable**, adaptive techniques for adjusting the structure and parameters of the transceivers to allow them to demonstrate the best performance in a variety of particular situations;

- **robust** techniques, which can demonstrate reasonable (required) performance in a variety of the unspecified situations.

The first approach assumes that the particular scenario can be identified, the optimal solution is known, and the required transceiver configuration can be provided. MIMO receivers capable of reconfiguring themselves by switching automatically between a 'beamforming' and a 'spatial multiplexing', Figure 6.17, can be considered as an example of the first approach [119]. It is currently known that there is a fundamental trade-off between spatial diversity and multiplexing in MIMO channels [120], but still there is no general approach to achieve this limit in a wide range of scenarios. An example of designing MIMO transceivers capable of adapting to varying propagation conditions, namely, to varying spatial correlation, is presented in the following section.

The second approach can be illustrated by 'short-burst' systems, which allow avoiding nonstationarity tracking. Another example are MIMO and interference cancellation techniques based on semi-blind estimation algorithms without explicit estimation of the propagation channel for all signals received by an antenna array [109, 110]. A general problem of these techniques is the fact that they must operate with a low number of observations. These situations are especially unsuitable for traditional array-processing tools, which are designed to give a good performance when the number of observations is high. General statistical analysis tools based on random matrix theory can be very useful in order to operate in low sample size situations.

The current vision of systems 'beyond 3G' (everything is connected everywhere at anytime with high frequency reuse and high mobility) suggests that complicated system/interference scenarios will be important, which cannot be reduced to a fixed set of the situations with the known optimal solutions. Thus, a combination of reconfigurable and robust techniques is recommended. In this context, algorithms able to gracefully adapt to varying degrees of knowledge of, for example CSI, are expected to achieve optimal performance.

6.3.3.1 MIMO Transceiver Reconfigurability with Linear Precoding

The design of reconfigurable MIMO transceiver based on linear precoding consists in introducing a linear filter at the MIMO transmitter that allows for the utilization of certain knowledge about the channel conditions and/or properties in order to improve performance of the transmission scheme with respect to a selected performance criterion.

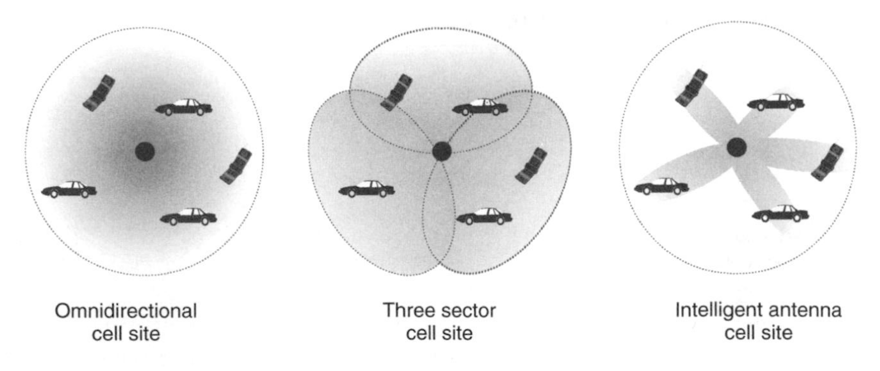

Figure 6.17 Adaptive beamforming principle

Linear precoding schemes for Space-Time Block Coded systems have been recently proposed [115, 121], which achieve reconfigurability to antenna correlation and CSI reliability by employing a transmission scheme adaptive to channel conditions.

Recent studies have shown that high fading correlations reduce MIMO channel capacity and system performance. An optimal linear precoder that assumes knowledge of the TX antenna correlations can improve the performance of a space-time coded system by forcing transmission on the nonzero eigenmodes of the TX antenna correlation matrix. The main advantage of this precoder is that it does not have to track fast-fading; it only tracks the slowly varying antenna correlations. These can be fed back to the transmitter using a low-rate feedback link or can be derived based on UL channel estimation.

Let us consider a communication system that consists of M-transmit and N-receive antennas.

The transmitter is assumed to have some knowledge of the second-order statistics of the channel. These statistics can be obtained from the feedback bits sent by the mobile station (MS), or they can be derived based on UL channel estimation. As depicted in Figure 6.18, the space-time block encoder Z maps the input data sequence x = (x1, x2, .. x Q) onto an M × Q matrix Z of codewords that are split into M parallel sequences. These codewords are then transformed by an M × M linear transformation L in order to adapt the code to the available channel information. The resulting sequences, encompassed in an M × Q matrix C = LZ, are sent on the M-TX antennas during Q time intervals. Transmitted data are recovered at the receiver by means of a ML receiver.

The signal received at the mobile is assumed to be a linear combination of several paths reflected from local scatterers, which result in uncorrelated fading across the RX antennas and therefore uncorrelated rows of matrix H. However, limited scattering at the BS can result in antenna correlation and, therefore, in correlated columns of matrix H. In such a case, as explained in [121], it is possible to use a geometry-based stochastic channel model in which the probability density function (PDF) of the geometrical location of the scatterers is prescribed. However, in the following analysis, a correlation-based channel model will be used instead for the sake of simplicity. According to the correlation-based model, the channel H can be written as follows:

$$\mathbf{H} = \mathbf{H}_W \mathbf{R}_T^{1/2}, \tag{9}$$

where HW is an N × M independently and identically distributed (i.i.d) complex matrix and RT is the MxM TX antenna correlation matrix. The received signal Y is corrupted by additive white Gaussian noise denoted by the N × Q matrix Σ with covariance matrix σ2IN:

$$\begin{aligned} \mathbf{Y} &= \mathbf{HC} + \Sigma \\ &= \mathbf{HLZ} + \Sigma \end{aligned} \tag{10}$$

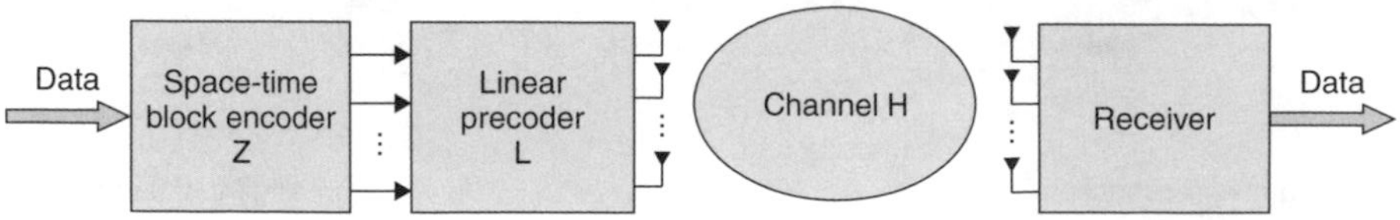

Figure 6.18 Reconfigurable transmission scheme combining space-time block codes and a linear transformation designed with reference to channel knowledge available at the transmitter

A block-fading model in which the channel remains constant over a number of symbol periods (spanning a space–time codeword) and then changes in an independent fashion in the following realization is assumed. The antenna correlation at the transmitter and, therefore, the computed precoding matrices used to compensate for antenna correlation are assumed stationary over time.

The linear transformation L is determined so as to minimize a given criterion, such as an upper bound on the Pairwise Error Probability (PEP) of the codeword. The PEP is defined as the error probability of choosing in favor of the codeword Zl instead of the actually transmitted codeword Zk.

If $\tilde{\mathbf{E}}(\mathrm{k, l, t}), = \mathbf{Z}^{\mathrm{k}}(\mathrm{t}) - \mathbf{Z}^{1}(\mathrm{t})$ denotes the code error matrix, maximum diversity is obtained when $\tilde{\mathbf{E}}(\mathrm{k, l, t})$ has full rank for all $^{\mathrm{k,l}}$. The coding gain is dominated by the minimum distance code error matrix, which is defined as $\mathbf{E} = \arg \min_{\tilde{\mathrm{E}}(\mathrm{k,l,t})} \det[\tilde{\mathbf{E}}(\mathrm{k, l, t})\tilde{\mathbf{E}}^{\mathrm{H}}(\mathrm{k, l, t})]$. It can be shown that the minimization of an upper bound on the average PEP is equivalent to the following optimization problem:

$$\max_{\mathrm{L}} \det\left(\mathbf{I} + \frac{E_s}{\sigma^2}\mathbf{R}_T^{1/2}\mathbf{L}\mathbf{E}\mathbf{E}^H\mathbf{L}^H\mathbf{R}_T^{1/2^H}\right)$$
$$s.t : Trace(\mathbf{L}\mathbf{L}^H) = P_0 \tag{11}$$

where E_{s} is the symbol energy and P_0 reflects a desirable power constraint.

The solution of this optimization problem is

$$\mathbf{L} = \mathbf{V}_{\mathbf{r}}\Phi_{\mathbf{f}}\mathbf{V}_{\mathbf{e}}^H$$
$$\mathbf{\Phi}_{\mathbf{f}}^2 = \left(\gamma\mathbf{I} - \left(\frac{E_{\mathrm{s}}}{\sigma^2}\right)^{-1}\mathbf{\Lambda}_{\mathbf{r}}^{-2}\mathbf{\Lambda}_{\mathbf{e}}^{-1}\right)_+ \tag{12}$$

where $\mathbf{R}_{\mathbf{T}}^{1/2} = \mathbf{U}_{\mathbf{r}}\mathbf{\Lambda}_{\mathbf{r}}\mathbf{V}_{\mathbf{r}}^H$ and $\mathbf{E}\mathbf{E}^H = \mathbf{U}_{\mathbf{e}}\mathbf{\Lambda}_{\mathbf{e}}\mathbf{V}_{\mathbf{e}}^H$ and $\gamma > 0$ is a constant computed from the trace constraint and (.)+ stands for max (.,0).

When orthogonal space–time codes are considered [121], the minimum distance code error matrix is such that $\mathbf{E}\mathbf{E}^H = \zeta\mathbf{I}$, where ζ is a scalar, $\mathbf{\Lambda}_{\mathbf{e}} = \zeta\mathbf{I}$, and $\mathbf{V}_{\mathbf{e}} = \mathbf{I}$. In this case the precoder becomes:

$$\mathbf{L} = \frac{1}{\sqrt{\mathrm{M}}}[\mathbf{w}_1\mathbf{w}_2\ldots\mathbf{w}_{\mathrm{M}}]\begin{bmatrix} \sqrt{1+\beta_1} & 0 & 0 & \ldots & 0 \\ 0 & \sqrt{1+\beta_2} & 0 & \ldots & 0 \\ 0 & 0 & \ddots & 0 & \vdots \\ \vdots & \vdots & 0 & \ddots & 0 \\ 0 & 0 & \ldots & 0 & \sqrt{1+\beta_{\mathrm{M}}} \end{bmatrix} \tag{13}$$

where w1, w2, ..., wM are the eigenvectors of the matrix, $\mathbf{R}_{\mathrm{T}}^{1/2}$, $\beta_i = \left[\left(\frac{1}{\lambda_{r,1}^2} - \frac{1}{\lambda_{r,i}^2}\right) + \ldots + \left(\frac{1}{\lambda_{r,M}^2} - \frac{1}{\lambda_{r,i}^2}\right)\right] / \left(\frac{\mathrm{E_S}}{\sigma^2}\right)$ and $\lambda_{\mathrm{r,1}}, \lambda_{\mathrm{r,2}}, \ldots, \lambda_{\mathrm{r,M}}$ are the eigenvalues (with $\lambda_{\mathrm{r,1}}, \geq \lambda_{\mathrm{r,2}}, \geq \ldots, \geq \lambda_{\mathrm{r,M}}$) of $\mathbf{R}_{\mathrm{T}}^{1/2}$.

When the antenna correlation is zero, the eigenvalues of $\mathbf{R}_T^{1/2}$ are equal and, therefore, $\beta_i = 0$ and the matrix of the eigenvectors equals the identity matrix. In this case, the precoder becomes $\mathbf{L} = \mathbf{V}_e^H$ and is an orthogonal transformation equivalent to STBC. When the antenna correlation is one, only one eigenvalue of $\mathbf{R}_T^{1/2}$ is nonzero and the precoder is equivalent to a beamformer.

In Figure 6.19, simulation results are presented for a 2 × 1 Universal Mobile Telecommunication System (UMTS) FDD system with spreading factor equal to 128, convolutional coding of rate 1/3 and data rate equal to 12.2 kbps [122]. The channel correlation matrix is assumed perfectly known at the transmitter. Performance results are depicted for Alamouti STBC, beamforming and the proposed reconfigurable design in terms of the required Eb/No for 1 % Frame Error Rate (FER) as a function of antenna correlation. As expected STTD performance degrades when antenna correlation increases, whereas antenna correlation is beneficial for beamforming performance. The proposed scheme performs similarly to STTD for low antenna correlations and becomes equivalent to beamforming for high antenna correlations. For correlation values between the two extremes, the proposed approach outperforms both STTD and beamforming.

The sensitivity of the proposed reconfigurable design to CSI errors at the transmitter was investigated in [122] both analytically and via simulations and its robustness was demonstrated.

The methodology of designing linear precoders presented in this section can be extended to nonorthogonal STBC [123] and can be generalized to any form of matrix modulation, of which STBC is just a special case. Moreover, it can be applied so as to optimize criteria

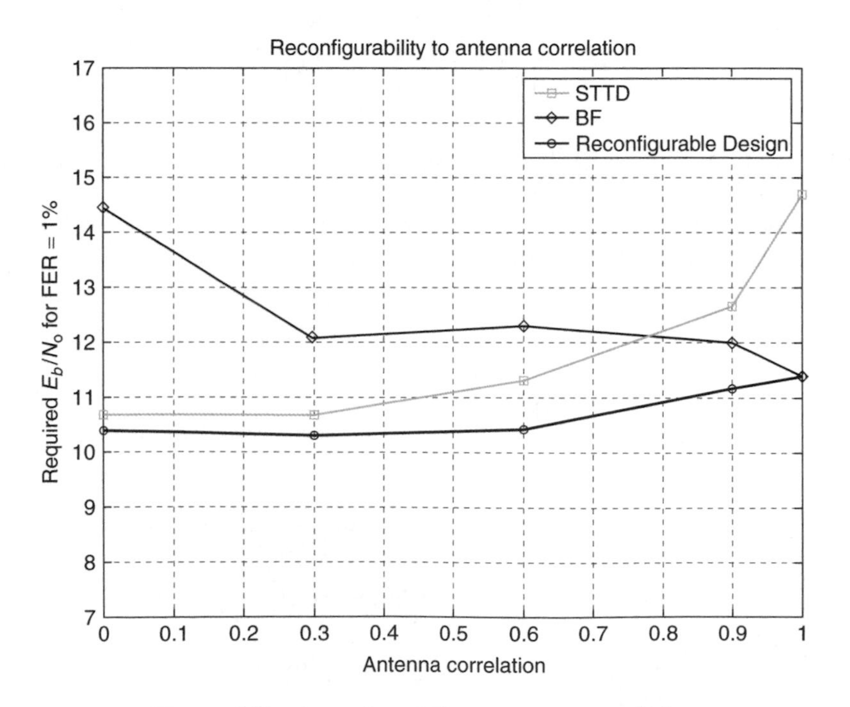

Figure 6.19 Reconfigurability to antenna correlation

other that the average PEP, but it is not always possible to achieve a closed form expression for the linear precoding matrix, and numerical methods need to be implemented.

6.3.4 Multiuser MIMO Downlink Precoding

SDMA promises high gains in the system throughput of wireless multiple antenna systems. If SDMA is used on the downlink of a MU MIMO system, either long-term or short-term CSI has to be available at the BS to facilitate the joint precoding of the signals intended for different users. Precoding is used to efficiently eliminate or suppress MUI via beamforming or by using 'dirty-paper' codes. It also allows us to perform most of the complex processing at the BS, which leads to a simplification of the MTs. In this section, we provide an overview of efficient linear and nonlinear precoding techniques for MU MIMO systems. The performance of these techniques is assessed via simulations on statistical channel models, and on channels generated by the IlmProp, a geometry-based channel model that generates realistic correlations in space, time, and frequency.

6.3.4.1 Linear Precoding Techniques

Block diagonalization (BD) is a linear precoding technique for the downlink of MU MIMO systems [124]. It decomposes a MU MIMO downlink channel into multiple parallel orthogonal single-user MIMO channels. The signal of each user is preprocessed at the transmitter using a modulation matrix that lies in the null space of all other users' channel matrices. Thereby, the MUI in the system is efficiently set to zero. BD is attractive if the users are equipped with more than one antenna. However, the zero MUI constraint can lead to a large capacity loss when the users' subspaces significantly overlap. Another technique also proposed in [124], named *successive optimization* (*SO*), addresses the power minimization and the near – far problems. It can yield better results in some situations but its performance depends on the power allocation and the order in which the users' signals are preprocessed. The zero MUI constraint is relaxed and a certain amount of interference is allowed. Minimum mean-square-error (MMSE) precoding improves the system performance by allowing a certain amount of interference especially for users equipped with a single antenna. However, it suffers a performance loss when it attempts to mitigate the interference between two closely spaced antennas, situation always occurring when the user terminal is equipped with more than one RX antenna. In [125] the authors proposed a new algorithm called *successive minimum mean squared error* (*SMMSE*), which deals with this problem by successively calculating the columns of the precoding matrix for each of the RX antennas separately. The complexity of this algorithm is only slightly higher than the one of BD but it can provide a higher diversity gain and a larger array gain than BD.

6.3.4.2 Nonlinear Precoding Techniques

It is well known that linear equalization suffers from noise enhancement and hence has poor power efficiency in some cases. The same drawback is experienced by linear precoding, which combats noise by boosting the TX power. This disadvantage of linear equalization at the receiver side can be avoided by DFE, which is, for instance, used in Very-High-Data-Rate Blocked Asynchronous Transmission (V-BLAST) [126].

Tomlinson-Harashima precoding (THP) is a nonlinear precoding technique developed for single-input, single-output (SISO) multipath channels. THP can be interpreted as moving the feedback part of the DFE to the transmitter. Recently it has been also applied for the pre-equalization of MUI in MIMO systems [127], where it performs spatial pre-equalization instead of temporal pre-equalization for ISI channels. Thereby, no error propagation occurs. Hence, the precoding can be performed for the interference-free channel. MMSE precoding in combination with THP is proposed in [128]. MMSE balances the MUI in order to reduce the performance loss that occurs with zero interference techniques while THP is used to reduce the MUI and to improve the diversity. SO THP proposed in [129] combines SO and THP in order to reduce the capacity loss due to the cancellation of overlapping subspaces of different users and to eliminate the MUI. After the precoding, the resulting equivalent combined channel matrix of all users is again block diagonal. This also facilitates the definition of a new ordering algorithm. Unlike in [130] and [128], this technique allows more than one antenna at the MTs and has no performance loss due to the cancellation of interference between the signals transmitted to two closely spaced antennas at the same terminal.

6.3.4.3 Scheduling

All precoding techniques suffer from two major drawbacks: They can only spatially multiplex a limited number of users in any given resource slot. Secondly, their performance largely degrades when serving spatially correlated users. In a real system, it is reasonable to assume that there exist a larger number of users than those a BS can simultaneously TX to. Therefore, there is a need for a scheduling algorithm, responsible to decide which users are to be spatially multiplexed at any given time and frequency.

The scheduler should take its decisions avoiding to group spatially correlated users, maximizing the system performance while remaining fair. In fact, for spatially correlated users, the MUI is very large because of the overlapping signal spaces. This, of course, degrades the performance of nonzero MUI techniques. Forcing the interference to zero, as the zero MUI techniques do, does not help much since it would be inevitable to use inefficient modulation matrices. Therefore, this situation should be avoided by the scheduler. Finally, fairness is very important in a real system since it assures that all users are served. Without it, the scheduler would simply pick the users with the best channels and transmit only to those. These prerequisites render the design of an efficient scheduler a challenging task. Examples of spatial correlation aware schedulers are the ones presented in [131, 132].

6.3.4.4 Simulation Environment

In order to assess the robustness of the algorithms in a realistic scenario we propose simulations results obtained with the help of the IlmProp channel model [133]. The IlmProp, includes a geometrical representation of the environment surrounding the experiment, as depicted in Figure 6.20. Three users employing a 2-element Uniform Linear Array (ULA) move at approximately 50 km/h around a BS mounting a 4-element ULA. All arrays are characterized by $\frac{\lambda}{2}$ SVD antennas, operating in the 5-GHz band. The mobiles had a Line-Of-Sight (LOS) component and several scattering clusters around them that

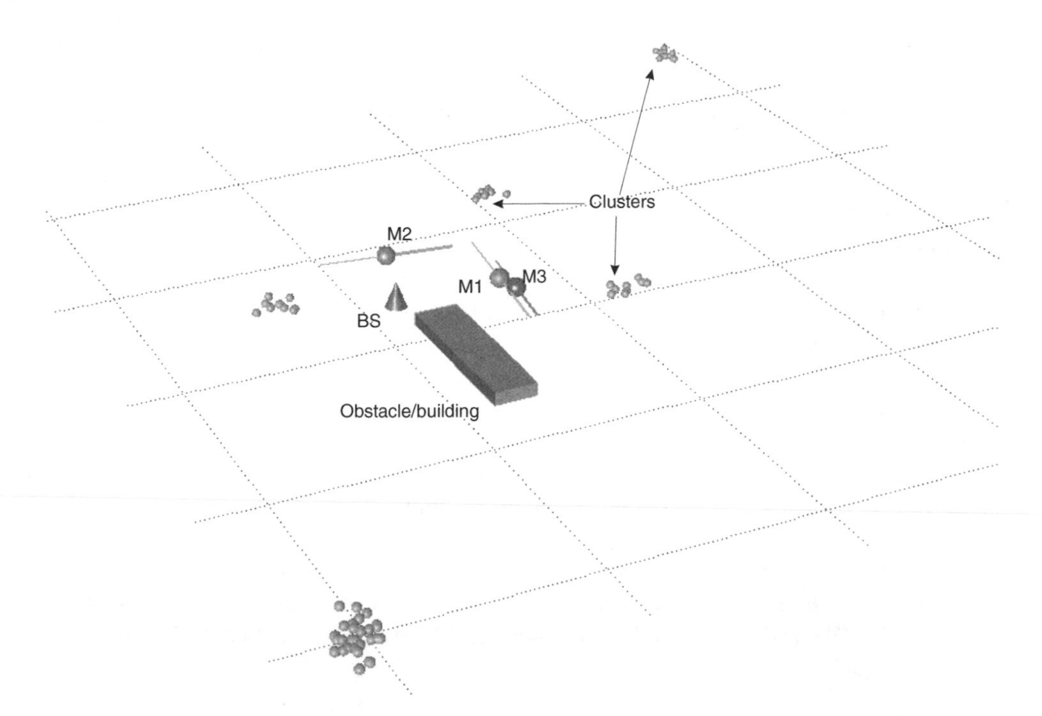

Figure 6.20 IlmProp channel model. Three mobiles (M1, M2, and M3) move on linear trajectories around a BS. Scattering clusters provide a delay-spread of 1 μs

provide delay spreads up to 1 μs. The channels have been computed for 64 subcarriers, separated by 150 kHz, sampled in time with a sampling interval equal to 70 μs. Users 1 and 3 move on a similar trajectory, remaining spatially correlated with one another for the total time of the simulation, while user 2 moves on a different path. The building depicted in Figure 6.20 obstructs the paths toward one of the far cluster for users 1 and 2 only. In addition to the LOS component and the paths originating from the clusters, the channels have been enriched with a Dense Multipath Component (DMC) [134]. The latter is modeled as a complex white noise whose power-delay profile exhibits the trend of decaying exponentials as described in [134]. The DMC represents approximately 20 % of the total channel's power.

6.3.4.5 Simulation Results

In this section, we will compare the performance of BD, SO THP, and SMMSE precoding. To do so, we take into account a purely stochastic channel HW and two channels $H^{I^{(1)}}$ and $H^{I^{(2)}}$, partly deterministic, generated by the IlmProp, a geometry-based channel model for MU time variant, and frequency-selective MIMO systems [133, 135]. The channel HW is assumed to be frequency-selective with the power-delay profile as defined by IEEE802.11n - D with Nonline of Sight (NLOS) conditions. The IlmProp channels $H^{I^{(1)}}$ and $H^{I^{(2)}}$ represent two typical outcomes of a scheduler that is not spatial correlation aware, in the first case, and one where the users' spatial correlation has been considered, for the second channel. We consider a MU MIMO downlink channel where *MT* TX antennas are located at the BS, and *MR*i RX antennas are located at the i-th MS. We will use the notation $\{MR1, \ldots, MRK\} \times MT$ to describe the antenna configuration of the system. In order to take into account channel estimation errors, we use a 'nominal-plus-perturbation' model. The estimated combined channel matrix can be represented as, $\hat{\boldsymbol{H}} = \boldsymbol{H} + \boldsymbol{E}$, where H denotes the flat-fading combined channel matrix of all users, and E is a complex random Gaussian matrix distributed according to $CN(0_{MR \times MT}, M_R^{\sigma_e^2} I_{MT})$. In case of OFDM, we model the channel estimation error on each SC in this way. The real advantage of SMMSE can be seen from the BER performance shown in Figure 6.21. By introducing MUI, SMMSE provides a higher diversity and a higher array gain than BD. SO THP does not have the same diversity gain as BD, that can be explained with the influence of the modulo operator used in THP. Next, we will investigate the influence of scheduling on the performance of the MU MIMO precoding techniques using the IlmProp channel model. In Figure 6.22, we show the 10 % outage capacity as a function of the ratio of total TX power P_T and additive white Gaussian noise at the input of every antenna σ_n^2. In the next Figure 6.23, we investigate the influence of scheduling on the BER performance of BD and SMMSE. As it can be seen in these two figures, the scheduling has a great impact on the capacity of the system. The capacity loss is about one bps/Hz and the SNR loss is about 2 dB. We also see that SMMSE clearly outperforms BD.

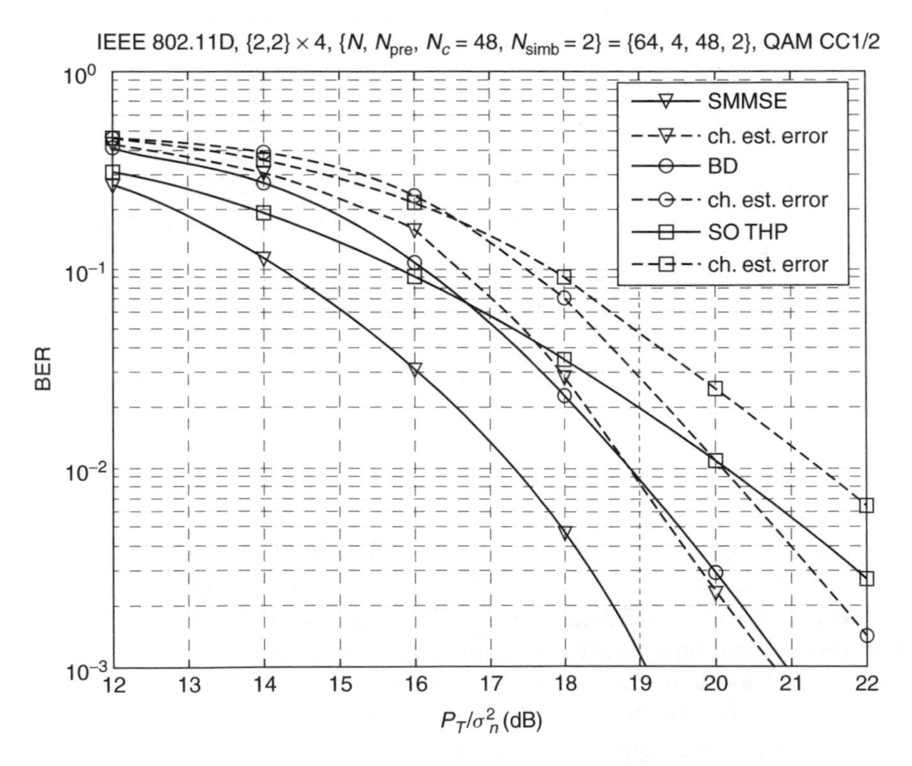

Figure 6.21 BER performance comparison of BD, SO THP and SMMSE in configuration $\{2, 2\} \times 4$

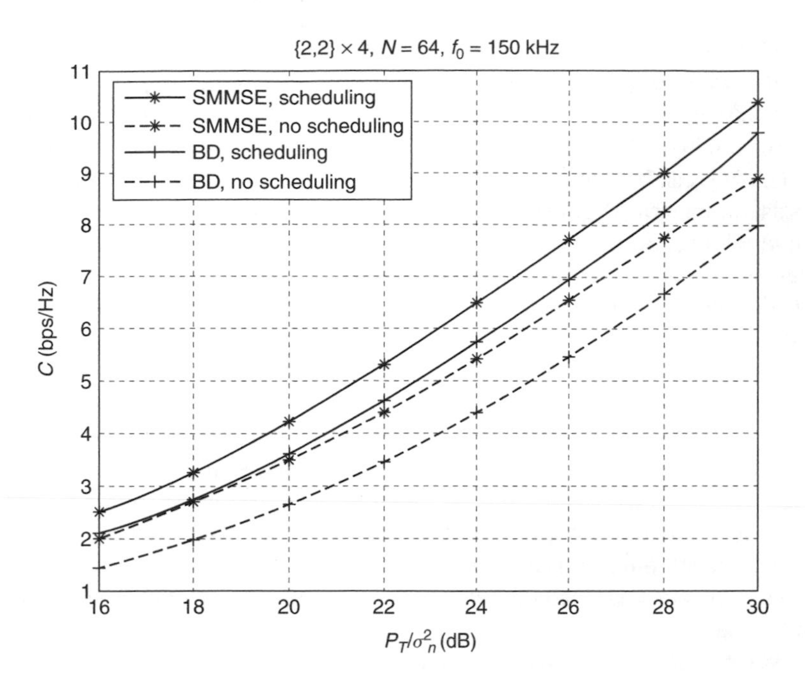

Figure 6.22 10 % outage capacity performance of BD and SMMSE, with and without scheduling

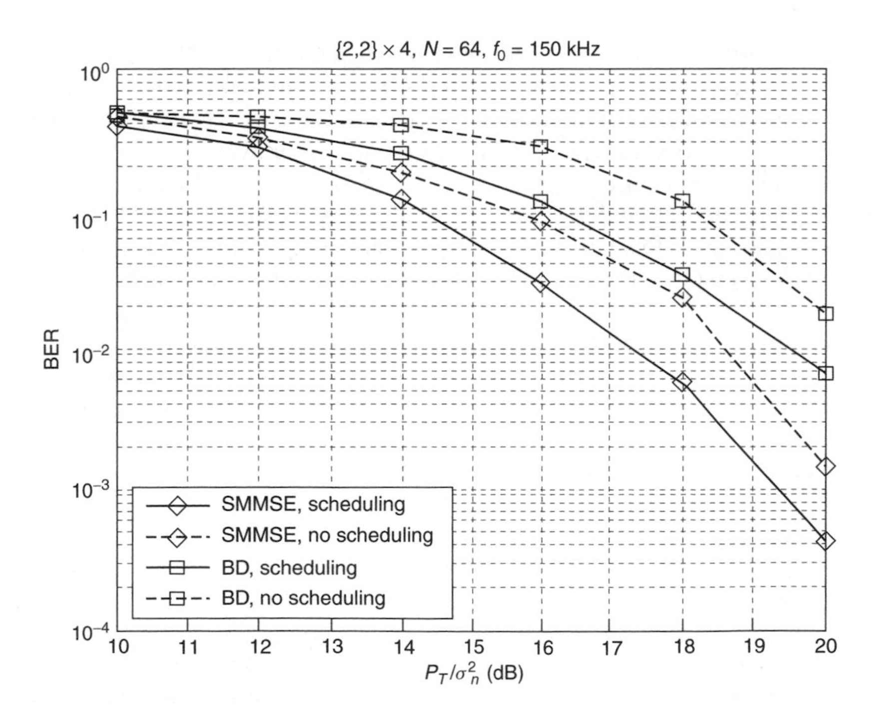

Figure 6.23 BER performance comparison of BD and SMMSE in configuration $\{2, 2\} \times 4$, with and without scheduling

6.3.5 Smart Antenna Cross-layer Optimization

The overall system performance can be enhanced further by finding ways to account for the upper layers in the Open Systems Interconnection (OSI) stack. For instance, resource management functions like scheduling, admission control, and Hybrid Automatic-Repeat-Request (HARQ) can benefit from the array-processing algorithms at the physical layer. In other words, an integrated system design approach must be employed if one is to achieve the best overall level of system performance. This means that the antenna array-processing techniques are developed in close collaboration with designs at the physical, link (MAC, DLC, scheduling, etc.) and network layers (radio resource management, routing, transport, etc.), that is, in a cross-layer fashion rather than attempting to optimize the designs in isolation of one another. In subsequent paragraphs, several important issues to be addressed when dealing with cross-layer designs will be further detailed.

As for the information to be exchanged among those functionalities residing in different OSI layers, a three-fold classification can be established as follows:

- CSI, that is, estimates for CIR, location information, vehicle speed, signal strength, interference level, interference modeling, and so on;
- QoS-related parameters, including delay, throughput, BER, PER measurements, and so on; and
- PHY-layer resources made available in the corresponding node, such as MU reception capabilities, spatial processing schemes, number of antenna elements, battery depletion level, and so on.

It is also important to carefully consider the optimization criteria. When applying MIMO techniques to wireless packet-switched networks, it is common practice to select transmission parameters with the goal of maximizing channel capacity. However, specific coding schemes being used, MAC strategies, scheduling policies, or even the performance of protocols in the upper layers of the protocol stack (such as Transport Control Protocol (TCP)) also have a major impact on link quality. Thus, all these constituent blocks should alternatively be designed so as to attain the highest possible throughput (i.e. maximum number of data packets successfully delivered) instead of the highest data rate (capacity-maximizing approaches).

For delay-tolerant services, the combination of MIMO architectures, such as V-BLAST, with HARQ schemes (Go-Back N, Selective Repeat, Stop&Wait) turn out to be a fertilized field for research. Cross-layer designs encompassing HARQ methods (i.e. combinations of Frequency Estimation for Channel (FEC) with ARQ strategies) have also attracted considerable attention, in particular, type-II HARQ methods that address soft joint decoding of successive packet retransmissions. By undertaking cross-layer designs, novel strategies for mapping TX data on specific streams, alternative methods to adjust frame size (as is the case of the HSDPA (High-Speed Downlink Packet Access) standard), or new antenna selection methods either on the TX or RX side can be derived. Apart from that, the identification of methods aimed at dynamically adjusting rates and powers on TX antennas is also needed.

In a MU context, the so-called opportunistic approaches have recently attracted considerable attention. In short, this is based on the idea of multiplexing users by granting the channel to those with higher chances of completing a successful transmission. In the end, overall (rather than individual link) throughput maximization is pursued. For highly correlated space channels, opportunistic beamforming approaches will decide pointing at the user with the highest SNIR out of those present in the system. On the contrary, in rich scattering scenarios where space channels exhibit low correlation (e.g. indoor), opportunistic approaches will implicitly exploit fading by granting access to those with highest instantaneous capacity (while possibly maintaining fairness among users, i.e. not causing excessive delay). This clearly departs from traditional strategies aimed at stabilizing individual links against channel (or interference) fluctuations by using multiple antennas. Moreover, artificial fading has to be induced for slowly time-varying channels on the transmit side by randomly changing transmit weights. More sophisticated multiple-access can be derived by combining space and code diversities (e.g. V-BLAST with DS-SS). In that case, additional issues like efficient user grouping arise.

In summary, the approaches mentioned in the last paragraph go beyond the link-level and jointly exploit MU diversity as a complement to code, time, frequency or space diversity. This clearly has an impact on the design of MAC protocols, which are forced to abandon the collision-avoiding paradigm (CSMA, Aloha, etc.) and evolve toward MU MAC schemes. MU capabilities of the system will clearly depend on specific physical-layer technique (optimum combining, TX beamforming, SVD-based, etc.) this, again, reinforcing the need for cross-layer designs.

In previous paragraphs, it has been shown how a tighter interaction of smart antenna techniques with upper layer functionalities can bring remarkable technical benefits. Furthermore, this could also entail further adoption of array-processing schemes in future wireless communication systems. That is, a layer-isolated approach where no cross-layer interaction is envisaged could possibly prevent most of the physical-layer innovations in the array-processing field from being extensively recognized at the standardization level.

6.3.6 Realistic Performance Evaluation

The adoption of Smart Antennas in future wireless systems relies on two main investigation approaches to be performed: (I) consideration of the Smart Antennas 'features' early in the design phase of future systems (top-down compatibility) and (II) realistic performance evaluation of Smart Antenna techniques according to the critical parameters associated with the future systems requirements (bottom-up feasibility). The latest trends that fall in the former approach have been discussed in the previous paragraphs of this section. The latter investigation approach consists in performance evaluation based on suitable simulation methodology with accurate modeling.

Realistic performance evaluation [136], both at the link and the system-level, of the techniques described in the beginning of Section 6.3 will be based on measurement data–based simulations as well as model-based simulations in different deployment scenarios (macro-cellular, micro-cellular, in-building, into building, ad hoc, etc.).

The measurement data–based performance evaluations should rely on link-level simulations and system-level simulations. Both are off-line simulations using multidimensional channel sounding field measurement data, but their objectives are different [137, 138]:

a. Link-level simulations provide an assessment of the implementation losses, and evaluate the ST-equalizer link-performance in terms of BER in various propagation environments. They focus on analyzing/evaluating the impact of ST-processing and/or coding configurations, parameters, and algorithms on performance.

b. System-level simulations provide an assessment of the overall multicell system performance, and evaluate performance metrics such as overall capacity, throughput, SINR and BER in terms of either geographical distributions in the area of interest or outage probability [139]. The main focus is on interference scenarios, which depend on cell design including frequency reuse and user distribution.

The efficiency of Smart Antenna techniques depends on the characteristics of a highly variable environment in terms of propagation, antenna array configuration, user behavior, terminal resources, minimum QoS demands, reliability of CSI, traffic patterns, service profiles, and interference scenarios. It is therefore critical to develop realistic spatio-temporal models for the characterization of the MIMO channel in a highly variable environment [140, 141], and of representative interference models for adequate evaluation of baseband signal processing techniques in multiple antenna, MU, multiservice, multi-technology radio networks. Furthermore, it is necessary to extend the operation of these models to enable their use in simulating multitechnology radio environments including ad hoc networks, such that it is possible to utilize differing models from different systems in a consistent manner. Of particular importance are any time-correlated effects (e.g. shadowing), which need to be applied across the models in a coherent manner. Since frequency diversity is the traditional tool to combat these phenomena, it is necessary to develop hybrid technology processing schemes where spatial diversity processing is combined with frequency, time, and code.

MIMO channel modeling is required in order to sufficiently assess the performance of Smart Antenna techniques at a link-level, but QoS evaluation in a multiuser, multiservice and multitechnology environment requires adequate modeling of the interference as well. An appropriate characterization of the spatio-temporal properties of these issues is important, not only to obtain realistic simulation results but also to establish a theoretical framework describing the link-level behavior from an analytical perspective. In this sense, new mathematical tools such as random matrix theory could be the key to a wider understanding of the MIMO channel characteristics [142, 143]. These novel techniques provide a very simple way of analyzing and studying the performance of MIMO channels with fading correlation at the transmitter or the receiver. Despite the fact that these models are asymptotic by nature, a good agreement with the actual (nonasymptotic) channel response is observed in practice.

As for the characterization of the global performance of the communications link, system-level simulators have usually been employed. These simulators generate certain traffic patterns within a network of cells and statistically measure interference in terms of Signal-to-Interference Ratio (SIR).

The realistic performance evaluation of Smart Antenna techniques will therefore rely on the following modeling issues [144–146]:

- MIMO channel modeling based on statistical formulation of the scatterers' distribution and real-time MIMO measurement campaigns followed by link-level simulations for a variety of antenna array configurations, air-interfaces, user mobility patterns, and service profiles;
- realistic interference modeling based on system-level simulations taking under consideration the intra- and inter-cell impact of smart antenna techniques, nonuniformity of

traffic (e.g. hot spots), mixed services scenarios, and interoperability between different air-interfaces [116];
- accurate mapping between link and system-level results based on more realistic interference models [113, 144, 147, 148];
- realistic modeling of implementation losses such as channel estimation errors, feedback quantization and delay, and so on.

6.3.7 Deployment of Smart Antennas in Future Systems – Implementation Issues

According to recent studies, Smart Antenna technology is now deployed in one of every 10 BSs in the world. The study predicts that the deployment of Smart Antenna systems will grow by a factor of 2 in the next five years. It was shown in the same study that Smart Antenna technology has been successfully implemented for as little as 30 % more cost than similar BS without the technology. However, implementation costs can vary considerably and still cost-effective implementation is the major challenge in the field.

The application of sophisticated smart antenna processing techniques in future systems will impose stringent demands upon both BS and terminal implementations, and this increased functionality will also be expected to be achieved at low cost. It is therefore important that the viability of the proposed smart antenna processing techniques is addressed, and novel technologies applied wherever appropriate.

At the BS of particular importance is the development of improved antenna structures (possibly employing Micro Electro Mechanical Systems (MEMS) technology, for example, micro-switches, or left-handed materials), improved cabling structures, and efficient low-cost RF/DSP architectures.

At the terminal the application of Smart Antenna techniques can have a significant impact, not only in terms of system performance but also in terms of cost and the terminal's physical size. It is therefore important to examine the viability of such terminals, which are likely to be both multimode and multiband in nature, and available in a wide variety of forms. Particular areas that need to be examined are efficient MIMO/diversity antenna system designs, small low power RF structures, use of RF combining techniques [149], and viable low power digital signal processing (DSP) implementations. Terminal cost requirements may lead to the use of nonperfect RF/analogue components and DSP algorithms for compensating this should be investigated.

In addition to this, for both the BS and the terminal, an important consideration is how it may be possible to effectively develop the antenna structures, RF architectures, and DSP implementations to enable them to operate efficiently within a wide variety of air-interface scenarios, both separately and in parallel. To this end, innovative development flow methodologies jointly covering the RF and the baseband parts of complex wireless System on Chip (SoC) should be studied. This methodology effort will allow a short time development of integrated smart antenna systems starting from high-level algorithm descriptions and going up to efficient implementation optimized for the scenarios through reusability of IP blocks. This should allow taking advantage of the high integration level of latest digital technology to improve 4G terminals in terms of complexity, cost and power consumption, thus allowing a mass introduction of new smart devices. In general, the implementation feasibility of different concepts and algorithms should be evaluated, considering analog and digital component technology that evolves with time and finally leads to a 'feasibility roadmap', which indicates the viability of different concepts over time.

A key output of this area of study is an understanding of the base technologies that are required to make the future use of Smart Antennas viable. The financial impact of the deployment of Smart Antenna technologies in future wireless systems has been studied in [150] for the cases of CDMA2000 in the United States and UMTS in Europe. The results of this study showed that 'smart antenna techniques are key in securing the financial viability of the operator's business, while at the same time allowing for unit price elasticity and positive net present value. They are hence crucial for operators that want to create demand for high data usage and/or gain high market share'. On the basis of this type of analysis, technology roadmaps along with their associated risks can be concluded, which will enable appropriate technology intercept points to be determined, resulting in the development of technologies appropriate for each application area.

6.3.8 Summary

The adoption of Smart Antenna and MIMO technologies in future wireless systems will have a critical impact on the achievement of higher data rates in a number of challenging scenarios (varying propagation conditions, multiservices, multitechnology networks) and it is expected to play a significant role in the realization of end-to-end reconfigurability and transparent operation across networks of different technologies. The array-processing techniques have to be taken into account in the original system design to provide the basis for scalable, high data rate as well as high capacity system solutions by achieving an optimum combination of array gain, diversity gain, interference reduction gain, and spatial multiplexing gain on the RX as well as the transmit side.

Multiuser MIMO systems that enable SDMA are particularly challenging. Here, we have investigated several linear and nonlinear precoding techniques suitable for the downlink of multiuser multiple-input multiple-output (MU MIMO) systems using the IlmProp channel model with realistic correlations in space, time, and frequency. These precoding techniques represent good candidates for a practical implementation because of the good compromise between performance and complexity. Nonlinear techniques like SO THP can theoretically achieve a high capacity. SO THP is especially attractive for users with multiple antennas since, it does not experience any capacity loss due to the cancellation of interference between closely spaced antennas at the same MT. SMMSE also addresses this problem. This technique is a modification of MMSE precoding, which optimizes the performance of users equipped with multiple antennas. SMMSE provides the best BER performance regardless of the number of RX antennas per user. The complexity of this technique is also reasonable, which makes it a potential candidate for implementation in future wireless systems. It outperforms other linear techniques (such as BD) and even nonlinear techniques like SO THP. Furthermore, the total number of RX antennas at MTs can be greater than the number of TX antennas at the BS. Moreover, it relies on linear processing and it is less sensitive to channel estimation errors than nonlinear techniques.

6.4 Duplexing, Resource Allocation and Inter-cell Coordination

Coexistence of different access technologies, hierarchical cellular deployment, a wide variety of data services, requirements for transparent operation across different technologies, adaptivity to varying network conditions and mobility, and QoS constraints introduce a number of challenges in the design of future-generation systems and the specification of new air-interfaces, such as efficiency and flexibility in the utilization of spectrum, dynamic resource allocation and exploitation of the MU diversity, and reconfigurable interference management and inter-cell coordination.

In this section, three critical issues for the design of next generation systems are addressed: (i) duplexing, (ii) scheduling and resource allocation, and (iii) interference and inter-cell coordination. A number of research directions are presented, which consist promising potential candidates for next generation systems specification.

Traditionally, the spectrum allocation process, which determines the scope of the duplexing mechanism and the selection of the appropriate air-interface technology are sequential actions. However, the spectrum allocation in turn impacts the duplexing choice, which significantly impacts the air-interface design and places limitations on the choice of both the air-interface features as well as the access control and resource management mechanisms to deliver optimum system performance.

This section first addresses the issue of duplexing and the related impact of duplexing on the services to be offered and the support of the air-interface features to maximize the performance. In Section 6.4.3, the design strategies that the system relies on to deliver high throughput with low outage are addressed from the perspective of distributed single cell solutions. Then, in Section 6.4.4, interference management schemes relying on coordination through explicit information exchange requiring fast inter-BS signaling or through some type of implicit messaging, are discussed from the perspective of a multicell environment. Finally, Section 6.4.5 contains conclusions and a summary of recommendations for the design of next generation systems.

6.4.1 Duplexing

Traditionally, the decision on how to partition the degrees of freedom available for communication between uplink and downlink has boiled down to two clear possibilities: either time or frequency separation. Each contending option has accepted advantages as well as clear drawbacks. Moreover, an implicit understanding developed over the years that TDD was attractive in low mobility micro-cellular scenarios while FDD was preferred in high mobility wide-area deployments.

In this section the duplexing issue is considered in the context of a new air-interface design, in an effort to go beyond the traditional paradigm where only pure FDD and TDD are evaluated, bringing other options into consideration.

6.4.1.1 Paired versus Unpaired Spectrum

The allocation of spectrum is a highly political issue, performed by certain national and international agencies and mostly beyond the control of equipment manufacturers. However, the allocation of spectrum strongly conditions the duplexing choice because, if unpaired spectrum is allocated, FDD cannot be used. Although unpaired spectrum is indeed easier to find, the historical tendency has been nonetheless to assign paired spectrum for wide-area systems. In paired spectrum, both FDD and TDD options are possible.

6.4.1.2 Link Asymmetry

An immediate advantage of TDD is that it enables an asymmetric allocation of degrees of freedom between uplink and downlink although in general not dynamically on a cell-by-cell basis, but rather on a system-wide basis. A cell-by-cell allocation is not an attractive option, as different uplink and downlink patterns in adjacent cells may result in catastrophic mobile-to-mobile and base-to-base interference, in addition to the common mobile-to-base and base-to-mobile interference encountered in FDD. A similar asymmetry in FDD would require uneven spectrum blocks, highly unlikely and very rigid. The relevance of link asymmetry, nonetheless, is still unclear. The UMTS forum forecasts a 2.3 ratio of downlink over UL for 2010 but this will depend largely on emerging and yet-to-be-envisioned applications. In that sense, TDD may be considered as the favored option but not decisively so because of the difficulty in predicting whether the asymmetry ratio will be significantly different from unity.

One of the main drawbacks of TDD systems, that may degrade the advantage of the inherent flexibility to cope with time-varying asymmetric traffic patterns, is, however, the possible interference between uplink and downlink. Since both uplink and downlink share the same frequency in TDD, the signals of the two transmission directions can interfere with each other. This kind of interference occurs if the BSs are not synchronized, but even with synchronized cells the interference can be present if the division between uplink and downlink is different in adjacent cells. The uplink and downlink signals on adjacent frequencies can interfere with each other and, therefore, the interference between uplink and downlink signals can take place also between two operators.

In order to eliminate this interference, it is usually considered that the whole system or at least the cells belonging to the same radio network controller use the same switching point between UL and DL. The switching point will be eventually changed in the long-term according to the traffic variations. Such an approach leads to inefficiencies if there is a wide variability of traffic patterns between neighboring cells, and proposals have been made where in some situations adjacent cells use transmit opposite slots or even randomization [151]. This is a topic that requires further investigation, and basically involves the following issues:

- system synchronization, which may not be the most critical owing to the availability of moderate cost given paired spectrum (GPS) solutions;
- devising efficient and complex radio resource allocation algorithms.

6.4.1.3 Link Reciprocity

Link reciprocity is usually regarded as the most attractive feature of TDD, which naturally enables it, at least for low-to-moderate normalized Doppler spreads. As a result of reciprocity, sophisticated TX processing schemes that necessitate instantaneous channel information become feasible [152]. The lack of reciprocity in FDD, in turn, makes these schemes dependant on the relay of CSI through feedback, which tends to incur unacceptable delays if conventional transmission techniques are employed. It was recently shown for flat-fading, that CSI transfer for FDD is actually feasible: the time occupied is typically a factor of 2 or 3 compared to the time required for TDD CSI transfer (e.g. simple use of up-pilots).

Two questions arise concerning the link reciprocity issue: (i) whether the majority of users fall within the Doppler range where reciprocity is sufficiently exact and if yes, (ii) what is, quantitatively, the value of channel information availability at the transmitter. In a cell operating individually, the broadcast advantage is sizeable if the number of

BS antennas is sufficient. In terms of cooperative network operation, the benefits need to be evaluated.

6.4.1.4 Link Budget

Let us consider an FDD system that is radiating a steady power level P. Consider now TDD. If the power level during the active part of the duplex is kept at P (same amplifiers used), there is a 3 dB increase in the thermal noise floor because of the doubling in bandwidth. With a pathloss exponent of 3.8, for instance, this results in a 17 % reduction in range. In exchange, the 'average' TX power is halved, which has no impact on the BS but would extend the battery life in a MT. If, in contrast, we desire to keep the range unchanged with respect to the FDD case, the power radiated during the active part must be doubled. This would require bigger amplifiers and it would also result in the same 'average' TX power and thus the same battery life.

6.4.1.5 Synchronicity and Guards

In FDD, uplink and downlink are orthogonal in frequency provided there is sufficient separation between the corresponding blocks. Each side of the link requires guard bands to accommodate filter roll-offs (Figure 6.24).

In TDD, temporal orthogonality is only possible if cells have synchronized uplink and downlink switch patterns plus guard times to account for propagation delays. This is in addition to guard bands roughly equal to those in FDD (Figure 6.25).

Orthogonality is essential in wide-area systems for otherwise catastrophic interference may take place (If the system has a micro-cell component with street-level or indoor bases, these may not need to be synchronous.).

The overhead represented by the guard times can be made as small as desired by extending the duplex time (time that either link is active). Guard times are an issue only because we desire to make the duplex time short in order to minimize physical-layer

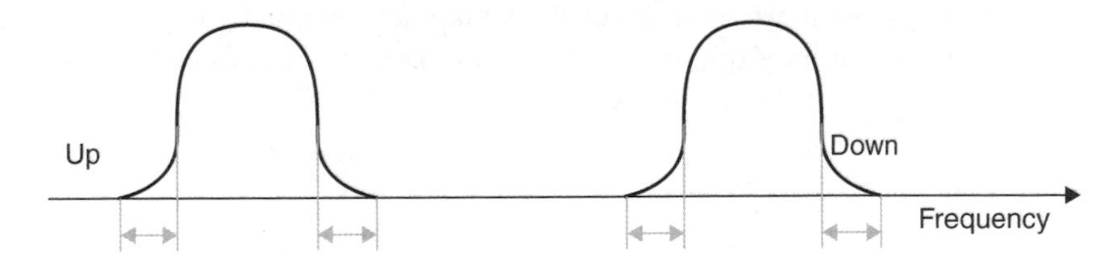

Figure 6.24 Guard bands in FDD

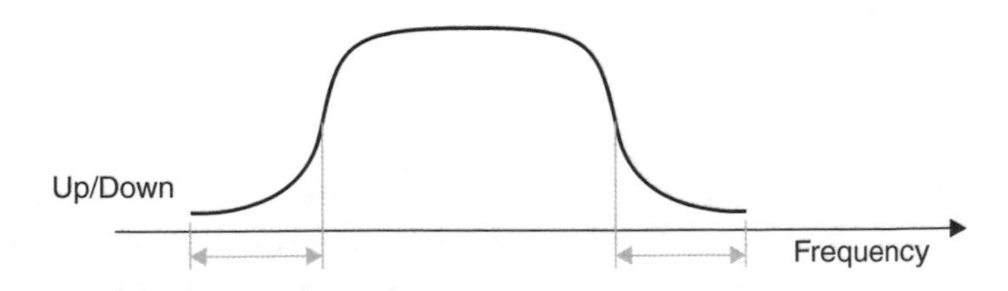

Figure 6.25 Guard bands in TDD

contribution to latency and also to ensure channel reciprocity over the widest possible range of Doppler spreads.

Synchronicity is essential in TDD. This is an inconvenience with respect to FDD but not insurmountable given the availability of low-cost GPS technology. A guard time evaluation needs to be performed in order to provide quantitative analysis.

6.4.1.6 Link Continuity

A drawback of TDD comes from the periodic interruptions in the links, which are active only for part of the time (usually but not necessarily 50 %). Interestingly, this issue did not exist in circuit-switched voice systems and thus has not been part of the traditional discussions on FDD versus TDD duplexing. This is a new issue that is caused by, and central to, packet-switched data traffic (This also includes packet-switched voice, which may be a replacement for circuit-switched voice.).

Besides higher bandwidth efficiencies, one of the central goals in the design of future-generation systems is to achieve an order-of-magnitude reduction in latency. This is being recognized as a necessary condition for the support of certain envisioned applications (such as gaming). With discontinuous links, no message – not even a 1-bit acknowledgement – can be relayed back with a delay inferior to the duplex time. This implies that the time taken by a basic roundtrip at the physical-layer level cannot go below a few msec and thus the aggregate delay experienced by a packet running through a scheduler and subject to ARQ can easily be on the order of 10 msec. This latency propagates through the protocol stack posing serious problems to the upper layers and causing bottlenecks. As a result, some of the throughput improvements enabled at the physical layer may not be realized, an issue that becomes increasingly important as data rates grow.

In summary, the primary issues on which the choice of a duplexing scheme rests appear to be link reciprocity and link continuity, each of which favors a different choice. Similarly, the remaining issues (synchronicity, link budget, symmetry, etc.) do not point to a clear preferred choice either. In light of these facts, the question that naturally arises is whether it is possible to combine FDD and TDD in such a way that the best of each is preserved. Along these lines, two novel duplexing schemes will be presented in the sections to follow, after a brief discussion of the complexity issues associated with TDD and FDD presented in the subsequent text.

6.4.1.7 Terminal Complexity Issues

The main goal of a terminal implementation in future wireless systems is low complexity with very low power consumption at the lowest possible costs. The basic operation in the terminal needs to be split into the baseband part and the RF part.

Considering the baseband part and from the algorithm point of view the difference between an FDD and a TDD system is very small. Most of the signal processing algorithms envisioned for future wireless systems are not dependant on a specific duplex mode and identical algorithms can be used in FDD or TDD systems.

The main differences are introduced in the architecture and the implementation level of the SoC. Here the burst characteristic of the TDD mode can lead to a simplification in the memory management and the scheduling processes of the different building blocks. Memory blocks can be reused for transmit and receive. By having the possibility to switch on and off the different processing blocks, the overall power consumption can be optimized. The processing complexity can easily be adapted to the needed data rate, depending on the number of used slots to be processed.

With reference to the RF building blocks, the TDD mode has some advantage. In FDD, the TX and RX processes are coincident in time, which requires a coupled pair of RF filters, known together as a duplexer, to prevent the co-located transmitter from degrading receiver performance. Because of the high Q requirements, a duplexer may be a moderately expensive component.

As in TDD, the TX and RX processes are separated in time, the overall integration of the RX and the TX modules on one chip is simpler to realize. No crosstalk between TX and RX need to be considered thus leading to relaxed constrains on the signal isolation between the TX and the RX chain. At the antenna, a simple two-way switch replaces the duplexing unit to separate the RX and the TX signals. The absence of the leakage problem may also result in less stringent linearity requirements for a TDD transmitter when compared to an FDD one. In this latter case, inter-modulation products may fall within the RX band, which will obviously not be removed by the duplexer.

6.4.1.8 Band Switching Duplexing

Band switching duplexing has been proposed in [153] and can be described as follows. GPS blocks, instead of reserving a block for uplink and the other for downlink, alternate their use every T sec, as depicted in Figure 6.26. With this scheme reciprocity is achieved and the channel can be estimated in each band when it is used for UL and then exploited when it is used for downlink. Synchronicity and guard times are still needed, as in TDD. Both links are always active (except on guard times).

Note that Band Switching is both TDD and FDD. It is TDD because every unit of bandwidth is used, alternatively, half of the time for uplink and half of the time for downlink; it is FDD because, at every point in time, half the spectrum is used for uplink and half for downlink.

Implementation of this duplexing scheme may introduce some challenges that need to be assessed. Nevertheless, it provides the best set of trade-offs and at the same time the TDD alternative remains an option in case unpaired spectrum is allocated, as many parameters, such as guard times, synchronicity, and so on are reusable. If link reciprocity fails to provide the expected gains, Band Switching can be easily reduced to conventional FDD.

6.4.1.9 Hybrid Division Duplexing (HDD)

The time division duplex (TDD) scheme is more suitable for data services in a short-range wireless communications system such as wireless LAN. The FDD scheme, on

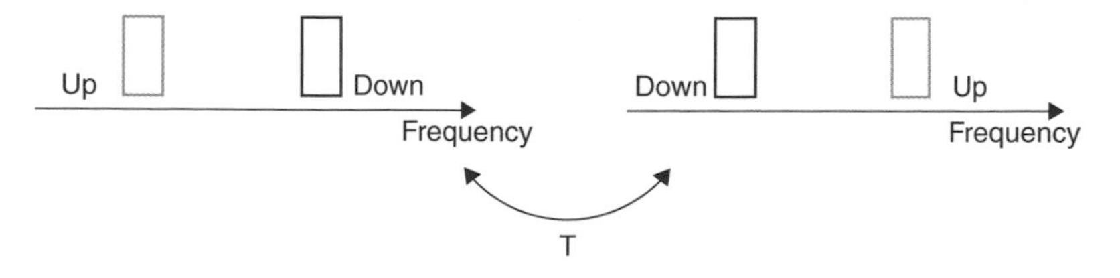

Figure 6.26 Band switching duplexing

the other hand, is more suitable for voice services in a wide-area mobile communications environment. However, future mobile communications systems will be required to provide multimedia service with both voice and data in a wide coverage area. Hybrid Division Duplexing (HDD) system has been proposed in [154] that aims to combine advantages of both TDD and FDD schemes and to increase the flexibility and efficiency of mobile communication systems. The HDD can be described as follows.

The frequency plan and the cell structure of HDD systems are depicted in Figure 6.27 and Figure 6.28, respectively. The HDD system has two frequency bands; one frequency band is for both uplink and downlink communications, which is identical to TDD mode, and the other band is devoted to UL communication only. We denote the two frequency bands as the TDD band and the FDD UL band, respectively. Note that the downlink communication is available only in the TDD DL, while UL traffic can be sent in either TDD UL or FDD UL bands. In HDD systems, the cell area and users are divided into two zones or virtual groups. Conceptually, terminals at the inner zone send UL signal in TDD UL, while those at the outer zone send signal in FDD UL. As a result, a terminal at the inner zone communicates in TDD mode, while a terminal in the outer zone communicates in FDD mode. As the terminal moves inwards or outwards within a cell, it needs to change from FDD mode to TDD mode or vice versa. This requires the BS to keep track of the location of each terminal. If location information is not available at BS, other information such as received power or received SNR at terminals can be used instead.

Terminals crossing the cell boundary are operating in FDD mode, so conventional handover schemes for FDD mode can be readily used. Once the handover is completed and the terminal is within a cell, the BS has a complete control of the terminal. So, the BS can direct the terminal to send UL data in either TDD UL or FDD UL based on the location. This is simply a matter of resource allocation of time and frequency for the BS, and is different from the handover situation in which two different BSs are involved.

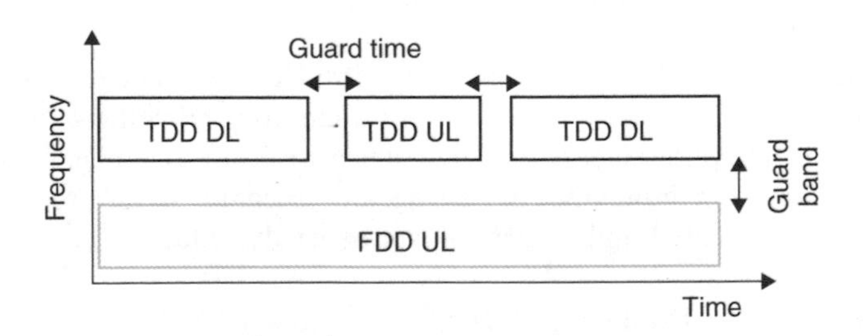

Figure 6.27 Frequency planning

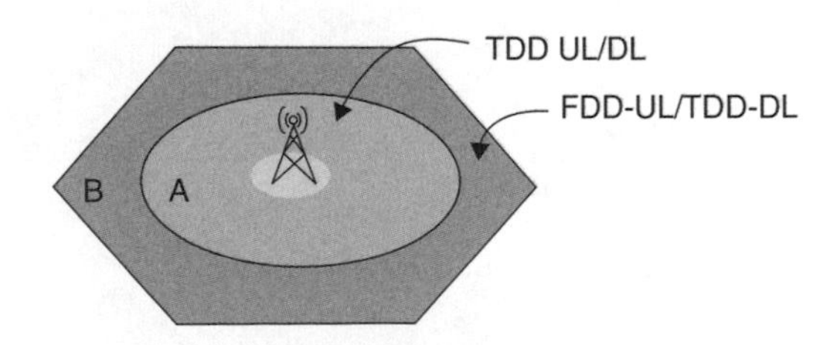

Figure 6.28 Cell structure of HDD systems

However, this does mean a change in resource allocation for the terminal, and can cause a signaling overhead. Thus, when a terminal is moving at a very high speed, it can be more beneficial to operate in FDD mode only throughout the entire cell area in order to reduce signaling overhead.

In summary, the HDD combines the advantages of TDD mode and FDD mode. Therefore, the HDD-based system could be a solution for providing a high data rate and asymmetric service with a nomadic user in a TDD micro-cell, and a reliable service to a high-speed mobile user in the FDD micro-cell.

6.4.2 Scheduling and Resource Allocation within a Cell

The efficient utilization of spectrum requires dynamic resource allocation strategies with the flexibility to adapt to varying wireless network conditions, user requirements and QoS constraints, and will be one of the major criteria for the design of a new air-interface. In this section, several promising candidates for efficient resource allocation, resource usage optimization, and scheduling are discussed.

6.4.2.1 Resource Allocation within the Context of an OFDM Air-interface

The benefit of scheduling multiple users in a single OFDM symbol by assigning different tones to different users with optimal power allocation across the tones has been studied in [155]. This scheme requires feedback on the signal-to-interference on a per sub-channel (a group of tones that have nearly similar channel frequency response) basis. The study showed that the benefit for delay-tolerant services is about 50 to 100 % depending on the TX power level, number of users and level of multipath, under ideal feedback conditions. In the delay-constrained scenario, the gains are expected to be even larger. However, the effect of feedback delay, feedback errors, and signaling overhead can diminish the gains somewhat. Future studies should consider allocation of resources for low bit-rate trickle channel type services as well.

6.4.2.2 Minimizing Over-the-Air Messaging in OFDM Systems

An OFDMA system using different coding and modulation schemes (link modes) on different subcarriers requires the signaling of the downlink quality on the different subcarriers to the BS and the signaling of the link mode for the different subcarriers to the terminal. This signaling may consume considerable air-interface resources, and can be easily minimized by introducing terminal-based decision schemes [156]. The terminal is now responsible for measurement of the channel and the communication of the related decision about the link mode parameters to be used by the BS for its communication with the terminal, based on a suitable table of link mode parameter sets available to both entities. The terminal may simply signal a step up or down from the current link mode, which takes much fewer bits than the current model of signaling back and forth between BS and terminal. Similar techniques can also be applied to manage power level changes across the subcarriers in an OFDM system.

Further optimization in signaling can be achieved [157] by (1) dividing the broadband channel into appropriate sub-bands on the basis of the observed frequency-selective fading, (2) measuring at the terminal the SIR conditions for each of the sub-bands, (3) determining at the terminal the appropriate sub-bands to be used and their optimum link modes, (4) signaling from the receiver to the transmitter the selected sub-bands (or only the incremental changes to the sub-bands) using a channel map and the optimum link modes with step up/down commands.

6.4.2.3 Region Division-based Fractional Loading Method for OFDMA Cellular Systems

Combining frequency-hopping (FH) with OFDMA can provide interference averaging and frequency diversity. Such useful features of the FH-OFDMA technique make it possible to achieve a cellular system with a frequency reuse factor (FRF) of one. A simulation study shows that the interference averaging method without inter-cell coordination does not satisfy the system performance requirements especially at high cell load. Moreover, it also shows that an interference management by controlling the cell loading imposes a fundamental limitation on the system capacity. To enhance the performance, therefore, we suggest a region division method combined with fractional loading technique [158], such that the overall system throughput and the spectral efficiency can be increased and UL Cross-Channel Interference (CCI) can be alleviated considerably by enhancing signal-to-interference level on a per sub-channel basis depending on its belonged region and allocated loading factor. From the simulation results, it is shown that the proposed method outperforms the conventional methods in the context of outage probability and throughput performance.

6.4.2.4 Near Capacity Multi-antenna Multiuser Communication

The use of multiple antennas to communicate with many users simultaneously (MU or 'broadcast' link) has recently received a great amount of attention, especially in wireless local area network (WLAN) environments such as Institute of Electrical and Electronics Engineers (IEEE) 802.11, where channel conditions change slowly and there is sufficient time for all parties to learn their channel conditions.

The capacity of an M-transmit, N-receive antenna link grows linearly in a Rayleigh-fading environment with the minimum of M and N when the receiver knows the channel [159]. It is also shown in [159] that K users, each with one antenna, can transmit to a single receiver with M antennas and the sum-capacity (total of transmission rates to all K users) grows linearly with the minimum of M and K. In a broadcast link case where the M antennas are used to transmit to the K users, it was shown in [94] that the sum-capacity grows linearly with min (M, K), provided both the transmitter and receivers know the channel.

Transmission schemes achieving the sum-capacity in multi-antenna MU links have been recently proposed, based on the 'dirty-paper coding' concept. Dirty-paper coding [160] is introduced for Gaussian interference channels, where it is shown that the capacity of a channel with interference, where the interfering signal is known at the transmitter is the same as in the case of the channel with no interference. In this approach, the interference is seen as 'dirt' and the desirable signal as ink. This information-theoretic solution is not to combat the dirt, but to use a code that aligns itself as much as possible with the dirt. Dirty-paper techniques are natural candidates for achieving sum-capacity in multi-antenna MU links because the transmitted signal for one user can be viewed as interference for another user, and this interference is known but out of the control of the transmitter.

Sphere-Encoded Multiple Messaging [161] allows for a basestation that is equipped with an M-element TX antenna array to send simultaneously and selectively K messages to K autonomous single-antenna terminals. The technique can also be used when M BSs, each equipped with a single TX antenna, are 'wired-together' to function as an M-element TX array. The total throughput, for a constant total power and spectrum, is proportional to min (M, K). When K becomes comparable to M, this technique significantly outperforms simple channel inversion with diagonal loading. In its current form the expected SNRs of the M users are equal, the transmission rates to the users are equal, and the scheme is designed for flat-fading. It performs within 4 dB of sum-capacity. Three extensions need to be considered: (a) handle users with different SNRs and permit unequal rate transmissions, (b) handle delay-spread channels through an OFDM version, (c) introduce scheduling considerations. Fast CSI, which gives the transmitter knowledge of the complex downlink propagation matrix in near real-time, is the essential enabler for Sphere-Encoded Multiple Messaging. Fast CSI transfer is relatively straightforward for TDD [152] where reciprocity holds. We recently found that FDD may still be a feasible alternative to TDD, requiring two to three times as much transmission time as TDD for the same quality CSI.

6.4.2.5 Cross-layer Optimization for Resource Usage Efficiency

There are two divergent approaches in MAC layer evolution. A MAC layer design, which is independent of the physical layer, may be preferred for the support of a wide range of air-interface alternatives. Such a design will enable the implementation of low latency seamless handoff among different access systems, with layer 2 context transfer across access systems, so as to benefit transfer of delay-sensitive services. On the other hand, cross-layer design has been proposed to improve an access system's efficiency when multimedia services are introduced into the system [162]. It is important that MAC layer design addresses both objectives in the evolution of future integrated access solutions.

6.4.2.6 QoS-optimized Access Network Technology

In future systems, different radio technologies may be used for different applications in order to optimize the resource usage and to meet the QoS requirements. For example, for a streaming audio service, which can tolerate packet delay, ARQ retransmission can be used, while for voice conversation, which cannot tolerate delay, fast retransmission techniques such as HARQ may need to be employed along with robust FEC. Similarly, a voice conversation could be efficiently served using a power-controlled subsystem to ensure a constant rate to the user, while audio streaming can be efficiently served with a rate-controlled system.

System design to match each and every application that emerges in the future is not practical, since the statistical characteristics of the applications are not known a priori. The best approach [163] is to design the system based on a 'toolbox' of air-interface technologies, which would provide a flexible mapping to an internal Radio QoS Class (RQC), and cater to a range of network QoS requirements, including the most stringent ones. This would enable minimum changes to the air interface when new applications with different QoS requirements emerge. When the network QoS classification is changed, there is no need to update the RQC since it is already flexible to adapt to any new QoS mapping. Only an update mapping the new network QoS into the appropriate RQC as a table entry at the wireless edge node is required. Further optimization of the RQC is possible by supplementing it with a set of secondary QoS parameters (e.g. packet delay), which includes per-packet based characterization, thus providing a finer granularity in the support of QoS requirements.

6.4.2.7 QoS Based Adaptive MAC States for Resource Usage Efficiency

MAC states are introduced in wireless systems to conserve terminal power and for efficient use of resources when bursty data is transmitted over wireless systems: Moving a wireless link to an intermediate (e.g. standby) state will result in minimizing the resources allocated to the link, such that transmission can be resumed with little delay. For next generation systems, since services with diverse QoS requirements are to be supported, there is a need for greater flexibility to adapt the MAC state transitions to suit the services' latency requirements, availability of data for transmission, and system load [163]. For instance, a transition from an intermediate state to an active state may need to be triggered earlier than normal if the traffic flow has a higher QoS. On the other hand, for a service with regular packet arrivals such as a streaming service, the MAC state changes may not be necessarily depending on the MAC state transition time and the delay requirement. For TCP-based flows, MAC state design also needs to minimize the delay as seen by the upper layer to avoid falsely triggering the congestion avoidance algorithm of TCP.

6.4.2.8 Scheduling Algorithm to Reduce Interference in Hybrid Division Duplexing

Since HDD contains a feature of TDD scheme, it is also prone to the interference problem as the TDD scheme. Figure 6.29 depicts different types of interferences that may occur in HDD systems. Cell 1 is in the downlink transmission period while the adjacent cell 2 is in the UL period. The worst scenario is when the terminal 1 (MS1) in cell 1 is receiving signal from BS1 at the cell boundary between cell 1 and cell 2, and the terminal 2 (MS2) is transmitting signal at the edge of the inner cell of cell 2 closest to MS1. Note that the terminals in cell 2 closest to MS1 at the cell boundary must TX in FDD UL, so that they cannot interfere with MS1. Therefore the interference between terminals in HDD systems is less than that in TDD systems. However, BS2 in cell 2 RXs interference from BS1, just as in TDD systems. As in TDD systems, having fixed uplink and downlink transmission periods can eliminate the problem. In order to relax this requirement and allow a variable UL/downlink transmission ratio, a scheduling method is proposed [164].

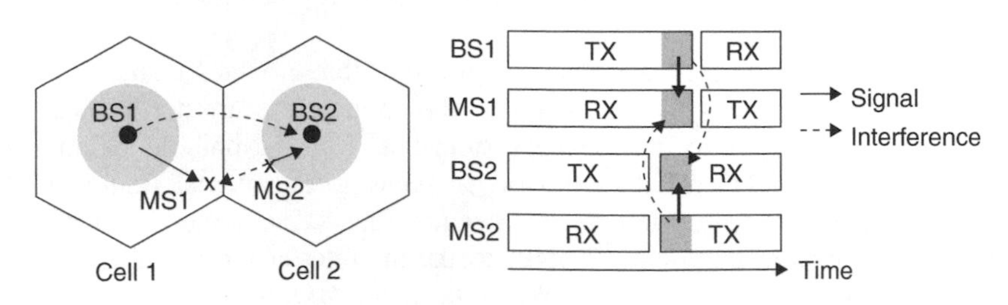

Figure 6.29 Interferences in HDD systems

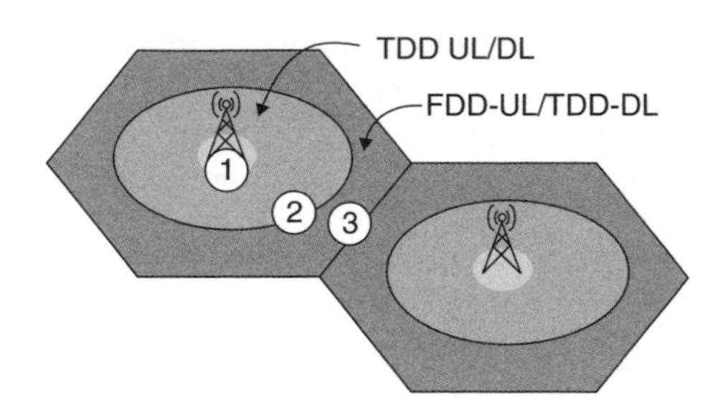

Figure 6.30 Terminal locations within a cell

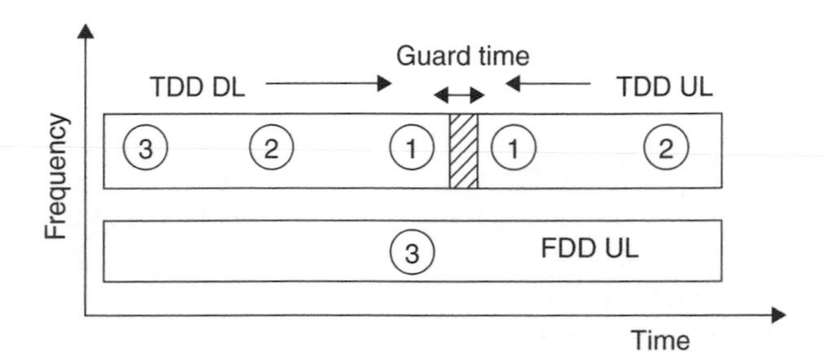

Figure 6.31 Scheduling algorithm based on user locations

Figure 6.30 and Figure 6.31 illustrate the scheduling method based on user location for reducing interferences in HDD technology. Here, the terminals are numbered from 1 to 3 in the order of increasing distance from the BS. The MS1 is located nearest to the BS in the inner zone, the MS2 is located second nearest to the BS, and the MS3 is in the outer zone. The uplink and downlink transmissions for the terminals are illustrated in Figure 6.31. In the downlink, the assignment principle is to allocate time slots to terminals furthest away from the BS first and to terminals nearest to the BS last. Time slots closest to the guard time, or time slot switching point are allocated to MS1. Time slots next closest to the guard time are allocated to MS2, and so on. In the UL, terminals nearest to the BS are allocated first, terminals nearest to the time slot switching point, and those next nearest to the BS are allocated next in TDD UL. Terminals in the outer zone of the cell are allocated to FDD UL for UL transmission. This time the slot scheduling method based on the user location reduces inter-cell interference when the time switching point is allowed to change within a certain limited range from a nominal position. Therefore, this scheduling algorithm improves the flexibility in transmission by allowing a certain limited amount of interference. This scheduling algorithm can be applied to the TDD-based cellular systems scheme as well.

6.4.2.9 Distributed Scheduling in Wireless Data Networks with Service Differentiation

In recent years, WLANs have been rapidly deployed all over the world. An important design issue in such data networks is that of distributed scheduling. The lack of centralized control leads to multiple users competing for channel access. This leads to significant throughput degradation. Existing approaches, such as Slotted Aloha and IEEE 802.11 DCF, also fail to provide differentiated service to users. The upcoming IEEE 802.11e Enhanced DCF aims to address these issues, but by supporting only strict priority classes, it is unable to provide dynamic service differentiation. A class of distributed scheduling algorithms, Regulated Contention Medium Access Control (RCMAC), has been proposed in [165], which provides dynamic prioritized access to users for service differentiation. Furthermore, by regulating MU contention, RCMAC achieves higher throughput when traffic is bursty, as is typically the case. In addition to WLANs, the basic concepts of RCMAC also have applications in multihop cellular networks, mesh networks, and sensor networks.

6.4.2.10 Jointly Opportunistic Beamforming and Scheduling (JOBS) for Downlink Packet Access

Jointly Opportunistic Beamforming and Scheduling (JOBS) [166] is a simple and robust downlink packet access technique for mobile scenarios where fading is expected. It is particularly suited for delay-tolerant data services to mobiles moving at moderate and high velocities. JOBS combines channel aware scheduling with 'dumb' beamforming, first proposed in [167], where it was shown that a significant multiuser gain can be achieved by this combination. The proposed scheme is particularly attractive, because no additional over-the-air signaling is required and the additional processing required at the mobile is minimal. However, the issue of delay and delay jitter was not considered in [167], which is also a crucial parameter of a packet data access system. In fact, in packet data systems – like 1xEVDO and HSDPA – there is a trade-off between throughput and delay. When the throughput gains, the delay performance deteriorates. The delay performance can be dramatically improved and at the same time the throughput is also improved by employing the JOBS algorithm, which exploits all past mobile reports to 'learn' the current preferred beam of each mobile, and puts this information to use by providing priority to mobiles waiting longer for the next packet. Simulations demonstrated improved performance in several scenarios.

On the other hand, several open issues remain that require further investigation, such as the effects of system latency. As in the case of 1xEVDO or HSDPA systems, the performance of fast moving mobiles degrades rather fast with system latency. In order to limit the built-in latency, a variety of schemes need to be investigated following a passive or active approach. An example of an active scheme is predicting the future state of the channel of the selected mobile, based on its past reports. An example of a passive scheme is using a certain back off, that is, forcing the BS to reduce its TX data rate to allow the mobile to RX the packet, even when the channel has deteriorated somewhat by the time the transmission actually takes place. Moreover, signaling overhead requirements, for example a certain pilot requirement must be evaluated.

6.4.2.11 Optimized Scheduling for QoS Support

As various new applications with stringent requirements emerge, allocating radio resources to multiple users with different signal conditions and QoS requirements is one of the key challenges in future systems. This allocation should be done to maximize the capacity/throughput while meeting the QoS and fairness requirements to the satisfaction of the user. In addition, the scheduler in future systems will operate to jointly optimize transmission in a multidimensional time-frequency-MIMO layer-space, and dynamically assign the users to different beam realizations created using smart antennas.

A channel and QoS adaptation mechanism is proposed in [168]. In order to efficiently support users with diverse channel conditions and QoS needs, channel adaptivity as well as QoS adaptivity should be supported [162]. While optimal scheduling prescribes that a user transmit when he/she has a good channel (during an upfade), the stringent QoS requirements (e.g. low delay) of the data awaiting transmission may override the channel condition criterion. The radio QoS requirements may be obtained from the application QoS specified in the service level agreement and the scheduler needs to support the soft statistical guarantees of these requirements. It is, however, expected that the widespread deployment of MIMO schemes in future wireless networks may reduce the benefits of scheduling on the basis of channel quality (up fades) since fluctuation of channel quality due to fast-fading are reduced [169].

It should be noted that, for the UL it might sometimes be necessary to TX data without waiting for the access node to schedule resource grants to the user. In such cases, autonomous scheduling can be employed where the mobile transmits data in a pre-allocated transmission resource. This form of scheduling may be particularly suited to services with low latency needs and constant bit-rate type of traffic characteristics. Both fast and autonomous scheduling approaches may be employed in conjunction to allow for highly efficient utilization of radio resources.

6.4.2.12 Fairness Considerations in Scheduling

The fairness requirement is also a widely debated issue in industry forums, since rendering a better service to the users located in specific geographic locations is considered unfair even though a minimum service requirement is met for all users. For example, in the 1xEV-DV evaluation methodology [156], a minimum fairness requirement is specified for the system, and the scheduler should meet this criterion. For the future systems, a scheduler may need to flexibly change the fairness on a per service basis. The capacity of a system is highly dependent on the level of throughput fairness it provides to the users. Therefore, if the revenue/charging is based on a per bit basis, the operators would prefer to operate in a moderately fair region rather than operating within a full fair premise. On the other hand, if the charging is based on a per session/subscription basis, the objective is to increase the number of users, where throughput scheduling schemes that are equal are preferred. Therefore, the flexibility of operating in different fairness regions in an optimal manner is an important design consideration for the scheduler.

Another important point to note is that the coverage provided by the system should be adequate in order for the scheduler to have full flexibility in all the required fairness scenarios. A service coverage concept where coverage is explained as a function of the operator's capability of offering a viable service has been proposed in [170]. Service coverage ubiquity can be measured, for example, by the availability of that service over the area for a minimum number of users. Such a standard performance evaluation methodology covering capacity, fairness, QoS requirements, and coverage need to be defined in order to assess and compare future system designs.

6.4.3 Interference and Inter-cell Coordination

The benefits of dynamic resource allocation and reconfigurable scheduling, presented in the previous sections, need to be validated in challenging wireless network setups, where coexistence of different access technologies, hierarchical cellular deployment, a wide variety of envisioned services, with increasing demand for data services, and mobility requirements result in challenging interference scenarios and introduce the requirement for adaptive and reconfigurable inter-cell coordination.

These coordination requirements introduce some constraints in the network architecture. Moreover, in order to facilitate the adoption of distributed architecture concepts, these coordination requirements need to be minimized and consideration should be given to implementing them using mobile-assisted coordination techniques or implicit coordination techniques.

In this section, we discuss several promising system and access control techniques and the associated coordination requirements.

6.4.3.1 Handoff

Current CDMA-based cellular systems employ soft handoff for voice and circuit data on the downlink and cell switching for packet data. Soft handoff is implemented on the UL for all services. While soft handoff is beneficial for current systems from the point of view of air-interface system capacity, it imposes severe constraints on the design of the backhaul network through timing requirements and QoS requirements. These stringent requirements make the access network complicated. Fast cell switching is a promising alternative to soft handoff for delay-sensitive packet service [171, 172]. Simulation results indicate that with reduced switching delays it is possible to achieve comparable performance on the downlink. The potential gain due to UL frame selection was also found to be small in the presence of incremental redundancy or other physical-layer HARQ techniques for error recovery. A flat All-IP network architecture becomes simpler to implement without soft handoff.

An alternate option is to enable the BS coordination required for soft handoff or interference avoidance schemes to be performed by the mobile through some simple over-the-air signaling rather than through a central entity in the network. This approach facilitates the implementation of coverage improvement techniques without the need for centralized resource control, thus significantly simplifying the access network architecture.

The complexity of seamless handoff processing, as a user migrates from one access network to another, increases with the expansion of the services to be supported and the integration of multiple-access networks in future wireless architectures. Seamless mobility solutions, which can accomplish handoff within milliseconds, within and across networks, are desirable [173].

6.4.3.2 Partial Handoff

MIMO links hold the potential to improve system capacity in future wireless communication systems. In such systems, the mobility management is more complicated than in existing cellular systems. The capabilities of MIMO communication allow for an additional handoff scheme called as *partial handoff* [174]. The basic idea is that in a region known as the *partial handoff region*, the MT communicates simultaneously with multiple BSs, as illustrated in Figure 6.32.

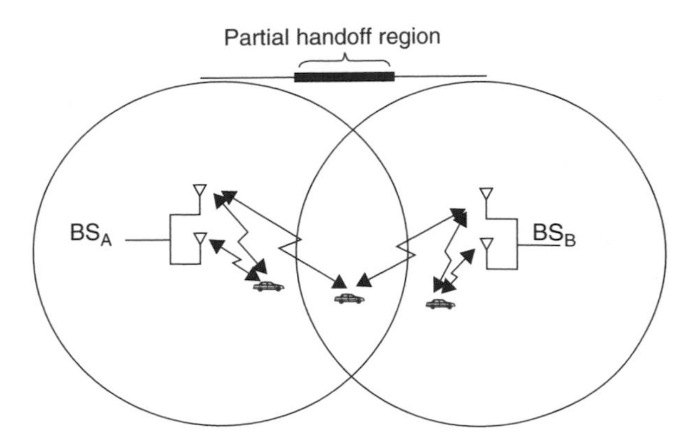

Figure 6.32 Illustration of the partial handoff region

The partial handoff does not necessitate the handoff of all the TX antennas from the serving BS to the target BS. Unlike in soft handoff for CDMA systems, the BSs are sending potentially different pieces of information and take advantage of the capabilities of MIMO communication. The advantage of partial handoff over conventional handoff is that it reduces the effect of TX antenna correlation by separating the TX antennas as far as possible. Hence, partial handoff can increase system capacity, link throughput, and other measures of performance by reducing the effect of spatial correlation.

6.4.3.3 Out-of-Cell Interference Considerations

The structure of out-of-cell interference, in the absence of inter-cell coordination, impacts the system performance significantly. Techniques used to improve performance can be broadly classified as interference avoidance, interference averaging, and interference smoothing. In [175], a dynamic packet assignment scheme is proposed for the downlink, which is based on a coordinated interference avoidance mechanism. The scheme shows significant throughput improvement. Other techniques used include beam forming, beam rotation, coordinated power management and dynamic interference removal.

The downlink and uplink interference environments are significantly different. For example, in a rate-controlled downlink, the mean out-of-cell interference caused to a stationary mobile is constant. This enables a mobile to identify the interferers and establish mechanisms to avoid or remove the interference. On the other hand, for a rate-controlled UL, interference caused to a particular BS varies from slot to slot since it depends on the power levels and the location of the mobiles selected for transmission in a given time slot, offering a greater challenge to minimizing or eliminating UL interference.

One can consider a system design where out-of-cell interference is bursty, less predictable or smoother by averaging over many users. A detailed simulation study comparing these options was carried out for the UL in [176] assuming the use of incremental redundancy schemes to deal with unpredictable interference that showed that interference averaging is not always essential. One can also consider investigating a coordinated power management scheme for the UL where selected mobiles are scheduled for transmission and their respective power levels determined so as to minimize overall interference.

While most of the interference avoidance schemes discussed above require explicit coordination among BSs, it is worthwhile considering techniques, which can be implemented without this requirement. In these systems, the interfering devices (e.g. BS or the mobile) may need to adhere to some generic noninterference policy, such as a maximum power limitation or slow coordinated power change mechanism. For example, in [177] an interference smoothing technique is proposed where a transmitter (BS for the downlink, mobile for the UL) is not allowed to increase or decrease its power levels abruptly. The start of high rate transmissions could abruptly change the interference it causes to other devices and the interfered links would not have time to adjust their power levels in order to continue communication. A power/rate ramp up/down process is defined, in which the increase or decrease is done slowly in small steps giving sufficient time for the interfered link to adjust its power levels or rates accordingly.

6.4.3.4 Superposition Coding for Unknown Interference

Superposition coding has been proposed [154] and shown to be a throughput-optimal encoding technique for compound channels in which the random state of the channel is unknown to the transmitter. Such a scenario arises in particular in a wireless network when the out-of-cell interference from neighboring BSs, and therefore the signal-to-noise ratio at the receiver, is unknown and time varying. The main idea behind superposition coding is to make sure that some data are correctly received even when the interference is large; additional information can be received when the interference is small. In the case of transmission to K receivers under total power constraint P, when deploying superposition coding the transmitter generates K codebooks with certain powers and rates, and transmits the sum of K codewords encoding the symbols intended for the K receivers. The receiver with the worst channel decodes the codeword with the corresponding rate and all the other codewords appear as noise to this receiver. The next-worst receiver is able to decode the rate corresponding to the worst and subtract it from the received signal. Then, it decodes the codeword of its corresponding rate, while it sees all other codewords as additional noise, and so on, until the best receiver decodes all codewords.

The benefit of applying superposition coding was investigated under the realistic scenario in which the transmission rates of the superposition codes have to be chosen from a fixed and pre-determined rate set. The maximum achievable throughput was explicitly calculated and compared to that of single-rate coding [178]. Superposition coding was found to achieve a finer granularity than a single-rate code, thereby improving throughput in situations where a higher-rate code cannot be used.

Initial results with infinite rate granularity on comparisons to retransmission techniques, such as incremental redundancy with a limited number of retransmissions, indicate that the benefit of superposition coding over incremental redundancy may not be that significant for delay-constrained applications. Further conclusions on the comparison between incremental redundancy and superposition coding need to be based on case studies with a fixed and finite rate set, in particular, conducted in the context of a power-controlled versus a scheduled mode of transmission in the UL.

6.4.3.5 Load Balancing among OFDM Base Stations

In a conventional OFDM system, the carrier-to-interference ratio (CIR) measured on pilot subcarriers is used to determine the choice of the serving BS as well as to determine the link mode (coding/modulation). It is proposed in [179] that the power of the pilot subcarriers may be decreased in a loaded BS for the purpose of load balancing. With the reduction of pilot powers, the loaded BSs CIR will appear lower than that of a neighbor, which has not reduced its pilot power, and therefore direct the terminal to the alternate BS. However, since the pilot CIRs are also used to determine the link mode, it is important that a compensation factor be applied so as not to bias the link mode calculations. Furthermore, since the data subcarriers are sent at maximum power, the BS coverage area will be unchanged. Thus, the proposed scheme will only encourage some terminals to start selecting less loaded BSs than their serving BS. This helps to balance the load between the different BSs, which increases the system throughput. In the case where adjacent BSs are all loaded, they may all reduce their pilot power level. The choice of the serving BS is based on the relative powers of the pilots. This will continue to ensure that the coverage area is still maintained and no coverage holes are created.

6.4.3.6 Self-organization of Base Stations

Significant expense is incurred in the configuration and management of BSs. In the future it is envisioned that BSs will probe the environment around them and adjust accordingly a number of parameters, such as their antenna configuration and TX power. Algorithms and protocols for such self-organization schemes are currently under investigation [180]. Centralized control (bunching) of BSs, intelligent relaying, dynamic cell sizing and intelligent handover are only a few examples of self-organization procedures. The information used to enable these procedures can be classified into four categories:

- Geographical Information (e.g. location-related parameters and propagation characteristics);
- Spatial/temporal information (e.g. available coverage, capacity and interference patterns);
- Network information (e.g. services offered, traffic);
- Contextual information (e.g. user profile).

Logical sensors located in the corresponding parts of the network would probe and buffer a number of parameter values associated with the above categories and compute the required adjustments in a dynamic fashion.

6.4.4 Summary

In these contributions, three critical issues for the design of next generation wireless systems have been addressed, namely, duplexing, resource allocation, and inter-cell interference coordination. The objective was to understand the major requirements and the challenges involved in the design of future-generation networks and present a number of research directions that appear to be promising potential candidates.

The major benefits and drawbacks of TDD and FDD were presented in terms of link reciprocity, link budget, synchronicity, and guard requirements. Terminal complexity issues were discussed. A new scheme, Band Switching Duplexing, was proposed, which flexibly alternates between the two and combines their features. Similarly, a novel hybrid duplex scheme that combines features of TDD and FDD was proposed, also aiming to combine advantages of both schemes.

Some insight on resource allocation and fractional loading across cells was given in the context of an OFDM air interface. Considering resource allocation within the cell, Sphere-Encoded Multiple Messaging, a near capacity multi-antenna MU transmission scheme, RCMAC, a distributed scheduling scheme supporting service differentiation, Joint Opportunistic Beamforming and Scheduling, a scheme combining the benefits of space and MU diversity, and a scheduling algorithm specially designed for HDD scheme were presented as potential candidates for future systems design. Cross-layer and QoS-optimized resource allocation and scheduling were discussed.

In an inter-cell coordination context, the delay sensitivity of handoff algorithms was addressed along with the benefits of fast cell switching as an alternative to soft handoff. Furthermore, the partial handoff idea was presented for increasing capacity and throughput by reducing the effect of spatial correlation. Superposition coding, as a throughput-optimal encoding technique in the presence of unknown interference, and load balancing among OFDM BS were proposed as possible candidates for interference management in future systems.

Finally, self-organization of BSs is believed to play an important role in achieving adaptive and reconfigurable operation in future wireless networks.

6.5 Multidimensional Radio Channel Measurement and Modeling

Future wireless communication systems are expected to employ higher carrier frequencies (2–6 GHz, up to 60 GHz in short range) and wider bandwidths (10–100 MHz, 500 MHz and more for ultra-wideband (UWB)) than currently deployed systems. Moreover, dual-polarized antenna arrays will be employed for improved coverage and higher data rates. It is obvious that to fully utilize the emerging technologies a better understanding of the behavior of wideband (WB) radio channels at new frequency bands is needed. Furthermore, the development and standardization toward future WB wireless communication systems require that system performance can be evaluated using realistic assumptions about the multidimensional radio channel.

The characteristics of multi-antenna radio channels are not well understood in the new bands of future WB wide-area and short-range communications. Up to the present extensive radio channel measurement campaigns have been carried out mostly in frequency bands of 800 to 2.5 GHz. Those measurements are typically narrowband (NB) and only a few of them include multiple antennae at both the transmitter and the receiver sides. Existing radio channel measurement campaigns and modeling activities have been reviewed, for example in [181–191]. Commonly used channel models, such as those specified by ITU (International Telecommunication Union) are mostly based on NB Ultra High Frequency (UHF) measurement campaigns. Indeed, realistic, measurement-based radio channel models do not exist for future-generation wireless communication systems, which are aiming at very high data rates and high spectral efficiency using multi-antenna techniques. Multiple-Input-Multiple-Output (MIMO) systems require more accurate channel models even for 2 to 2.5 GHz band, for which the 3GPP and 3GPP2 standardization bodies have recently specified the so-called *SCM* (*Spatial Channel Model*) model for the UMTS evolution [192].

Channel models can be classified into three groups: NB, WB, and UWB channels. The definition of WB is that the bandwidth of interest is higher than coherence bandwidth of the channel. In other words, the channel is frequency-selective. Current UWB definition according to Federal Communications Commissions [193, 194] is that the relative bandwidth is more than 20 % or more than 500 MHz as shown by

$$\frac{2(f_H - f_L)}{f_H + f_L} > 0.2 \tag{14}$$

or

$$f_H - f_L > 500 \text{ MHz}. \tag{15}$$

where f_H and f_L are the higher and lower −10-dB bandwidths, respectively. Some channels at very low frequency can be UWB and NB at the same time. Moreover, mm-wave systems may be categorized as UWB even if the relative bandwidth is very small (Figure 6.33).

To enhance the coverage and capacity of future wireless systems multiple TX and RX antennas are required in sectorization, beam steering, diversity, or MIMO mode. It is also reasonable that the above modes can be employed in an adaptive manner in which the maximum data rate can be traded for improved coverage. To be effective, the selection of the best mode has to be based on the instantaneous radio channel conditions.

Several scientific papers have shown that the capacity of MIMO systems increases linearly as a function of the number of antenna elements if the channel realizations are uncorrelated between the antenna elements, as is the case in rich scattering environments [195]. In addition, the signal-to-noise ratio has a major effect to the system capacity. Most of the MIMO techniques assume far-field conditions between antenna arrays and environment. Realistic terminals are subject to near-field disturbance and are not covered by the exiting far-field investigations [196]. The near-field effects and the actual radio channel conditions may increase the inter-antenna correlation significantly, which degrades the achievable spectrum efficiency. It remains unclear how different multi-antenna modes should be applied flexibly in various channel propagation conditions to achieve best overall system performance. Future radio transmitters adapt to the radio

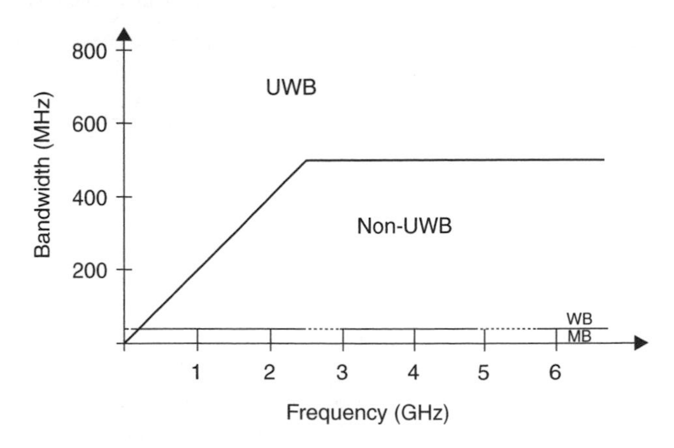

Figure 6.33 NB, WB, and UWB channels

environment by employing feedback information about the actual radio channel state from the receiver. Understanding the critical parameters associated with accurate radio channel modeling is therefore of critical importance for system engineering [197].

The radio channel model formulation consists of seven phases (Figure 6.34). First, the major requirements for system design such as capacity and coverage requirements, propagation environment, frequency spectrum, maximum transmitter power, and mobile velocity are needed. The second phase defines generic models, while the third and fourth phases cover planning and conducting a measurement campaign.

Data post-processing will be completed in the fifth phase, and channel modeling in the sixth phase. The seventh phase consists of building the simulation model, where the channel is combined with the communication system simulation. It is worth emphasizing that the first and the last phases require strong cooperation between the propagation research and the communication research.

6.5.1 State of the Art

6.5.1.1 Multichannel Modeling Methods

Radio channel models with MIMO ports are typically applied in spatial multiplexing mode, in which parallel data streams are transmitted simultaneously to achieve high data rates for a single user. Figure 6.35 illustrates a system model for the MIMO communication system. The task of the transmitter is to match the signal waveform to the MIMO radio channel and the task of the receiver is to extract MIMO channel parameters for efficient detection of the transmitted information.

In general, it can be stated that the modeling of the MIMO radio channels should be based on physical propagation environments. Synthetic models could be beneficial for MIMO modem functionality testing. Thus, extensive measurement campaigns are needed

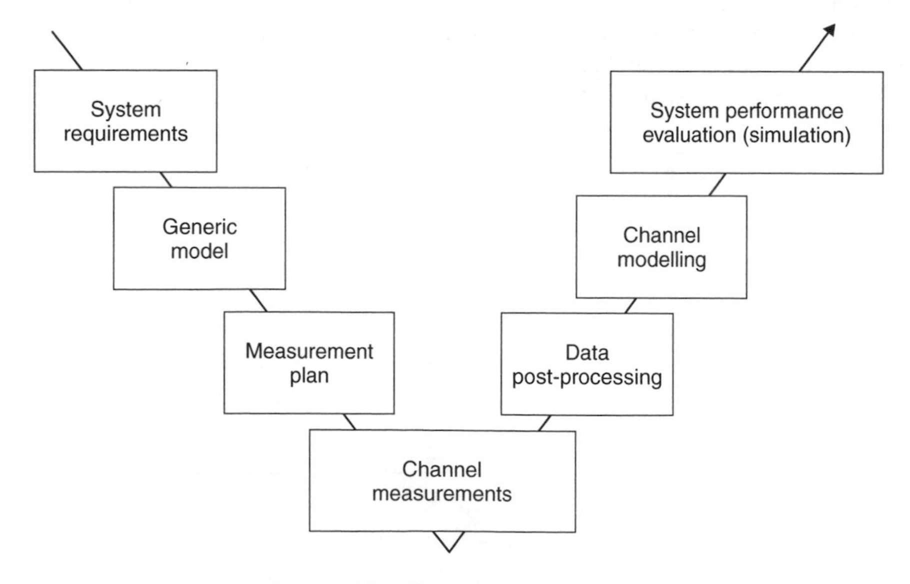

Figure 6.34 Channel modeling process

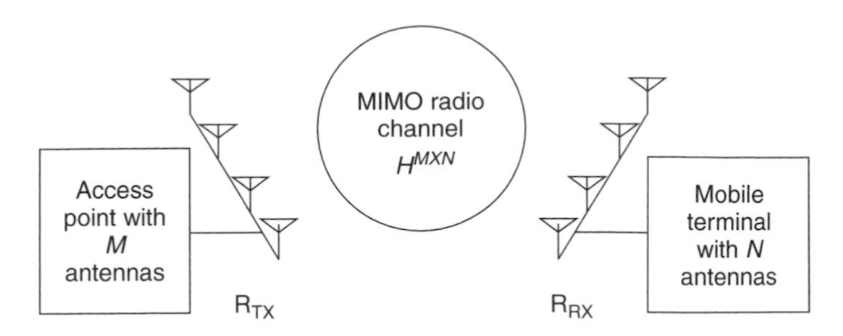

Figure 6.35 Communication system with M-transmit and N-receive antennas and a $M \times N$ MIMO radio channel. Spatial correlation matrices RTX and RRX define the inter-antenna correlation properties

to create channel models for new frequency bands. Basically, the radio channel modeling can be of deterministic or stochastic nature. Deterministic modeling involves a detailed reproduction of the actual physical wave propagation process. Alternatively, ray-tracing methods can be utilized to create a model for a specific environment. Thus, the geometry of the specific environment and the characteristics of the reflectors and scatterers therein define the channel properties. In stochastic modeling, a large number of measurements are conducted in order to define a statistical representation of the channel. For example, main categories often consist of macro, micro, and pico cells in the cellular system and several subcategories usually exist within each main category. A good representative of the stochastic channel models is the MultiElement Transmit and Receive Antenna (METRA) model [145] in which the measured correlation matrices of the transmitter and receiver end are combined in a simple way to generate a MIMO channel model.

The stochastic channel modeling is often based on the assumption of space–time stationarity. It may hence not be adequate in order to describe:

- the channel's time-varying dynamics;
- the performance of reconfigurable techniques that attempt to adjust to possibly abrupt statistical changes of the environment/interference;
- particular channel conditions (such as 'keyholes') that, even though rare from a statistical point of view, require successful handling.

In such cases, a more accurate deterministic description of the channel may be required, which is the situation for nearly all investigations relating to handset terminals. On the other hand, the deterministic model has its limitations too, arising from its inherent inability to fully characterize the environment. In attempting to do so, deterministic models may result in high numerical complexity. They also become site-specific, taking away from the model's generality. The choice of a proper set of parameters that depict adequately the environment avoiding over-parameterization should then be sought.

In order to achieve a reasonable compromise between sufficient detail and complexity, a geometry-based stochastic channel model can be used. This model allows capturing geometry-specific channel and interference attributes. In a Geometry-based Stochastic Channel Model, the PDF of the geometrical location of the scatterers is prescribed. For each channel realization, the scatterers' locations are taken randomly from this PDF, and the multipath delays and directions are calculated by a ray-based approach assuming that each scatterer carries one multipath component. The PDF must be selected in such a way that the resulting Power-Delay Profiles (PDP) and Power-Azimuth Profiles (PAS) agree reasonably with the measured values. Measurement results show that it is realistic to assume that scatterers are grouped into clusters in space. An open issue of this approach is the correct parameterization of these scattering clusters to the considered scenario. Currently, cluster parameters are chosen empirically. To improve the model's accuracy, algorithms have to be developed that are able to identify multipath clusters in measurement data and extract the necessary model parameters.

The COST259 model [141] is an example of geometry-based stochastic modeling, which has a three-level top-bottom structure. It is a model framework enabling to tune the three-level parameter set of a specific simulation environment. As another example, the 3GPP and 3GPP2 standardization bodies have defined the SCM, which utilizes geometry-based stochastic modeling. The model is intended for the evaluation of the system and link-level MIMO performance of 3G radio communication systems. Also parametric double-directional stochastic models have been developed. These models can be implemented as a tapped delay line, in which each of the N resolvable multipath components has specific parameters such as complex gain, delay, polarization, and path directions. Recent research has shown that besides the specular path model also diffuse scattering has to be considered [115, 116, 141, 145, 146, 192, 198–200].

6.5.1.2 Standardized Channel Models

Standardized channel models provide a full description of the parameterization and stochastic properties along with a number of test cases for certain values of the model parameters. These test cases offer a number of scenarios to be considered for simulations and aim at establishing comparability in terms of channel models for calibration and performance evaluation purposes. Conseil d'Europe projects (EC), Cooperation in Scientific and Technical Research (COST), Code-Division Test bed (CODIT) and METRA provided channel modeling methodology without proposing specific test cases. The newly developed SCM and IEEE 802.11n models include also specific model parameters for outdoor and indoor scenarios, respectively.

3GPP SCM

The scope of the 3GPP-3GPP2 Spatial Channel Modeling AdHoc Group (SCM AHG) was to develop and specify parameters and methods associated with the spatial channel modeling that are common to the needs of the 3GPP and 3GPP2 organizations [192]. The model includes two sub-models:

- calibration (link-level) model, and
- simulation (system-level) model.

Calibration Model

The SCM link-level channel model considers a MIMO link, where a single transmitter sends signal through a radio channel to a single receiver. It is a spatial extension of the International Telecommunication Union-Radio Sector (ITU-R) Rec. M.1225 model. The spatial extension includes (PAS), angle spread (AS) and propagation path directions in three environments: urban macrocell, urban microcell and suburban macrocell.

Simulation Model

The SCM system-level model follows COST259. The model is based on geometry, but a subset of the parameters is stochastic. The bandwidth is 5 MHz and the center frequency is around 2 GHz. System-level simulations often include multiple BSs and multiple MTs. A snapshot-based model is assumed for the generation of the clusters of scatterers. During a snapshot, the channel undergoes fast-fading according to the mobile movement. However, delays are kept constant. Two consecutive snapshots are independent and include randomly located clusters, which make the channel model discontinuous.

The channel modeling methodology of SCM can be described by the following three steps:

- Specify an environment: suburban macro, urban macro, or urban micro.
- Obtain the parameters to be used in simulations, associated with that environment.
- Generate the channel coefficients based on the parameters.

The channel model includes several parameters, such as number of paths, number of sub-paths, mean angular spread (AS) at BS and MS, AS per path at BS and MS, Angle of Arrival (AoA)/Angle of Departure (AoD) distributions, variable delay-spread, path loss, shadowing. These parameters are fixed in the specification of a number of test cases. The number of paths in all test cases is six.

IEEE 802.11n

The IEEE 802.11n model has been developed for indoor WLAN high throughput applications. Different environments from small office to large open space exist (models A–F). Both LOS and NLOS cases are included in the models. It is assumed that the propagation environment can be modeled via clusters. The number of clusters and multipath components vary between models. The maximum number of delay positions considered is 18. The 802.11n model uses tap specific AoA and AoD characteristics and antenna geometry. Transmitter and receiver correlation matrices are calculated from the geometry. MIMO correlation matrix is obtained via the Kronecker product of the spatial correlation matrices at transmit and receiver antenna arrays.

Channel Modeling in Ongoing Research Projects

Currently, several ongoing collaborative research projects carry out studies on channel modeling. In Europe, COST is continuing its long tradition in COST273. Wireless World Initiative (WWI, [201]) in Europe has a WINNER project (Wireless World Initiative New Radio [202]) which is working toward specifying next generation air- interface and enabling technologies. The channel modeling work of WINNER is to provide a suitable MIMO channel model – based on measurement data – according to the system assumptions on frequency bands, bandwidth, scenarios, and candidate MIMO algorithms. EC project Pervasive Ultra-wideband Low Spectrum Energy Radio Systems (PULSERS) studies Ultra-Wideband Systems. In MAGNET project extensive WB and UWB radio, MIMO radio propagation investigations are carried out for personal area network (PAN) and body area network (BAN) scenarios [203]. NEWCOM [204] is a Network of Excellence type of cooperation in Europe.

Simulation Aspects

Algorithm research and evaluation require realistic measurement-based channel models. The models need to be supported by an efficient channel simulator. Channel simulators can be implemented in SW (e.g. MatLab), or hardware, or a combination of both. While the models should accurately reflect real environments, they need to be simplified to reduce simulation complexity. Complexity is a very critical issue in SW simulations, especially for broadband multi-antenna systems. With increasing complexity the simulation time becomes a problem for SW simulators while the hardware cost becomes a limiting factor in real-time HW simulators. Therefore, reducing simulation complexity without decreasing the modeling accuracy significantly is a key research topic.

Measurement-based channel modeling approach reflecting the above requirements can be found in [205, 206]. These papers propose a low complexity subspace-based channel model suitable to represent frequency-selective time variant MIMO channels. The model allows the generation of new random channels with the same spatial, time, and frequency correlations of a reference channel.

Field tests of wireless communication systems are cumbersome and expensive. Although field tests provide realistic results they can only be performed for a limited number of environments. Moreover, the results may vary even for the same environment because of changes in, for example, weather conditions, traffic, and vegetation. Radio channel simulation has been found efficient in transceiver algorithm design and development, verification and validation of wireless systems. Usually, the far-field conditions of the radio channel can be adequately modeled but the realization of near-field conditions for terminals is rather difficult.

The main challenges of real-time HW simulators have been the implementation of real-time filtered noise, uncorrelated fading, and real-time convolution. Requirements on large bandwidth, high dynamic range, accuracy, number of channels, capability of simulating measured data and usability will become more and more important in future applications. For SW simulators, the optimization of simulation parameters to ensure reasonable simulation time and accurate modeling are the major challenges.

One interesting research topic is playback simulation of WB multidimensional measured data ([207] and [208]). The main idea is to bypass the channel modeling phase. It makes the process from measurement to simulation much faster. Playback simulation employs directly the measured data and is very realistic but there are still many challenges in reducing the effects of undesired spurious signals and noise. However, this is a very viable solution when working with near-field, user-proximity terminal scenarios.

System-level simulation is a relatively new research area and it is not as well understood as the link-level simulation. Realistic modeling of MU interference in a multicell environment is a big challenge. System-level simulation faces the problem that simplified modeling is needed in order to avoid excessive complexity.

6.5.1.3 Radio Channel Sounders

From a historical perspective, the first sounding experiments have been carried out by using single tone Carrier Wave (CW) signals. This was enough as long as only the NB channel behavior was of interest. Single tone CW sounding, however, gives us no information to resolve path time delays. For that, we need a bandwidth, which is roughly the inverse of the desired delay resolution. Sequential sounding at a number of different frequencies is the easiest approach to achieve a delay resolution, which may even be very high since standard vector network analyzer techniques can be applied. The drawback is the resulting huge measurement time that precludes mobile measurement. The only way around this is to keep the environment fixed during one series of frequency sampling measurements. This actually has its equivalent in the sequential sampling of the antenna array geometry and may be considered as an equivalent to the synthetic antenna aperture approach in the frequency domain. Sustained measurement along some longer trajectory is clearly prohibitive. The network analyzer application also requires a cable connection between the TX and the RX sites, which limits the link distances.

Short duration repetitive pulses together with envelope detectors have been used in early broadband real-time sounding experiments. The main drawbacks of this method are the high peak-to-mean power ratio at the transmitter and that only power-delay profiles can be measured. To achieve maximum signal-to-noise ratio at the receiver, excitation signals are required, which have a minimum crest factor. The crest factor is given by the ratio of the peak value of the signal to its root mean square (rms) amplitude. Minimum crest factor signals are distinguished by a constant magnitude envelope in the time domain. At the same time, they must have a constant spectrum that leads to a short autocorrelation function.

The probably most well-known examples of those excitation signals are periodic pseudorandom binary signals (PRBS). PRBS can be very easily generated by a linear feedback shift register, since only digital circuits are required. This makes it possible to generate very broadband excitation signals, even suitable for ultra-wideband sounding. Another excitation signal concept is known as the *periodic multi-sine signal*. In communication engineering terms, this signal may be called a *multicarrier spread spectrum signal* (*MCSSS*). The MCSSS is defined by its complex Fourier coefficients. Thus, the spectrum of the TX signal can easily be shaped and even adaptively modified. However, a D/A converter is required at the TX. The real-time bandwidth of existing broadband sounders is about 100 to 120 MHz.

In principle, we have the following three main options for MIMO sounding:

- sequential measurement (antenna switching at both TX and RX);
- semi-sequential measurement (antenna switching at TX, parallel measurement at RX) [209];
- parallel measurement at both TX and RX.

Currently, the most advanced radio channel sounders (i.e. the highest number of channels) apply antenna switching techniques for multichannel measurements or combine parallel and switched channels [210]. Figure 6.36 depicts the basic TX/RX structure of a switched MIMO sounder.

Antenna Arrays

The spatial dimension of the channel response is accessed by antenna arrays. This mainly relates to 'true' arrays but can also include synthetic aperture arrays. The latter arrays consist of a sequentially sampled spatial aperture where only one antenna (or a subset) of the respective array is physically deployed. The angular resolution capability of any array depends on the effective aperture size and shape as seen from the respective wave direction. Sophisticated antenna architecture design is required to achieve high Angle of Arrival (AoA)/Angle of Departure (AoD) resolution. This has to go along with low antenna element coupling, mechanically and electrically stable construction, and precise

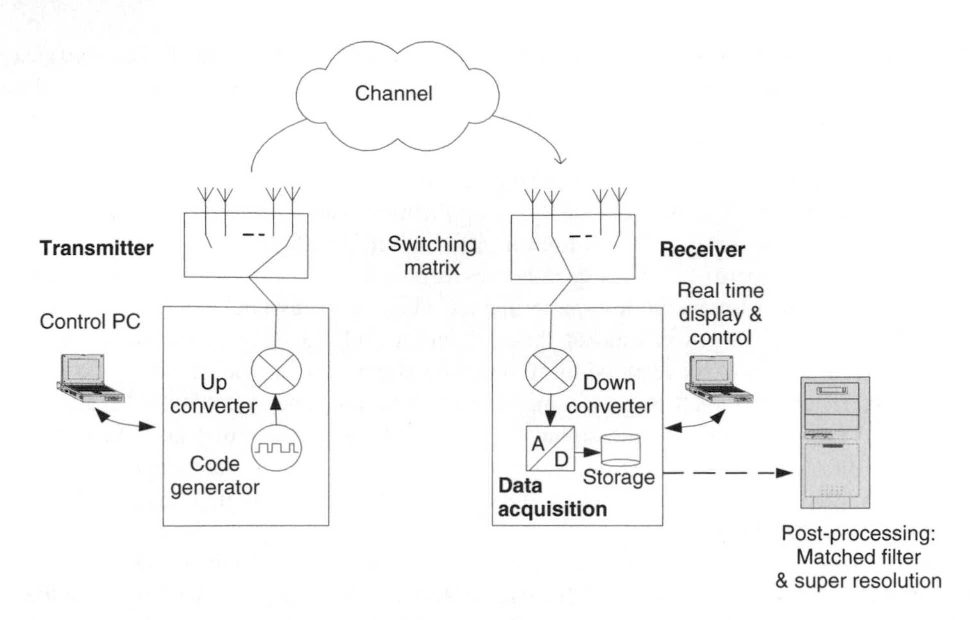

Figure 6.36 Block diagram of a switched multidimensional radio channel sounding system. Source: Elektrobit Testing Ltd.

calibration. This has also to include the antenna switches and feeder cables. Particularly, the feeder cables can be problematic for measurement with small terminals, so nongalvanic RF optic solutions have to be employed [196, 211]. Since there is always a trade-off between various specifications including resolution, measurement time, availability and costs, there is a wide variety of useful antenna array architectures.

Figure 6.37 illustrates a state-of-the-art 3D antenna array with 25 dual-polarized patch elements. The antenna elements are positioned in a way that allows omni-directional channel probing in azimuth direction and almost full measurement in the elevation angle. Only a small cone in space angle along the supporting pole of the array cannot be covered. The antenna can be applied as the mobile antenna since it is transmitting omni-directionally. Its 3D structure enables the study of roof reflections in indoor environments. The antenna array in Figure 6.38 represents a BS antenna with limited azimuth aperture. It is a typical patch antenna with 16 dual-polarized elements. It enables good resolution both in azimuth and elevation directions.

The uniform circular array in Figure 6.39 consists of 32 sleeve antennas, which do not require a ground plane. Here a 2 W power switch is included to support the application as a TX antenna. The stacked polarimetric uniform circular patch array (SPUCPA) in Figure 6.40 is a wide aperture, very high-resolution cylindrical array. It comprises of four stacked rings of 24 polarimetric patches. So it has 192 output ports in total. The switch is arranged inside of the cylinder. The cylindrical architecture gives a maximum resolution in azimuth for low elevation paths and also a considerable resolution in elevation.

A spherical antenna has been introduced in [212] and [213] and two different implementations are shown in Figure 6.41 and Figure 6.42. The spherical array is meant for spatial radio channel measurements at the MS and it covers the full 4π solid angle. The array has 32 dual-polarized elements with an inter-element distance of 0.641 R or 0.714 R

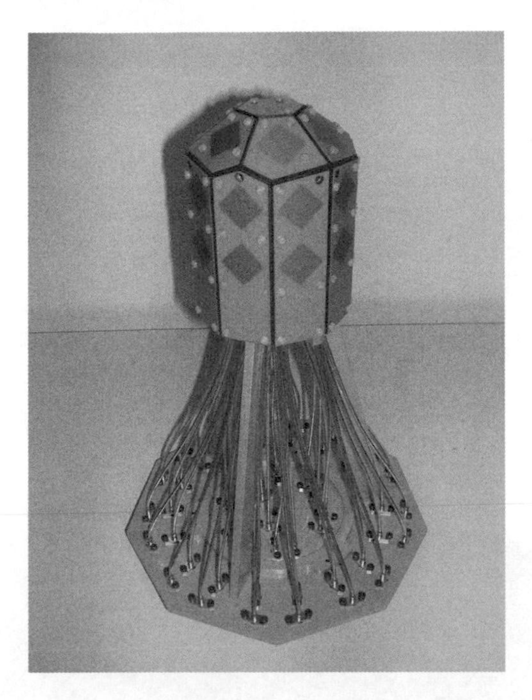

Figure 6.37 3-D Dual-polarized antenna array for 5.25 GHz. Source: Elektrobit Testing Ltd.

Figure 6.38 Dual-polarized 4×4 antenna arrays for 5.25 GHz. Source: Elektrobit Testing Ltd.

Figure 6.39 Vertical polarized circular dipole array (UCA32). Source: Technische Universität Ilmenau.

(R is the radius of the sphere). The elements are mounted on the surface of an aluminum sphere and pointed toward the normal of the sphere. Their orientation is such that the two polarization vectors are parallel to unit vectors $u\theta$ and $u\phi$. The 64-channel RF switching unit is placed inside the sphere together with its control electronics.

Figure 6.43 shows an antenna array consisting of monopole elements distributed on a sphere. This array has the advantages of being useful in the range of 3 to 10 GHz in addition to covering a mixture of polarizations and directions [214].

Another practical question is an efficient description of the antenna patterns for the geometry-based or geometry-based stochastic channel modeling. The EADF (Effective Aperture Description Function) [215] seems to be a promising technique.

6.5.1.4 Recent Results in Radio Channel Measurements and Modeling

WINNER Project

WINNER project is focusing on beyond 3G (B3G) radio system using a frequency bandwidth of up to 100 MHz [202]. Work package 5 (WP5) of the WINNER project is developing double-directional channel models for the WINNER system. In the beginning

Figure 6.40 Stacked polarimetric uniform circular patch array (SPUCPA4 × 24). Source: Technische Universität Ilmenau.

of the project, WP5 selected initial channel models from the literature for early link-level and system-level simulations. Then METRA [145] based IEEE 802.11n model [199] was selected for indoor simulations and 3GPP SCM [192] was selected for outdoor simulations. The SCM model has a bandwidth of 5 MHz and center frequency of 2 GHz, but required bandwidth and center frequency in the WINNER project are up to 100 MHz and 2 to 6 GHz, respectively. Thus, WP5 performed some modifications and implemented a new model, called *SCME* (*SCM Extension*) [216]. However, in spite of the modification, the initial channel models are not sufficient for the WINNER system simulations due to the different requirements. Therefore new models will be developed within WINNER. The new models are described in Deliverable D5.4, which will be published in October 2005. Some examples of WINNER measurement campaigns are shown in Figures 6.44 and 6.45.

Figure 6.46 shows the cross-correlation between shadow-fading and angle spread. It is clearly shown that angle spread is higher when the shadow loss is high.

SIMO and MIMO Channel Measurements for Coverage Investigation

One of the drawbacks of moving toward higher carrier frequencies is that the path loss increases and consequently the coverage range becomes small.

In order to compare the indoor coverage range of a WLAN based on 802.11a with that of a Digital Enhanced Cordless Telecommunication (DECT) system, two measurement campaigns have been carried out in an old office building [217]. We consider DECT as the specification of a professional high quality wireless telephone system and its coverage range as a basis for comparison with Voice over wireless local area network (VoWLAN)

Figure 6.41 Spherical array with 32 dual-polarized elements. Source: Helsinki University of Technology.

systems based on IEEE 802.11a. The building is a typical office environment with long corridors, in our case, 38 m long, and middle-sized rooms. First, channel measurement with co-located antennas has been performed. Five independent wireless nodes (RACooN lab) [218, 219] which can TX and RX signals in the operation band of 5.1to 5.9 GHz and are synchronized via a Rubidium clock are used. A SIMO system with four co-located antennas as receivers and one mobile node as the transmitter has been set up.

We concentrate on NLOS. By applying RX beamforming, the range was increased from 12.4 to 21.5 m. It is obvious that in order to achieve the coverage performance for voice traffic we should be able to apply these techniques in both directions: uplink and downlink. This results in a MIMO system.

In a second measurement campaign, we investigate the effect of the antenna spacing on the coverage. In these measurements, a 2×4 MIMO system with locally distributed antennas has been considered. TX nodes were the mobile nodes and the distance between the TX antennas was about 110 cm. Each of the RX antennas was located in the edge of a rectangular trolley. Distance between RX antennas was in one direction 10λ (57 cm) and in the other direction about 110 cm. Again channel measurements have been performed using RACooN nodes and in the same location as previous measurements.

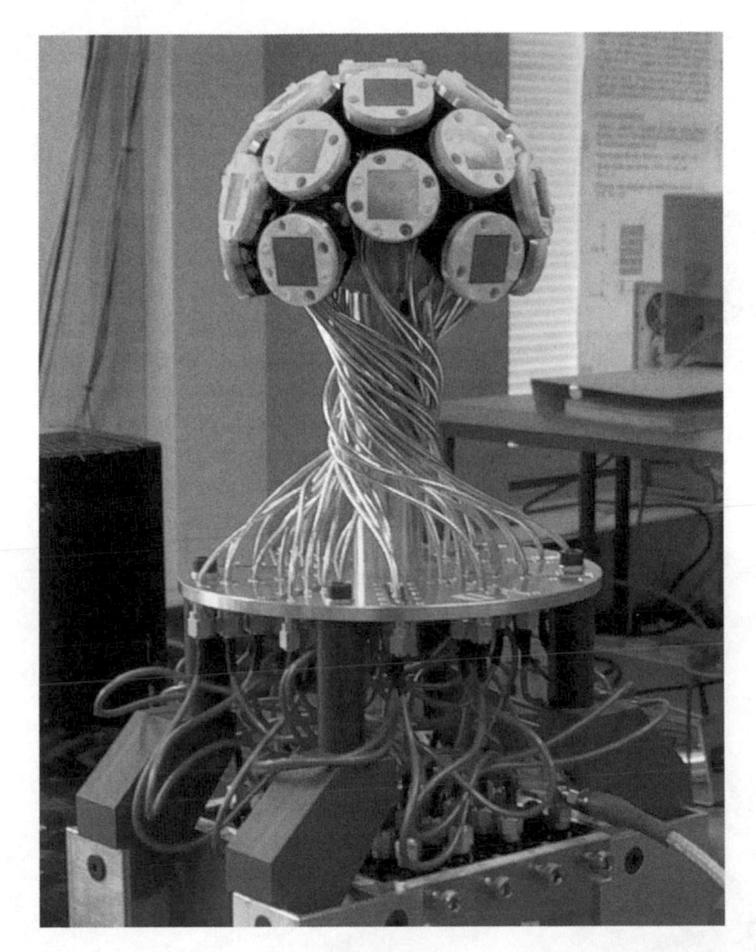

Figure 6.42 Spherical array. Sources: Helsinki University of Technology and Elektrobit Microwave Ltd.

In this measurement as we expected, the correlation between channels was reduced and we benefit more from space diversity and perform better against shadowing. For the 2×4 MIMO case with RX beamforming and selection between two TX antennas, the coverage range was increased from 12.4 to 31.8 m, which shows an 8 m improvement compared to the SIMO case and almost a 19 m enhancement compared to the SISO case.

In the simulation, we use an NLOS indoor channel model from [220] with RMS delay-spread of 100 ns. This model has been recommended for HiperLAN/2 but due to the similarity of the Physical layers (PHY) in HiperLAN2 and IEEE 802.11a we have employed this model. The model consists of an 18-tap delay line.

By comparing the measurements and simulation results, we modified our channel model used in the simulation. Applying MMSE criterion to the measured and simulated SNR versus distance we obtained the path loss exponent of 3.3. Using the new parameter instead of the path loss exponent of 3, we achieved a model that conforms better to the real channel in our scenario compared to the previous model. Results from simulations

Figure 6.43 Wideband spherical array of monopole elements. Source: Aalborg University/Antennas and Propagation.

Figure 6.44 Rural area measurements in Tyrnävä, Finland. Source: Elektrobit Testing Ltd.

Figure 6.45 Urban macrocell measurements in Stockholm, Sweden. Source: Royal Institute of Technology, Stockholm.

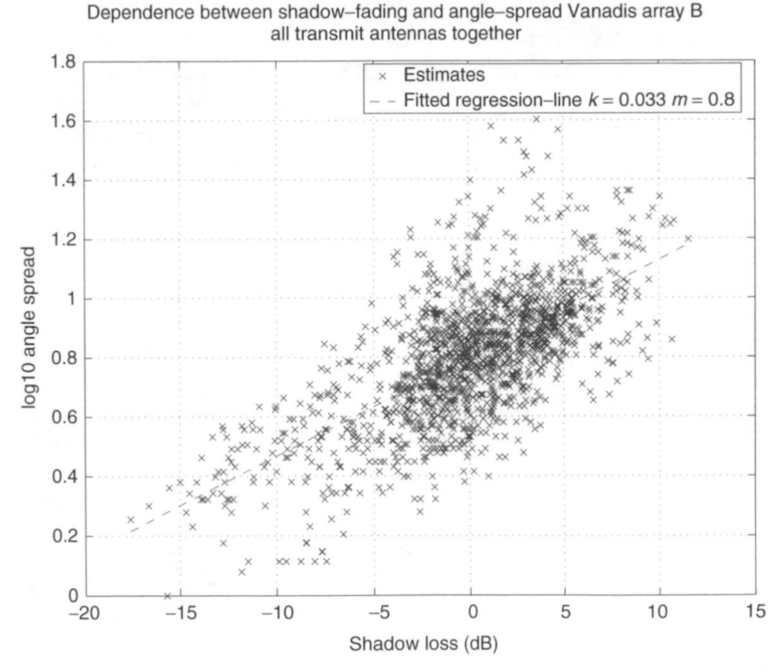

Figure 6.46 Dependence between shadow-fading and angle spread (Stockholm, Sweden). Source: Royal Institute of Technology, Stockholm.

confirm the measurement results and show that by applying diversity techniques we can increase the coverage range of each AP and consequently the number of required Access Points (APs) reduces. In our scenario in the SISO case, we need at least 2 APs to cover the whole floor. This number can be reduced to one AP by employing diversity techniques that provide both diversity gain and array gain.

Indoor MIMO Channel Modeling

A new accurate channel model at 5.8 GHz for indoor MIMO communications has been tested with the sounder described in [210]. Optic RF connected handsets have been used to emulate the mobile units [196]. The model takes into account the parts of the environment not seen directly by the antennas in a double bounce model with a coupling matrix between the end environments [221, 222]. The model uses a new richness measure to compare model and environments.

Polarization Investigation

In the basic MIMO approach, spatial multiplexing is applied in order to increase the spectral efficiency of the radio link. For this purpose, polarization diversity provides a competitive alternative to spatial diversity. In [223], MIMO channel measurements in an indoor environment in 5.25-GHz band by applying dual-polarized antennas in both transmitter and receiver ends are considered. The goal in the measurement campaign was to find out how well the polarization properties of the channel are preserved in the indoor environment. The channel is measured in fixed points and locations with both LOS and non-LOS (NLOS) scenarios being considered [223].

The angle of departure and the angle of arrival versus delay, and the polarization of resolved paths are given in [224]. The results also showed that the SISO may suffer from polarization mismatch, see Figure 6.47 [223].

Comparisons of MISO, SIMO, diversity MIMO, and information MIMO are represented in Figure 6.48. Results show that the information MIMO provides clearly the best approach. Diversity MIMO is robust against polarization mismatches since it combines all four channels. MISO and SIMO may suffer from polarization mismatch but since

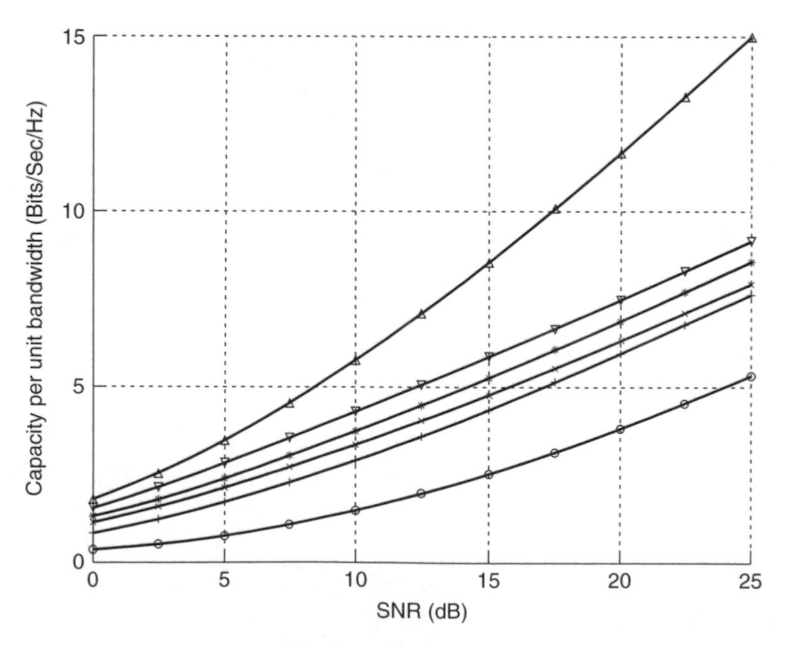

Figure 6.47 Information rates at example location for SISO (vertical to vertical 'x', vertical to horizontal 'o'), MISO ('+'), SIMO ('*'), diversity MIMO ('∇') and information MIMO ('Δ')

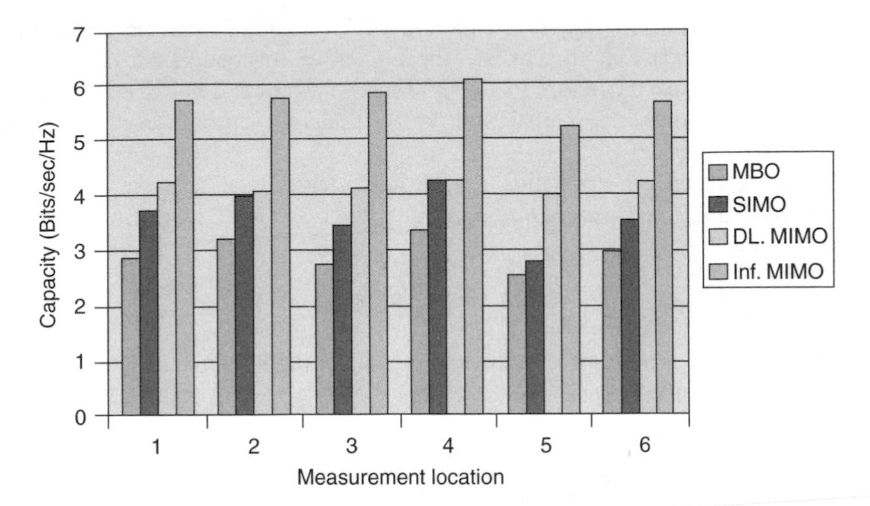

Figure 6.48 Comparison of MISO, SIMO, diversity MIMO and information MIMO

combination over two channels is carried out, the losses are not significant as may be the case with SISO.

The paper [223] showed the utilization of polarization as a diversity technique in a real indoor environment. The results indicated the clear benefit of using polarization as a diversity technique.

Indoor UWB MIMO Channel Modeling

An ultra-wideband, 3 to 6 GHz MIMO channel sounder was developed at Antennas Propagation, Aalborg University for investigations in realistic short-range personal and body area network (PAN/BAN) scenarios [203]. The channel sounder is based on swept time-delay correlation performed in hardware and has two fully parallel transmitters (or four with switching) and four parallel RX branches. The equipment allows dynamic measurements including user movements and body-proximity scenarios. A large variety of small to medium-size hand-held and body-worn type of terminal antennas can be employed using an optic RF connection setup [211].

Novel UWB PAN and BAN indoor channel models have been proposed, which take into account the effects of realistic small-size antenna systems, the user dynamics and user body-proximity effects [203].

Another MIMO UWB sounder is being developed within the PULSERS project [225]. This real-time sounder covers the bandwidth from (near) zero to 3.5 GHz or from 3.5 to 10.5 GHz. The air interface is based on custom specific SiGe circuits and the TX signal is a binary periodic pseudo-random sequence. The sounder architecture comprises a scalable number of sequential transmitter and parallel receiver modules, which are connected by Ethernet to the data concentrator module. So the setup can emulate a distributed UWB antenna system or a sensor network.

Playback Simulation

In playback simulation, we try to bring radio channel into the lab. The process is described in [226]. The general idea is to measure the channel, generate channel models from the measurements, and then simulate the channel in a HW emulator to create repeatable tests. By this process, we can see in real life the effects of the channel to the receiver in question. The equivalent procedure that uses a software-based simulation environment is described [227]. It allows very realistic link-level simulation, optimization, and evaluation of transceiver architectures.

Figure 6.49 represents a simulation result. It is clearly seen that we can mimic the nature with reasonable accuracy. Only spot 1 shows reasonable divergence. This is due to the antenna characteristics.

6.5.2 Channel Modeling Process

This section depicts a generic process for radio channel modeling. The subsections below describe each phase of the process.

6.5.2.1 System Requirements

In B3G systems, the main goal is to improve spectral efficiency and provide high data rates and high QoS. The intended data rates for the B3G systems are in the range of 100 Mbit/s for wide-area coverage and 1 Gbit/s for hot spots. Moreover, it is possible that variable bandwidths are applied depending on user location and service requirements. It has been stated that the criterion for the success is the achieved system capacity in Erlang/(MHz $\times$ km^2), which indicates that the performance evaluation of the candidate air-interface

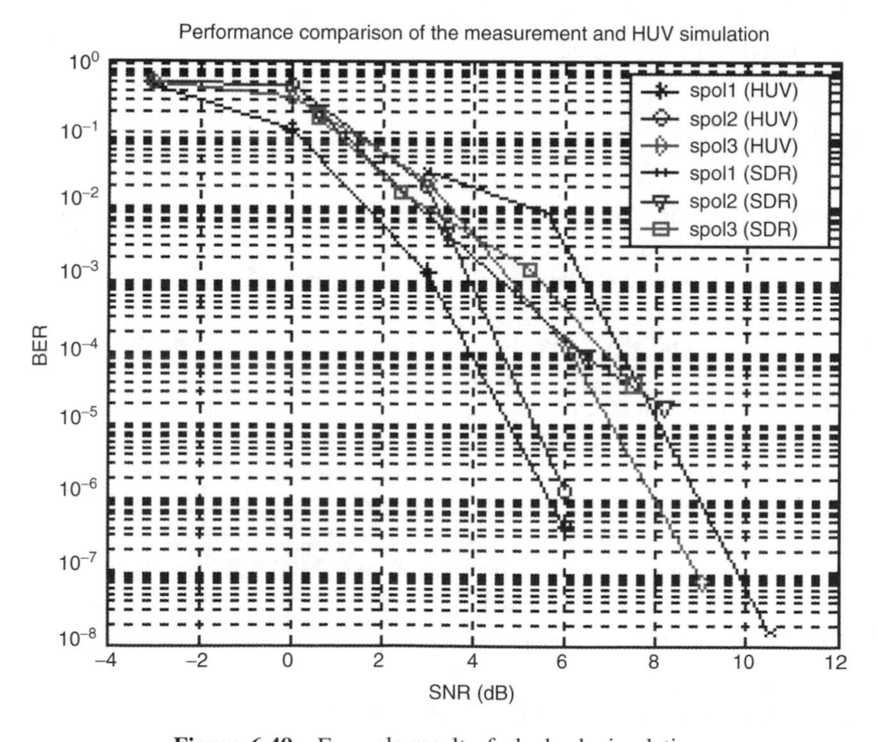

Figure 6.49 Example result of playback simulation

concepts have to be performed both at link and system levels. For example, the impact of new scheduling and interference coordination methods in the system-level need to be studied along with the potential coexistence of multiple radio access systems [228]. In addition, it is expected that some new location services set extremely tight requirements for the accuracy of radio channel measurements and models including absolute position of the mobile, mobile route, and long-term evolution of the channel parameters.

In the Wireless World Research Forum (WWRF) White Paper on smart antennas [195] the potential of multi-antenna systems as an enabler to address the goals of next generation system design in terms of capacity and spectrum efficiency are discussed. Both conventional beam forming and advanced MIMO approaches, such as space–time coding and spatial multiplexing, are addressed. It is also emphasized that realistic performance evaluation, both at the link and the system-level, of the proposed multi-antenna techniques will be based on measurement data–based simulations as well as model-based simulations in different deployment scenarios (macro-cellular, micro-cellular, in-building, into building, ad hoc, and multihop networks, etc.) and will require the following:

- realistic MIMO channel modeling;
- realistic interference modeling;
- accurate mapping between link and system-level results;
- realistic modeling of implementation losses;
- realistic terminals and user handlings.

Although some of these topics have been addressed in earlier projects ([115, 116, 118, 144, 145]), an extensive effort is still required in order to agree on link and system-level modeling of B3G multi-antenna systems. For example, many of the link-level properties of MIMO radio channels are not fully characterized, especially in high carrier frequencies and wide bandwidths where future systems design tends to be more environment-specific.

6.5.2.2 Generic Model

A generic channel model should be selected so that it supports all the system requirements, which can be foreseen. For example, COST259 or the generic WINNER model could serve as the generic model. However, it can be difficult to encompass in these models realistic near-field terminals and user-handling scenarios.

6.5.2.3 Measurement Plan

Multidimensional radio channel modeling also sets stringent requirements for the radio channel measurement campaigns, both in terms of specifications of the measurement equipment and of the measurement experiments planning, in order to provide data of suitable accuracy and statistical efficiency. Measurements have to be planned carefully to create sufficient statistics avoiding at the same time extremely large data files.

In measurement plans the most critical issue is to select a number of specific test environments, which represent reasonably well typical network deployment and user scenarios. Also, different network architectures – such as cellular, ad hoc, peer-to-peer, relay-based, hybrid, tandem, distributed radio, distributed MIMO, cooperative techniques – should be taken into account. The selection of antenna configurations for the terminals and BSs including beam forming and diversity antennas, different sectorization schemes, distributed antenna systems and polarization, and sensor networks has to be done. Other important aspects are mobility scenarios (for the terminal and BS), and user terminal size, and handling conditions. Typical user environments include, for example, outdoor, indoor, outdoor-to-indoor, and PANs.

Outdoor scenarios can be further divided into the following types: Dense urban: macro, micro, Urban: macro, micro, Suburban/residential: macro, Rural: macro, Suburban Hilly: macro, Rural Hilly: macro, and Highway: macro, micro. Indoor scenarios can be office, home, airport/shopping malls, or some specific environments such as factory halls. Outdoor-to-indoor measurements should be done for dense urban, urban, suburban/residential, university campus, and business park environments, both macro and microcells.

The different scenarios have to be described in detail including the environment maps, photographs at the measurement locations, and so on. Measurement plan usually also includes the general measurement parameters such as the number of TX and RX antennas and their heights, terminal location/mobility, TX powers and so on, which are addressed in the following.

6.5.2.4 Channel Measurements

On the basis of the current assumptions on the expected features of B3G systems, a number of key parameters can be identified for the characterization of the multidimensional MIMO radio channel (see Table 6.1). These parameters determine the requirements for the specification of measurement devices as well as a list of interesting radio environments to be investigated.

Channel sounding for Gbit/s WLAN access systems will require a bandwidth larger than 1 GHz since the influence of the scattering and shadowing objects in the near antenna range at 60 GHz can only be correctly evaluated for simulation purposes if the temporal resolution is high enough. 1 GHz corresponds to 30 cm, which is still about 80 times wider than the wavelength at 60 GHz. So the measurement bandwidth should even be higher, for example, 5 GHz.

UWB-sounding requires a minimum bandwidth of at least 500 MHz within a frequency range between 3.1 to 10.6 GHz. This huge bandwidth of totally 7.5 GHz creates absolutely new challenges in propagation measurement and parameter estimation. We are faced with a paradigm shift from narrowband to broadband processing (note that existing 'broadband' sounder systems of 100 to 240 MHz bandwidth at, e.g. at 5 GHz, are still narrowband in terms of relative bandwidth). Also, in the mm-wave band we need some GHz of sounding bandwidth in order to resolve the relevant objects of the propagation environment, which is required to deduce realistic channel models. Joint DoA/DoD estimation capability is also required for mm-wave and UWB.

Novel UWB signal processing will allow reconstructing the actual propagation environment in terms of location, shape, and size of the scatterers from CIR data that are recorded along some route by moving the MS around. The method takes advantage of the excellent range resolution of UWB signals. It is of specific interest for indoor propagation studies, which often suffer from near range effects and too low spatial resolution. Additionally, the antenna element is the dispersive element itself and it is inherently included in the channel

Table 6.1 Measurement parameters

Parameter	Value
Carrier frequency	2–6 GHz for cellular and WLAN applications, 3.1–10.6 GHz UWB and up to mm-Wave for short range, BAN, and PAN.
Bandwidth	10–100 MHz or more (scalable), 500 MHz–7 GHz for UWB and mm-Wave
Sampling	I & Q (phase and amplitude)
Data storage	High-speed multichannel data recording, large memory
Antenna configuration	Beam forming, distributed, diversity, MIMO, hybrid, flexible measurement configurations with respect to the number of TX and RX antennas, antenna types, antenna constellations, antenna mountings, etc.
Number of TX antennas	High enough to enable antenna-independent models and to ensure the high spatial resolution (e.g. 8–32 dual-polarized, depends on measurement scenario and accuracy requirement)
Number of RX antennas	High enough to enable antenna-independent models and to ensure the high spatial resolution (e.g. 8–32 dual-polarized, depends on measurement scenario and accuracy requirement)
Multiple sounder operation	Synchronization and coordinated sounder air-interface operation to emulate distributed multiple link systems.
Mobility	On-body 0–0.5 m/s, pedestrian 0–5 km/h, mobile 20–250 km/h, fast trains >250 km/h
User device	Type and handling
Delay resolution	~10 ns, 200 to 300 ps for UWB
Max delay range	1–100 microseconds, up to 500 ns for UWB and mm-wave
Dynamic range of the measurement	10–50 dB
Post-processing	Efficient post-processing tools for data analysis and automated signal feature extraction.
Specific topics	Multiband measurements System-level measurements, e.g. 4×4, 8×8 MIMO Measurements with application-specific antennas (body-worn, hand-held, etc.)
Other tools	Location control (GPS etc.), scene video
Other requirements	Easy-to-use and flexible user-interface
Accessories	Measuring vehicle, lifts, masts, and other supporting mechanics

model. Thus generic UWB channel models are difficult to generate [211]. Further applications are for investigation of localization principles in cooperative UWB sensor networks. UWB radio channel models have already been reported, for example, in [229–237].

Future channel sounders are also expected to support distributed MIMO systems. This will require cooperation between multiple sounders and distributed antenna deployment. These measurement systems will not only allow to emulate MU systems, but also more advanced schemes such as cooperative distributed downlink transmission systems, relay and multihop links, distributed and cooperative sensor networks, and also MT navigation features (infrastructure-based and stand-alone).

Furthermore, a large variety of antenna arrays and elements are employed aimed at different applications, ranging from very high resolution in DoA/DoD/Azimuth/Elevation/ Polarization approaches, which may require a large number of antenna array elements to antenna arrays for hand-held devices, which are designed to exploit space diversity but are subject to cost and available space constraints. UWB an mm-wave antenna and antenna array design for high-precision spatial measurement is still an open task.

6.5.2.5 Data Postprocessing

For accurate radio channel modeling, multiple parameters from the measured data have to be extracted reliably. Therefore, advanced post-processing algorithms are required for multidimensional detection and parameter evaluation. The key motivation of the accurate radio channel characterization is to understand the basic link and system-level limitations for advanced high data rate services for future wireless communication systems. The main emphasis has to be given to the spatial properties of multi-antenna radio channels. The MIMO radio channel studies are related mainly to the estimation of the MIMO channel matrix and its statistical properties. The spatial correlation properties at the transmitter and at the receiver are important when comparing the performance in terms of channel capacity (maximum achievable performance bound) and/or link quality (FER) of different multi-antenna schemes. For example, the relative performance of diversity and beam forming schemes depends on the angular spread at the transmitter and the receiver, as well as the distribution of the co-channel users in the network. Another important topic related to MIMO performance is the potential advantage of polarization diversity as an additional dimension of diversity to be exploited along with space, time, frequency, and code diversity.

For the joint high-resolution multidimensional parameter estimation of the interested channel parameters of interest (DoA, DoD, Time delay, Doppler and polarimetric path weights), various algorithms have been proposed, including the multidimensional Unitary ESPRIT (Estimation of Signal Parameters via Rotational Invariance Techniques) algorithm [238, 239] and the SAGE (Space Alternating Generalized Expectation Maximization) method [238] which, essentially, is an EM (Expectation Maximization) based simplified ML parameter estimation procedure. The SAGE approach, for example, may be applied to a large variety of antenna array architectures [240–247]. ML parameter estimation procedures are in general more flexible to cope with these requirements, whereas ESPRIT is restricted to arrays that show a shift invariant behaviour. SAGE is essentially a consecutive parameter-wise or parameter-subset-wise search. A gradient based-multidimensional ML channel parameter estimation framework, called *RIMAX*, has been described in [248–250]. The algorithm is based on EM and on gradient-based non-linear least squares (Gauss–Newton/Levenberg-Marquardt). The resulting estimates are the deterministic parameters of the specular propagation paths as well as the delay distribution of diffuse scattered components. The estimator exhibits robust adaptive model order control and high-resolution performance of closely spaced coherent propagation paths in all dimensions and provides reliability information of the estimated parameters. Recent investigations have shown that the SAGE algorithm combined with suitable switching modes of the TX and RX arrays allows for estimation of Doppler frequencies with magnitude up to half the rate with which the sub-channels of the MIMO system consisting of the TX array, the propagation channel, and the RX array are sounded, and especially is independent of the number of elements in these arrays [246, 247].

For near-field and user-proximity conditions, the high-resolution algorithms will have higher challenge to perform than in the case of traditional far-field conditions.

6.5.2.6 Channel Modeling

Because of wide bandwidths, use of multiple antennas and new carrier frequencies accurate radio channel models do not exist for B3G systems. Further requirements for channel modeling may lead to practical simulation problems since conventional tap-delay line modeling may become rather challenging due to the large number of taps. Thus, one important issue in channel modeling is to allow simple and accurate implementation on a simulator. A proper choice of the model structure and dimension can dramatically reduce the computational complexity. Some mixture of deterministic and stochastic modeling approaches seems to be adequate for WB modeling. The modeling of MIMO radio channels makes the identification of efficient simulation models even more challenging especially for system performance simulations of MU and multicell scenarios. Furthermore, future ubiquitous systems where the coexistence of several air-interface standards is envisioned, new network architectures bring additional demands for the modeling.

For short-range indoor networks, sensor networks, PANs, BANs, and so on, we are faced with completely new modeling requirements in terms of near range behaviour, antenna architectures, bandwidth and application scenarios, and so on. In the huge UWB range below 10 GHz, we can expect a strong frequency-selective channel response because of the different structural- and material-dependent scattering and transmission characteristics. Completely different modeling paradigms arise in the mm-wave band because of high propagation attenuation, low material penetration, and stronger influence of object roughness. Also, the influence of shadowing objects will be completely different. Investigation of localization principles in cooperative UWB sensor networks will require channel models that are geometrically correct.

In addition to accurate MIMO channel characteristics (correlation, DoA, DoD, polarization, diffuse and specular components), inter-path, MU, inter- and intra-cell, adjacent channel, heterogeneous systems interference and noise models are needed. Effects of interference coordination with system-level scheduling and different feedback schemes as well as MU, multicell scenarios (cellular, fixed wireless, ad hoc, multihop, relay and hybrid networks) have to be modeled. Channel models should also support dynamically changing environments, which are unavoidable in realistic near-field and user-proximity scenarios.

6.5.2.7 System Performance Evaluation (Simulation)

For realistic evaluation of system performance, it is required that the relevant global statistics of certain well-defined propagation environments are reproduced by the channel model. This may include large-scale changes of structural and statistical parameters and transition between different scenarios. The system-level simulations introduce a number of challenges associated with the identification of the right balance in the complexity versus accuracy trade-off. To this end, a number of modeling methodologies are potential candidates:

- new approaches to generating MIMO channel models from measurement data;
- antenna-independent channel measurement and modeling that allows antenna embedding and de-embedding (it requires to include polarization);
- channel measurement and modeling with embedded antenna system effects for realistic user-proximity scenarios (possibly the only viable solution in UWB);
- techniques for efficient MIMO channel modeling to enable feasible simulation times, including efficient link to system-level interface methodologies;
- feasibility study of deterministic/stochastic/semi-deterministic modeling.

Performance of wireless techniques/technologies for cellular mobile communications systems have to be evaluated in terms of three categories: link-level, network-level, and system-level. Performance figures of these levels are summarized as follows:

- Link-level performance figures are related to efficiencies of radio signal transmission techniques/technologies such as modulation/demodulation schemes, radio access schemes, coding and decoding techniques, signal processing algorithms, and interference scenarios.
- Network-level performances are related to efficiencies of the joint operation of radio signal transmission and resource allocation schemes such as joint radio access and dynamic frequency/time slot allocation, adaptive modulation, and autonomous segregation as well as dynamic power allocation based on the water-filling criterion.
- System-level performances indicate spectrum efficiency of overall operation of the link-level and network-level techniques/technologies. System-level performance indicates how much traffic can be carried by the system considered per-Hz per-square meters. This figure is referred to as Erlang Capacity.

Link-, network, and system-level performances can be estimated by conducting simulations for the algorithms, techniques, and technologies to be evaluated, assuming multiple users in the cell layout considered. Model-based propagation channels have long been used for the simulations where all propagation paths between multiple users and multiple BSs have to be simulated. However, considering the fact that in order to support higher information bit rates, future mobile communication systems have to use broader bandwidth than their ancestral systems, accuracy of evaluations through model-based multicell MU simulations is limited.

If CIR data representing channel transfer function between transmitter's each antenna element to receiver's each antenna element can be measured, the spatial and temporal spreads are inherently expressed by the data set. Such measurement data can be usually obtained through multidimensional channel sounding field measurement campaign, The received composite signal's waveform can be calculated by convoluting the transmitted waveform with the CIR data element-by-element. If this process is performed for each of the simultaneous users located in cells considered, and if the calculated waveforms are added all together, the received composite signal waveform comprising desired and interference users' and base stations' received signals, can be calculated. Then, the algorithms for the techniques/technologies to be evaluated can be run, and their performances can be evaluated in a quantitative manner. This method provides us with much more realistic performance estimates/prediction in real fields than the model-based simulations because the data represents realistic propagation scenarios. Objectives of link, network, and system-level simulations are as follows:

- Link-level simulations aim at evaluating techniques/technologies link-performances in terms of BER and FER in various propagation environments. They focus on analyzing/evaluating the impact of the system configurations, parameters, and algorithms that are related to signaling schemes over radio on performance.
- Network-level simulations aim at evaluating techniques/technologies network-performances in terms of throughput, blocking rate, and call loss rate in various propagation environments. They focus on analyzing/evaluating the impact of the network configurations, random access schemes, resource allocation methods, parameters, and algorithms, that are related to network layer aspects including handoff and power control, on performance.
- System-level simulations aim at obtaining system-level performance figures such as signal-to-interference plus noise power ratio (SINR) and BER in terms of geographical performance distributions in the area of interest and outage probability. The main focus is on interference scenarios, which depend on cell design including frequency reuse and user distribution.

Measurement data sets obtained through real-time MIMO channel sounding in various propagation environments are used to evaluate performances of broadband MIMO systems as well as to estimate the multipath characteristics in considered scenarios. By correlating both results, the performance dependency and sensitivity of the transmission and signal detection algorithms to propagation conditions can be identified. Goals of this phase of the research should be the realistic evaluation and cross-layer issues, spanning from the physical layer to the system layer, for broadband MIMO broadband mobile communications systems. Focus is given to link-, network- and system-level issues, taking into account the imperfections due to restrictions for practical implementations.

For efficient system-level performance evaluation the following topics have to be considered: approximation methods (approximation of analytical models, reducing the complexity, BER simulations), channel simulation techniques (SW/hardware simulation), and system-level simulation techniques (multicell simulation aspects, MU simulation aspects).

6.5.3 Open Issues and Research Topics

Although considerable progress has been made in the area of radio channel modeling there are a number of open issues that require further investigations.

6.5.3.1 Open Issues in Multichannel Radio Channel Characterization

Open Issues Related to Propagation

Propagation phenomena and channel models for high carrier frequencies (3–6 GHz) need to be further studied in terms of path loss as a function of frequency, fading characteristics, channel correlation properties, eigenstate properties, and correlation properties between interfering base stations and interfering users.

Furthermore, diffuse scattering and cluster statistics require realistic modeling.

Multidimensional channel modeling (e.g. indoor 3D channels and outdoor 2D/3D channels), has not yet been adequately addressed.

Dynamic and transition scenarios (e.g. outdoor to indoor, corner effects, user-proximity effects) and indoor/outdoor positioning scenarios have not received significant attention and require further analysis.

Finally, cross-correlation between small-scale and large-scale effects is an important aspect in determining the accuracy of a channel model but remains relatively unknown.

Open Issues Related to System Parameters

The introduction of new system assumptions imposed by the design requirements for next generation wireless networks translates into a number of new channel modeling challenges associated with the modeling for higher and scalable bandwidths (10 to 100 MHz or more), multiband operation, and UWB systems. New transceiver architectures, such as multi-antenna mode selection including MIMO, diversity and beamforming, and network structures introduce a number of requirements in channel modeling that cannot be addressed on the basis of methodologies followed for current systems.

Open Issues Related to Measurements and Efficient Modeling

Antenna-independent field prediction techniques, including antenna embedding and de-embedding, and antenna modeling of mutual coupling and dispersive behavior, remains an open issue.

In the case of nonconventional access architectures for example, multihop, ad hoc, cooperative transmission, and so on, modeling of the overall (or equivalent) channel between source and destination would give us a better insight to understand the equivalent link behavior.

In MU MIMO system, time and frequency (and space) scheduling are fundamental for an effective use of the radio resources. The interaction between this multidimensional scheduling and the channel behavior in these scenarios should be further investigated and understood. Knowing the multidimensional channel behavior will help designers to better predict its evolution in time and ultimately, to exploit efficiently the available radio resources. Moreover, efficient methods for generating MU MIMO channels should be investigated in order to reduce the evaluation time of network and system performance.

Another interesting aspect that should be investigated results from the fact that B3G wireless communication systems will be highly heterogeneous, encompassing different kinds of networks and a broad range of different terminals (with different sizes and capabilities). From the MIMO standpoint it means that one cannot necessarily assume a fixed MIMO antenna configuration, for the entire period during which a communication link is active. It is in fact fixed for the time that a given terminal is connected to a given BS but after a handover (horizontal or vertical) the MIMO channel may end up with a different configuration. How this time-variable MIMO configuration affects the link, network, and system performance remains also an open point for further research.

How handling of the terminal, body-proximity, and nearby obstacles impact on the characteristics of the MIMO channel need to be better understood as, for instance, the antenna element coupling resulting from the closeness of objects to the array may reduce the rank of the channel matrix.

In several situations, in ad hoc networks, for instance, the physical dimensions of the MIMO system could be very small resulting in communications over a near-field MIMO channel.

Finally, MIMO channel reciprocity, both short- and long-term, plays a decisive role in the design of MIMO transceivers – in terms of the required signaling overhead–and the evaluation of the conditions under which channel reciprocity holds will be extremely valuable in that respect.

6.5.3.2 Proposed Research Topics

From the above considerations, it can be stated that the development and standardization of evolving and future WB wireless communication systems require system performance evaluation using realistic channel modeling assumptions. Indeed, efficient exploitation of the radio channel requires that the transmitter and the receiver be designed to 'match' the instantaneous radio propagation environment.

There could be several propagation scenarios for each environment depending on the BS location, mobile user movement, frequency, and bandwidth assumptions. The following procedure for channel modeling should be completed for each propagation scenario:

- extensive multidimensional WB measurement campaigns including multi-antenna transmission and reception as well as polarization measurement. Several campaigns are needed for each propagation scenario to increase the generality of the results;
- extensive measurement campaigns including realistic user terminal and device handling;
- development of refined, efficient estimation techniques for channel parameter estimation;
- collection and creation of data libraries for different environments and user scenarios;
- development of multidimensional radio channel models for link-level simulations of future wireless systems;
- development of multidimensional radio channel models for system-level simulations of future wireless systems;
- verification of the link and system-level channel models.

6.5.4 Summary

In this section, the state of the art on radio channel measurements, modeling and simulation methodology are presented. Because of the complexity of future wireless systems, propagation research efforts will require close cooperation and efficient work sharing between academia and industry, and also between propagation and communication researchers. This section focused mostly on the wide-area communications in frequency range of 2 to 6 GHz. Future work will address UWB and microwave frequencies (up to 60 GHz).

It should be noted that designing next generation wireless MIMO systems without knowing the operational frequency bands is a difficult if not an impossible task. The World Radio Conference (WRC) is expected to decide on the spectrum allocation for B3G systems in 2007. Producing conclusive results on the relative merits between different bands or critical issues concerning certain bands could have a significant impact on the final decision.

6.6 Acknowledgment

The following individuals contributed to this chapter: Reiner S. Thomä, Marko Milojevic (Technische Universität Ilmenau), Bernard H. Fleury, Jørgen Bach Andersen, Patrick C. F. Eggers, Jesper Ø. Nielsen, István Z. Kovács (Aalborg University, Denmark), Juha Ylitalo, Pekka Kyösti, Jukka-Pekka Nuutinen, Xiongwen Zhao (Elektrobit Testing, Finland), Daniel Baum, Azadeh Ettefagh (ETH Zürich, Switzerland), Moshe Ran (Holon Academic Institute of Technology, Israel), Kimmo Kalliola, Terhi Rautiainen (Nokia Research Center, Finland), Dean Kitchener (Nortel Networks, UK), Mats Bengtsson, Per Zetterberg (Royal Institute of Technology – KTH, Sweden), Marcos Katz (Samsung Electronics, Korea), Matti Hämäläinen, Markku Juntti, Tadashi Matsumoto, Juha Ylitalo (University of Oulu/CWC, Finland), Nicolai Czink (Vienna University of Technology, Austria).

References

[1] ITU Recommendation ITU-R M.1645, "Framework and Overall Objectives of the Future Development of IMT-2000 and Systems Beyond IMT-2000", 2003.

[2] S. Kaiser *et al*, *"WWRF/WG4/Subgroup on New Air Interfaces White Paper: New Air Interface Technologies - Requirements and Solutions"*, draft version, August, 2002.

[3] W. Mohr, R. Lüder amd K-H. Möhrmann, "Data Rate Estimates, Range Calculations and Spectrum Demand for New Elements of Systems Beyond IMT-2000", *Proc. 5th Symposium Wireless Personal Multimedia Communications WPMC02*, Honolulu, HI, Oct., Vol. 1, 2002, pp. 27–30.

[4] Z. Wang and G. B. Giannakis, "Wireless Multicarrier Communications – Where Fourier Meets Shannon", *IEEE Signal Process. Mag.*, Vol. 17, May, 2000, pp. 29–48.

[5] P. Frenger, "A Framework for Future Wireless Access", *Proc. WWRF13*, Working Group 4, Jeju, Korea, Mar., 2005.

[6] L. J. Cimini Jr. and N. R. Sollenberger, "Peak-to-Average Power Ratio Reduction of an OFDM Signal Using Partial Transmit Sequences", *IEEE Commun. Lett.*, Vol. 4, No. 3, March, 2000, pp. 86–88.

[7] R. van Nee and R. Prasad, *OFDM for Wireless Multimedia Communications*, Artech House, 2000.

[8] K. Fazel and S. Kaiser, *Multi-Carrier and Spread Spectrum Systems*, John Wiley & Sons, 2003.

[9] D. Falconer, S. L. Ariyavisitakul, A. Benyamin-Seeyar and B. Eidson, "Frequency Domain Equalization for Single-Carrier Broadband Wireless Systems", *IEEE Commun. Mag.*, Vol. 40, No. 4, April, 2002, pp. 58–66.

[10] H. Sari, G. Karam and I. Jeanclaude, "Frequency-Domain equalization of Mobile Radio and Terrestrial Broadcast Channels", *Proc. Globecom '94*, San Francisco, CA, Nov.–Dec., 1994, pp. 1–5.

[11] I. Koffman and V. Roman, "Broadband Wireless Access Solutions Based on OFDM Access in IEEE 802.16", *IEEE Commun. Mag.*, Vol. 40, No. 4, April, 2002, pp. 96–103.

[12] H. Sari, Y. Levy and G. Karam, "An Analysis of Orthogonal Frequency-Division Multiple Access", *Proc. of IEEE GLOBECOM '97*, Phoenix USA, pp. 1635–1639.

[13] S. Kaiser, "MC-FDMA and MC-TDMA Versus MC-CDMA and SS-MC-MA: Performance Evaluation for Fading Channels", *Proc. 5th International Symposium On Spread Spectrum Techniques and Applications*, Oulu Finland, Vol. I, 1998.

[14] S. Kaiser, "OFDM Code Division Multiplexing in Fading Channels," *IEEE Trans. Commun.*, Vol. 50, No. 8, August, 2002, pp. 1266–1273.

[15] B. Muquet, M. de Courville, P. Duhamel and G. Giannakis, "OFDM with Trailing Zeros Versus OFDM with Cyclic Prefix: Links, Comparisons and Application to the Hiperlan/2 System", *Proc. ICC 2000*, New Orleans, LA, June, 2000, pp. 1049–1053.

[16] M. Muck, M. de Courville, M. Debbah and P. Duhamel, "A Pseudo Random Postfix OFDM Modulator and Inherent Channel Estimation Techniques", *Proc. Globecom 2003*, San Francisco USA, Dec., 2003.

[17] M. Muck, M. De Courville and P. Duhamel, "Postfix Design for Pseudo Random Postfix OFDM Modulators", *Proc. International OFDM Workshop*, Dresden, Germany, Sept., 2004.

[18] M. Muck, M. De Courville, X. Miet and P. Duhamel, "Iterative Interference Suppression for Pseudo Random Postfix OFDM Based Channel Estimation", *IEEE International Conference on Acoustics, Speech and Signal Processing*, Philadelphia USA, Mar., 2005.

[19] M. Muck, A. R. Dias, M. De Courville and P. Duhamel, "A Pseudo Random Postfix OFDM-Based Modulator for Multiple Antennae Systems", *Proc. ICC 2004*, Paris, France, May, 2004.
[20] K. Fazel and L. Papke, "On the Performance of Convolutionally-Coded CDMA/OFDM," *Proc. PIMRC '93*, Yokohama, Japan, Sept., 1993, pp. 468–472.
[21] D. Lacroix, J. P. Javaudin and N. Goudard, "IOTA, an Advanced OFDM Modulation for Future Broadband Physical Layers", *Wireless world Research Forum (WWRF) Meeting #7*, Eindhoven, Holland, Dec., 2002.
[22] A. Gusmão, R. Dinis, J. Conceição and N. Esteves, "Comparison of Two Modulation Choices for Broadband Wireless Communications", *Proc. VTC 2000 Spring*, Tokyo, Japan, May, Vol. 2, 2000, pp. 1300–1305.
[23] H. Witschnig, T. Mayer, A. Springer, L. Maurer, M. Huemer and R. Weigel, "The Advantages of a Known Sequence versus Cyclic Prefix in a SC/FDE System", *Proc. 5th Symposium Wireless Personal Multimedia Communications WPMC02*, Honolulu, HI, Oct., Vol. 3, 2002, pp. 1328–1332.
[24] D. Falconer and S. L. Ariyavisitakul, "Broadband Wireless Using Single Carrier and Frequency Domain Equalization", *Proc. of the 5th International Symposium on Wireless Personal Multimedia Communications*, Honolulu, HI, Oct. 27–30, 2002.
[25] L. Brühl and B. Rembold, "Unified Spatio-Temporal Frequency Domain Equalization for Multi- and Single-Carrier CDMA Systems", *Proc. VTC 2002 Fall*, Vancouver, Canada, Oct., 2002, pp. 676–680.
[26] R. F. H. Fischer and J. B. Huber, "On the Equivalence of Single-and Multicarrier Modulation: A New View", *Proc. ISIT 1997*, Ulm, Germany, July, 1997.
[27] F. Horlin, F. Petré, E. Lopez-Estraviz, F. Naessens and L. Van der Perre, "A Generic Transmission Scheme for Fourth Generation Wireless Systems", presentation at *WWRF Meeting #11*, Oslo, Norway, June, 2004.
[28] U. Sorger, I. De Broeck and M. Schnell, "Interleaved FDMA – A New Spread-Spectrum Multiple-Access Scheme", *Proc. ICC98*, Atlanta, GA, June, 1998, pp. 1013–1017.
[29] D. Falconer, R. Dinis, C. T. Lam and M. Sabbaghian, "Frequency Domain Orthogonal Signature Sequences (FDOSS) for Uplink DS-CDMA", presentation at *WWRF Meeting #10*, New York, Oct., 2003.
[30] D. Falconer, R. Dinis, T. Matsumoto, M. Ran, A. Springer and P. Zhu, *''WWRF/WG4/Subgroup on New Air Interfaces White Paper: A Mixed Single Carrier/OFDM Air Interface for Future-Generation Cellular Wireless Systems"*, Aug., 2003.
[31] C.-M. Chang and K.-C. Chen, "Frequency-Domain Approach to Multiuser Detection in DS-CDMA Communications", *IEEE Commun. Lett.*, Vol. 4, No. 11, Nov., 2000, pp. 331–333.
[32] R. Dinis, D. Falconer, C. T. Lam and M. Sabbaghian, "A Multiple Access Scheme for the Uplink of Broadband Wireless Systems", *Proc. Globecom, 2004*, Dallas, TX, Dec., 2004.
[33] D. Galda and H. Rohling, "A Low Complexity Transmitter Structure for OFDM-FDMA Uplink Systems", *Proc. VTC Spring 2002*, Birmingham, Alabama, USA, 2002, pp. 1737–1741.
[34] P. Xia, S. Zhou and G. B. Giannakis, "Bandwidth-and Power-Efficient Multi-Carrier Multiple Access", *IEEE Trans. Commun.*, Vol. 51, No. 11, Nov. 2003, pp. 1828–1837.
[35] Y. Goto, T. Kawamura, H. Atarashi and M. Sawahashi, "Variable Spreading and Chip repetition Factors (VSCRF)-CDMA in Reverse Link for Broadband Wireless Access", *Proc. PIMRC 2003*, Beijing, China, 2003.
[36] N. Maeda, Y. Kishiyama, K. Higuchi, H. Atarashi and M. Sawahashi, "Experimental Evaluation of Throughput Performance in Broadband Packet Wireless Access Based on VSF-OFCDM and VSF-CDMA", *Proc. PIMRC 2003*, Beijing, China, 2003.
[37] D. A. Wiegandt, Z. Wu and C. R. Nassar, "High Throughput, High Performance OFDM via Pseudo-Orthogonal Carrier Interferometry Spreading Codes", *IEEE Trans. Commun.*, Vol. 51, No. 7, July, 2003, pp. 1123–1134.
[38] Z. Wang and G. B. Giannakis, "Complex-Field Coding for OFDM over Fading Wireless Channels", *IEEE Trans. Inf. Theory*, Vol. 49, No. 3, Mar., 2003, pp. 707–720.
[39] N. Benvenuto and S. Tomasin, "On the Comparison Between OFDM and Single Carrier with a DFE Using a frequency Domain Feedforward Filter", *IEEE Trans. Commun.*, Vol. 50, No. 6, June, 2002, pp. 947–955.
[40] J. Tubbax, L. Van der Perre, S. Donnay and M. Engels, "Single Carrier Communications Using Decision-Feedback Equalization for Multiple Antennas", *Proc. ICC 2003*, Anchorage, AK, May, Vol. 4, 2003, pp. 2321–2325.
[41] G. Caire, G. Taricco and E. Biglieri, "Bit-Interleaved Coded Modulation", *IEEE Trans. Inf. Theory*, May, 1998, pp. 927–946.

[42] J. Louveaux, L. Vandendorpe and T. Sartenaer, "Cyclic Prefixed Single Carrier and Multicarrier Transmission~: Bit Rate Comparison", *IEEE Commun. Lett.*, Vol. 7, No. 4, April, 2003, pp. 180–182.
[43] N. Yee, J.-P. Linnartz and G. Fettweis, "Multi-Carrier CDMA in Indoor Wireless Radio Networks," *Proc. PIMRC 1993*, Yokohama, Japan, Sept., 1993, pp. 109–113.
[44] N. Zervos and I. Kalet, "Optimized Decision Feedback Equalizer Versus Optimized Orthogonal Frequency Division Multiplexing for High Speed Data Transmission over the Local Cable Network", *Proc. International Conference On Communication*, Boston, USA, June, Vol. 2, 1989, pp. 1080–1085.
[45] S. Kaiser and J. Hagenauer, "Multi-Carrier CDMA with Iterative Decoding and Soft-Interference Cancellation", *Proc. GLOBECOM 97*, Phoenix, AZ, Nov., 1997.
[46] D. Kivanc, G. Li and H. Liu, "Computationally Efficient Bandwidth Allocation and Power Control for OFDMA", *IEEE Trans. Wireless Commun.*, Vol. 2, No. 6, Nov., 2003, pp. 1150–1158.
[47] C. Y. Wong, R. S. Cheng, K. B. Letaief and R. D. Murch, "Multiuser OFDM with Adaptive Subcarrier, Bit and Power Allocation", *IEEE Sel. Areas Commun.*, Vol. 17, No. 10, Oct., 1999, pp. 1747–1758.
[48] P. Montezuma and A. Gusmão, "A Pragmatic Coded Modulation Choice for Future Broadband Wireless Communications", *Proc. VTC 2001 Spring*, Rhodes, Island, May, Vol. 2, 2001, pp. 1324–1328.
[49] Z. Wang, X. Ma and G. B. Giannakis, "OFDM or Single-Carrier Block Transmissions?", *IEEE Trans. Commun.*, Vol. 52, No. 3, Mar., 2004, pp. 380–394.
[50] N. Benvenuto and S. Tomasin, "Block Iterative DFE for Single Carrier Modulation", *IEE Electron. Lett.*, Vol. 39, No. 19, Sept., 2002.
[51] R. Dinis, A. Gusmão and N. Esteves, "On Broadband Block Transmission over Strongly Frequency-Selective Fading Channels", *Proc. Wireless 2003*, Calgary, Canada, July, 2003.
[52] P. Schniter and H. Liu, "Iterative Equalization for Single Carrier Cyclic Prefix in Doubly Dispersive Channels", *Conference. Record Asilomar Conference on Signals, Systems and Computers*, Nov., 2003.
[53] T. Abe, S Tomisato, T Matsumoto, "Performance Evaluations of a space-time Turbo Equalizer in Frequency Selective MIMO Channels Using Field Measurement Data", *IEE Workshop on MIMO Communication Systems*, London, England, Dec., 2001, pp. 21/1–21/5.
[54] K. Kansanen and T. Matsumoto, "Frequency Domain Turbo Equalization for Broadband MIMO Channels", presentation at *COST 273 Meeting, TD(04) 143*, Göteborg, Sweden, June 9–10, 2004.
[55] X. Li and J. Ritcey, "Bit-Interleaved Coded Modulation with Iterative Decoding", *IEEE Trans. Commun.*, Vol. 45, No. 11, Nov., 1997, pp. 169–171.
[56] H. Atarashi, S. Abeta and M. Sawahashi, "Variable Spreading Factor-Orthogonal Frequency and Code Division Multiplexing (VSF-OFCDM) for Broadband Packet Wireless Access", *IEICE Trans. Commun.*, Vol. E86-B, No. 1, Jan., 2003, pp. 291–299.
[57] S. Abeta, H. Atarashi, M. Sawahashi and F. Adachi, "Performance of Coherent Multi-Carrier/DS-CDMA and MC-CDMA for Broadband Packet Wireless Access", *IEICE Trans. Commun.*, Vol. E84-B, No. 3, Mar., 2001, pp. 406–414.
[58] R. M. Buehrer, N. S. Correal and B. D. Woerner, "Simulation Comparison of Multiuser Receivers for Cellular CDMA", *IEEE Trans. Veh. Technol.*, Vol. 49, No. 4, July, 2000, pp. 1065–1085.
[59] G. B. Giannakis, "Filterbanks for Blind Channel Identification and Equalization", *IEEE Signal Process. Lett.*, Vol. 4, No. 6, June, 1997, pp. 184–187.
[60] B. Muquet, Z. Wang, G. B. Giannakis, M. De Courville and P. Duhamel, "Cyclic Prefixing or Zero Padding for Wireless Multicarrier Transmissions?", *IEEE Trans. Commun.*, Vol. 50, No. 12, Dec., 2002, pp. 2136–2148.
[61] B. Le Floch, M. Alard and C. Berrou, "Coded Orthogonal Frequency Division Multiplex", *Proc. IEEE*, Vol. 83, No. 6, June, 1995, pp. 982–996.
[62] J. P. Javaudin, C. Dubuc, D. Lacroix and M. Earnshaw, "An OFDM Evolution to the UMTS High Speed Downlink Packet Access", *Proc. VTC 2004 Fall*, Los Angeles, CA, Sept., 2004.
[63] T. Pollet, M. Van Bladel and M. Moeneclaey, BER Sensitivity of OFDM Systems to Carrier Frequency Offset and Wiener Phase Noise", *IEEE Trans. Commun.*, Vol. 43, No. 2/3/4, Feb./Mar./April, 1995, pp. 191–193.
[64] P. Robertson and S. Kaiser, "Analysis of the Effects of Phase-Noise in Orthogonal Frequency Division Multiplex (OFDM) Systems," *Proc. ICC '95*, Seattle, WA, June, 1995, pp. 1652–1657.
[65] P. Robertson and S. Kaiser, "The Effects of Doppler Spreads in OFDM(A) Mobile Radio Systems", *Proc. VTC 99*, Amsterdam, Netherlands, pp. 329–333.
[66] H. Steendam and M. Moeneclaey, "The Effect of Carrier Frequency Offsets on Downlink and Uplink MC-DS-CDMA", *IEEE Sel. Areas Commun.*, Vol. 19, No. 12, Dec., 2001, pp. 2528–2536.

[67] H. Steendam, M. Moeneclaey and H. Sari, "The Effect of Carrier Phase Jitter on the Performance of Orthogonal Frequency-Division Multiple-Access Systems", *IEEE Trans. Commun.*, Vol. 46, No. 4, April, 1998, pp. 456–459.
[68] A. M. Tonello, N. Laurenti and S. Pupolin, "Analysis of the Uplink of an Asynchronous Multi-user DMT OFDMA System Impaired by Time Offsets, Frequency Offsets and Multipath Fading", *Proc. VTC 2000 Fall*, Boston USA, pp. 1094–1099.
[69] J. Armstrong, "Analysis of New and Existing Methods of Reducing Intercarrier Interference Due to Carrier Frequency Offset in OFDM", *IEEE Trans. Commun.*, Vol. 47, No. 3, March, 1999, pp. 365–369.
[70] G. Leus, I. Barhumi, O. Rousseaux and M. Moonen, "Direct Semi-Blind Design of Serial Linear Equalizers for Doubly-Selective Channels", *Proc. of 2004 IEEE International Conference on Communications*, Paris France, June 20–24, Vol. 5, 2004, pp. 2626–2630.
[71] G. Carron, R. Ness, L. Deneire, L. Van der Perre and M. Engels, "Comparison of Two Modulation Techniques Using Frequency Domain Processing for In-House Networks", *IEEE Trans. Consumer Electron.*, Vol. 47, No. 1, Feb., 2001, pp. 63–72.
[72] P. Struhsaker and K. Griffin, Contribution to IEEE 802.16.3c-01/46, "Analysis of PHY Waveform Peak to Mean Ratio and Impact on RF Amplification", Mar. 6, 2001.
[73] C. Rapp, "Effects of HPA Nonlinearity on a 4DPSK/OFDM Signal for a Digital Sound Broadcasting System", *Proc. 2nd European Conference on Satellite Communications*, Liege, Belgium, Oct., 1991, pp. 179–184.
[74] P. J. Sinderbrand, "WCA Coalition Proposal to the FCC in the Matter of Amendment of Parts 1,21, 73, 74, and 101 of the Commission's Rules to Facilitate the Provision of Fixed and Mobile Broadband Access, Educational and Other Advanced Services in the 2150–2162 and 2500–2690 MHz Bands", Sept. 8, 2003.
[75] S. H. Müller and J. B. Huber, "OFDM with Reduced Peak-to-Average Power Ratio by Optimum Combination of Partial Transmit Sequences", *Electron. Lett.*, Vol. 33, No. 5, 27th Feb., 1997, pp. 368–369.
[76] S. H. Müller and J. B. Huber, "A Comparison of Peak Power Reduction Schemes for OFDM", *Proc. Globecom '97*, Phoenix, USA, 1997.
[77] P. Zhu, A. Khandani and W. Tong, "Scrambling-Based Peak-to-Average Power Reduction Without Side Information", *Proc. WWRF8bis*, Beijing, Feb., 2004.
[78] M. Lampe and H. Rohling, "Reducing Out-of-Band Emissions Due to Nonlinearities in OFDM Systems", *Proc. VTC Spring '99*, Houston, USA, 1999.
[79] A. E. Jones and T. Wilkinson, "Combined Coding for Error Control and Increased Robustness to System Nonlinearities in OFDM", *Proc. VTC '96*, Atlanta, GA, May, 1996.
[80] A. E. Jones, T. A. Wilkinson and S. K. Barton, "Block Coding Scheme for Reduction of Peak to Mean Envelope Power Ratio of Multicarrier Transmission Scheme", *Electron. Lett.*, Vol. 30, No. 25, Dec., 1994, pp. 2098–2099.
[81] J. Armstrong, "Peak-to-Average Power Reduction for OFDM by Repeated Clipping and Frequency Domain Filtering", *Electron. Lett.*, Vol. 28, Feb., 2002, pp. 246–247.
[82] R. Dinis and A, Gusmão, "A Class of Nonlinear Signal Processing Schemes for Bandwidth-Efficient OFDM Transmission with Low Envelope Fluctuation", *IEEE Trans. Commun.*, Vol. 52, No. 11, Nov., 2004, pp. 2009–2018.
[83] H. Ochiai, "Power Efficiency Comparison of OFDM and Single-Carrier Signals", *Proc. VTC, Fall*, Vancouver Canada, 2002, pp. 899–903.
[84] X. Li and L. J. Cimini Jr., "Effects of Clipping and Filtering on the Performance of OFDM", *IEEE Commun. Lett.*, Vol. 2, No. 5, May, 1998, pp. 131–133.
[85] C. Van den Bos, M. H. L. Kouwenhoven and W. A. Serdijn, "Effect of Smooth Nonlinear Distortion on OFDM Symbol Error Rate", *IEEE Trans. Commun.*, Vol. 49, No. 9, Sept., 2001, pp. 1510–1514.
[86] FCC web site on cognitive radio: www.fcc.gov/oet/cognitiveradio/
[87] S. Hijazi, B. Natarajan, M. Michelini, Z. Wu and C. R. Nassar, "Flexible Spectrum Use and Better Coexistence at the Physical Layer of Future Wireless Systems via a Multicarrier Platform", *IEEE Wireless Commun.*, Vol. 11, No. 2, April, 2004, pp. 64–71.
[88] P. Höher, S. Kaiser and P. Robertson, "Pilot-Symbol-Aided Channel Estimation in Time and Frequency", *Proc. GLOBECOM '97 (CTMC)*, Phoenix, AZ, Nov., 1997.
[89] Y. Li, L. J. Cimini and N. R. Sollenberger, "Robust Channel Estimation for OFDM Systems with Rapid Dispersive Fading Channels", *IEEE Trans. Commun.*, Vol. 46, No. 7, July, 1998, pp. 902–915.

[90] Y. Li, "Simplified Channel Estimation for OFDM Systems with Multiple Antennas", *IEEE Trans. Commun.*, Vol. 50, No. 1, Jan., 2002, pp. 67–75.
[91] D. Lacroix, N. Goudard and M. Alard, "OFDM with Guard Interval Versus OFDM/Offset QAM for High Data Rate UMTS Downlink Transmission", *Proc. VTC 2001, Fall*, Atlantic City, NJ, Oct., 2001, pp. 2682–2686.
[92] R. W. Heath, M. Airy and A. J. Paulraj, "Multiuser Diversity for MIMO Wireless Systems with Linear Receivers," *In Proc. 35th Asilomar Conf. on Signals, Systems, and Computers*, IEEE Computer Society Press, Pacific Grove, CA, Nov., 2001.
[93] Q. H. Spencer, C. B. Peel, A. L. Swindlehurst and M. Haardt, "An Introduction to the Multi-User MIMO Downlink," *IEEE Commun. Mag.*, Vol. 42, No. 10, Oct., 2004, pp. 60–67.
[94] S. Vishwanath, N. Jindal and A. J. Goldsmith, "On the Capacity of Multiple Input Multiple Output Broadcast Channels," *Proc. of the IEEE International Conference on Communications (ICC)*, New York, April, 2002.
[95] Z. Pan, K. K. Pan and T. Ng, "MIMO Antenna System for Multiuser Multi-Stream Orthogonal Space Time Division Multiplexing," *Proc. of the IEEE International Conference on Communications*, Anchorage, Alaska, May, 2003.
[96] K. K. Wong, "Adaptive Space-Division-Multiplexing and Bit-and Power Allocation in Multi-User MIMO Flat Fading Broadcast Channel," *Proc. of the IEEE 58th Vehicular Technology Conference*, Orlando, FL, Oct., 2003.
[97] M. Haardt, C. Mecklenbräuker, M. Vollmer, and P. Slanina, "Smart Antennas for UTRA TDD," *Eur. Trans. Telecommun. (ETT)*, special issue on Smart Antennas, J. A. Nossek and W. Utschick, guest editors, Vol. 12, No. 5, 2001, pp. 393–406.
[98] A. Paulraj, R. Nabar and D. Gore, *Introduction to Space-Time Wireless Communications*, Cambridge University Press, 2003.
[99] A. Hottinen, O. Tirkkonen and R. Wichman, *Multi-Antennas Transceiver Techniques for 3G and Beyond*, Wiley, 2003.
[100] G. J. Foschini and M. J. Gans, "On Limits of Wireless Communications in a Fading Environment when using Multiple Antennas", *Wireless Pers. Commun.*, Vol. 6, 1998, pp. 311–335.
[101] V. Tarokh, N. Seshadri and A. R. Calderbank, "Space-Time Block Coding for Wireless Communication: Performance Criterion and Code Construction", *IEEE Trans. Inf. Theory*, Vol. 44, No. 2, Mar., 1998, pp. 744–765.
[102] V. Tarokh, H. Jafarkhani and A. R. Calderbank, "Space-Time Block Coding for Wireless Communication: Performance Results", *IEEE Sel. Areas Commun.*, Vol. 17, No. 3, Mar., 1999, pp. 451–460.
[103] V. Tarokh, A. Naguib, N. Seshadri and A. R. Calderbank, "Combined Array Processing and Space-Time Coding," *IEEE Trans. Inf. Theory*, Vol. 45, May, 1999, pp. 1121–1128.
[104] E. Telatar, "Capacity of Multi-Antenna Gaussian Channels", *European Transactions on Telecommunications*, Vol. 10, No. 6, Nov/Dec, 1999, pp. 585–595.
[105] D. Chizhik, F. Rashid-Farrokhi, J. Ling and A. Lozano, "Effect of Antenna Separation on the Capacity of BLAST in Correlated Channels", *IEEE Commun. Lett.*, Vol. 4, No. 11, Nov., 2000, pp. 337–339.
[106] J. Bach Andersen, "Array Gain and Capacity for Known Random Channels with Multiple Element Arrays at Both Ends," *IEEE J. Sel. Areas Comm.*, Vol. 18, Nov., 2000, pp. 2172–2178.
[107] F. R. Farrokhi, G. J. Foschini, A. Lozano and R. A. Valenzuela, "Link-Optimal Space-Time Processing with Multiple Transmit and Receive Antennas," *IEEE Commyn. Lett.*, Vol. 5, Mar., 2001, pp. 85–87.
[108] J. C. Liberti and T. S. Rappaport, *Smart Antennas for Wireless Communications*, Prentice Hall, 1999.
[109] D. Hatzinakos (Editor), "Signal Processing Technologies for Short Burst Wireless Communications", Special Issue of Signal Processing, Elsevier, *Signal Processing*, Vol. 80, No. 10, Oct., 2000.
[110] M. Kuzminskiy and D. Hatzinakos, "Multistage Semi-Blind Spatio-Temporal Processing for Short Burst Multiuser SDMA Systems", *32nd Asilomar Conference on Signals, Systems and Computers*, Pacific Grove, CA, 1998, pp. 1887–1891.
[111] T. Abe and T. Matsumoto, "Space-Time Turbo Detection in Frequency Selective MIMO Channels with Unknown Interference", *Proc. WPMC01*, Aalborg, Demmark, 2001.
[112] T. Abe and T. Matsumoto, "Iterative Channel Estimation and Signal Detection in Frequency Selective MIMO Channels", *Proc. VTC2001-Fall*, Atlantic City, NJ, 2001.
[113] G. G. Raleigh and J. M. Cioffi, "Spatio-Temporal Coding for Wireless Communication", *IEEE Trans. Commun.*, Vol. 46, Mar., 1998, pp. 357–366.

[114] H. Sampath and P. Stoica, "A Paulraj: Generalized Linear Precoder and Decoder Design for MIMO Channels Using the Weighted MMSE Criterion", *IEEE Trans. Commun.*, Vol. 49, Dec., 2001, pp. 2198–2206.
[115] IST-FITNESS Project (http://www.ist-fitness.org), 2003.
[116] IST-FLOWS Project (http://www.flows-ist.org), 2003.
[117] IST-STRIKE Project (http://www.ist-strike.org)
[118] IST-MATRICE Project (http://www.ist-matrice.org)
[119] A. Lozano, F. R. Farrokhi and R. Valenzuela, "Asymptotically Optimal Open-Loop Space-Time Architecture Adaptive to Scattering Conditions", *IEEE 53rd Vehicular Technology Conference*, VTC 2001, Spring, Vol. 1, pp. 73–77.
[120] L. Zheng and D. N. C. Tse, "Diversity and Multiplexing: a Fundamental Tradeoff in Multiple Antenna Channels", *IEEE Trans. Inf. Theory*, Vol. 49, No. 5, May, 2003, pp. 1073–1096.
[121] A. Alexiou and C. Papadias, "Reconfigurable MIMO Transceivers for Next-Generation Wireless Systems", *Bell Labs Tech. J., Future Wireless Commun. Issue*, Vol. 10, No. 2, July, 2005, pp. 139–156.
[122] A. Alexiou, M. Qaddi, "Robust Linear Precoding to Compensate for Antenna Correlation in Orthogonal Space-Time Block Coded Systems", *3rd IEEE Sensor Array and Multichannel Signal Processing Workshop*, July, 2004.
[123] A. Medles, A. Alexiou. "Linear Precoding for STBC to Account for Antenna Correlation in Next Generation Broadband Systems", *IEEE PIMRC 2005*, Berlin, Germany, Sept., 2005.
[124] Q. Spencer and M. Haardt, "Capacity and Downlink Transmission Algorithms for a Multi-User MIMO Channel," *Proc. 36th Asilomar Conference on Signals, Systems, and Computers*, IEEE Computer Society Press, Pacific Grove, CA, Nov., 2002.
[125] V. Stankovic and M. Haardt, "Multi-User MIMO Downlink Precoding for Users with Multiple Antennas," *Proc. of the 12th Meeting of the Wireless World Research Forum (WWRF)*, Toronto, Canada, Nov., 2004.
[126] P. W. Wolniansky, G. J Foschini, G. D. Golden and R. A. Valenzuela, "V-BLAST: An Architecture for Realizing Very High Data Rates Over the Rich-Scattering Wireless Channel," *Proc. ISSSE 98*, Pisa Italy, Sept., 1998.
[127] G. Cinis and J. Cioffi, "A Multi-User Precoding Scheme Achieving Crosstalk Cancellation with Application to DSL Systems," *Proc. Asilomar Conference on Signals, Systems, and Computers*, Nov., Vol. 2, 2000, pp. 1627–1637.
[128] M. Joham, J. Brehmer and W. Utschick, "MMSE Approaches to Multiuser Spatio-Temporal Tomlinson-Harashima Precoding," *Proc. 5th International ITG Conference on Source and Channel Coding (ITG SCC'04)*, Erlangen Germany, Jan., 2004, pp. 387–394.
[129] V. Stankovic and M. Haardt, "Successive Optimization Tomlinson- Harashima Precoding (SO THP) for Multi-User MIMO Systems," *Proc. IEEE International Conference on Acoustics, Speech, and Signal Processing (ICASSP)*, Philadelphia, PA, Mar., 2005.
[130] G. Caire and S. Shamai, "On the Achievable Throughput of a Multi Antenna Gaussian Broadcast Channel," *IEEE Trans. Inf. Theory*, Vol. 49, No. 7, July, 2003, pp. 1691–1706.
[131] M. Fuchs, G. Del Galdo and M. Haardt, "A Novel Tree-Based Scheduling Algorithm for the Downlink of Multi-User MIMO Systems with ZF Beamforming," *Proc. IEEE International Conference on Acoustics, Speech, and Signal Processing (ICASSP)*, Philadelphia, PA, Mar., 2005.
[132] G. Del Galdo and M. Haardt, "Comparison of Zero-Forcing Methods for Downlink Spatial Multiplexing in Realistic Multi-User MIMO Channels," *Proc. IEEE Vehicular Technology Conference 2004-Spring*, Milan, Italy, May, 2004.
[133] G. Del Galdo, M. Haardt and C. Schneider, "*Geometrybased Channel Modelling of MIMO Channels in Comparison with Channel Sounder Measurements*," Advances in Radio Science – Kleinheubacher Berichte, 2003.
[134] A. Richter, "Estimation of Radio Channel Parameters: Models and Algorithms", Ph.D. Thesis, Ilmenau University of Technology, 2005.
[135] http://tu-ilmenau.de/ilmprop.
[136] K. Conner, D. Das, S. Gollamudi, J. Lee, P. Monogioudis, A. L. Moustakas, S. Nagaraj, A. Rao, R. Soni and Y. Yuan, "Intelligent Antenna Solutions for UMTS-Algorithms and simulation Results", *IEEE Commun. Mag.*, Vol. 42, No. 10, Sept., 2004, pp. 28–39.
[137] T. Yamada, T. Matsumoto, S. Tomisato and U Trautwein, "Results of Link-Level Simulations Using Field Measurement Data for an FTDL-Spatial/MLSE-Temporal Equalizer", *IEICE Trans. Commun.*, Vol. E84-B, 7, July, 2001, pp. 1956–1960.

[138] T. Yamada, T. Matsumoto, S. Tomisato and U Trautwein, "Performance Evaluation of FTDL-Spatial/MLSE-Temporal Equalizers in the Presence of Co-channel Interference – Link-Level Simulation Results Using Field Measurement Data – ", *IEICE Trans. Commun.*, Vol. E84-B, 7, July, 2001, pp. 1961–1964.
[139] A. Czylwik and A. Dekorsy, "System Level Simulations for Downlink Beamforming with Different Array Topologies," *Proc. of the IEEE Global Telecommunications Conference (GLOBECOM 2001)*, San Antonio, TX, 2001, pp. 3222–3226.
[140] COST 259, (http://www.lx.it.pt/cost259)
[141] L. M. Correia (Editor), *Wireless Flexible Personalised Communications*, John Wiley & Sons, 2001.
[142] R. Müller, "A Random Matrix Model of Communication Via Antenna Arrays", *IEEE Trans. Inf. Theory*, Vol. 48, No. 9, Sept., 2002, pp. 2495–2506.
[143] A. Lozano and A. M. Tulino, "Capacity of Multiple-Transmit Multiple-Receive Antenna Architectures", *IEEE Trans. Inf. Theory*, Vol. 48, No. 12, Dec., 2002, pp. 3117–3128.
[144] IST-ASILUM Project (http://www.ist-asilum.org)
[145] IST-METRA Project (http://www.ist-metra.org)
[146] IST-SATURN Project (http://www.ist-saturn.org), 2003.
[147] U. Rehfuess, K. Ivanov and C. Lueders, "A Novel Approach of Interfacing Link and System Level Simulations with Radio Network Planning", *GLOBECOM'98*, Vol. 3, 1998, pp. 1503–1508.
[148] J. Pons and J. Dunlop, "Enhanced System Level/Link Level Simulation Interface for GSM," *IEEE VTS 50th Vehicular Technology Conference*, Amsterdam Netherlands, Vol. 2, 1999, pp. 1189–1193.
[149] ESPRIT-ADAMO Project (http://www.cordis.lu/esprit)
[150] D. Avidor, D. Furman, J. Ling and C. Papadias, "On the Financial Impact of Capacity-Enhancing Technologies to Wireless Operators", *IEEE Wireless Commun.*, Vol. 10, No. 4, Aug., 2003, pp. 62–65.
[151] H. Haas, B. Wegmann and S. Flanz, "Interference Diversity Through Random Time Slot Opposing (RTO) in a Cellular TDD System", *Proc. of the IEEE Vehicular Technology Conference, 2002. VTC 2002-Fall*, 2002 IEEE 56th, Vancouver Canada, Sept. 24–28, Vol. 3, 2002.
[152] R. Venkataramani and T. L. Marzetta, "Reciprocal Training and Scheduling Protocol for MIMO Systems", *Proc. 41st Annual Allerton Conference on Communication, Control, and Computing*, Monticello, IL, Oct. 1–3, 2003.
[153] P. Bosch and S. Mullender, "*Band Switching for Coherent Beamforming in Full Duplex Wireless Communication*", patent application filed.
[154] T. M. Cover and J. A. Thomas, *Elements of Information Theory*, John Wiley, 1991.
[155] S. Das and H. Viswanathan, "Dynamic Frequency Assignment in a Multi-User OFDM System" *IEEE Vehicular Technology Conference, Fall 2004*, Los Angeles, CA, 2004.
[156] 1xEV Evaluation Methodology, Addendum (V6), 1xEV-DV WG5 Evaluation AHG, July 25, 2001. ftp://ftp.3gpp2.org/TSGC/Working/2001/TSG-C_0108/TSG-C-0801-Portland/WG5/.
[157] B. Hashem, D. Steer and S. Periyalwar, "*Dynamic Sub-Carrier Assignment in OFDM Systems*", US Patent 6,721,569, April 13, 2004.
[158] S. Y. Park, Y. Lee and S. Yun, "Region Division Based Fractional Loading Method for Uplink FH-OFDMA Cellular Systems," *Proc. of WWRF 11*, Oslo, Norway, June, 2004.
[159] I. E. Telatar, "Capacity of Multi-Antenna Gaussian Channels", *Eur Trans. Telecommun.*, Vol. 10, Nov./Dec., 1999, pp. 585–595.
[160] M. Costa, "Writing on Dirty Paper", *IEEE Trans. Inf. Theory*, Vol. 29, May, 1983, pp. 439–441.
[161] C. Peel, B. Hochwald and L. Swindlehurst, "A Vector-Perturbation Technique for Near-Capacity Multi-Antenna Multi-User Communication – Part I: Channel Inversion and Regularization", and "A Vector-Perturbation Technique for Near-Capacity Multi-Antenna Multi-User Communication – Part II: Perturbation", *IEEE Trans. Commun.*, submitted to, see also website http://mars.bell-labs.com.
[162] M.-H. Fong, G. Wu, Z. Hang, D. K. Yu, A. R. Schmidt, "*Automatic Retransmission Request Layer Interaction in a Wireless Network*", US Patent 6,760,860, July, 2004.
[163] S. Periyalwar, B. Hashem, G. Senarath, K. Au and R. Matyas, "Future Mobile Broadband Wireless Networks: A Radio Resource Management Perspective", *Wireless Commun. Mobile Comput.*, Vol. 3, 2003, pp. 803–816.
[164] W. S. Jeon and D. G. Jeong, "Comparison of Time Slot Allocation Strategies for CDMA/TDD Systems," *IEEE J. Sel. Areas Commun.*, Vol. 18, July, 2000, pp. 1271–1278.
[165] P. Gupta, Y. Sankarasubramaniam and A. Stolyar, "Distributed Scheduling in Wireless Data Networks with Service Differentiation," to be presented at the *2004 IEEE International Symposium on Information Theory*, Chicago, Ill, June 27 – July 2, 2004.

[166] D. Avidor, J. Ling and C. Papadias, "Jointly Opportunistic Beamforming and Scheduling (JOBS) for Downlink Packet Access," *IEEE International Conference on Communications (ICC 2004)*, Paris, France, June 20–24, 2004, to appear.

[167] P. Viswanath, D. N. C. Tse and R. Laroia, "Opportunistic Beamforming Using Dumb Antennas," *IEEE Trans. Inf. Theory*, Vol. 48, No. 6, June, 2002, pp. 1277–1294.

[168] Z. Haas (Guest Editorial), "Design Methodologies for Adaptive and Multimedia Networks", *IEEE Commun. Mag.* (Special Issue), Vol. 39, No. 11, Nov., 2001. pp. 106–107.

[169] A. G. Kogiantis, N Joshi and O. Sunay, "On Transmit Diversity and Scheduling in Wireless Packet Data, *IEEE International Conference on Communications*, Helsinki Finland, 2001, pp. 2433–2437.

[170] G. Senarath, S. Periyalwar, M. Smith and R. Matyas, "Coverage Performance Evaluation Methodology for Next Generation Wireless Systems", *WWRF-12*, Toronto, Canada, Nov., 2004, submission to.

[171] S. Das, W. M. ManDonald and H. Viswanathan "Delay Sensitivity Analysis of CDMA Downlink Handoff Algorithms", *Globecom*, San Francisco, USA, 2003.

[172] S. Das, H. Viswanathan and G. Rittenhouse, "Dynamic Load Balancing Through Coordinated Scheduling in Packet Data Systems", *Proc. of IEEE Infocom 2003*, San Francisco, CA, 2003.

[173] L Morand and S Tessier, "Global Mobility Approach with Mobile IP in "All IP" Networks", *IEEE ICC*, Vol. 4, 2002, pp. 2075–2079.

[174] R. W. Heath Jr., S. Cho and C.-M. Wang, "Partial Handoff in MIMO-OFDM Cellular Systems", *WWRF 12*, Toronto, Canada, Nov., 2004.

[175] J. Chuang and N. Sollenberger, "Beyond 3G: Wideband Wireless Data Access Based on OFDM and Dynamic Packet Assignment", *IEEE Commun. Mag.*, Vol. 38, July, 2000, pp. 78–87.

[176] S. Das and H. Viswanathan, "On the Reverse Link Interference Structure for Next Generation Cellular Systems" *IEEE Globecom 2004*, Dallas, TX, 2004.

[177] G. Senarath and A. Abu-Dayya, *"Method and Apparatus for Enabling the Smooth Transmission of Bursty data in a Wireless Communications System"*, US Patent Number: 6,778,499, August 17, 2004.

[178] S. Das, T. E. Klein and S. Mukherjee, "Maximum Throughput for the Additive Gaussian Noise Channel with a Pre-Determined Rate Set and Unknown Interference" *Proce. of CISS 2004*, Princeton, NJ, 2004.

[179] B. Hashem, S. Periyalwar and A. Kotov, *"Method and System for Providing Load-Balanced Communication"*, US patent 6,748,222, June, 2004.

[180] WWRF-SIG3, "Self-Organization in Wireless World Systems", http://www.wireless-world-research.org/

[181] H. Hashemi, "The Indoor Radio Propagation Channel", *Proc. IEEE*, Vol. 81, No. 7, July, 1993, pp. 943–968.

[182] J. Bach Andersen, T. S. Rappaport and S. Yoshida, "Propagation Measurements and Models for Wireless Communication Channels," *IEEE Commun. Mag.*, Vol. 33, Jan., 1995, pp. 42–49.

[183] B. H. Fleury and P. E. Leuthold, "Radiowave Propagation in Mobile Communications: An Overview of European Research," *IEEE Commun. Mag.*, Vol. 34, Feb., 1996, pp. 70–81.

[184] R. B Ertel, P. Cardieri, K. W. Sowerby, T. S. Rappaport and J. H. Reed, "Overview of Spatial Channel Models for Antenna Array Communication Systems", *IEEE Pers. Commun.*, Vol. 5, Feb., 1998, pp. 10–22.

[185] J. Bach Andersen, "A Propagation Overview", *5th International Symposium on Wireless Personal Multimedia Communications (WPMC)*, Honolulu, Hawaii, USA, Oct. 27–30, Vol. 1, 2002.

[186] T. K. Sarkar, Z. Ji, K. Kim, A. Medouri and M. Salazar-Palma, "A Survey of Various Propagation Models for Mobile Communication", *IEEE Antenn. Propag. Mag.*, Vol. 45, No. 3, June 2003, pp. 51–82.

[187] A. F. Molisch, "A Generic Model for MIMO Wireless Propagation Channels in Macro- and Microcells", *IEEE Trans. Signal Process.*, Vol. 52, No. 1, Jan., 2004, pp. 61–71.

[188] J. P. Kermoal, L. Schumacher, K. I. Pedersen, P. E. Mogensen and F. Fredriksen, "A Stochastic MIMO Radio Channel Model with Experimental Validation," *IEEE J. Sel. Areas Commun.*, Vol. 20, No. 6, Aug., 2002, pp. 1211–1226.

[189] T. Jämsä, M. Haardt and R. Thomä, "Radio Propagation and Network Planning", *WWRF/WG4 Contribution*, Oct. 5, 2001.

[190] K. Yu and B. Ottersten, "Models for MIMO Propagation Channels, A Review", *Wiley J. Wireless Commun. Mobile Comput.*, Vol. 2, No. 7, Nov., 2002, pp. 653–666.

[191] M. A. Jensen and J. W. Wallace, "A Review on Antennas and Propagation for MIMO Wireless Communications", *IEEE Trans. Antennas Propag.*, Vol. 52, No. 11, Nov., 2004, pp. 2810–2824.

[192] "Spatial Channel Model for Multiple Input Multiple Output (MIMO) Simulations", Technical Report, Release 6, 3GPP TR 25.996 V6.1.0 (2003-09), http://www.3gpp.org/.

[193] Federal Communication Commission, FCC, "*The First Report and Order in the Matter of Revision of Part 15 of the Commission's Rules Regarding Ultra-Wideband Transmission Systems*", FCC 02–48, ET Docket No. 98–153, USA, 2002.

[194] Federal Communication Commission, FCC, http://www.fcc.gov/Bureaus/Engineering_Technolgoy/News_Releases/2002/nret0203.html, 2002.

[195] M. Haardt and A. Alexiou, (Editors) WWRF White Paper "*Smart Antennas and Related Technologies (SMART)*", WWRF/WG4/Subgroup on Smart Antennas, version 1.5, Mar. 26, 2003.

[196] W. A. Th. Kotterman, "Characterisation of Mobile Radio Channels for Multiantenna Terminals", Ph.D. Thesis, ISBN 87.90834-68-2, Aalborg University, July 2004.

[197] E. Bonek, M. Herdin, W. Weichselberger and H. Özcelik, "MIMO – Study Propagation First!", *Proc. of IEEE International Symposium on Signal Processing and Information Technology (ISSPIT'03)*, Darmstadt, Deutschland, Dec. 14–17, 2003, invited paper (Special Session).

[198] http://www.lx.it.pt/cost273/

[199] IEEE 802.11-03/940r2 "IEEE P802.11 Wireless LANs, TGn Channel Models", Jan. 9, 2004.

[200] IEEE 802.15-02/368r5-SG3a, "IEEE P802.15 Wireless Personal Area Networks, Channel Modeling Sub-Committee Report", Dec., 2002.

[201] http://www.wireless-world-initiative.org/

[202] http://www.ist-winner.org/

[203] IST 2004–507102, "My Personal Adaptive Global Net (MAGNET)", Deliverable D3.1.2a: "PAN Radio Channel Characterisation (Part 1 and 2)", October 2004 and June 2005 (www.ist-magnet.org).

[204] http://www.ismb.it/newcom/index.html

[205] G. Del Galdo, M. Milojevic, M. Haardt and M. Hennhöfer, "Efficient Channel Modeling for Frequency-Selective MIMO Channels," *Proc. IEEE/ITG Workshop on Smart Antennas*, Munich, Germany, Mar., 2004.

[206] G. Del Galdo, M. Haardt and M. Milojevic, "A Subspace-Based Channel Model for Frequency Selective Time Variant MIMO Channels," *Proc. 15th International Symposium on Personal, Indoor, and Mobile Radio Communications (PIMRC)*, Barcelona, Spain, Sept., 2004.

[207] J. Kolu, T. Jämsä and A. Hulkkonen, "Real Time Simulation of Measured Radio Channel", *IEEE VTC 2003 Fall*, Orlando, FL, Oct. 06–09, 2003.

[208] J. Kolu, J-P. Nuutinen, T. Jämsä, J. Ylitalo and P. Kyösti, "*Playback Simulations of Measured MIMO Radio Channels*", TD (04) 110, COST273, Gothenburg, Sweden, June, 2004.

[209] COST273 project internal temporary document, TD-04-079, "Semi-sequential MIMO channel measurements in indoor environments", http://193.136.221.5/cost273/.

[210] J. Ø. Nielsen, J. B. Andersen, P. C. F. Eggers, G. F. Pedersen, K. Olesen, E. H. Sørensen and H. Suda, "Measurements of Indoor 16 × 32 Wideband MIMO Channels at 5.8 GHz", *IEEE International Symposium on Spread Spectrum Techniques and Applications (ISSSTA)*, Sydney, Australia, Sept., 2004.

[211] I. Z. Kovacs, P. C. F. Eggers and G. F. Pedersen, "Body-Area Networks", in M. G. Benedetto, C. Politano, T. Kaiser, A. Molisch, I. Oppermann and D. Porcino (Editors), *UWB Communication Systems – A Comprehensive Overview*, EURASIP Book, 2005.

[212] K. Kalliola, H. Laitinen, L. I. Vaskelainen and P. Vainikainen, "Real-Time 3D Spatial-Temporal Dual-Polarized Measurement of Wideband Radio Channel at Mobile Station", *IEEE Transa. Instrum. Meas.*, Vol. 49, No. 2, 2000, pp. 439–448.

[213] K. Kalliola, "Experimental Analysis of Multidimensional Radio Channels", doctoral dissertation, Helsinki University of Technology, Feb., 2002.

[214] J. B. Andersen, J. Ø. Nielsen, G. F. Pedersen, K. Olesen, P. C. F. Eggers, E. H. Sørensen and S. Denno, "A 16 by 32 Wideband Multichannel Sounder at 5 GHz for MIMO", *Antennas and Propagation Society Symposium, 2004*, IEEE Vol. 2, June 20–25, 2004, pp. 1263–1266.

[215] M. Landmann and G. Del Galdo, "Efficient Antenna Description for MIMO Channel Modeling and Estimation," *Proc. European Conference on Wireless Technology (ECWT 2004)*, Amsterdam, The Netherlands, Oct., 2004.

[216] D. S. Baum, J. Salo, G. Del Galdo, M. Milojevic, P. Kyösti and J. Hansen, "An Interim Channel Model for Beyond-3G Systems", *Proc. IEEE Vehicular Technology Conference 2005 Spring*, Stockholm, Sewden, May, 2005.

[217] A. Ettefagh, M. Kuhn, B. Cheetham and A. Wittneben, "Comparison of Distributed and Co-located Antenna Diversity Schemes for the Coverage Improvement of VoWLAN Systems", *PIMRC 2005*, Berlin, Germany, Sept., 2005.

[218] http://www.nari.ee.ethz.ch/wireless/research/projects/racoon/introduction.html refer to this page for information on the RACooN testbed of the Swiss Federal Institute of Technology.

[219] S. Berger and A. Wittneben, "Experimental Performance Evaluation of Multiuser Zero Forcing Relaying in Indoor Scenarios", *IEEE Vehicular Technology Conference, VTC Spring 2005*, Stockholm, Sweden, May 2005.

[220] M. Debbah, J. Gil, P. Fernandes, J. Venes, F. Cardoso, G. Marques, L. M. Correia, FLOWS project "*Final Report on Channel Models*", 2004, Online available: http://www.flows-ist.org/main/outputs/list.htm

[221] J. Bach Andersen and J. Ø. Nielsen, "Modeling the Full MIMO Matrix Using the Richness Function", *WSA2005*, Duisburg, Germany, April 2005.

[222] J. Bach Andersen, "Propagation Aspects of MIMO Channel Modeling", in H. Bolcskei, C. Papadias, D. Gesbert and A.-J. van der Veen (Editors), *Space-Time Wireless Systems: From Array Processing to MIMO Communications*, Cambridge University Press, 2005.

[223] J. Hämäläinen, R. Wichman, J.-P. Nuutinen, J. Ylitalo and T. Jämsä, "Analysis and Measurements for Indoor Polarization MIMO in 5.25 GHz Band", *Proc. of IEEE Vehicular Technology Conference*, Stockholm, Sweden, May, 2005.

[224] J. Ylitalo, J.-P. Nuutinen, J. Hämäläinen, T. Jämsä, M. Hämäläinen, "Multi-Dimensional Wideband Radio Channel Characterisation for 2–6 GHz Band", *WWRF11 Meeting*, Oslo, Norway, June, 2004.

[225] J. Sachs, M. Kmec, P. Peyerl, P. Rauschenbach, R. S. Thomä and R. Zetik, "A Novel Ultra-Wideband Real-Time MIMO Channel Sounder Architecture," *XXVIIIth URSI General Assembly*, New Delhi, India, Oct. 23–29, 2005.

[226] J. Kolu, P. Kyösti, J.-P. Nuutinen, T. Jämsä, "*Playback Simulation of Measured MIMO Radio Channels, COST273, TD-04-110*", Gothenburg, Sweden, 2004.

[227] U. Trautwein, C. Schneider and R. Thomä, "Measurement Based Performance Evaluation of Advanced MIMO Transceiver Designs," *EURASIP J. Appl. Signal Process.*, Vol. 2005, No. 11, 2005, pp. 1712–1724.

[228] The Book of Visions, WWRF, Dec., 2001, http://www.wireless-world-research.org/.

[229] R. A. Scholtz and J.-Y. Lee, "*Problems in Modeling UWB Channels*", *Conference Record of the Thirty-Sixth Asilomar Conference on Signals, Systems and Computers*, Pacific Grove-California, USA, Nov., 2002.

[230] J. M. Cramer, R. A. Scholtz and M. Z. Win, "On the Analysis of UWB Communication Channels", *Proc. IEEE Military Communications Conference*, Nov. 1999.

[231] M. Z. Win and R. A. Scholtz, "Characterization of Ultra-Wide Bandwidth Wireless Indoor Channels: a Communication-Theoretic View", *IEEE J. Sel. Areas Commun.*, Vol. 20, No. 9, Dec., 2002, pp. 1613–1627.

[232] J. Foerster, IEEE P802.15-02/490r1-SG3a, "Channel Modeling Sub-committee; Final Report", Mar., 2003.

[233] J. Foerster, M. Pendergrass and A. Molisch, "A Channel Model for Ultrawideband Indoor Communications", *Proc. the 6th International Symposium on Wireless Personal Multimedia Communications*, Yokosuka, Japan, Oct. 19–22, Vol. 2, 2003, pp. 116–120.

[234] S. S. Ghassemzadeh and V. Tarokh, "UWB Path Loss Characterization in Residential Environments", *Proc. 2003 IEEE Radio Frequency Integrated Circuits (RFIC) Symposium*, June, 2003, pp. 501–504.

[235] V. Hovinen, M. Hämäläinen and T. Pätsi, "Ultra Wideband Indoor Radio Channel Models: Preliminary Results", *IEEE Conference on Ultra Wideband Systems and Technologies, UWBST2002*, Baltimore, MD, 2002.

[236] V. Hovinen and M. Hämäläinen, "Ultra Wideband Radio Channel Modeling for Indoors", *COST273 Workshop*, Helsinki, Finland, 2002.

[237] D. Cassioli, M. Z. Win and A. F. Molisch, "The Ultra-Wide Bandwidth Indoor Channel: from Statistical Model to Simulations", *IEEE J. Sel. Areas Commun.*, Vol. 20, No. 6, Aug., 2002, pp. 1247–1257.

[238] B. H. Fleury, M. Tschudin, R. Heddergott, D. Dahlhaus and K. Pedersen, "Channel Parameter Estimation in Mobile Radio Environments Using the SAGE Algorithm", *IEEE J. Sel. Areas Commun.*, Vol. 17, No. 3., Mar., 1999, pp. 434–450.

[239] M. Haardt, R. S. Thomä, and A. Richter, "Multidimensional high-resolution parameter estimation with applications to channel sounding" in Y. Hua, A. Gershman and Q. Chen, (Editors), *High-Resolution and Robust Signal Processing*, Marcel Dekker, New York, NY, 2003, Chapter 5, pp. 253–338.

[240] B. Fleury, X. Yin, P. Jourdan and A. Stucki, "High-Resolution Channel Parameter Estimation for Communication Systems Equipped with Antenna Arrays", *Proc. 13th IFAC Symposium on the System Identification (SYSID 2003)*, Rotterdam, The Netherlands Aug., 2003.

[241] X. Yin, B. H. Fleury, P. Jordan and A. Stucki, "Polarization Estimation of Individual Propagation Paths Using the SAGE Algorithm", *Proc. the 14th IEEE International Symposium on Personal, Indoor and Mobile Radio Communications (PIMRC2003)*, Beijing, China, Vol. 2, Sept., 2003, pp. 1795–1799.

[242] B. H. Fleury, P. Jourdan and A. Stucki, "Performance of a SAGE-Based High-Resolution Scheme for Estimation of Bidirection Dispersion in the Radio Channel", *Proc. Vehicular Technology Conference Spring 2002*, Birgmingham, AL, May, 2002.

[243] B. H. Fleury, P. Jourdan and A. Stucki, "*High-Resolution Channel Parameter Estimation for MIMO Applications Using the SAGE Algorithm*", International Zurich Seminar, Feb., 2002.

[244] B. H. Fleury, P. Jourdan, A. Stucki, "High-resolution channel estimation for MIMO systems using the SAGE algorithm", *XXVIIth General Assembly of the International Union of Radio Science (URSI)*, Maastrich, The Netherlands - invited paper, Aug., 2002.

[245] B. H. Fleury, X. Yin and A. Kocian, "Impact of the Propagation Conditions on the Properties of MIMO Channels", *International Conference on Electromagnetics in Advanced Applications (ICEAA 03)*, Torino, Italy, invited paper, Sept., 2003, pp. 783–786.

[246] X. Yin, B. H. Fleury, P. Jourdan and A. Stucki, "Doppler Frequency Estimation for Channel Sounding Using Switched Multiple Transmit and Receive Antennas", *Proc. IEEE 2003 Global Communications Conference, (Globecom'03)*, San Francisco, CA, Dec., Vol. 4, 2003, pp. 2177–2181.

[247] T. Pedersen, C. Pedersen, X. Yin, B. Fleury, R. Pedersen, B. Bozinovska, A. Hviid, P. Jourdan and A. Stucki, "Joint Estimation of Doppler Frequency and Directions in Channel Sounding Using Switched Tx and Rx Arrays", *IEEE Globecom 2004 Signal Processing for Communications*, Dallas, TX, Dec., 2004, accepted.

[248] R. S. Thomä, M. Landmann, A. Richter and U. Trautwein, "Multidimensional high-resolution channel sounding measurement", EURASIP Book Series on Signal Processing and Communications, Volume 3 "SMART ANTENNAS-STATE OF THE ART", ISBN 977-5945-09-7, pp. 241–271.

[249] A. Richter, M. Landmann and R. S. Thoma, "Maximum Likelihood Channel Parameter Estimation from Multidimensional Channel Sounding Measurements," *Proc. IEEE VTC2003-Spring*, Jeju, Korea, April, 2003.

[250] R. S. Thoma, M. Landmann and A. Richter, "RIMAX – a Maximum Likelihood Framework for Parameter Estimation in Multidimensional Channel Sounding," *2004 International Symposium on Antennas and Propagation*, Sendai, JP, August 17–21, 2004.

7

Short-range Wireless Communications

Edited by Gerhard Fettweis (Vodafone Chair, TU Dresden), Ernesto Zimmermann (Vodafone Chair, TU Dresden), Ben Allen (King's College London), Dominic C. O'Brien (University of Oxford) and Pierre Chevillat (IBM Research GmbH, Zurich Research Laboratory)

7.1 Introduction

Probably the largest portion of the practical applications of short-range communication takes the form of Wireless Local Area Network (WLAN), Wireless Personal Area Network (WPAN) and Wireless Body Area Networks (WBAN), covering ranges from tens of meters down to sub-meter communications.

Data rates for wireless cellular and local area networks (LAN) have been steadily increasing in recent years, with an approximate fivefold increase in throughput every four years. With new applications such as wireless multimedia and the replacement of cables for communication purposes in home, office, and public access scenarios, we can anticipate this trend to continue. Three challenging requirements arise from the call for higher data rates in next generation wireless systems: we have to increase spectral efficiency, design systems for larger bandwidths with the aim of reducing the costs per bit.

Four important candidate techniques for short-range communications are considered in this chapter. These are as follows:

- MIMO–OFDM in time division duplexing (TDD)
- Ultra-wideband (UWB)
- Optical communications
- Wireless sensor networks (WSN).

Technologies for the Wireless Future – Volume 2 Edited by Rahim Tafazolli

7.2 MIMO–OFDM in the TDD Mode

The increasing demand for higher data rates in next generation wireless systems translates into three challenging requirements for the physical layer development: We have to increase both the spectral efficiency and the bandwidth and reduce the costs per bit. A combination of multiple-input multiple-output (MIMO), orthogonal frequency division multiplex (OFDM) and TDD may be suitable to meet these requirements. Since a substantial amount of spectrum is reserved for TDD systems across the world, this combination could be a promising candidate for the air interface in the next generation of mobile communications systems.

Exploiting the rich scattering typical of indoor and urban environments [1], MIMO systems allow for sound gains in the spectral efficiency, thus facilitating the transmission at high data rates (HDRs) in a spectrum which is usually limited by regulation and other factors.

With increasing bandwidth, more and more echoes are resolved in the channel, calling for efficient equalization techniques. OFDM is a very attractive option for solving this problem, and the combination of MIMO and OFDM allows a substantially reduced complexity of the spatio-temporal processing [2]. This is due to the fact that the OFDM pre- and postprocessing transforms the multipath channel into multiple flat fading channels for which well-known MIMO detection schemes can be used.

But we still have to exploit the plethora of diversity offered by the multipath MIMO channel, which is not trivial with MIMO–OFDM. Note that the maximal diversity order in the multipath MIMO channel is given as the product of the numbers of received antennas and resolved echoes, when the spatial multiplexing technique is used. In general, space-frequency codes must be designed to realize this diversity, which may be different from the well-known designs for space–time codes [3].

Alternatively, one may attain the capacity by water filling and related methods, based on channel state information (CSI) available at the transmitter. With channel-aware preprocessing and adaptive modulation, data are loaded onto the subcarriers and spatial channels according to the postprocessing signal-to-interference and noise ratio (SINR). The error bursts typical for fading channels can be avoided prior to the transmission. Channel coding is no longer required to realize the multipath diversity. Consequently, the interleaver size and the corresponding delay can be reduced.

The basic challenge is to make CSI available at the transmitter. This can be achieved by sending feedback information over the reverse link, but there is a more efficient method in the TDD mode. The latter uses the same carrier frequency alternately for transmission and reception. Because of channel reciprocity, we may transmit training sequences in the uplink (UL) direction before the data transmission in the downlink (DL). The reciprocal CSI is then used to preprocess the transmitted data, as well as for optimal scheduling in the space-frequency domain (see Figure 7.1).

From the system point of view, we must consider what the ultimate gain for the whole system is. In the above example, the CSI obviously helps in the downlink optimization. But at the same time, the training sequences eat up valuable resources in the (capacity in the channel) battery power. There is an optimal trade-off point somewhere that needs to be assessed in future research.

We feel that this simple principle is the key to realizing the above requirement to reduce the costs per bit while providing a HDR. With CSI at the transmitter, we gain

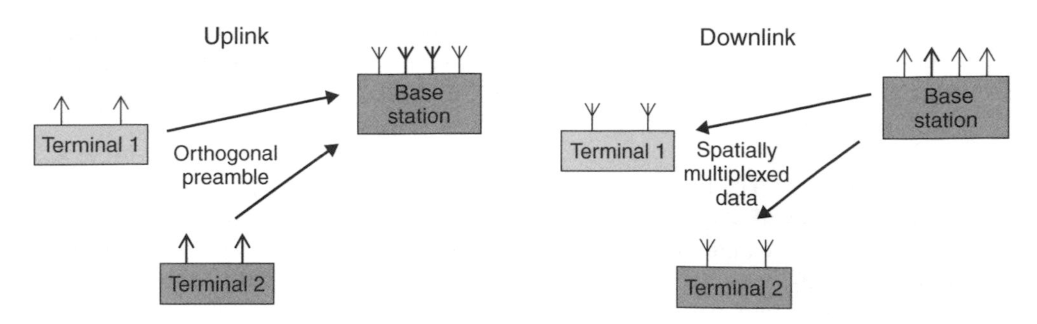

Figure 7.1 In the time-division duplex mode, the channel information for the downlink can be made available at the transmitter by using the reciprocal channel information from the uplink, or vice versa

the flexibility to shift the signal processing effort almost completely to the base station (BS) in both uplink and downlink. This allows a simple terminal design. Alternatively, we can share the effort between two mobile terminals (MTs) in the ad hoc mode. For each carrier, a simple matrix-vector multiplication at both stations, combined with rate and power control, is sufficient to attain the full capacity. (This is called 'eigenmode signaling' [4].)

The aim of this section is to identify open research challenges concerning the combination of MIMO, OFDM, and TDD, which have not been previously addressed, to be able to eventually realize the system concept. The section is organized as follows: In Section 7.2.2, application scenarios and requirements for future wireless systems are described. The system concept for the physical layer is introduced in Section 7.2.3. The following Sections 7.2.4 to 7.2.6 summarize the state of the art and analyze open research items in the three component technologies of the combined approach proposed here. Section 7.2.7 highlights that a cross-layer optimization is required to achieve the optimal system performance and it addresses the optimal space-frequency scheduling strategy. Section 7.2.8 is concerned with the real-time implementation. Finally, conclusions are drawn in Section 7.2.9.

7.2.1 Application Scenarios and Requirements

The past years have been marked by two parallel developments that significantly transformed the way people work and live – the advent of the Internet and the widespread introduction of personal mobile communications. It is widely anticipated that these two digital industries, will converge in the next decade, with wireless networks replacing cable-based solutions in home, office, and hot-spot environments. Seamless interoperation with other communication infrastructure will facilitate user mobility by providing the desired information anywhere at any time. This trend is already visible in the rapidly rising number of Wireless LAN access points (APs) in hotels, lounges, and offices around the world.

Examples of applications in this context comprise access to email, the Intranet, and the Internet, the distribution of multimedia content in home and hot-spot environments, replacement of Ethernet networks in offices, and wireless peripheral interfaces for digital equipment. Naturally, customers expect the same high quality of service (QoS) and ease of use as known from wire-based solutions. This translates into a multitude of challenges: high-capacity reserves, in order to support the extreme peak data rates, reconfigurability and multiband operation in order to provide service in a heterogeneous environment, and plug-and-play network configuration in order to enable fast and flexible set-ups.

- *Home Environment*: The massive use of high quality multimedia applications (streaming audio and video, with data rates in excess of several tens of Mbps) for numerous users is one reason for the anticipated need of very high bit rates. Other key requirements in this area are self-configuration and zero maintenance features as well as low transmit powers, to minimize exposure of humans to electromagnetic radiation.
- *Office and Enterprise Environment*: WLAN solutions available to date already enable office staff to work detached from their desks. However, with high numbers of users accessing a single AP and 100 Mbps Ethernet being state of the art, peak data rates can be expected to be of the order of 1 Gbps. Moreover, core business applications such as Voice over Internet Protocol (IP) and videoconferencing demand very high QoS (including encryption features).
- *Hot-Spot/Public Access Environment*: A large scale coverage in a future heterogeneous wireless access network will be provided by next generation cellular networks, whereas HDR access in urban and hot-spot environments will be provided by short-range wireless systems. Expected high variations in user data rates and differing service requirements call for a highly flexible MAC. In order to enable user mobility, the system will have to interoperate with other B3G standards.

7.2.2 Operating Principle of the Air Interface

The above application scenarios require an air interface that offers a high peak rate, makes the best possible use of the available spectrum, and enables wireless terminals at low cost. Since in the future a significant quantity of applications will be driven by the continuous communication between users rather than by the burst-type internet access, the end-to-end delay becomes increasingly important for the acceptance of new services. In order to meet these requirements, we need an air interface with high spectrum efficiency, which operates as close as possible to the information-theoretic limits, but allows a high-bandwidth communication in real time.

From the information-theoretic point of view, there are two approaches with almost equivalent performance limits: There is the classical 'blind' transmission scheme, which sometimes overloads the wireless fading channel and repairs the resulting error bursts at the receiver by interleaved channel coding, which unfortunately introduces additional delay between transmission and reception. The second concept is an adaptive scheme based on channel side information at the transmitter avoiding these errors already in advance. We feel that the adaptive scheme fits the above requirements; better it also has another interesting impact. From the core network perspective it is desirable that only error-free data may enter through the wireless-to-wired gateway. With the classical scheme, there is an unavoidable possibility of error over the wireless link. With the adaptive scheme, the throughput may vary in time, according to the time-varying capacity of the wireless channel. Provided the channel variation can be tracked sufficiently fast at the transmitter, it can be guaranteed that the variable-rate data are free of error when entering the gateway, even under extreme bit-error rate (BER) constraints in the core network.

The physical and MAC layers are shown in Figure 7.2, in the UL of the infrastructure mode. We assume that each MT as well as the AP features multiple antennas.

Uplink: The MT starts by transmitting pilot sequences for the channel estimation. According to the channel and interference situation at the AP, the scheduler in the AP assigns the required resources in the space-frequency domain to the MT such that the desired QoS requirements (error rates, delay) can be met. The adaptive flow control distributes the data accordingly. Each spatial and frequency channel should at least have individual modulation, steered by the link adaptation unit. Individual coding could be used as well, as indicated in Figure 7.2, but might be too complex, eventually. Power control is realized using the joint space-frequency preprocessing after modulation. Finally, the signals are passed through an inverse fast Fourier transform (IFFT), modulated onto the carrier and transmitted over the air. In principle, the corresponding units are used in reverse order at the AP. As in the classical system, it is easy to shift the signal processing effort almost completely to the AP in the UL, except for power and rate control, which are performed at the terminal but steered from the AP. The (usually complex) MIMO signal processing is performed at the AP.

Downlink: At first, the AP requests the MT to transmit pilot sequences for the channel estimation in the UL direction and additional information of the current interference scenario at the MT. Using channel reciprocity, the CSI from the UL is then correspondingly used to steer the space-frequency precoding and to schedule the data transport in the space-frequency domain at the AP, to maximize the throughput and to minimize the latency of the individual users. The data are precoded, which means that the MIMO processing is already done at the AP, so that only minor postprocessing is required at the MT, such as scaling, demodulation and decoding. The data streams to be multiplexed in space will arrive spatially separated at the MT antennas.

Ad hoc mode: Here the effort is shared between the stations, based on the eigenmode signaling, which results in the minimal total effort for the MIMO processing as well. Initially, the MTs exchange pilot signals for the channel estimation. The space-frequency preprocessing at the transmitting MT is then used to couple the data optimally into the MIMO channel and the postprocessing at the receiving MT is used to couple them out. The throughput is optimized by the channel-adaptive space-frequency power and rate control.

Summarizing the above statements, we see that the TDD mode offers the freedom to shift the signal processing effort to wherever it is desired, since it allows us to reuse the CSI from the UL for the DL, and vice versa, owing to channel reciprocity. This particularly enables the design of MTs at reasonable costs, even for very HDRs. Moreover, optimal link adaptation becomes feasible since near-perfect channel side information is available at the transmitter.

7.2.3 MIMO

7.2.3.1 Broadband MIMO Channels

The mean capacity of an independent and identically distributed (i.i.d.) Rayleigh-fading MIMO channel is well known to scale linearly with the minimum of the numbers of transmit and receive antennas [5]. This is perfectly confirmed by the measurement data in Figure 7.3 for an indoor non-line-of-sight (NLOS) scenario. When the line-of-sight (LOS)

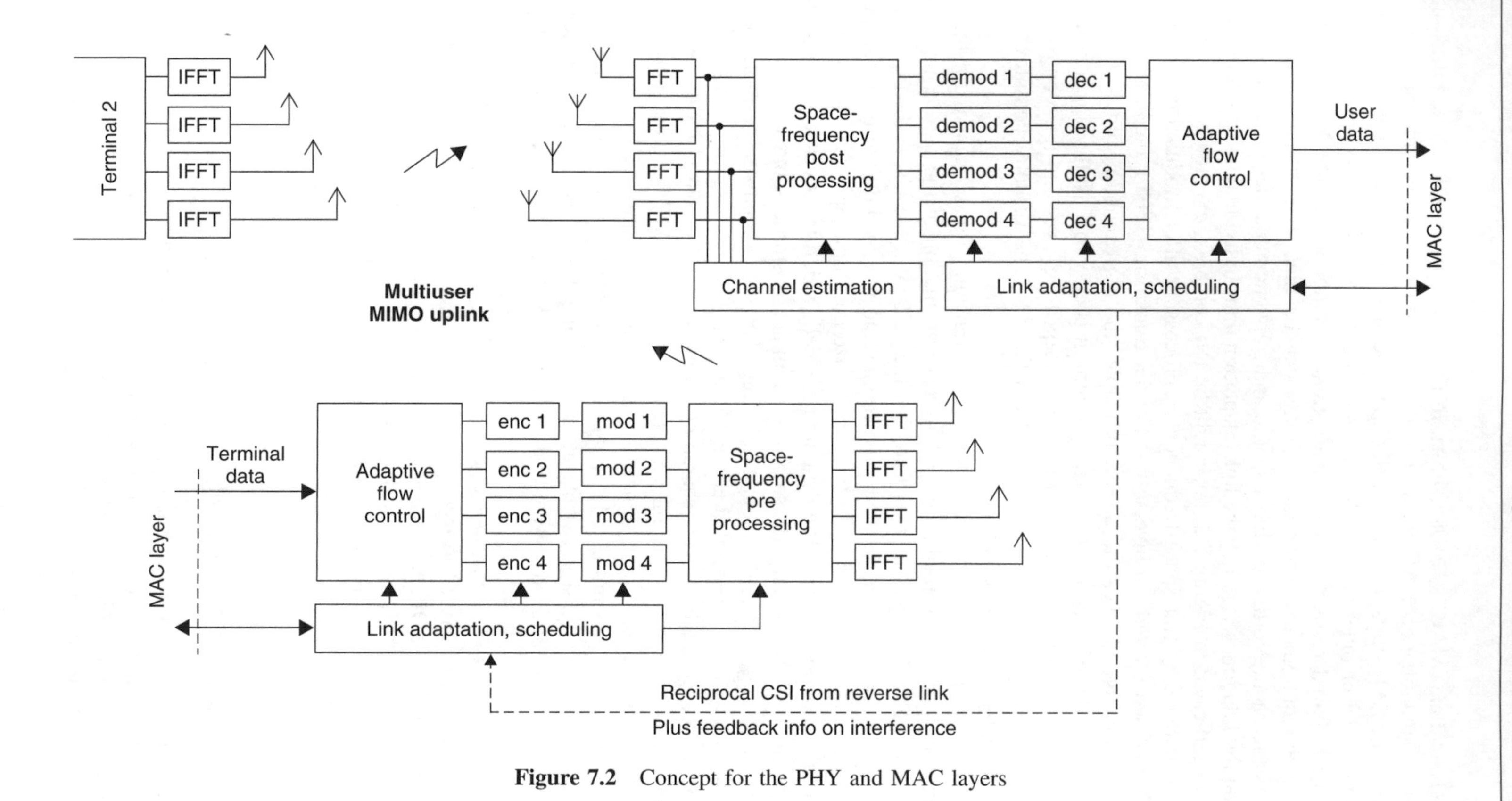

Figure 7.2 Concept for the PHY and MAC layers

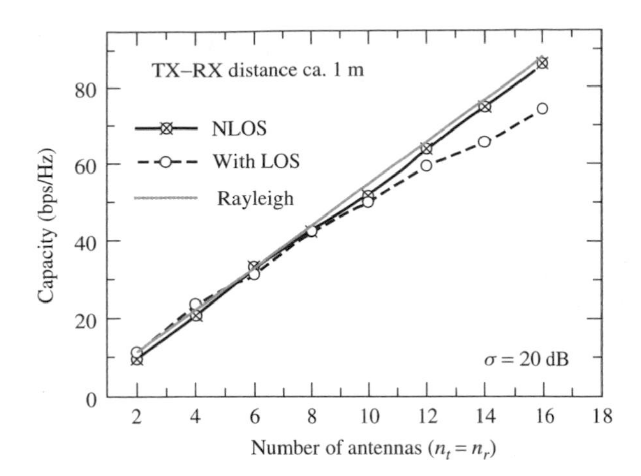

Figure 7.3 Measured indoor MIMO capacities versus the numbers of antennas (from [229])

is added, a minor degradation is observed at large numbers of antennas, indicating a minor rank loss of the channel matrix. The huge measured capacities (more than 80 bps/Hz with 16 antennas at 20 dB SNR) are related to the observation that the LOS component is normally not as important compared with the scattered signal, at least indoors with omni-directional antennas. In the general case of MIMO rice channels and non-omni-directional antennas, one has to take into account the trade-off between the multiplexing gain (due to the NLOS scattering environment) and the path loss gain due to the LOS component. The trade-off is mainly a function of the eigenvalue distribution of the mean matrix and the Tx–Rx distance [6, 7].

Future wireless systems may occupy bandwidths ranging from several 10 MHz to even 100 MHz. With increasing bandwidth, more and more taps are resolved and the channel offers a high degree of multipath diversity. When spatial multiplexing is used, the achievable diversity order is equal to the product of the numbers of receive antennas and taps. Broadband spatial multiplexing has then two implications on the channel capacity: Firstly, the mean capacity increases with the number of antennas (see Figure 7.3). Secondly, the more the taps that are resolved, the sharper becomes the capacity distribution (see Figure 7.4), which is a result of the huge amount of diversity offered by the channel [8]. Therefore, broadband MIMO systems not only provide higher capacity, but also have higher link reliability and promise almost wirelike QoS. If we anticipate that the slope of the capacity pdf in Figure 7.4 (which is simulated over 10^4 random i.i.d. channel realizations) can be continued (dashed line), then the outage probability becomes negligible even when it is measured against figures of merit in the core network.

On the other hand, the spatial structure of the MIMO channel may change as we go from indoor to outdoor scenarios. In Figure 7.5, the ordered singular values (SVs) for three scenarios (plotted versus frequency) are compared with the theoretical distributions for the i.i.d. Rayleigh-fading channel (The SVs can be considered as the amplitude gains of the spatially multiplexed streams.). The random matrix theory predicts an almost equal spacing between the mean SVs. Indoor data agree fairly well with this expectation, and a similar distribution is found at an urban crossing, as well. But on an almost empty

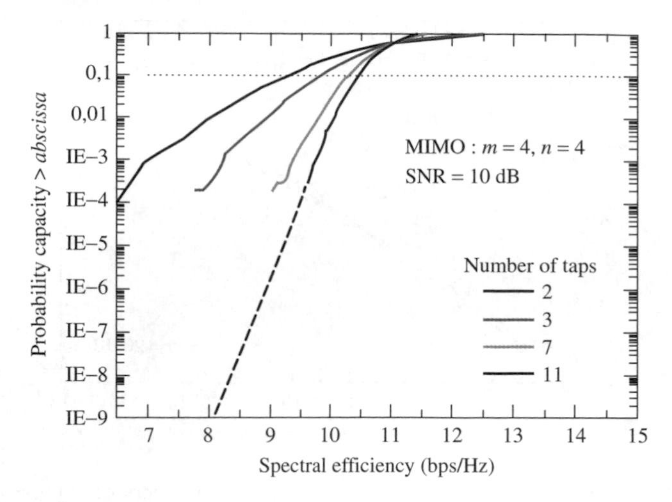

Figure 7.4 Broadband MIMO systems promise wire-like quality of service (see text)

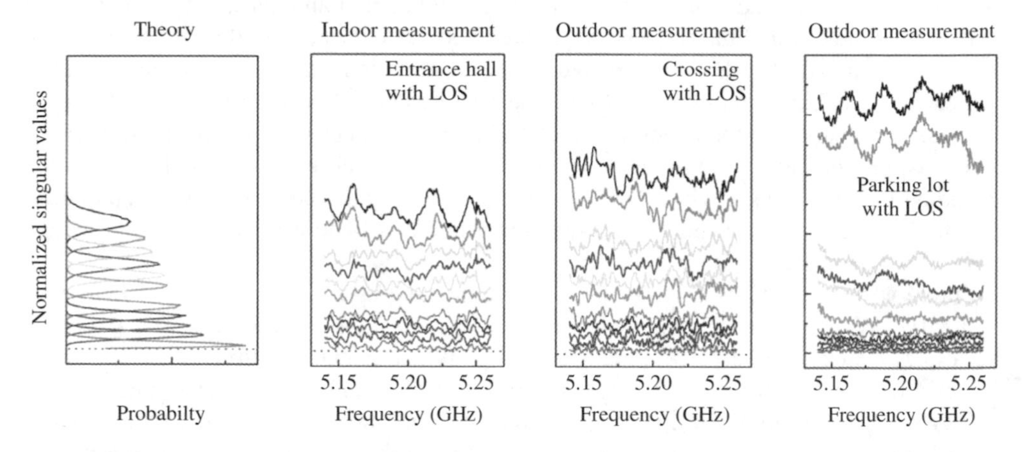

Figure 7.5 Singular value distributions in three scenarios, compared with the distribution in theory

parking lot, which is a rather extreme outdoor scenario where the scattering is significantly reduced, the two largest SVs start to separate from the others. Note that different field directions are addressed in these measurements by four sets of triple antennas addressing different field directions. The two channels with significantly enhanced SVs are due to polarization, which allows opening multiple spatial channels even in the worst-case LOS scenario.

Note that the gains of the spatial channels change from frequency to frequency, which is obvious from Figure 7.5, and sometimes they become zero for one or more spatial channels at certain frequencies. In order to transmit the data optimally over such arbitrarily conditioned MIMO channels, we need an individual link adaptation at each frequency bin and on each spatial channel. Such a concept may be realistic if we exploit the channel knowledge from the reverse link direction, and when the channel is not changing too rapidly. Of course, we need to signal what transmission mode is used in order to reduce receivers processing load. Then the granularity of this adaptation (in the frequency domain, the number of different modulations, coding rates etc.) becomes an issue as too fine a granularity consumes too much capacity. A proper trade-off between adaptation and signaling granularity is required here.

Channel Modeling Issues

The above results point out the necessity of developing a better understanding of the underlying propagation conditions, and more generally the features of the channel critical to communication systems operating in the broadband MIMO–OFDM setting for short-range and long range scenarios. Of particular importance among these features are the parameters characterizing the eigenmodes/singular modes of the propagation channel, the time-frequency variant structure, the effect of polarization and correlation. In [9], a broadband MIMO model nicknamed Maxent MIMO Model based on information-theoretic considerations (maximum entropy principle) was developed and shown to be capacity complying. It includes the time-variant structure as well as the angles of arrival and departure. The maximum entropy principle is a general method, which is consistent with the state of one's knowledge to construct maximum entropy-based models. One of the interesting features of the models is their ability to take into account the state of one's knowledge and leave in an unconstrained space all the information not provided. The resulting models are mainly based on the product of random matrices with Gaussian entries (due to information based on the mean and the variance, otherwise this is not the case) with some pattern mask depending on the relations in the environment. The Maxent model encompasses Sayeed's virtual representation [10], the Kronecker model [11], and the so-called keyhole channels [12] as special cases. However, the effect of polarization, path loss (for the link budget), the effect of transmit and receiver filters, the effect of the antenna pattern, and correlation are still open issues and need to be addressed in the light of the transmission techniques to be used in MIMO–OFDM TDD mode.

MIMO Algorithms for TDD

One basic advantage of MIMO–OFDM is that we can reuse the well-known algorithms from the flat fading channel on each carrier. In general, we wish to shift as much signal processing as possible to the AP, to allow economic MTs at HDRs. Next we distinguish between the infrastructure and ad hoc mode.

Infrastructure uplink: Linear MIMO detection schemes, as the minimum mean squared error (MMSE) detector, may have the worst performance at first sight, but they have the proven potential to be implemented in real time for larger numbers of antennas, using parallel multiply-and-accumulate structures realizable in field programmable gate arrays (FPGAs) and application-specific integrated circuits (ASICs). The loss of capacity is not as large when a suitable link adaptation is used [13], which is surprisingly efficient also with linear schemes [14]. These adaptive linear algorithms may be extended to the case of MIMO–OFDM [5] where the gap in the optimal schemes reduces further, eventually because of the multipath diversity. Hence, the linear schemes may serve as a good basis to introduce the new technique in the field. Later, the detectors may be replaced by better schemes, such as successive interference cancellation (SIC), lattice-aided detection, sphere decoding or maximum-likelihood detection (MLD). Note that for each of these schemes, a suitable link adaptation must be developed in order to satisfy QoS requirements.

Infrastructure downlink: Recent information theory shows that the DL capacity region is the same as in the UL, provided the channel coefficients are known at the transmitter. This fundamental result is called the UL/DL duality [15]. So it is not surprising that the well-known UL processing schemes have corresponding counterparts for the DL. For instance, the DL scheme for the well-known zero-forcing algorithm in the UL is called *channel inversion* [16]. Similarly, dual algorithms based on the SIC (Tomlinson-Harashima precoding [17]), on modulo-lattice reduction [18], and sphere precoding have recently been reported [19]. The dual scheme for the MLD is the classical 'dirty paper' precoding, which states, in principle, that known interference signals are not relevant for the channel capacity. But the realization is not yet known. As for the UL, the better the performance, the higher is the effort.

Ad hoc mode: Here, the optimal eigenmode signaling can be directly used. It requires simple matrix-vector operations both at the transmitter and at the receiver. There are hints that eigenmode signaling is sensitive in time-varying channels, since the CSI at the transmitter and receiver may differ from each other and this could eventually lead to a confused stream assignment [20]. These effects can be avoided by passing the pilot signals through the precoder, which is then considered as a part of the effective channel. So the stream assignment can be restored at the receiver [21].

7.2.4 OFDM

7.2.4.1 Motivation and Principle

Proposals for bandwidth usage in current research [22] and standardization efforts [23] as well as large amounts of spectrum allocation for unlicensed operation in the high GHz range are clear indicators that future wireless systems will use bandwidths ranging from several tens of MHz up to several hundreds of MHz, for wireless broadband applications (i.e. 'Gigabit Wireless').

However, increasing the transmission bandwidth simultaneously increases the sampling frequency of the channel. As the sampling interval becomes shorter, the paths impinging from different scatterers are eventually temporally resolved, the (discrete sampled) channel impulse response is, in this case, no longer a single Dirac impulse ('tap') but a sequence of taps; a tapped delay line (TDL), usually described by a power delay profile (PDP) [24].

Traditional single carrier (SC) systems will suffer from strong intersymbol interference (ISI) when transmitting over a frequency-selective channel, which makes equalization at the receiver cumbersome. For complexity reasons, it is therefore advantageous to perform equalization in the frequency domain. Two Fourier transforms are required – one of which is done at the transmitter, the other one at the receiver. The aim of this multicarrier (MC) modulation is to have a number of parallel subcarriers high enough so that the bandwidth of one subcarrier is significantly lower than the channel's coherence bandwidth and the frequency-selective channel is hence transformed into several narrowband frequency flat fading channels. In the time domain, this means that for each individual subcarrier, the symbol time is raised much above the maximum channel delay such that the effects of ISI are negligible.

A commonly used and spectrally efficient [25] technique arising in this context is OFDM. By means of an IFFT, the transmitter transforms the frequency-domain data samples on several subcarriers (which are equidistantly distributed in the frequency domain) into the time domain, adds a prefix/postfix and transmits the resulting signal over the channel. The receiver cancels the prefix/postfix and uses a fast Fourier transform (FFT) to transform the received signal back into the frequency domain. We will dwell on the details of pre-/postfixes in the following section. The number of subcarriers is usually a power of 2, to allow for efficient implementation of the IFFT/FFT.

7.2.4.2 OFDM System Design

FFT Size

One major challenge for the design of an OFDM system is the selection of an appropriate FFT size (number of subcarriers). In order to avoid OFDM intersymbol-interference, the length of the guard interval T_{GI} must be larger than the maximum channel impulse response (Note that the length of the guard interval should be designed to take into account the length of the channel impulse response of the *full channel*, i.e. the concatenation of transmitter filters, channel, and receiver filters.). However, the guard interval is an additional overhead that should not exceed a certain threshold $\eta_{\max}$ (usually, 25 % of the OFDM symbol length). Otherwise, the spectral efficiency would be reduced too much. On the other hand, the OFDM symbol must be short enough such that the temporal variations resulting from relative movements of the transmitter, receiver, and/or scatterers do not result in a time-varying channel within one OFDM symbol, which would result in inter-carrier interference (ICI) and thus significantly deteriorate system performance.

Since subcarrier spacing and symbol duration are inversely proportional, the former requirement sets a lower bound on the number of subcarriers: $N_{sc} > B \times T_{GI}/\eta_{\max}$, where B is the bandwidth of the channel. It is independent of both carrier frequency and user mobility. The latter requirement sets an upper bound on the number of subcarriers since at any fixed bandwidth, the OFDM symbol length scales linearly with the number of subcarriers. If we assume, for example, a Jakes spectrum and design the OFDM symbol to be shorter than the 99 % channel coherence time, then the number of subcarriers is upper bounded by $N_s < 0.032B/(f_{d,\max}(1 + \eta_{\max}))$, where $f_{d,\max} = f_c v/c$ is the maximum Doppler frequency. The upper bound obviously depends on the carrier frequency as well as on the user mobility. While OFDM carriers are orthogonal, their spectrums do overlap. Doppler shift or, equivalently, insufficient frequency synchronization thus introduces ICI. Doppler spread, or phase noise introduces ICI as well (the effect is a sum of shift effects). Concerning the effect of synchronization on the OFDM performance, the reader is referred to [26]. Usually, a Doppler spread (or phase noise) equivalent to 3 to 5 % of the carrier spacing may be acceptable. This gives another lower bound on the carrier spacing (and hence an upper bound on the number of subcarriers). In Wi-Fi, the carrier spacing is rather high to account for cheap oscillators. For future wireless LAN systems in the 60-GHz range, Doppler spread and phase noise considerations may put different requirements.

Prefix/Postfix Design

After fixing the FFT size and the guard interval length, one needs to decide which type of prefix/postfix should be used. The 'standard' approach is the use of a cyclic prefix (CP). The L last samples of the OFDM symbol are taken and appended to the beginning of the OFDM symbol, where L is the number of samples in the guard interval. The addition of a CP ensures that the convolution of the transmitted signal with the channel impulse response is circular and can hence be diagonalized on an FFT basis. Equalization (and MIMO detection) can thus be effectively performed in the frequency domain on a set of flat channels.

Zero-Padding (ZP) [27] follows a different approach by appending to the transmitted data sequence a guard interval containing L zero entries instead of pre-pending the OFDM symbol with redundant signal copies. Note that the length of the guard interval remains the same and intersymbol-interference between adjacent OFDM symbols is prevented, as for CP–OFDM. The main advantage is that channel equalization by a simple inversion of the channel is possible even when the channel is badly conditioned, that is the channel transfer function has zeros close to or on subcarriers. The drawback is a higher receiver complexity since the FFT used in conventional CP–OFDM receivers needs to be replaced by a bank of finite impulse response (FIR) filters. However, this complexity increase may be reduced [27].

Recent work [28] proposes to use a pseudo-random-postfix (PRP) as guard interval. The aim is to facilitate low-complexity channel estimating for highly mobile environments. The inserted postfix contains a fixed data sequence weighted by a pseudo-random factor in order to avoid stationary effects. Channel estimation can then be assisted by averaging the (unweighted) received postfix sequence over several OFDM symbols and de-convoluting the resulting vector. Once the channel is known, the postfix can be subtracted from the received data sequence and conventional ZP-OFDM receiver algorithms can be applied to gain knowledge of the transmitted data.

Pulse Shaping

The first drawback of the standard OFDM waveform is the rectangular pulse shape in the time domain. The corresponding sinc-Spectrum has a bad localization in the frequency domain. This waveform can be modified by an appropriate windowing. But in a conventional OFDM setup, this leads to the need for either a postprocessing equalizer [29, 30] or to the addition of an extra guard interval [31], in order to remove the resulting ISI. Another way to introduce a windowing leading to good frequency localization is to use oversampled filter banks [32, 33]. The oversampling increases the spacing between the subcarriers, and it produces the equivalent of the time-domain guard interval in the frequency domain. So again we have, as for the standard OFDM, a loss in the spectral efficiency.

Another way to introduce the pulse shaping without loss of orthogonality is to include a time-offset when modulating each subcarrier. For instance, the offset quadrature amplitude modulation (OQAM) modulation format [34] has been used in [35] to get, in the real field, a set of orthogonal waveforms named *Isotropic Orthogonal Transform Algorithm* (*IOTA*). IOTA has the nice property of being nearly optimal with regard to the time-frequency localization criterion, that is an appropriate criterion when considering transmission over time and frequency dispersive channels. As any OFDM/OQAM scheme, IOTA can reach a maximum spectral efficiency, which is not the case of standard OFDM with CP or ZP. Furthermore, like any other modulated transforms, it can be easily and efficiently implemented, thanks to FFT algorithms. However, compared to classical OFDM, an extra cost is required in order to implement poly-phase pre- and post-filters to generate a waveform that is longer than the simple rectangular window. However, as shown in the filter bank implementation described in [36], nearly optimal results can be obtained with very short waveforms.

On the basis of a theoretical analysis [37] it appears that, for transmission over time and frequency dispersive channels, orthogonal waveforms are no longer the best choice and that non orthogonal waveforms have to be used instead. A bi-orthogonal generalization of OFDM/OQAM has already been investigated [38, 39] that could be a candidate for the fourth generation of mobile communication systems.

7.2.4.3 Implementation

Peak-to-average Power Ratio (PAPR)

The time-domain signal in an OFDM system is the superposition of the signals of a large number of subcarriers. This results in an approximately Gaussian distribution of the I- and Q-components of the complex base-band signal. Consequently, OFDM systems require transmit and receive signal-processing blocks with a high dynamic range. This leads to more costly Radio Frequency (RF) components since amplifiers with a larger linear dynamic range are less efficient than the 'switching' amplifiers used in the previous mobile systems, for any given supply voltage level [40].

There are two approaches to reduce this problem: we can either avoid large peak-to-average power ratio (PAPRs) (PAPR reduction) or live with the clipping effects that otherwise occur in the high power amplifier (HPA) of the transmitter. The latter effect is twofold – on the one hand, it distorts the transmitted signal waveforms and thus leads to an increased BER, especially for higher-order modulation schemes. On the other hand, the clipping leads to a broadened spectrum of the transmitted OFDM signal – which is the far more damaging effect, since the emission is usually restricted by a spectrum mask.

There has been active research in recent years in the area of preprocessing the OFDM signals for PAPR reduction by coding. However, the additional redundancy of PAPR reduction coding causes a reduction in user data rate and most schemes also lack flexibility. Another approach is to avoid 'bad' OFDM words by changing them for equivalent ones. The equivalence is given by changing pseudo-randomly the transmitted word, which however requires channel side information, at least intrinsically. An extensive overview of PAPR reduction schemes is given in [41].

Whenever a broadened spectrum of the OFDM signal is acceptable, one may relax the requirements on RF front-end components, allowing for cheaper power amplifiers to be used. To appropriately detect the non-linearly distorted signal at the receiver, digital base-band compensation techniques can be employed [42, 43].

Phase Noise

The performance of an OFDM system can be strongly degraded by the presence of random phase noise in oscillators, especially if a system targets HDRs at very high carrier frequencies (for instance at 60 GHz). Phase noise causes constellation rotation (common phase error, CPE), and ICI. Several methods have been proposed to compensate the effects of phase noise. After addressing the problem of estimating the CPE [44, 45], more advanced algorithms were presented, which focus on suppressing the resulting ICI also [46]. However, for the carrier frequencies and modulation formats currently envisaged for next generation WLAN systems, phase noise is not (yet) a limiting factor, as long as the subcarrier spacing remains in the order of several hundred kilohertz (e.g. 312.5 kHz as in the Institute of Electrical and Electronics Engineers (IEEE) 802.11a/n systems).

IQ Imbalance

The mismatch between the in-phase and quadrature (IQ) components of the time-domain base-band signals after up-conversion to the radio frequency must not be overlooked. It is not as widely communicated in the scientific community as other topics, although it is critical for system performance. The FFT processing in the conventional OFDM receiver requires that the IQ mismatch between I- and Q-paths in OFDM transceivers is either absent or sufficiently reduced.

The IQ imbalance is typical for low-cost direct conversion transceivers, where the 90° hybrid for the local oscillator may not be perfect (see Figure 7.6, right, where the interior of an IQ modulator is shown). The origin of the phase imbalance is well understood if we compare the precision at which the two mixers can be placed on a printed circuit board (which may be 1 mm), with the carrier wavelength (which is 6 cm in free space at 5 GHz). This would result in a phase error of 6°. Moreover, there are amplitude imbalances, due to the different mixer efficiencies in both branches, which are typically in the order of 1 to 2 dB. Both effects are narrowband in their nature, since only the local oscillator phase is affected, which has a fixed frequency. But there may be a difference also in the two paths towards the summation point, after the mixers. It becomes critical particularly in broadband OFDM systems. The resulting phase mismatch then becomes frequency-selective, and this error is referred to as the broadband imbalance in the following.

There are several ways out. A simple narrowband method would be calibration and correction of the time-domain signals. The IQ mismatch is estimated against a perfect normal and corrected individually, both at the transmitter and at the receiver, to pre-compensate or to restore the complex base-band signals, respectively. This method is reliable as long as the RF parameters can be held sufficiently constant (oscillator and modulation powers).

A second method is to estimate the effect of the IQ imbalance in the frequency domain. The latter causes a cross talk between equally indexed carriers in the upper and in the lower sideband (LSB), depending on the wireless channel as well [47]. Therefore, one may

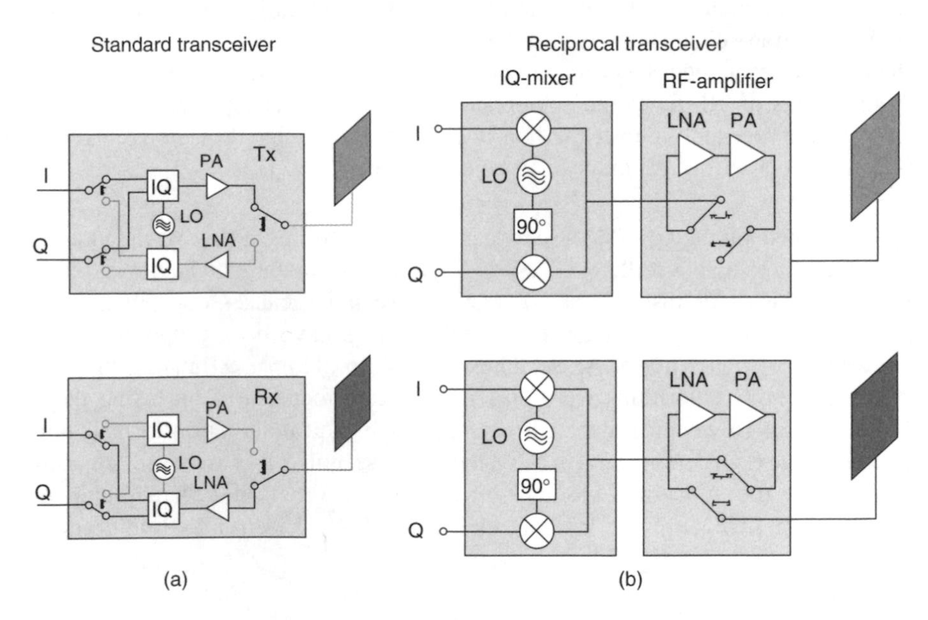

Figure 7.6 Standard transceivers lose the reciprocity in the base-band (a). A reciprocal transceiver may reuse all components in both link directions, using a transfer switch (b)

use revised training sequences for channel estimation. For instance, one could transmit a training symbol only in the upper sideband (USB), and estimate both the correct channel coefficients in the USB and the cross talk coefficients in the LSB. Then a second symbol is transmitted only in the LSB and the estimation is repeated. With this information, one may perform a joint detection of the equally indexed carrier signals in the USB and LSB. But in the case of MIMO–OFDM, both the dimensions of the channel matrices and the effort for the reconstruction of the data signals are then increased, and the channel-tracking rate is reduced as well.

The problems associated with this approach [47] arise from the fact that the estimation and compensation of the IQ mismatch is done on the basis of known training symbols within the OFDM signal. The dependence on these training symbols can be avoided, if a blind estimation of the IQ mismatch parameters is performed [48]. In this approach, the estimation and compensation of the quasi-static IQ mismatch is separated from the tracking of the time-variant channel coefficients. Furthermore, the IQ mismatch compensation can be performed for each receive antenna independently, also in the case of MIMO–OFDM. The computational effort no longer scales with the number of transmit antennas.

Both, the narrow- and broadband imbalance can be avoided with digital up- and down-conversion. But this results in a relatively high sampling frequency for the modulated intermediate frequency (IF) signal (>100 MHz), at least at the transmitter, which must provide a sufficient gap, for the purpose of filtering, between the desired and the image frequency after the IF signal is up-converted to the RF domain. The digital technique is truly broadband, since the analogue origins of the imbalance are no longer present. With commercial components, digital conversion can currently be applied up to 25-MHz bandwidth.

For even higher bandwidth, one must still use the analogue techniques. The broadband imbalance can be avoided by integrating both modulators onto a single chip, which removes the most critical path difference after the mixers. The narrowband imbalance may then either be calibrated out as described above or almost reduced, by a well-designed 90° hybrid on the same chip. Such components are commercially available, up to carrier frequencies of 2.7 GHz [49] and their performance is calibrated using a multitone test signal. The above-mentioned cross talk between equally indexed carriers is then specified at a mirror frequency distance of −36 dB.

7.2.5 TDD

7.2.5.1 Reciprocity

Before we start, it should be noted that the Personal Handyphone System (PHS) in Japan already operates in the TDD mode, and it exploits the channel reciprocity for the array processing at the BS. Unfortunately, the technology is driven by a single company (Arraycom) and PHS is not an internationally supported standard. Moreover, the PHS is a narrowband system and it does not apply multiple antennas at the MT. Many questions concerning the antenna calibration must be reconsidered for the broadband case until the reciprocity-based technique is applicable to OFDM signals.

For dealing with reciprocity, we must consider two points: Firstly, it requires special (calibrated) RF front ends to be realized. Secondly, the reciprocity may get lost because of the terminal movement, so that the CSI taken over from the reverse link becomes outdated. Both issues must be considered, separately.

Transceiver Calibration

The major advantage of the TDD system is the inherent channel reciprocity, and the presented system concept is fully based on it. But the reciprocity holds only at the antennas, and normally it gets lost in the base band after the IQ mixer. This becomes obvious from Figure 7.6(a). One usually uses different IQ mixers, amplifiers, and path lengths in the separate RF chains at the transmitter and at the receiver.

There has been previous work concerning the calibration of the PHS BS array antennas in order to exploit channel reciprocity [50]. In principle, one of the BS antennas is used as a reference transmitter, and the test signal transmitted from that antenna over the air is measured at the receivers of all other BS antennas for the purpose of calibration. On the basis of the results of such measurements, a calibration procedure is developed, which allows the desired reciprocal processing of the base-band signals of all antenna elements.

Papers from the scientific community recently proposed self-calibration techniques [51–54]. For instance, a noise source is used as a reference [52], while the antenna is disconnected from the transceiver. But then the effort in the RF chain is increased, because of additional switches and directional couplers, all of which must be perfectly matched. This is likely to increase the costs of the terminal (note that reciprocity is also required at the terminal, at least in the ad hoc mode).

More recently, a reciprocal transceiver structure has been proposed. The idea is shown in Figure 7.6(b). In principle, one may reuse the IQ mixer and the low-noise and power amplifiers in the TDD mode, since the terminal does not transmit and receive simultaneously. By using an RF transfer switch, the link direction of the in-line, low-noise, and power amplifiers can be reversed. Moreover, the IQ mixer is reused, both as a modulator and demodulator. In a hybrid setup, which has already been tested in the lab, a calibration procedure is still required. It directly estimates all the parameters contributing to the residual non-reciprocity (phase errors, amplitude imbalances), based on a precise transceiver model. From the estimated parameters, individual calibration matrices are formed for the transmitter and receiver modes. The error vector magnitude of the residual non-reciprocity has been reduced down to −30 dB. Currently, the calibration is narrowband, and it is useful in a 20-MHz bandwidth. But we would like to point out that a careful integration of the amplifier and the transfer switch, combined with digital up- and down-conversion or an analogue IQ mixer with negligible imbalance could make these transceivers calibration-less. This is desirable, at least at the MT [55].

Guaranteeing reciprocity in wireless communication devices can pose additional constraints from the point of view of joint antenna and printed circuit board design. This is clear from the fact that in the event that a nonnegligible amount of signal energy enters the receiver at points other than the front end (i.e. behind the antennas, switches, low-noise amplifiers (LNAs) and power amplifiers) the composite channel will not be reciprocal. This could easily happen in RF subsystems, which separate the front end from the up/down-conversion circuits for the ease of deployment of the equipment.

Time-variant Channels

Of course, the wireless channel must itself be reciprocal as well. It is easy to prove this with two antennas and a network analyzer in the lab, by alternately measuring the channel in both directions. The reciprocity is given even in indoor scenarios, where the scattering is rich. But as soon as antennas move faster than the channel is updated, time variation destroys the channel reciprocity. Hence, time variation must be negligible while the channel coefficients are estimated in the UL; the precoded data are then transmitted in the DL.

At low mobility, this requires a simple change in the OFDM burst structure, where the pilot signals for the channel estimation are initially transmitted from the MT to the BS. At higher mobility, the changes in the link direction must be realized very fast. The challenge is then to still identify the signals from different antennas, to estimate the channel with good quality, and to still transmit a useful amount of data in a fraction of the channel coherence time. But these requirements may not be much stricter than in conventional wireless systems, which use the channel information only at the receiver. We need to transmit the pilot signals only from the terminal towards the BS, either in the UL (as in the conventional system) or prior to the DL and receive the data within the coherence time. Such a roundtrip needs two times the propagation delay (2 μs for a distance of 300 m), which is shorter than one OFDM symbol if the number of subcarriers and the length of the guard interval are properly designed as described in Section 7.2.5.

7.2.5.2 Interference

Interference is the true headache in cellular TDD systems. It arises from many reasons (uplink-downlink synchronization, intra- and intercell interference) and it is handled with additional interference management functionality.

Synchronization-related interference: A major promise of the TDD mode is the flexible load distribution between UL and DL. However, the results in [56] clearly indicate that the switching point between up- and downlink must be synchronized in adjacent cells, particularly in the DL, while the UL is less affected.

Intracell interference: When a continuous Code Division Multiple-Access (CDMA) signal is transmitted over a multipath fading channel, the spreading codes are no longer orthogonal, which causes multiple-access interference as well as interference between consecutive CDMA symbols. Both can be removed with rather complex receiver structures (see e.g. [57]), which is critical at the MT.

The OFDM signaling waveforms remain orthogonal in multipath channels also. Provided a suitable guard interval is used, and the synchronization of the MTs is well established, there may be no intracell interference, at least in the DL.

In the UL, each terminal must be individually synchronized. This is straightforward, when the terminals are operated in the Time Division Multiple-Access (TDMA) mode, as in the HiperLan/2 standard. With orthogonal frequency division multiple-access (OFDMA) in the UL, the synchronization becomes an issue to be addressed, particularly when multiple users must be supported at a SC, as suggested by the information theory.

Intercell interference: Reusing the carrier frequency in adjacent cells became popular with the CDMA systems where it was claimed that all interference can be suppressed by the processing gain. But the exemplary results in [56] indicate that this approach might be wrong, since no cooperation between adjacent BSs is realized. The performance in a given cell depends strongly on the load in the adjacent cells, and it is obvious that the intercell interference is a limiting factor when the system load is high, even with optimal receiver structures.

Interference management: Cooperation between the BSs might be a principal way out. Joint detection and transmission techniques have been discussed in the context of BS antenna arrays since a decade or so. Multiple base-stations may form a virtual array by means of optical fiber or microwave/free-space optical interconnects between them. The drawback of this approach is that the infrastructure costs rise significantly. But any sort of cooperation is economical for a cluster of cells, where the system is frequently used at high load.

In view of reducing the complexity of the system management as well as the signaling overhead between nodes in the radio access network (either BS or radio network controllers), algorithms allowing mitigation of intercell interference in a decentralized mode are of particular interest. Such algorithms may exploit two fundamental features of wireless data access communications: First, data traffic is not continuous but bursty. In order to prevent buffer overflow, BSs typically operate at a fraction of the nominal physical layer capacity. At the physical frame/slot level, this translates into a utilization ratio of less than 100 % of the physical resource (time slots and frequency subcarriers in a TD-OFDMA access system). It is possible to exploit this property with the aim to reduce interference by using a smart 'frame filling algorithm' [58]. One example is the so-called left–right algorithm whereby a given cell fills up the temporal frame from the left-most slot toward the right, while the neighboring cell fills up the frame with traffic from the right to the left. On an average, several slots are unused by one of the BSs, and thus are not subject to any interference. Joint power control and antenna beamforming can be used to further enhance interference mitigation, by finding rules to allocate interference-prone slots to strong users and interference-mitigated slots to weakest users. More general frame filling rules can be researched that will lead to a reduced interference level.

7.2.6 Cross-layer Design

7.2.6.1 Multiple-access for MIMO: Single versus Multiple Link Optimization

In wireless single user links, one applies multiple antennas in order to increase the spectral efficiency and the performance. Many well known results and techniques can achieve a certain performance and can guarantee a certain spectral efficiency [5]. Furthermore, the properties of the multiple-antenna channel and their impact on the performance was analyzed [59].

On the other hand, in multiuser scenarios, the MIMO multiple-access channel appears in the uplink transmission from multiple users to the multi-antenna BS. In general, the optimal transmit strategies depend crucially on the applied performance metric. Different power constraints can be imposed. The sum power of all users can be constrained to limit intercell interference. The transmit powers of each user can be constrained or the transmit power of each antenna of each user can be bounded. In addition to the power constraint, the type of CSI at the mobiles and at the base plays an important role. In the proposals for Highspeed Uplink Packet Access (HSUPA) or High-speed Downlink Packet Access (HSDPA), the multiuser transmit as well as receive strategy depend on the channel quality and on the QoS requirements of the users. To apply results for the MIMO multiple-access channel to the MIMO broadcast channel, the duality theory in [60, 61] can be used. The capacity regions of both multiuser systems are equal. The perfect CSI assumption is crucial for the duality theory and for the following results.

Each user who participates in the multiple-access channel or broadcast channel, has its own performance, which depends on its own transmit strategy as well as on the transmit strategies of all other users. A popular overall performance metric is the sum of all individual performances of all users. This leads to the sum capacity, which can be used as a measure of total throughput. For the single-antenna multiuser case, the sum capacity has been analyzed in [62], and it turned out that the optimal strategy is to allocate power to only the user with the best channel quality. The corresponding optimal multiple-access scheme is a simple TDMA. The result has led to the development of opportunistic downlink scheduling algorithms [63]. If multiple antennas are applied, the sum capacity optimization problem is solved by iterative waterfilling [64] and power allocation [65]. For fixed channel matrices and for a fixed sum power constraint, the algorithm proposed in [65] provides the optimal transmit covariance matrices and their respective transmit powers of all users. The higher the SNR, the more active the users. In general, it is necessary to support more than one user at a time to achieve the sum capacity [66]. This result necessitates the development of multiple-access schemes that distribute the available temporal and spectral resources not exclusively to one best user, but to a set of active users. The important point to note here is that multiple-access schemes that are optimal for single-antenna systems can be highly suboptimal in multiple-antenna multiple-user systems. The multiuser and spatial diversity offered by the underlying multiple antenna physical layer (PHY) has to be taken into account when designing the multiple-access scheme.

In contrast to the sum performance optimization, there has been some effort to analyze the complete performance region of the MIMO multiple-access channel and broadcast channel [66]. Since each user has his individual performance, the set of performances that are simultaneously achievable can be used to design multiple-access strategies, which support a certain point in the performance region. The boundary of the performance region is of major interest because of its power efficiency. In [67], the capacity region of the single-antenna multiple-access channel has been derived and its intrinsic polymatroid structure has been explored. It turns out, that the optimal transmit strategy that achieves a certain point on the boundary of the performance region requires more than one user to be active, that is, TDMA is suboptimal, except at the point that achieves the maximum sum-rate. The suboptimality of TDMA even for small SNR values was observed in [68].

7.2.6.2 Queuing Theoretic Analysis: Stability and Capacity

With the expected increasing proliferation of new services, which require low delay and high-rate uplinks, like for example image/video-upload, is associated a need for flexible uplink transmission design. Hence, the issue of efficient uplink scheduling is gaining increasing importance in the design of for example, HSUPA radio link scheme proposed for future use. Recently, a number of multiple-access scheduling policies based on combined optimization of the bursty data-packet traffic of the data link layer (DLL) and the information flow of the PHY has been presented [69]. Such scheduling design constitutes a distinct field of cross-layer design of the network communication stack [70]. The idea behind cross-layer design is to allow for the combined optimization of different objectives throughout the communication stack, which can result in increased network efficiency. In [71], the combination of information theory and queuing theory has been demonstrated. The need for a kind of cross-layer optimization was recognized in [72]. The goal of cross-layer scheduling policies is to choose such a physical layer transmit strategy, which optimizes a desired measure of efficiency of packet traffic in the multiple-access DLL. Hence, the goal can be for example, the minimization of mean or maximal bit queue length or achieving stability (finite length at any time) of bit queues at all transmitters.

The scheduler can optimize different objective criteria. An important criterion from the DL point of view is the stability of the bit queues of the users. In [73], it is shown that the stability region of arrival rates corresponds to the ergodic capacity region of the MIMO multiple-access channel. All arrival rate vectors that lie inside the capacity region of the MIMO multiple-access channel can be supported without infinite waiting time (or bit queue size). Furthermore, the optimal scheduling algorithm was developed in [73]. Interestingly, the optimal SIC order depends only on the bit queue length and not on the channel realizations. Other important properties of the capacity region based on the polymatroidal theory [67] lead to a complete characterization of the optimal transmit strategy. Currently, the computational efficient implementation of the optimal scheduling algorithm is under investigation.

7.2.6.3 Minimizing Dependency on Channel State Information

Optimal multi-access transmission strategies require a complete CSI to be available at the transmitter/scheduler to form the correct precoding matrices, and power allocation. In view of the uncertainty surrounding the channel estimation at the transmitter due to the non-full reciprocity of the TDD channel, it is of interest to investigate scheduling techniques that either are more robust to errors in transmit channel state information (TCSI) or reduce the need for it. Recently, a reduced feedback scheme for multiuser diversity in single transmit antenna systems [74, 75] was proposed, which reduces the feedback needs by as much as 90 % while maintaining 90 % of the system capacity. The technique is based on a channel quality thresholding principle used to discard certain users from the competition in accessing the channel. These types of techniques are currently being investigated for extension to the MIMO case with Space Division Multiple-Access (SDMA).

7.2.6.4 Dynamic Resource Allocation for Deterministic Channel Use

A key advantage of wideband OFDM(A) systems is the possibility of performing multiuser waterfilling both in time and frequency. Although the ergodic capacity region is not increased by the wideband resources [67], the additional dimensions potentially allow for a more fair use of the channel due to the increased randomness in the system. This randomness can be beneficial if constraints are placed in order to guarantee a certain instantaneous bandwidth. As mentioned earlier, the latter are particularly important for today's circuit-switched applications (e.g. voice, real-time video) if they are to be run effectively on wireless packet networks. In the context of a cross-layer view, we are interested in wideband resource allocation strategies guaranteeing the peak queue length as opposed to average queue length for a given link. In [76, 77] orthogonal allocation and power control strategies guaranteeing a deterministic channel use (i.e. guaranteed instantaneous bit rate) are considered for parallel (e.g. OFDMA) slowly fading channels with multiple antennas. Although clearly suboptimal from the point of view of the delay-limited capacity region [78], which to date remains an open problem for frequency-selective multi-antenna channels, it is shown that reasonably simple orthogonal allocation strategies can yield both multiuser diversity and spatial multiplexing, without the need for phase information at the transmitter. These strategies are thus very appealing for slowly fading TDD systems since exploiting amplitude reciprocity does not pose significant constraints on electronics design. The achievable rates approach those of the ergodic sum-rate, however, with strict guarantees on channel use.

7.2.7 Real-time Implementation

The system concept in Figure 7.1 is based on full channel information at the transmitter. Since the wireless channel may change rapidly, in particular, when the terminal is moving, we need to adapt to the current channel realization already prior to transmission. The long-term channel variation is not predictable in rich scattering scenarios. This might be possible only in cases where the LOS component is dominant. So the adaptation to the channel must take place in a fraction of the coherence time. Typical values both for Wireless LANs and cellular systems are of the order of a few milliseconds, even though a higher mobility is required for cell phones. Note that the proper frame length depends on the SNR at which a system operates. After an initial estimate of the channel, the time variation causes an interference rising with a slope of 20 dB per decade in time. Once this interference becomes comparable to the noise, a new estimate is needed. Wireless LANs operate at much lower noise and interference levels than cellular systems, and so they may be similarly sensitive to the time variation of the channel, even if the mobility is much lower [79].

All operations concerning the transmitted signal (spaced-frequency, precoding, and scheduling) must be properly adapted in this short time. So the algorithms must strictly satisfy real-time constraints while performing close to optimal capacity, which is the basic challenge in the system concept considered here.

Most principle problems in the real-time implementation of MIMO–OFDM have already been solved separately, but the system integration has not yet been fully completed. Regarding the UL, the definition of orthogonal preambles, the corresponding low-complexity channel estimation, the fast weight calculation, and at least the linear data reconstruction have already been integrated and successfully tested over the air in a number of prototypes, almost simultaneously [80–83].

In the following, we refer to [80]. Figure 7.7(a) shows part of the reconstructed data signals (I and Q signals for two antennas) after the spatially multiplexed transmission over the air at 5.2 GHz, estimating the channels for all 48 used subcarriers, calculating the weight matrices and reconstructing the data signals using the linear MMSE algorithm, which all of is done in real time. The minor cross talk between the I and Q branches is due to the IQ imbalance in the three receiver chains. Figure 7.7(b) shows the effect of the multipath fading, which is resolved in the frequency domain with OFDM. Using the channel-aware MIMO–OFDM concept described above, the corrupted carriers in the USB would not be loaded with data, while the LSB will carry most of the traffic. Concerning the complexity, the 3-antenna receiver for 2 transmitted data streams fits into a Virtex II/8000 FPGA, where the channel estimation and data reconstruction are performed. The limitation comes from the dedicated BlockRAMs needed to store the channel and weight coefficients close to either the correlation circuits for the channel estimation or to the multipliers in the matrix-vector multiplication unit used for data reconstruction, respectively. An additional TI 6713 Digital Signal Processor (DSP) is occupied with tracking the channel sufficiently

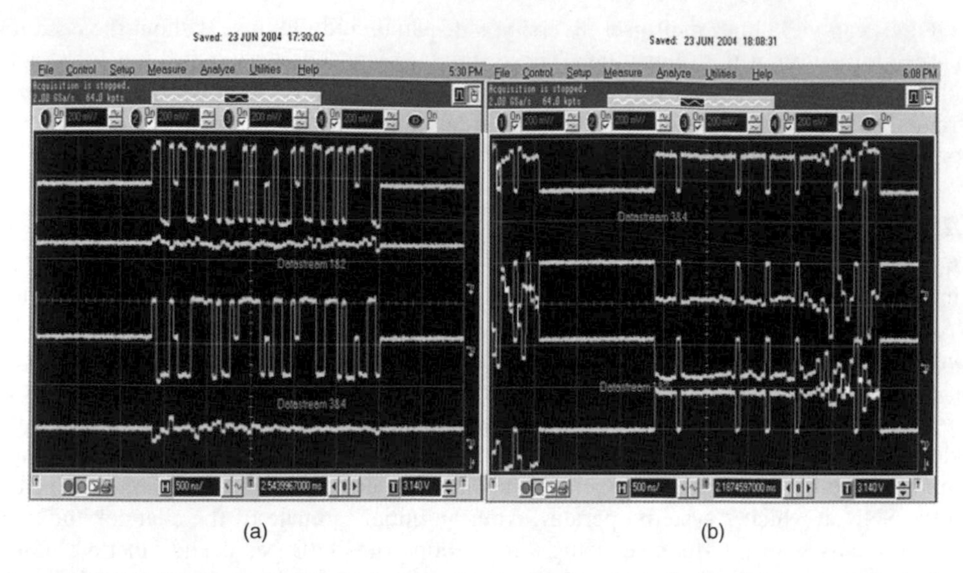

Figure 7.7 (a) Reconstructed data signals after reordering them in the frequency domain (top: antenna 1, bottom: antenna 2). (b) In the captured channel, the multipath fading corrupts the signals in the USB, while it provides good conditions in the LSB

fast. The OFDM signals are continuously transmitted and reconstructed, and the data transmission is only interrupted by the inserted preambles for the synchronization and channel estimation.

The combination of adaptive modulation and linear MIMO detection has recently been demonstrated with a flat fading MIMO detector. It may be straightforward to implement the adaptation for multiple carriers as well [84]. The DL is in principle very similar. For a SC, the linear precoding and the interference-free signal reception at multiple MTs has already been demonstrated, based on the channel side information obtained over a perfect feedback link [85]. The principle of a calibration-free TDD transceiver, which is reciprocal in the base band, has recently been reported (Section 7.2.6). The challenge is now to integrate all these techniques.

7.2.8 Summary and Main Research Challenges

It was shown that a combination of MIMO, OFDM, and TDD could be a promising way to fulfill the requirements of the next generation of wireless systems. Customers increasingly expect the same high QoS and ease of use as known from wired solutions, which implies a high bandwidth, ease of use, robustness in fading environments, negligible latency, and low-cost terminals. It has been illustrated that the broadband MIMO channel promises wire-like QoS over the wireless link, as long as the user is inside the coverage area. There are plenty of MIMO measurements in the literature but we need more broadband data also at longer ranges, which are relevant for cellular applications. In order to reduce the delay and to approach the channel capacity also in arbitrarily conditioned MIMO channels, we have proposed an adaptive system concept, based on channel side information at the transmitter. The latter is realistic in the TDD mode, due to the inherent channel reciprocity. A wide range of transmission and detection techniques are already known. The challenge here is the real-time constraint due to the channel-aware transmission. At least the simplest techniques have already been implemented and tested in real time, but we need further research to develop more efficient algorithms with the potential of real-time implementation. The OFDM system is highly developed, since it is already used in multiple standards and commercial products. Its inherent drawbacks are widely known and have mostly already been addressed. Alternative schemes not losing spectral efficiency due to the CP or guard interval may need further research, also in the context of MIMO. There are two principal ways to realize the reciprocity in the base band, but both the calibration and the removal of the IQ imbalance are not yet addressed satisfactorily, at least in the broadband case. The aspect of interference in such an adaptive concept must still be evaluated in a multicellular scenario. We may also need a guideline from information theory concerning the optimal scheduling and link adaptation in the multiuser broadband MIMO case. The dramatic progress, which is due to significant worldwide research efforts in this field, may indicate that the system concept described above can eventually be realized.

7.3 Ultra-wideband: Technology and Future Perspectives

UWB communications offers a radically different approach to wireless communication compared to conventional narrowband systems. Global interest in the technology is huge. Some estimates predict that the UWB market will be larger than the existing wireless LAN and Bluetooth markets combined by year 2007 [86]. This is due to the capability of these license-exempt wide bandwidth wireless systems to yield low-cost, short-range, extremely high-capacity wireless communications links. The actual achievable data rate naturally depends on the particular technology and propagation conditions; Figure 7.8 shows typically quoted data rates for UWB links, with 500 Mbps at 2-m range, 110 Mbps at 10 m used as conservative figures. The use of UWB has already been deregulated in the United States, Singapore is set to follow shortly with Japan, China; other countries are not far behind. The European position on deregulation, at the time of writing, is unclear although substantive work is being undertaken by the European Conference of

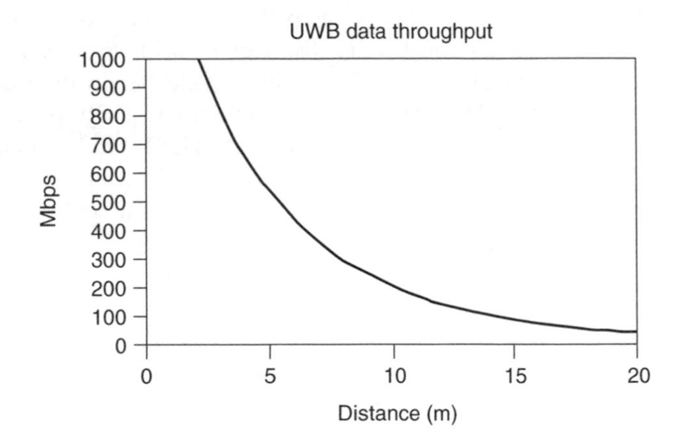

Figure 7.8 UWB data throughput – typically quoted system performance

Postal and Telecommunications (CEPT) and others [87] to produce a coordinated approach across Europe. However, Europe has considerable activity in UWB. In addition to European Telecommunications Standards Institute (ETSI) developing regulations, the European Union, as part of its IST (Information Society and Technologies) initiative, has funded a number of projects including UCAN (UWB Concepts of Ad hoc Networks), PULSERS (Pervasive Ultra Wide band Low spectral Energy Radio Systems) and ULTRAWAVES (ULTRA Wide band Audio Video Entertainment System). The description of this section is therefore highly opportune, bringing into focus the application and implementation of this exciting new technology.

7.3.1 Setting the Scene

The fundamental concept behind UWB communications, that of using a low peak power that is spread over a wide (typically an octave or more) bandwidth is hardly new; indeed the very earliest experiments of Hertz used a spark gap generator, in effect, producing UWB signals! Historically, wide bandwidth radio has used extremely narrow impulses (typically between 0.2 and 0.5 ns pulse width) to provide the communications link. These approaches are hence known as 'impulse radio' (IR) or 'carrier free' systems since a conventional carrier frequency is not present. More recently multiple sub-band systems have been developed using more conventional narrowband techniques occupying a number of sub-bands, which together utilize the available UWB spectrum. Even though not all modern UWB systems employ impulse transmission, it is appropriate to consider a 'birth date' of the technology in the early 1960s [88] when time-domain electromagnetics was being developed for circuit characterization using short-duration impulses. During the 1970s and early 1980s the emphasis was on using wideband, low power, impulse signals in radar and other defense-related applications. The low spectral power density makes these systems attractive to the military because of their inherent low probability of intercept. Equally, the broadband nature of the signal is effective in allowing radar analysis of complex environments such as those encountered by certain ground probing radars. More recently, UWB position location is being developed to determine the location, and to track moving objects within an indoor space to an accuracy of a few centimeters or less.

The application of UWB to ubiquitous commercial communication scenarios was not seriously considered in these early years principally due to the cost of implementation and an unclear market need. By the late 1980s, digital technology had improved to a point that the commercial practicality of low power wide bandwidth communications could be clearly demonstrated. The growth throughout the late 1990s of mobile multimedia communications gave a clear market imperative for HDR wireless communications. Current wireless LAN technologies are capable of up to 54 Mbps up to distances of 150 m (e.g. IEEE 802.11, also known as Wi-Fi). Similarly, Bluetooth (Frequency-hopping Spread-spectrum (FHSS)) provides a short-range 2 Mbps data rate. All these systems are inherently bandwidth limited. While improvements in technology can be anticipated, the inherent narrowband nature of the systems will limit capacity. Comparison with the 500 Mbps (As of March 2004, at least a 1 Gbps over 2 m appears technologically practical) over 2- to 4-m range of UWB systems shows why this technology is generating such interest. The low spectral power density of UWB technology in principle allows UWB to coexist with existing systems (with some caveats) and hence eases deregulation of their use, opening the way for UWB to become the short-range wireless technology of choice for many systems. As an example of application growth, a real-time UWB link between a handheld camcorder and plasma television was demonstrated in May 2004 [89].

The attractions of UWB in the HDR communications environment can be summarized as the potential to deliver ultra high speed data transmission (potentially 1 Gbps over short distances [90]; some experts even expect data rates of 10 Gbps in 2006), coexisting with existing electrical systems (due to the extremely low-power spectrum density) with low-power consumption using a low-cost one-chip implementation. UWB equipment can also be used in lower rate applications such as indoor location determination of people and assets, or other sensor network related applications. In this case, the available channel capacity is used to service a large number of lower data rate devices.

Currently, global growth is inhibited by the lack of a clear operational standard. Equally important, the timetable for deregulation of such systems in some countries is unclear; a key point here being the interoperability (or otherwise) of UWB with the existing users. The issue here is achieving agreement on the spectral emission mask necessary to avoid undesirable interference levels to and from UWB wireless systems. It is possible that different regions of the globe may require different emission standards.

7.3.1.1 UWB Communication Technologies

Modern UWB radio includes both impulse and multiband solutions. As noted above, the adoption of the Federal Communication Commission (FCC) emission regulations has provoked a number of differing technological solutions, all capable of meeting the FCC emission mask and hence using the UWB frequency ranges.

For IR, information is carried in a set of narrow duration pulses of electromagnetic energy. Approximately, the bandwidth required is inversely proportional to the pulse width – so for example, a 1-ns pulse has a bandwidth of 1 GHz. (Note this is only approximately true because of the actual shape of the transmitted pulse in practice.) The center frequency depends on the 'zero crossing rate' of the waves making up the pulse. This is the same type of pulse structure as is used in high-resolution radar for example. In UWB communications, information can be carried by the consideration of the various properties of subsequent pulses.

In multiband solutions, the allowable UWB spectra is split into a number of sub-bands allowing the use of a modified form of narrowband techniques within each sub-band. An important example of this is the proposed use of OFDM within the UWB context.

Examples of these classes of UWB systems include the following:

- *Proposed OFDM-Pulsed Multiband Approach*: In this approach, the full UWB spectrum is split into a number of sub-bands each of 528 MHz width. Each 528 MHz channel comprises 128 carriers modulated using Quadrature Phase-shift Keying (QPSK) on OFDM tones. The composite signal occupies the 528-MHz channel for approximately 300 ns before switching to another channel. In this way, seven time-frequency hopping OFDM signals occupy approximately 3.7 GHz of bandwidth, 14 would occupy 7.4 GHz, and so on.
- *Direct-Sequence Ultra-wideband (DS-UWB)*: Derived directly from secure military communications, a DSSS can be designed to occupy 1.5 GHz at the lower part of the UWB spectrum and 3.7 GHz at the high edge [91]. A combination of both bands

with M-ary Bi-Orthogonal Keying (MBOK)modulation and QPSK yields a data rate of 448 Mbps.

- *Time Modulated Ultra-wideband (TM-UWB)*: This uses extremely short pulses (less than one nanosecond) with a variable pulse-to-pulse interval. The interval variation is measured to produce information flow across the link, including the required information plus a channel code. A single bit of information may be spread over multiple pulse pairs and coherently added in the receiver. Since TM-UWB is based on accurate timing, it is well suited to both communications and distance determination. Systems are commercially available demonstrating bit rates of approximately 10 Mbps at 40 m.

7.3.1.2 Regulations and Standards

A major landmark in the development of UWB has been the publication in early 2002 of the FCC Order and Report deregulating the use of UWB systems in the United States. The FCC has defined a UWB system as having bandwidths greater than 20 % of the center frequency, measured at points 10 dB down from the peak level, or RF bandwidths greater than 500 MHz, whichever is smaller. There are a number of key points to the related emission regulations (US 47 CFR Part 15(f)). Firstly, to avoid inadvertent jamming of existing systems such as Global Positioning System (GPS) satellite signals, the lowest band edge for UWB for communication purposes is set at 3.1 GHz, with the highest frequencies at 10.6 GHz. Within this operational band, emission must be below −43 dBm/MHz EIRP – a limit the FCC have stated to be conservative. Importantly, the FCC deregulation does not specify or imply any particular implementation technology. There are currently at least five identifiable UWB technologies that have been designed to comply with the FCC legal limits (as indicated in the previous section). Each has its proponents and detractors, each has its own technical benefits and restrictions and none are mutually compatible! It should also be noted that the FCC has deregulated UWB technology for other applications as well as communications – for example, various types of imaging and vehicular applications. The defined allowable spectrum is different for different applications.

While this deregulation has obviously had a major impact on the United States' development of UWB, other countries are also advancing with respect to the regulatory framework. Of particular note is Singapore, which has established an aggressive schedule for deregulation. Experimental licenses have been granted for UWB with an EIRP limit 6 dB higher than the FCC specification. These will be assessed before the final deregulation standard is formalized. Japan has granted experimental licenses for the demonstration of UWB with deregulation slated for a 2004/2005 timeframe. The International Telecommunication Union (ITU) is preparing reports in a similar timeframe.

The legal situation in Europe is generally yet to be clarified. There is, however, very substantial activity. Of considerable concern has been the system coexistence issue of a potentially large number of UWB users operating with existing licensed spectrum users. It may be that the European deregulation will modify the spectral mask of UWB around fixed link operational frequencies (approximately 3 to 5 GHz) with a 20 dB reduction over the FCC mask in these areas.

With this as a background, UWB has generated an enormous interest various standard forums. A UWB physical layer is under development within the IEEE 802 LAN/Wide

Area Network (WAN) forum, group 3a of IEEE P802.15 where UWB has been considered for an alternative radio physical layer for the emerging WPAN standard. This is seen as a key standard and has been the subject of much controversy over competing technology; at the time of writing this standard has not yet been decided. Competing alternative technologies are very different. Each technology can provide the target data rates of at least 110 Mbps to 480 Mbps over a personal area network distance from 2 to 10 m. Which ever standard is eventually adopted, UWB systems can significantly out-perform competing wireless communications systems operating in license-free spectrum by virtue of the large available bandwidth despite the significantly lower EIRP levels. They also have unique and interesting propagation characteristics.

UWB systems can also provide opportunity for distance and position finding information. As such these are likely to be considered as possible physical layer definitions by the IEEE 802.15.4 a group. This is also under consideration, with longer range (up to 100 m) applications in mind, and includes WSN with high power efficiency requirements.

7.3.1.3 Future Challenges

Once the standards and regulation positions have been resolved, there is little doubt that deployment and growth in the use of UWB will be rapid. One area of future interest is to increase the data achievable rate at longer ranges (10–30 m). As would be expected, this is limited both by the total energy in a particular UWB pulse structure and by propagation channel characteristics.

For indoor application, any system can be dominated by multipath considerations. As an approximate guide for UWB, a propagation characteristic inversely proportional to the distance cubed is often quoted [92] (The validity of these models in cases where communication is not LOS (for example, through an office partition) is still under investigation.). The long delay spread caused by the multipath can have both positive and negative implications. On the positive side, the multipath arrivals will undergo less amplitude fluctuations (fading) since there will be fewer reflections that cause destructive/constructive interference within the resolution time of the received pulse. However, the multipath also causes the channel to be dispersive – that is, the output-received signal is significantly different from the input pulse. Figure 7.9 illustrates the received signal for a 2-ns input pulse width in a complex, NLOS indoor environment.

To produce higher data rates at these longer ranges will require a much greater understanding of the propagation characteristics. For example, the antenna characteristic (itself a dispersive structure) couples strongly with the propagation problem. It is possible that better control of the antenna characteristics (though still commensurate with low-cost implementation) will give significant performance enhancement.

UWB-based indoor networks capable of extremely HDRs are clearly technologically feasible. However, optimum routing and network configuration is still a research topic, complicated by the unique propagation characteristics of UWB. One can foresee the UWB-based indoor networks being evolved over the next few years.

Whether future regulators will allow further extension of the UWB band will depend, at least in part, on the results of the various system coexistence studies currently being performed. With the use of spectrum 'notching' techniques around critical areas, extension of bandwidth is a real possibility and, providing crucial technologies such as the antenna

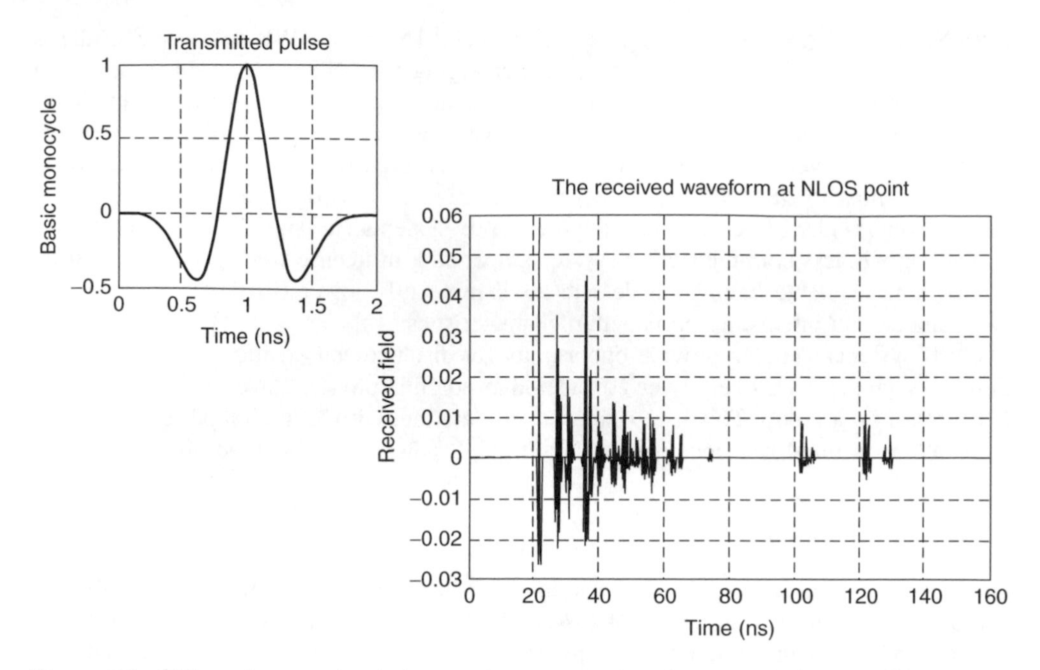

Figure 7.9 Effect of a complex indoor environment on impulse transmission over 30-m range, non–line-of-sight conditions

can be solved, should yield still further increase in short-range data rates. It remains to be seen just how the propagation channel will limit this with range.

UWB technology will improve with time. Advances in transmitter and receiver design, including the antenna, will improve efficiency and performance. In this regard, commercial UWB technology is still in its infancy – the future looks exciting.

7.3.1.4 Structure of the Section

The remainder of this section elaborates on the above discussion and is structured as follows. Section 7.3.2 describes perceived applications of UWB technology for communications. Section 7.3.3 then focuses on technological aspects of UWB. Medium Access Control (MAC) for UWB is discussed in Section 7.3.4. Section 7.3.5 describes spectrum issues that relate to the transmission of UWB signals. Some future perceptions of UWB technology are presented in Section 7.3.6, and is finally summarized in section 7.3.7.

7.3.2 Applications

Future wireless networks are envisaged to provide users with their desired information anywhere and at any time, by working seamlessly with other communication infrastructure. Examples of high-rate applications in this context comprise access to the information backbone and distribution of multimedia content in home and public access environments as well as the replacement of wire-based networks in single-office/small business environments. The inherent features of UWB make this technology a promising candidate for the above-mentioned scenarios. Moreover UWB is resilient against multipath fading, making it perfectly suitable for indoor environments. Coexistence with other wireless systems and robustness against jamming (or, in this context, interference from other spectrum users) helps to ensure easy deployment. Last but not least, the high-resolution position location capabilities enable new applications such as location-aware content provision in offices and public areas in general.

The above-mentioned notion of UWB implies the provision of HDRs. Conversely, the distinct potential of UWB in low data rate (LDR) applications where location estimation may be desired is owing to its inherent high temporal resolution because of the large signal bandwidth. This enables precise location determination and tracking. On the basis of these capabilities, new paradigms in protocol design, location-aware computing, location determination, and so on are feasible. This paves the way for innovative LDR applications, for example, the identification of assets and their exact positions are useful in logistics. WBAN is yet another feasible application, facilitating medical supervision, which may improve the quality of life of patients. Moreover, the potential for transceiver simplicity yields low-cost, low-power devices. This allows for a long lifetime of battery-operated nodes making large sensor networks for (indoor/outdoor) surveillance, smart homes, and security applications viable [93]. Such low-power implementation is considered feasible with careful signal and transceiver architecture design.

The origins of modern UWB radio technology date back to the late 1960s [94]. Since then its application range has been extended tremendously. At the beginning, radar technology pioneered a basic understanding of non-sinusoidal (pulse) signals and the development of working systems. Later, GPR (Ground Penetration Radar) became popular opening the way for detecting hidden objects. Applications never thought of before became possible: discovery of underground water resources, landmine detection, detecting cracks in rocks (a life-saving feature in mining) and so on. Extending the idea of discovering invisible targets, nowadays radar technology has developed further for imaging systems employed in security systems and for biomedical imaging applications. This technique can reveal objects or subjects, not perceptible to the human eye [95].

Wireless communications using UWB follows a completely different technical approach compared to traditional radar, even though in both applications large-bandwidth signals are used. In radar, the transmitted signal is known and the communication channel is unknown, that is, radar can be interpreted as estimating the channel and trying to extract distinct features of a channel. On the other hand, in communications, the transmitted signal is unknown and has to be estimated using at least certain knowledge of the channel. Thus, the extension of UWB to wireless communications is a completely new approach, yet just utilizing known concepts.

Considering already established wireless communications systems, such as narrowband and optical communications, UWB technology needs to verify that it can offer benefits that other systems cannot provide. Optical communications for example, usually require LOS for efficient operation, whilst narrowband radio communications systems would require an extremely high transmit power to operate at the envisaged high data rates that UWB can obtain.

In the future an abundant deployment of devices communicating with high-bandwidth signals will become common, penetrating our daily life at home, in industry, and in logistics – just to mention a few example areas. The remainder of this section focuses on communication-based applications, which forms the context of this section.

7.3.2.1 Low Data Rate Applications

Here, IR is considered as one technology accomplishing UWB operation. With careful signal and architecture design, the transmitter can be kept much simpler than with conventional narrowband systems, permitting extreme low energy consumption and thus long-life battery-operated devices, which are mainly used in LDR networks with low duty cycles. Nevertheless, the receiver design remains the major challenge for IR-based systems. As the number of resolvable multipath components is much higher than in narrowband signals, traditional Rake receivers provide too complex to be implemented in low-energy devices. Energy detection receivers are a promising approach to build simple receivers [96]. Energy management schemes may alleviate the strict energy bounds imposed by batteries [93].

Surveillance of areas difficult to access by humans can be achieved by the deployment of sensor networks [97]. Collecting difficult-to-gather data might lead to new insights and research topics in other research areas. The inherent noiselike behavior of UWB systems makes robust security systems highly feasible. They are not only difficult to detect, but also excel in jamming resistance. These characteristics are essential, not only for traditional security alarm systems, but also for WBANs, which are envisaged for medical supervision.

Because of the simple transceiver architecture and the thereby expected low costs of transceivers, the number of devices to be employed can be over dimensioned. This allows for highly redundant data sources, pushing new algorithms and paradigms in data aggregation, coding, and transmission reliability. With this approach, a certain percentage of nodes may fail (due to device failure, bad transmission conditions, and so on.) without affecting the functioning of the system as such. Deliberately designing devices with higher failure probability will again lower the cost of a single device. For complementing smart homes, actuators can be controlled by a central operator, making human intervention unnecessary.

Even though short range, LDR communications using alternative PHY concepts (i.e. UWB) are currently discussed in IEEE 802.15.4 (a working group) [98], a lot of research is necessary to actually bring those systems into our daily life on a large scale, where user acceptance and applications are central to large scale deployment as well as technological research.

The very distinct potential of UWB in low-rate applications is owing to its inherent temporal resolution because of large bandwidth, enabling positioning with previously unattained precision, tracking, and distance measuring techniques, as well as accommodating high node densities due to the large operating bandwidth [93]. Many routing protocols are known which reduce controlling overheads, using location information [99]. GPS is often unavailable in indoor environments. Hence, innovative methods for location determination are highly desired. Even in areas with good GPS reception, GPS transceivers may turn out to be too complex and too energy-hungry for long-life battery-operated devices. Today's indoor solutions use either infrared or ultrasonic approaches. The former requires line-of-sight propagation, which cannot be guaranteed, and the latter has the disadvantage of propagating with limited penetration. Simple UWB radio technology may fill this gap between demand and physical constraints, and is currently under development [100].

With proper position information, new paradigms in location-aware protocol design and computing come into play opening new opportunities for ad hoc network design. By reducing protocol overheads, this will again decrease the energy costs for data transmission.

For industrial needs, for example in the automotive field, distance measuring system is yet another example for the deployment of UWB systems as logistics will also profit from highly precise location determination.

7.3.2.2 High Data Rate Applications

HDR applications of UWB wireless technology initially drew much attention, since many of the applications are suited to the consumer market. Hence, commercial interest in technology development, standards and regulation has been, and still is very high.

The very definition of UWB [100] – a bandwidth exceeding 500 MHz (for carrier frequencies above 2.4 GHz) and an extremely low power spectral density (75 nW/MHz, according to FCC rules), along with inherent features such as operation in unlicensed frequency bands and resilience against multipath fading–makes UWB the perfect candidate technology for these kinds of scenarios. The problem of designing transceivers with reasonable complexity, also suitable for handheld devices, is one of the main challenges for high-rate applications. Robustness against jamming is also very important, as a large number of electrical devices emitting narrowband noise are usually found in home and office environments, as well as interfering signals from other wireless services operating in the section of the UWB bandwidth.

Within the context of HDR, the main application areas include the following:

- **Internet Access and Multimedia Services:** Regardless of the envisioned environment (home, office, hot-spot), very HDRs (>1 Gbps) have to be provided – either due to high peak data rates (download activity, streaming video), high numbers of users (lounges, cafés, etc.), or both. Because of their high algorithmic complexity, conventional narrowband systems with high spectral efficiency may not be applicable for low-cost and low-power devices (e.g. for small handheld devices). Using very large bandwidths at lower spectral efficiency and employing simple transceivers, UWB is an attractive solution for such applications.
- **Wireless Peripheral Interfaces:** A growing number of devices (laptop, mobile phone, personal digital assistant (PDA), headset and so on) are employed by users to organize themselves in their daily life. The interconnectivity of all these devices is increasingly important as a common database (contact information, calendars, emails, and documents) is held redundantly and in parallel on several devices. Users expect the required data synchronization and exchange to happen conveniently or even automatically. Standardized wireless interconnection is highly desirable to replace cables and proprietary plugs [101, 102]. It has to be emphasized, however, that wireless solutions in this context will be attractive mainly for battery-powered devices without the need for an external power supply.
- **Location-based Services:** In the context of high-rate data applications, location-based service provisioning is an increasingly important topic. To supply the user with the information he/she currently needs, at any place and any time (e.g. location-aware services in museums or at exhibitions), the user's position has to be accurately measured.

Especially in indoor environments, current solutions (such as GPS) cannot fulfill this demand. Here, UWB techniques may be used to accommodate positioning techniques and data transmission in a single system.

7.3.2.3 Home Networking and Home Electronics

One of the most promising commercial application areas for UWB technology is wireless connectivity of different home electronic systems. It is believed that many electronics manufacturers are investigating UWB as the wireless means to connect together devices such as televisions, DVD players, camcorders, and audio systems, which would remove some of the wiring clutter in the living room. This is particularly important when we consider the bit rate needed for high-definition television that is in excess of 30 Mbps over a distance of at least a few meters. An example of a possible home-networking setup using high-speed wireless data transfer of UWB is shown in Figure 7.10 [103].

Of course, UWB wireless connections to and from personal computers (i.e. wireless USB) are also another possible consumer market area, with products expected in the next few years.

A recent proposal is to use UWB as the wireless link in a ubiquitous 'homelink', which consists of an amalgamation of wired and wireless technologies [104]. The wired technology proposed by the authors is based on the IEEE 1394 standard. This is an attempt to effectively integrate entertainment, consumer communications, and computing within the home environment. The reason for the choice of IEEE 1394 is that it provides an isochronous mode, in which data are guaranteed to be delivered within a certain time frame after transmission has started. Bandwidth is reserved in advance, which gives a constant transmission speed. This is important for real-time applications, such as video

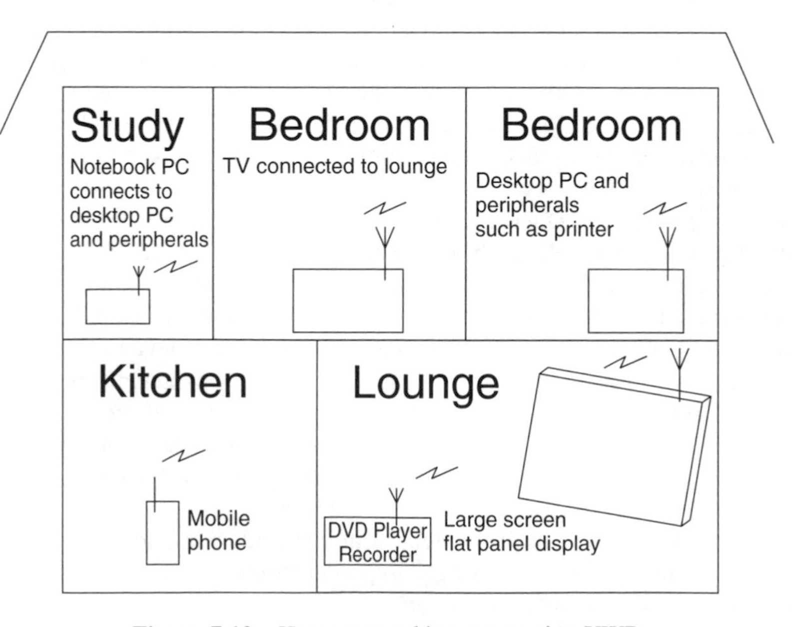

Figure 7.10 Home-networking setup using UWB

Table 7.1 Contents and requirements for home networking and computing

Service	Data rate (Mb/s)	Real-time feature
Digital Video	32	Yes
DVD, TV	2–16	Yes
Audio	1.5	Yes
Internet	>10	No
PC	32	No
Other	<1	No

broadcasts, to ensure that there is no break in the movie or television program for the viewer. Some possible services and required data rates are shown in Table 7.1.

Furthermore, IEEE 1394 has an asynchronous mode, in which data are guaranteed to be delivered, but bandwidth is not reserved and no guarantee is made about the time of arrival of the data.

IEEE 1394 provides scalable performance with 100, 200, and 400 Mbps, which is comparable with the target for UWB transmission speeds. IEEE 1394b is under consideration and will support 800 to 1,600 Mbps and may be extended to 3,200 Mbps. The considerations for home appliances can be described as economy, easy operation, flexibility, and high reliability. Particularly in the area of economy (i.e. cheap devices) and reliability, UWB can be expected to perform well. However, there are various other wireless systems that are targeting this application. In particular, the established IEEE 802.11a standard, which has data rates of up to 54 Mbps, is a strong contender. IEEE 802.11a chipsets have been reported to have speeds of greater than 70 Mbps and are expected to increase beyond 100 Mbps in the near future. Alternatively, wireless transceivers based on the 802.11 g standard may also provide an economical, if slightly lower data rate.

Bluetooth, HomeRF, and 802.11b standards are not strong contenders, because their maximum data rates are approximately 720 kbps, 1.6 Mbps, and 11 Mbps, respectively. Another possible alternative is Wireless 1394, a wireless extension to the wired 1394 system.

7.3.2.4 Wireless Body Area Networks (WBAN)

WBANs are another example of how our life could be influenced by UWB. Probably the most promising application in this context is medical body area networks. Because of the proposed energy-efficient operation of UWB, battery driven handheld equipment is feasible, making it perfectly suitable for medical supervision. Moreover, UWB signals are inherently robust against jamming, offering a high degree of reliability, which will be necessary to provide accurate information of the patient's health and reliable transmission of data in a highly obstructed radio environment.

The possibility to process a large amount of data and transfer vital information anytime and anywhere using UWB WBAN would enable tele-medicine to be the solution for future medical treatment of certain conditions. In addition, the ability to have controlled power levels would provide flawless connectivity between body-distributed networks.

UWB also offers good penetrating properties that could be applied to imaging in medical applications; with the UWB body sensors, this application could be easily reconfigured to adapt to the specific tasks and would enable HDR connectivity to external processing networks (e.g. servers and large workstations).

Currently, researchers are performing studies and analyses on potential UWB antenna systems that could be applied to body-centric networks. One of the topics is the characterization of on-body propagation channels with respect to UWB. As a result, statistical and deterministic propagation models should be developed, which will help to optimize future radio systems.

7.3.3 Technology

When we consider the trends and possibilities in UWB applications, it is pertinent to address the physical limitations of current technologies and the likely future developments in all areas of the transceivers. The high signal bandwidth places stringent conditions on the radiating and receiving structure of the antenna, the front end for both transmitter and receiver, and the transceiver chain (if there is one). The transceiver technology must be linear to avoid signal distortion, or at least be predictable over the whole band, so that predistortion and correction techniques can be reliably applied if needed. This forms the context of the subsequent discussion in this section, which starts by highlighting propagation-related issues.

7.3.3.1 Signal Propagation

Because of large signal bandwidths far exceeding the coherence bandwidth of the propagation environment, the UWB channel suffers from frequency-selective fading. However, sub-nanosecond temporal resolution allows for fine multipath resolution, resulting in reduced fade depths (about 5 dB) compared to narrowband systems (over 35 dB). In a typical UWB channel, a very large number of multipaths are observed, depending on the propagation environment and spreading bandwidth. Pulsed UWB systems count on temporal orthogonality of multipaths; as the pulses are narrow in time, their delayed reflections are often well separated and can therefore be easily resolved. Multicarrier (OFDM and CDMA) UWB systems rely on efficient energy distribution in the spectrum to prevent multipath and intersymbol interference, ensuring that sufficient information is always presented to the receiver for reliable demodulation. The low power spectral density, however, presents link budget and front-end receiver sensitivity challenges that impose restrictions on the operating range (although techniques such as antenna and processing gain are being considered as range extension technologies).

The average total received energy is distributed between a number of multipath arrivals. It has been reported that there are hundreds of paths for the 6-GHz bandwidth in a NLOS multipath propagation [105]. A measurement conducted in a typical modern office shows that the amount of energy captured by a UWB system increases rapidly until the number of multipath components is a few tens (perhaps up to 50 [106, 107]). The energy carried by higher-order multipath components is generally considered small. Nonetheless, the delay spread caused by even a few tens of multipath components can be substantial. Receivers must be efficient at gathering up this multipath energy, which therefore presents limitations to the achievable data rate. Research is continuing in this area. For a multiple sub-band scheme, the multipath manifests itself as narrowband fading within any sub-band. When fading affects a particular carrier, the information being carried is repeated on a different carrier, thus avoiding the fade. In a heavily multipath dominated environment this obviously significantly reduces throughput.

Rake reception makes use of high temporal resolution and multipath propagation to yield enhanced operation, providing a gain of up to 4 dB [108]. Similarly, two-branch transmit diversity using polarization orthogonality with space–time codes has been shown to provide a diversity gain of about 5 dB [109]. The use of multiple-antenna techniques is especially useful in UWB communications because of the small size of high-frequency antennas, enabling the compact placement of multiple-antenna elements within a mobile device to enhance the link reliability.

7.3.3.2 Electronics and A/D Conversion

The performance of A/D converters has a major impact on the choice of receiver architecture in a digital wireless system. In UWB systems, this phenomenon is particularly exacerbated by the large operating bandwidth. In pulsed systems, high-frequency A/D converters allow the implementation of correlation in the digital domain and enable new modulation and multiple-access concepts that exploit pulse shape. Lower-frequency converters that make use of analogue correlation are described in [110]. The dynamic range is usually relaxed, which ensures the feasibility of digital radio for UWB systems. These solutions have a promising future since they are particularly in line with the evolution of silicon technologies. Moreover, with the new concept of cognitive radio, all-digital architectures are becoming more and more attractive. However, the A/D converters that are required in such transceivers are not available off-the-shelf and need further development [111].

Similar to previous wireless technologies, the first UWB products will rely on RF-analogue oriented products with reduced performance A/D converters. New products are likely to appear with higher performance in terms of dynamic range and better immunity to narrowband interferers that could increase the integration level due to relaxed RF filtering requirements. Therefore, there is a real challenge in the implementation of high-frequency, medium dynamic range A/D converters. Among the various solutions, time interleaved converters [112] are a promising way of exploiting the performances of advanced technology with moderate power consumption. However, although the basic principle of such converters is well known, they have seldom been used in commercial products and will require adaptation to the constraints of UWB. Imbalance problems and propagation delays in this architecture [113] remain to be rectified. These developments should take into account frequency synthesizer architecture [114] to find optimal solutions.

In developing gigabit communication systems, the capability of electronic systems required to decode and process the data is becoming a significant limitation. Techniques for bandwidth maximization such as MIMO communications or software Rake receivers will necessitate the development of digital signal-processing technologies that far exceed current state-of-the-art performance. In fully exploiting the UWB channel, and particularly for future development in bands above 10 GHz, the need for faster processing technology is fundamental and is an enabling technology.

Recent developments in digital systems to clock speeds in excess of 20 GHz [115] have stretched the limits of semiconductor technology. Recently, single devices with clock speeds up to 1 THz have been demonstrated using extreme nano-scale lithographic fabrication [116]. Typical gate dimensions of the order of a few tens of nanometers are employed for this kind of performance. At present, lithography is not sufficiently reliable to enable VLSI with this level of performance. In addition, there are significant problems of heat loading in such extreme semiconductor devices. While these may one day be solved, the likelihood is that any devices operating with frequencies upwards of 100 GHz will require significant cooling. In the search for greater and greater processing speeds, a variety of new technologies are being pursued. Among these is SiGe hybrid logic technology, which uses Heterojunction Bipolar devices and has demonstrated 40 GHz performance [117].

The most advanced nonsemiconductor digital logic scheme currently under development is the superconducting scheme known as RSFQ (rapid single flux quantum) logic. RSFQ processes data in the form of flux quanta trapped within a superconducting matrix. The flux quanta are converted to and from voltage pulses whose time integrated voltage is $\sim 2 \times 10 - 15$ versus. Simple A/D converters [118] and microprocessors [119] have been demonstrated with simple devices shown to function up to and beyond 750 GHz [120]. The significant advantage of RSFQ is in process yield – typical integration scales for a 100-GHz system are greater than 1 micron in contrast to the extreme sub-micron requirements for silicon systems. The major drawback to RSFQ is often cited as the requirement for cryogenic cooling. SNR limitations require cooling below 30 K even when circuits are fabricated using high temperature cuprate superconductors (e.g. YBa2Cu3O6.94). While this is a problem for mobile systems, the cooling systems are not large and static installations are already in the field for superconducting filter systems.

Somewhat less well developed are nano-electronics solutions such as those based on carbon nanotubes. These macromolecules can be fabricated in both semiconducting and metallic forms and simple devices such as nano-FETs have been demonstrated [121]. At present, no realistic estimate can be made for the speed of these devices, but, based on scale alone the potential operating speed should be well in excess of 1 THz. Many issues remain to be resolved, including the not inconsiderable problem of how to fabricate a circuit from devices whose typical dimensions are a few nanometers.

While it is at present impossible to predict which of the many technologies under development may achieve dominance, the number of good prospects for ultrafast logic and systems bodes well for the future requirements of extreme signal-processing electronics. Consequently, the requirements for UWB digital wireless fall within this remit.

7.3.3.3 Single- and Multiband Techniques

As mentioned previously, IR was the initial form of UWB technology. Research effort is continuing in the investigation of UWB schemes based on IR, which uses a single wide-band signal. Impulse-based UWB systems have maintained their hold on niche markets, with applications ranging from imaging radars to location systems and WSN.

On the other hand, multiband techniques have reinterpreted the UWB concept to allow for the application of, in effect, multiple narrow or wide band approaches. Proposed more recently, they provide a natural evolution from the existing wireless signaling technology, as current narrowband and wideband designs can be readily adapted to operate in UWB configurations by utilizing a parallel architecture. They are based on conventional radio technologies, such as multicarrier modulation (also referred to as OFDM), DS-CDMA, and so on, and can be used to achieve bit rates of the order of 2 Gbps in an indoor dense multipath environment [122]. They also permit adaptive selection of bands to minimize interference and enhance coexistence with other wireless services. Both multiband approaches provide a natural evolution from narrowband to UWB without the need to devise novel signaling techniques, thus facilitating reuse of existing expertise for rapid production of marketable UWB products. Figure 7.11 illustrates the spectrum usage and sub-band division in MB-OFDM, which is based on frequency-hopping a 500-MHz wide OFDM block across the UWB spectrum.

7.3.3.4 Antenna Design and Performance

Microwave signal propagation requires smaller antennas than those required for typical RF communication, which makes it easier to produce compact devices. Multi-antenna systems can also be produced with acceptable sizes. Many special designs suitable for efficient UWB signal radiation have been proposed in the literature, such as planar elliptical dipoles, balanced antipodal Vivaldi, D*dot, TEM horn, and so on. These antennas have frequency-dependent radiation patterns. Many frequency-independent antennas can also be used with UWB, such as the biconical, discone, bowtie, horn, log-spiral, and trapezoidal designs. However, some of these antenna designs cannot be used directly for UWB systems, as they radiate dispersed signal waveforms due to phase variation as a function of frequency and look angle (i.e. they cause signal distortion). This problem occurs both in single-band and multiband systems, and any UWB system must compensate for this distortion.

The effect of the antenna can be modeled as a differentiation operation for the transient UWB waveform. Thus a radiated pulse waveform is the first derivative of the generated pulse in the case of a simple dipole, whereas loop antennas radiate the second derivative of the antenna current in the far field while also reducing the radiated energy. The signal

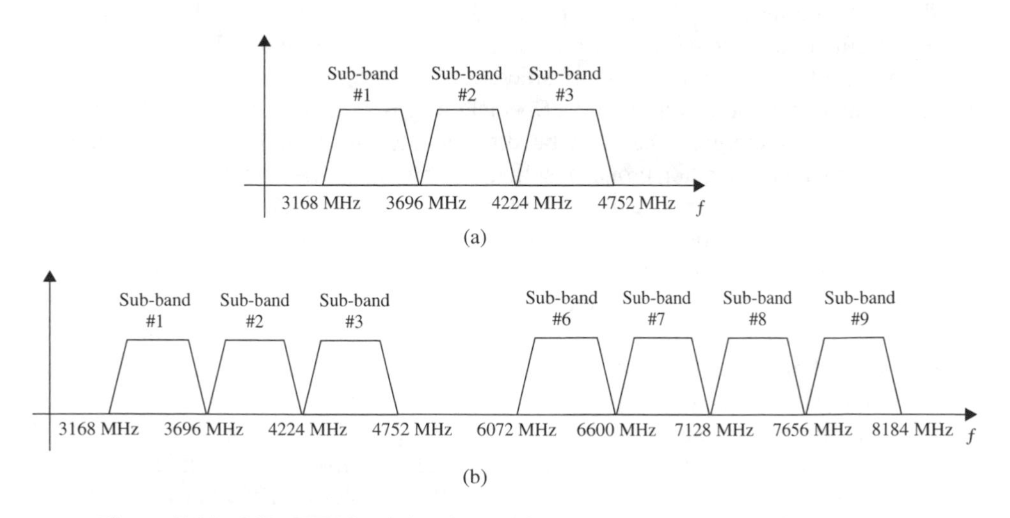

Figure 7.11 MB-OFDM sub-bands used in Mode 1 (a) and Mode 2 (b) devices

undergoes the same transformation at the receiving antenna, so that there is a multiple derivative operation. For a monopole or conical antenna used at the receiver, the output is the integral of the incident electric field. Such operations would only cause phase shifts in a narrowband sinusoidal signal, but the effect is more deleterious for nonsinusoidal UWB signals and requires careful signal design and power matching for distortion compensation. Some of the techniques that have been proposed for measuring the efficiency of UWB antennas differ from those used in the case of narrowband antennas.

Wideband antenna design is now a mature field, with a variety of high performance, compact, and cheap antennas readily available in the market. Existing multiple-antenna concepts can be adapted for UWB systems. Thus traditional space–time processing can be applied to achieve range extension, spatial filtering, and QoS benefits pertaining to these techniques. However, to ensure beneficial operation UWB signal propagation characteristics need to be taken into account when designing such systems.

Additionally, specialized UWB antennas can be designed; however, antenna properties depend strongly on frequency. The antenna causes frequency-dependent spatial filtering, which can be characterized by a directional linear time-invariant model. The transmitted waveform is filtered by the antenna structure. The ultra-wide bandwidth allows for resolving the fine structure of the transient transmitting and receiving behavior of the antenna. For the free-space propagation channel, the channel impulse response depends only on the antenna's filtering behavior and the distance. In the case of a LOS channel, the energy content of the channel impulse response is dominated by the direct path, which again is filtered by the transient responses of the employed antennas. For the analysis of environments with the condition of rich multipath, one has to take into account that the antenna is not only a frequency-domain filter but also has a spatial filtering aspect. This results in different filtering characteristics for different directions and leads to a weighted excitation of the different paths. A directional linear time-invariant (LTI) model can be used for characterizing this in terms of antenna impulse response or transfer function [123]. Adequate antenna measurement procedures have been developed for the measurement of the antenna impulse response [124, 125]. With such angularly dispersive properties displayed by UWB antennas, it is also important to deploy techniques for the normalization of antenna effects in channel measurements and in receiver design [126].

The choice of an antenna needs consideration of size and cost as well as radiation efficiency and dispersion characteristics. Special designs suitable for UWB operation have been proposed in literature. They can be differentiated by size, impedance bandwidth, directivity, gain, and dispersion properties. In this context, dispersion refers to the distortion introduced by frequency-dependent phase velocity as well as ringing owing to frequency-dependent movement of the antennas phase center. In particular, antennas consisting of multiple coupled resonators, such as logarithmic periodic antennas, exhibit poor time-domain dispersion qualities despite their flat gain in the frequency domain. Introducing lossy material can reduce the dispersion effects, but will also lower the radiation efficiency.

The near-field environment, particularly around small antennas, can influence the radiation characteristics, which may lead to additional dispersion. Therefore, the near-field environment has to be considered early in the design process of an integrated UWB antenna for a given application. The co-design of antenna, front-end amplifiers and signal properties is therefore an important topic for UWB technology. UWB is especially resistant against fast-fading effects when integrating over the entire bandwidth. Therefore the combinations with multi-antenna systems, which exploit spatial diversity for fading reduction, promise a very flat channel response resulting in increased efficiency and range. For location awareness, a system may use two or more antennas and calculate the angle-of-arrival by evaluating the signal delay between the two antennas. For impulse-modulated systems, time-domain beamforming is also of interest, since no 'grating lobes' emerge as long as the pulse repetition rate is low enough. However, for extended arrays there are directions where the pulses emanating from the single radiators do not overlap. In these directions, a train of single pulses is generated. Therefore, the radiation becomes more dispersive for directions away from boresight. This effect reduces with increasing number of elements.

7.3.3.5 Interference

Recent theoretical and experimental investigations show that UWB emissions do not cause significant interference to other devices operating in the vicinity [127]. Low-power spectral density causes UWB signals to lie below the unintentional emitter noise limits defined by the wireless regulations. As a consequence, a UWB system is deemed not to cause any more interference to a narrowband receiver than the spurious emissions from a computer, a microwave oven, or a car's ignition. Also, since current applications use UWB technology for very short ranges, any detectable interference can usually only be caused by a particularly high spatial density of UWB devices near the receiver. However, even in such a case, the interference would be localized by the transmission range of the UWB devices. Conversely, the effect of narrowband interference from other closely located emitters to UWB receivers is not significant unless a very high interference power is radiated as discussed in [128].

A conventional interference mitigation technique for narrowband systems is adaptive equalization at the receiver. Since adaptive equalizers require one or more filters for which the number of adaptive tap coefficients is of the order of the number of data symbols spanned by the multipath, they are not suitable for UWB indoor communications with over 100 Msym/s and ISI with more than 30 to 50 symbols. Frequency-domain equalization (FDE) is however a suitable candidate [129], which is the analogue of the conventional equalizer. This type of frequency-domain detector for indoor UWB IR communications has been shown to provide multiuser orthogonality in extremely frequency-selective environments [130].

7.3.3.6 Waveform Shaping

IR UWB communications may use any of a large family of signal waveforms that satisfy the regulatory spectral masks. More popular among them are the Gaussian pulse, Gaussian doublet, Gaussian monopulse (derivate of Gaussian), Mexican hat (2nd derivate of Gaussian), Morlet (modulated Gaussian), Rayleigh, Laplacian, prolate spherical wave functions, and Hermite families of waveforms [131]. The pulse length, rise time of the leading edge of the pulse, and the passband of radiating antenna determine signal bandwidth and spectral shape, while the pulse shape determines its center frequency. Gating, pulse repetition rate, modulation, and selection of dithering code are other factors that determine overall waveform shape. Note that the Fourier transform, autocorrelation, energy spectral density, and so on of the Gaussian pulse are also Gaussian, therefore none of the conventional filtering processes has any significant effect on its general shape, and only its amplitude and duration are affected. Its spectrum is asymmetric about the center frequency, which is the reciprocal of the monocycle's duration, while the 3-dB bandwidth is 116 % of the center frequency. A Gaussian monocycle has a single zero crossing. For additional derivatives of the Gaussian function, the relative bandwidth decreases, while center frequency increases, as does the number of zero crossings, which reduces the resolution of the system. The pulse shape is ideally preserved during propagation through a linear nondispersive medium such as free space. In practice, however, waveform distortion is caused due to physical propagation phenomena such as diffraction, scattering, and penetration through walls and other materials.

7.3.4 UWB MAC Considerations

As discussed earlier, UWB holds great potential for wireless ad hoc and peer-to-peer networks (i.e. networks with no fixed infrastructure or central controller). The enormous bandwidth of UWB systems means that very HDRs can be supported for short-range applications, or much longer distances can be supported with LDRs. The range of data rates may be from hundreds of kbps to Gbps. Systems designed to support such a wide range of data rates and specialized modulation schemes are being considered. Major medium access control (MAC) development activity is being undertaken in the IEEE standardization bodies [132] as well as European Union projects such as PULSERS [133].

Without dramatically changing the air interface, the data rate can be changed by orders of magnitude according to the system requirements. This means that HDR and LDR devices will need to coexist. The narrow time-domain pulse also means that UWB offers the possibility for very high location estimation accuracy. However, each device in the network must be 'heard' by a number of other devices in order to generate a position from a delay or signal angle-of-arrival estimate. These potential benefits, coupled with the fact that an individual low power UWB pulse is difficult to detect, offer some significant challenges for the multiple-access (MAC) design.

As highlighted in Section 7.3.2, UWB systems have been targeted at very HDR applications over short distances, as well as very LDR applications over longer distances. Classes of LDR devices are expected to be of very low complexity and very low cost, implying the MAC will also need to be of very low complexity. The potential proliferation of UWB devices means that a MAC must deal with a large number of issues relating to coexistence and interoperation of different types of UWB devices with different capabilities. The complexity limitations of LDR devices may mean that very simple solutions are required. HDR devices, which are expected to be of higher complexity, may have much more sophisticated solutions.

Research into UWB MAC design is at an early stage and no comprehensive solution for the unique difficulties for UWB systems have been proposed in the open literature. UWB operation is frequently associated with peer-to-peer operation in an ad hoc network configuration. Existing or emerging standards which allow direct peer-to-peer communications include the IEEE 802.15.3 [134] and ETSI HiperLAN Type 2 standards, although only in the context of a centrally managed network. This section will examine some of the difficulties facing the UWB MAC design and look at the impact of some of the key physical layer parameters on the MAC configuration.

7.3.4.1 Multiple-access for UWB Systems

MAC Objectives

The very wide bandwidth of UWB systems means that many potential solutions exist to control this resource. The devices may use all, or only a fraction of the bandwidth available in the 3.1 to 10.6 GHz operating band. The devices will still be classed as UWB provided they use at least 500 MHz. As reported earlier, candidates for the physical layer signal structure of UWB systems include IR, OFDM, multicarrier, and hybrid techniques. All of these possible techniques mean that different UWB devices may or may not be able to detect the presence of other devices. The main issues to be addressed by a UWB MAC include coexistence, interoperability, and support for positioning/tracking.

Coexistence

The potential proliferation of UWB devices of widely varying data rates and complexities will require coexistence strategies to be developed. Here, coexistence is in the context of coexisting UWB users, that is, it is assumed that coexistence between UWB and other spectrum users does not form part of this discussion. This is, however, of paramount importance to the success of UWB and, under some circumstances, a MAC designed to account for this could provide part or all of a solution. Other aspects providing such solutions are – physical layer signal design and transceiver architecture with temporal/frequency/spatial or polarization-based filtering.

Strategies for ignoring or working around other UWB devices of the same or different types based on physical layer properties will reflect up to the MAC layer. Optimization of the UWB physical layer should lead to the highest efficiency, lowest BER, lowest complexity transceivers. The assumptions of the physical layer will however have implications on MAC issues. These are, for example, initial search and acquisition process, channel access protocols, interference avoidance/minimization protocols, and power adaptation protocols. The quality of the achieved 'channel' will have implications on the link level, which may necessitate active searching by a device for better conditions, which is what happens with other radio systems.

Within this context, other noteworthy developments include the Common Signaling Mode (CSM) initiative, an approach allowing different types of UWB to communicate with each other over the same wireless network [135].

Interoperability

The most common requirement of MAC protocols is to support inter-working with other devices of the same type. With the potentially wide range of device types, the MAC design challenge is to be able to ensure cooperation and information exchange between devices with differing data rates, QoS class, or complexity. In particular, emphasis must be placed on how low complexity, LDR devices can successfully produce limited QoS networking with higher complexity, HDR devices.

Positioning/Ranging Support

Position estimation techniques are integrally linked to the MAC. This includes strategies for improving timing and therefore ranging accuracy and for exchanging timing information to produce positioning information. The provision of such support is a key requirement for many UWB wireless applications.

It is possible for any single device to estimate the arrival time of a signal from another device based on its own time reference. This single data point in relative time needs to be combined with other measurements to produce a 3D position estimate relative to some system reference. Exchange of timing information requires cooperation between devices. Being able to locate all devices in a system presents a variation of the 'hidden node' problem. The problem is further complicated for positioning because multiple receivers need to detect the signal from each node to allow a position in 2- or 3-dimensions to be determined.

Location tracking requires that each device can be sensed/measured at a suitable rate to allow a reasonable update rate. This is relatively easy for a small number of devices, but difficult for an arbitrarily large number of devices. Information exchange of timing and position estimates of neighbors (ad hoc modes) requires coordination between devices and calculation of position, and needs to be done somewhere (centralized or distributed) and the results fed to the information sink.

Finally, it is important to have the received signal as unencumbered by multiple-access interference as possible in order to allow the best estimation of time of arrival. As an illustration, every 3.3 ns error in delay estimation translates to a minimum 1 m extra error in position estimation.

All of the following issues require MAC support: information exchange, device sampling rate, node visibility, and signal conditioning. These pose significant obstacles to existing WLAN and other radio systems offering reliable positioning/tracking when added on to the MAC at the post-design. Hence, UWB system design has the opportunity to fully integrate positioning and tracking functionality much earlier in its design cycle.

Constraints and Implications of UWB Technologies on MAC Design

Some qualities of UWB signals are unique and can be used to produce additional benefits in terms of MAC design. For example, the accurate ranging capabilities associated with UWB signals may be exploited by the higher layers for location-aware services. Conversely, some aspects of UWB signals pose problems, which must be solved by the MAC design. For example, using carrier-less impulse signals does not assist in the implementation of the carrier sensing capability needed in popular approaches such as Carrier Sense, Multiple-access/Collision Avoidance (CSMA/CA) MAC protocols.

Another aspect that impacts MAC design is the relatively long synchronization and channel acquisition time in UWB systems. In [134], the performance of the CSMA/CA protocol is evaluated for an UWB physical layer. CSMA/CA is used in a number of distributed MAC protocols. It is also adopted in the IEEE 802.15.3 MAC and is being considered as the IEEE802.15.3a MAC.

The time required to achieve bit synchronization in UWB systems is typically high, of the order of a few milliseconds [134]. Considering that the transmission time of a 10 000 bit packet on a 100 Mbps rate link is only 0.1 ms, it is easy to understand the impact of synchronization acquisition on CSMA/CA-based protocols. The efficiency loss due to acquisition time can be minimized by using very long packets. However, this may impact performance in other ways.

Acquisition preambles are typically sent at a higher transmit power than data packets [136]. This impacts both the interference level and the energy consumption in highly burst traffic. This must be taken into account when determining the efficiency of the system.

The adoption of CSMA/CA as a distributed protocol must be jointly evaluated with the performance of the underlying UWB physical layer. In general, it may not be a suitable choice for an UWB MAC unless proper synchronization techniques are developed. One solution to this problem is the exploitation of the very low duty cycle of IR. Synchronization can be maintained during silent periods by sending low power preambles for synchronization tracking [136]. This approach is feasible only for communications between a single pair of nodes, which is not the case in peer-to-peer networks.

7.3.4.2 Location and Tracking Issues

MAC layer designs are often constrained by the distributed algorithms that use partial local knowledge to solve a global problem, such as network synchronization, positioning, node authentication, and data compression.

These algorithms require real-time feedback of data gathered by other cooperating terminals on particular information, such as the delay between the end of sent packets and an immediate ACK from the receiver. This information may be used for range evaluation. Similarly, real-time transmissions require a suitable implementation to support algorithms within the MAC layer for QoS support.

With this close physical layer to MAC layer coupling requirement, the Open Systems Interconnection (OSI) models should be reconsidered because bounds between PHY, MAC and upper layers become less clear. Some researchers propose to keep the OSI model but add a vertical 'service' scheme, as shown in the Figure 7.12 below. The services are related to the tasks of the distributed algorithms, where synchronization and localization are shown as examples.

Overview of Localization

A location-based application needs to have a minimum number of location nodes, which are able to measure the position in the space occupied by the positioned node. The geometrical constraints are such that at least three nonaligned nodes are required to obtain an estimation of the two-dimensional coordinates. This geometric calculus is based on the

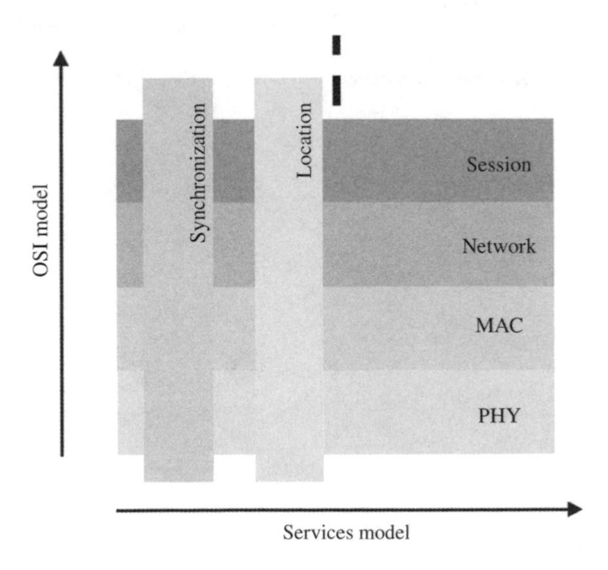

Figure 7.12 OSI layered communications model

assumption that the measured signal is free of perturbations. The main perturbations of the measured signal come from additive noise, interference, and multipath (resulting from angular spread). Consequently, a larger number of nodes must typically be employed to satisfy the location application requirements. This is illustrated in the figure below, where four coordinators are shown to reduce the estimation error compared to three.

Hence, for such a wireless network, the minimum number of nodes needed to provide an acceptable estimate of the position is N. Then, each node must be covered by at least N others. The working surface S in our network is assumed to be the one that provides at least N coordinators for every node position P. Thus, an analytical framework can be constructed to ascertain the location estimation performance for a given operating scenario, that is, particular node density, coverage, and angular spread. The information on the time-of-arrival estimates must be exchanged between nodes to produce a position estimate as shown in Figure 7.13 (i.e. cooperation), which requires MAC support.

Ranging

The position calculation may be performed using many different methods. One such method is based on the angle-of-arrival of the different channel paths. The most promising techniques are based on the signal transport delay between the emitter and receiver [136]. 'One way' ranging uses the assumption that the clocks of the two entities in communication are perfectly synchronized. However, this assumption is difficult to realize in practice. It is even more difficult under ad hoc network conditions where all the nodes do not share the same global reference clock. Then, we use the so-called 'two-way' ranging method, which avoids the requirement for clock synchronization.

The two-way ranging method can estimate the distance between two nodes without a priori knowledge of a shared clock reference. Using this technique, one node sends a

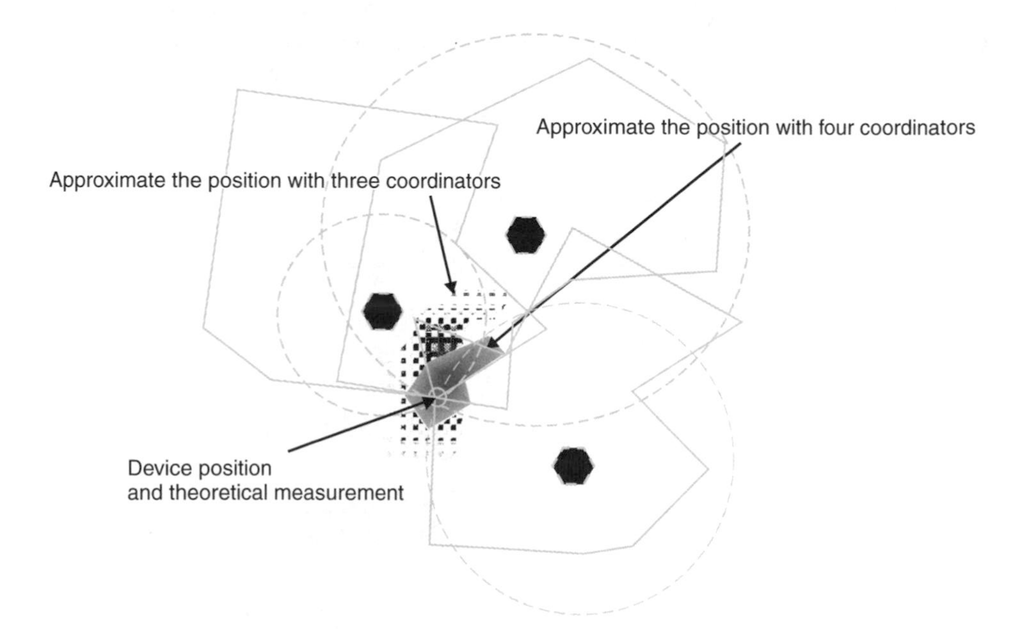

Figure 7.13 Illustration of position measurement error. Isotropic channels are described by circle and real channels by randomized geometric surface

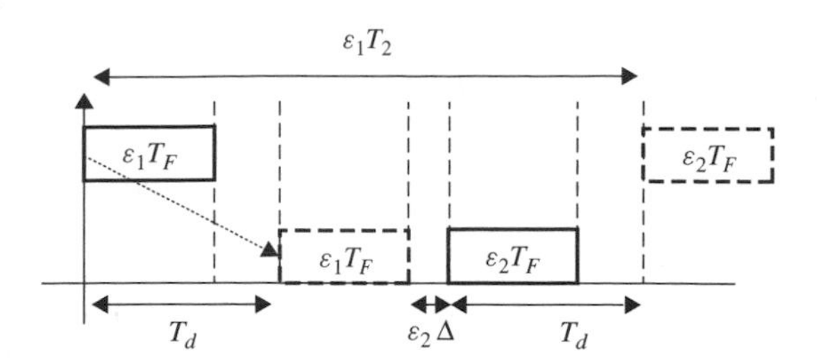

Figure 7.14 Principle of the two-way ranging

frame to another one. When receiving the message, and after a known delay (i.e. guard time), the second sends a response. The cumulative delay between the sent and received messages by the first node gives information on distance (as illustrated by Figure 7.14). On the basis of a theoretically perfect clock, the delay of the signal transport between the two devices is given by $T_d = \frac{d}{c}$, where d is the distance and c is the propagation velocity. It is assumed that each device has a reference clock with a nonnegligible bias with our reference clock. This bias is given by the parameter ε. The principal is shown in the figure below, where these parameters have been incorporated.

Thus, the transport delay estimator evaluated at device one is equal to $\hat{T}_d = \frac{T_2 - T_F - \varepsilon_{12}\Delta}{2}$ where $\varepsilon_{12} = \frac{\varepsilon_2}{\varepsilon_1}$ is the clock bias between the two nodes.

The evaluation of the position of the nodes uses the distributed algorithm detailed in [137]. This algorithm needs a number of measurement steps to converge. This generates an overhead for the network throughput. Therefore, reducing the need of this information suggests the use of data-packet structure to evaluate the time-of-arrival estimator.

7.3.4.3 Summary

At the time of writing, no standard exists for a UWB MAC. There is work to be done in standardization bodies to adapt and possibly optimize and enhance existing MAC's for use with UWB systems. The reason for not simply adopting an existing MAC is partly due to the inherent complications of using a physical layer signal, which is difficult to detect and difficult to synchronize for intended users. Efforts have been made to develop MAC's, which allow efficient operation for large numbers of devices; however, this work is still at an early stage.

An example of this is IEEE 802.15.3 MAC, where its critical points as a MAC suitable for UWB wireless systems have been outlined in this chapter. While this MAC has been developed for high-speed ad hoc applications and has been targeted for UWB by interested groups, its suitability for UWB, either impulse-based or for other physical layer UWB solutions, is yet to be proven.

Research on MAC solutions for UWB is needed and is ongoing in industrial research centers and in universities. However, the research rarely takes into account the peculiarities of UWB signaling. As shown, limitations due to specific characteristics of the UWB physical layer do exist and attention must be paid to these in the MAC design. The potential for UWB systems to offer flexible data rates, service a large numbers of users, and provide positioning services are critically dependent on the capabilities of the MAC, particularly when power efficiency and QoS constraints also need consideration.

7.3.5 Spectrum Landing Zones

7.3.5.1 Regulatory Bodies

One of the important issues in UWB communication deployment is the frequency allocation. Because of the implicit use of a very wide spectrum range, UWB systems are not planned to operate under any specific allocation but there are lots of existing and planned wireless systems operating under allocated bands within the UWB signal band. Some companies in the United States are working towards removing the restrictions from the FCC's regulations relating to the applications that are able to utilize UWB technology. These companies have established an Ultra Wide band Working Group (UWBWG) to negotiate with the FCC. Some other companies and organizations in the United States want existing narrowband allocated services to be well protected from possible interference generated by UWB systems or in their 'out of band' field. Similar discussions on frequency allocation and protection of other radio services from interference have also emerged in Europe. Currently, there are no dedicated frequency bands for UWB applications identified in the ETSI, in the ECC (European Communication Committee) decisions, or in the ITU Radio Regulation treaty.

UWB Regulation in the United States

Before the FCC's first Report and Order [138, 139], there was significant effort by industrial parties to convince the FCC to release the application of UWB technology under the FCC Part 15 regulation limitations and allow license-free use of UWB products. The FCC Part 15 Rules permit the operation of classes of RF devices without the need for a license or the need for frequency coordination (47 C.F.R. 15.1). The FCC Part 15 Rules attempt to ensure a low probability of unlicensed devices causing harmful interference to other users of the radio spectrum (47 C.F.R. 15.5). Within the FCC Part 15 Rules, intentional radiators are permitted to operate within a set of limits (47 C.F.R. 15.209) that allow signal emissions in certain frequency bands. They are not permitted to operate in sensitive or safety-related frequency bands, which are designated as restricted bands (47 C.F.R. 15.205). UWB devices are intentional radiators under FCC Part 15 Rules.

In 1998, the FCC issued a Notice of Inquiry (NOI) [140]. Despite the very low transmission power levels anticipated, proponents of existing spectrum users raised many claims against the use of UWB for civilian communications. Most of the claims related to the anticipated increase of interference level in the restricted frequency bands (e.g. TV broadcast bands and frequency bands reserved for radio astronomy and GPS). The Federal Aviation Administration (FAA) expressed concern about the interference to aeronautical safety systems. The FAA also raised concerns about the direction finding ability of UWB systems. The organizations that support UWB technology see large-scale possibilities for new innovative products utilizing the technology, as outlined earlier in this section.

When UWB technology was proposed for civilian applications, a definition for a UWB signal did not exist. The Defense Advanced Research Projects Agency (DARPA) provided the first definition for UWB signals based on the fractional bandwidth, Bf, of the signal. The first definition provided was that a signal can be classified as a UWB signal if Bf is greater than 0.25. The fractional bandwidth can be determined using the following formula [96]

$$B_f = 2\frac{f_H - f_L}{f_H + f_L}$$

where f_L is lower and f_H is higher than the -3 dB point in a spectrum, respectively.

In February 2002, the FCC issued the FCC UWB rulings that provided the first radiation limitations for UWB, and also permitted technology commercialization. The final report of the FCC First Report and Order [138, 139] was publicly available during April 2002. The document introduced four different categories for allowed UWB applications, and set the radiation masks for them. The prevailing definition has subsequently decreased the limit of Bf to the minimum of 0.20. Also, according to the FCC UWB rulings, the signal is recognized as UWB if the signal bandwidth is 500 MHz or more. In the formula above, f_H and f_L have also been redefined as the higher and lower -10 dB bandwidths, respectively. The FCC radiation limits are presented in Table 7.2 for indoor and outdoor (handheld UWB systems) data communication applications.

UWB Regulations in Europe

At the time of writing, regulatory bodies in Europe are awaiting further technical input on the impact of UWB on existing spectrum users. The European approach is somewhat more cautious than that of the United States, as Europe requires that a new technology must be shown to cause little or no harm to existing radio services. The European organizations have of course to take into account the FCC's decision, in full awareness of the potential commercial benefits of achieving globally compatible conditions of radio spectrum use for UWB.

Currently in Europe, the recommendations for short-range devices belong to the CEPT working group CEPT/ERC/SRDMG/REC 70-03 [141]. In March 2003, the European Commission has given a mandate [140] to the European Standardization Organizations, with the purpose of establishing a set of Harmonised Standards covering UWB applications. ETSI has subsequently established Task Group ERM TG31A to develop a set of Harmonised Standards for Short-Range Devices using UWB technology by December 2004, following the completion of spectrum compatibility studies by CEPT. The standardization working groups for UWB include ERM/TG31A covering generic UWB, and ERM/TG31B, which covers UWB for automotive applications operating in higher-frequency bands. These groups have developed two technical reports respectively on communication and on UWB radio location applications. In the related report [142], a preliminary slope-based spectrum mask is presented for UWB communication systems to operate within. This mask, referred below as the provisional CEPT/ETSI mask, is supported by the UWB industry within ETSI as the starting point for studies by CEPT.

Table 7.2 FCC radiation limits for indoor and outdoor (handheld UWB systems) communication applications (dBm/MHz)

	Frequency range (GHz)		
	$f < 3.1$	$3.1 < f < 10.6$	$f > 10.6$
Indoor mask	$-51.3 + 87\log(f/3.1)$	-41.3	$-51.3 + 87\log(10.6/f)$
Outdoor mask	$-61.3 + 87\log(f/3.1)$	-41.3	$-61.3 + 87\log(10.6/f)$

ETSI should further develop two standards on communication and on radio location UWB applications.

The European Commission has further given CEPT a Mandate [143] on the development of harmonized conditions of use for UWB devices in the European Union. This mandate asks for 'an in-depth analysis of the risks of harmful interference of UWB on all potentially affected radio services, including via aggregation effects, as well as taking into account the existing regulatory environment'.

Within CEPT, ECC has set up Task Group TG3 in March 2004 to undertake the task of responding to this Mandate. This group has produced a first report to be approved by the ECC in July 2004, which states that one key factor in the consideration of appropriate frequency bands for use by UWB is the compatibility with radio communication services using these bands in CEPT countries. CEPT's final report to the European Commission is to be approved by the ECC in April 2005.

CEPT has already undertaken a number of compatibility studies which indicate that limits for emissions from UWB devices for communication applications would need to be lower than the FCC limits in order to avoid harmful interference to some of the radio communication applications in CEPT countries. It can therefore be expected that ETSI/CEPT will follow some of the FCC's recommendations but will not necessarily directly adopt the FCC's regulations. Figure 7.15 shows the provisional CEPT/ETSI mask (Formula serve to establish proposal CEPT/ETSI mask) limits as well as the FCC masks.

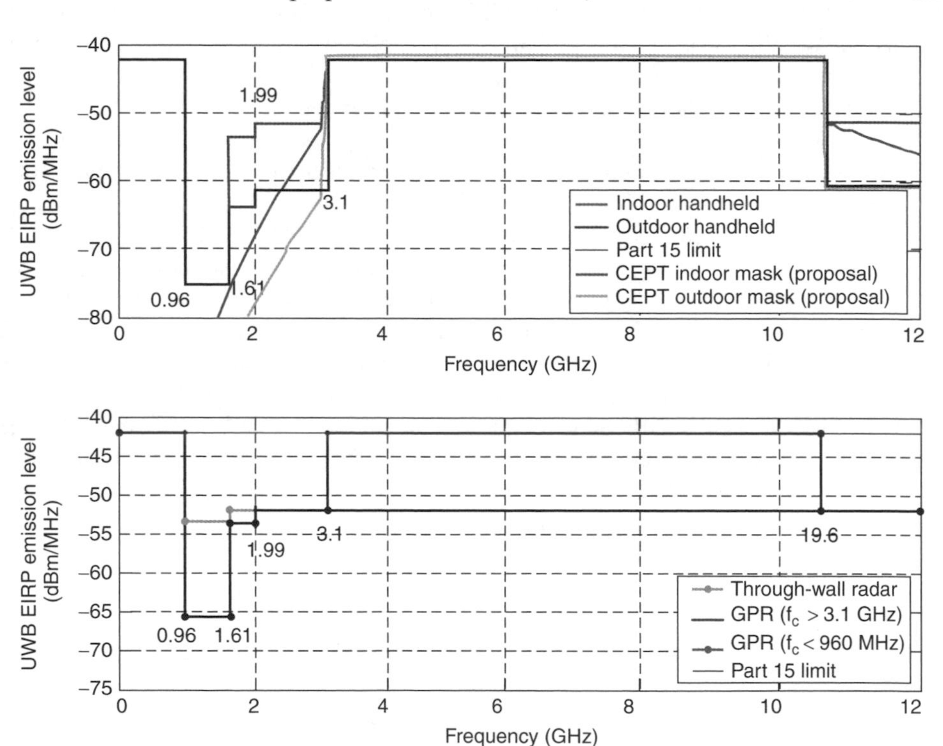

Figure 7.15 UWB radiation mask defined by FCC and provisional CEPT/ETSI proposal

The upper plot represents the masks for data communication applications for indoor and outdoor use. The lower plot gives the FCC radiation mask for radar and sensing applications. In all cases, the maximum average power spectral density follows the limit of FCC Part15 regulations [144]. Therefore, the limit will depend on compatibility studies results made by CEPT/TG3.

UWB Regulation within ITU

International Telecommunication Union – Radio Sector (ITU-R) Study Group 1 has established Task Group 1/8 to carry out relevant studies concerning the proposed introduction of UWB devices and the implications of compatibility with radio communication services. The mandate of TG 1/8 is focused on the study of:

- compatibility between UWB devices and radio communication services (Question ITU-R 227/1);
- spectrum management framework related to the introduction of UWB devices (Question ITU-R 226/1);
- appropriate measurement techniques for UWB devices.

7.3.5.2 UWB Standardization by IEEE

IEEE 802.15.3a

The IEEE established the 802.15.3a study group to define a new physical layer concept for short range, HDR applications. This ALTernative PHYsical layer (ALT PHY) is intended to serve the needs of groups wishing to deploy HDR applications. With a minimum data rate of 110 Mbps at 10 m, this study group intends to develop a standard to address such applications as video or multimedia links, or cable replacement. While not specifically intended to be a UWB standards group, the technical requirements very much lend themselves to the use of UWB technology. The study group has been the focus of significant attention recently as the debate over competing UWB physical layer technologies has raged. The work of the Study Group also includes analyzing the radio channel model proposal to be used in the UWB system evaluation.

The purpose of the study group is to provide a higher speed PHY for the existing approved 802.15.3 standard for applications, which involve imaging and multimedia [145]. The main desired characteristics of the alternative PHY are as follows:

- coexistence with all existing IEEE 802 physical layer standards;
- target data rate in excess of 100 Mbps for consumer applications;
- robust multipath performance;
- location awareness;
- use of additional unlicensed spectrum for high-rate WPANs.

IEEE802.15.4

The IEEE established the 802.15.4 study group to define a new physical layer concept for LDR applications utilizing a UWB air interface. The study group addresses new applications that require only moderate data throughput, but require long battery life such as low-rate WPANs, sensors, and small networks, as described in [93] and [133].

7.3.6 What is Next?

The majority of the discussion so far has focused on the state of the art of UWB technology as it is today and will be in the near future. This section, however, outlines perceptions of future UWB developments and drivers that may shape future evolutions of this technology.

- *Integral part of 4G communications networks*: 4G communications networks are commonly considered to be the fusion of personal, local/home, cellular, and wider area communications. Thus, with the application of UWB as a personal or local/home wireless access technology, it is within the remit of '4G'.
- *Very HDR requirements driven by the development of 3D imaging technology and applications*: 3D imaging technology has long been considered as a natural advance in visual communications and recently, progress has been made in display and signal technology. The wireless interconnection of such systems would naturally require a higher data rate compared to conventional image transmissions. With UWB applications including the provision of image transmission, it should consider the requirements for supporting 3D image transmissions.
- *Digital implementation solutions*: Currently, AD/DA conversion technology cannot support sampling rates or power efficiency requirements for an all-digital implementation of a UWB transceiver. Should such devices become available, the door would be opened for the implementation of advanced modulation and signal-processing techniques.
- *Super-high density networks*: The spatial density of a network's nodes and associated traffic models provide requirements for the multiple-access scheme design. With the increasing pervasiveness of wireless systems, the spatial density will increase. Thus, with this in mind, future UWB systems will require suitable spectrum recourses in order to support the demand for HDRs to an increased number of network nodes. UWB WSN is an example of a system that, if scaled, could fill a super-high density networking requirement.

7.3.7 Summary

This section presented an overview of the state of the art of UWB wireless technology and highlighted key application areas, technological challenges, spectrum operating zones, and future drivers.

The fact that UWB technology has been around for many years with a military context, and has been used for a wide variety of applications is strong evidence of the viability and flexibility of the technology. The simple transmit and receiver structures that are possible make this a potentially powerful technology for low complexity, low-cost communications. The physical characteristics of the signal also support location and tracking capabilities of UWB much more readily than with existing narrower band technologies.

The severe restrictions on transmit power (less than 0.5 mW maximum power, which is equivalent to -3 dBm for a 6.5 GHz wideband system with the FCC emission mask) have substantially limited the range of applications of UWB to short distance, HDR, or long distance, LDR applications. The great potential of UWB is allowing flexible transition between these two extremes without the need for substantial modifications to the transceiver. On the other hand, UWB technology emission masks outside the United States are still under discussion pending the outcome of compatibility studies with other radio services.

While UWB is still the subject of significant debate, there is no doubt that the technology is capable of achieving very HDRs and is a viable alternative to existing technology for WPAN, short range, HDR communications, multimedia applications, cable replacement, and WSN. Much of the current debate centers on which PHY layer(s) to adopt, the development of a standard, and issues of coexistence and interference with other radio services.

Despite the promising range of feasible applications, many questions are still open, and technical challenges need to be solved:

- Which UWB technology meets best a given application – IR, wideband CDMA, OFDM, or something different?
- Can antenna characteristics be optimized to provide efficient, distortion-free transmission and reception?
- What are the key requirements for a certain application? What MAC protocol fits best to a certain application?
- What does an energy-efficient, flexible, HDR MAC layer look like?
- How can we achieve integration into systems beyond 3G and coexistence with other wireless communications systems?
- Extreme energy efficiency calls for cross-layer optimization.

For a proper implementation, all these questions have to be answered well before starting a distinct design.

7.4 Wireless Optical Communication

It is commonly agreed that the next generation of wireless communication systems, usually referred to as 4G systems, will not be based on a single access technique but will encompass a number of different complementary access technologies. The ultimate goal is to provide ubiquitous connectivity, integrating seamlessly operations in most common scenarios, ranging from fixed and low-mobility indoor environments in one extreme to high-mobility cellular systems at the other extreme. Surprisingly, perhaps the largest installed base of short-range wireless communications links are optical, rather than RF, however. Indeed, 'point and shoot' links corresponding to the Infrared Data Association (IRDA) standard are installed in 100 million devices a year, mainly digital cameras and telephones. In this section, we hope to show that optical wireless (OW) communications has a part to play in the wider 4G vision. An introduction to OW is presented, together with scenarios where optical links can enhance the performance of wireless networks.

7.4.1 Introduction

The OW channel has THz of unregulated bandwidth, and characteristics that are distinct from that of radio. It should be noted that our aim is to show that the optical channel has complementary characteristics and can, in certain situations, add to, but certainly not replace the capability of a RF 4G wireless system. Together these media might provide a broad spectrum of channel characteristics and capabilities that radio alone would find difficult to meet.

The aims of this section are as follows:

- Introduce OW and the components and systems used.
- Summarize the state of the art, and the rich research community that exists.
- Compare the characteristics of OW with radio.
- Identify particular areas where OW can contribute to the 4G vision, and areas of future research.

7.4.2 Optical Wireless Communications as a Complementary Technology for Short-range Communications

In this section, an overview of OW Communication systems is presented. The main emphasis in this section is put on OW for indoor environments. Another important approach to OW, free-space optics (FSO), a point-to-point optical connection supporting very high rates in outdoor environments, will not be considered in this section. We start with a brief introduction and classification of OW systems and then continue with different engineering aspects, including transmitters, receivers, the optical channel, and other related issues. This introduction is based on the studies and reviews presented by [146–149]. Comparisons to conventional radio systems are presented to give the reader a broader perspective of the possible base-band technologies. An up-to-date account of different techniques, practical systems and standards related to OW as well as future research issues complete the section.

Basic System Configuration

Figure 7.16 shows a number of different OW configurations. There are two basic configurations; communications channels either use diffuse paths (Figure 7.16(a)) or LOS paths (Figure 7.16(b)) between transmitter and receiver. In a diffuse system an undirected source (usually Lambertian) illuminates the coverage space, much as it would be illuminated with artificial lighting. The high reflectivity of normal building surfaces then scatters the light to create an optical 'ether'. A receiver within the coverage space can detect this radiation, which is modulated in order to provide data transmission. Diffuse systems are robust to blocking and do not require that transmitter and receiver are aligned, as many paths exist from transmitter to receiver. However, multipath interference at the receiver can cause ISI and the path loss for most systems is high. The alternative approach is to use directed LOS paths between transmitter and receiver.

Wide LOS systems such as that shown in Figure 7.16(b) use ceiling-mounted transmitters that illuminate the coverage area, but minimize reflections from walls, ensuring that a strong LOS path exists. The wide beam ensures coverage. As the beams are narrowed path loss reduces and the allowed bit rate increases, albeit at the cost of coverage. Narrow beam systems therefore either require tracking to allow user mobility (Figure 7.16(c)), or some sort of cellular architecture to allow multiple narrow beams to be used (Figure 7.16(d)). A third class of systems also exists; quasi-diffuse systems minimize the number of multipaths by limiting the surface reflections, but allow robust coverage by directing radiation

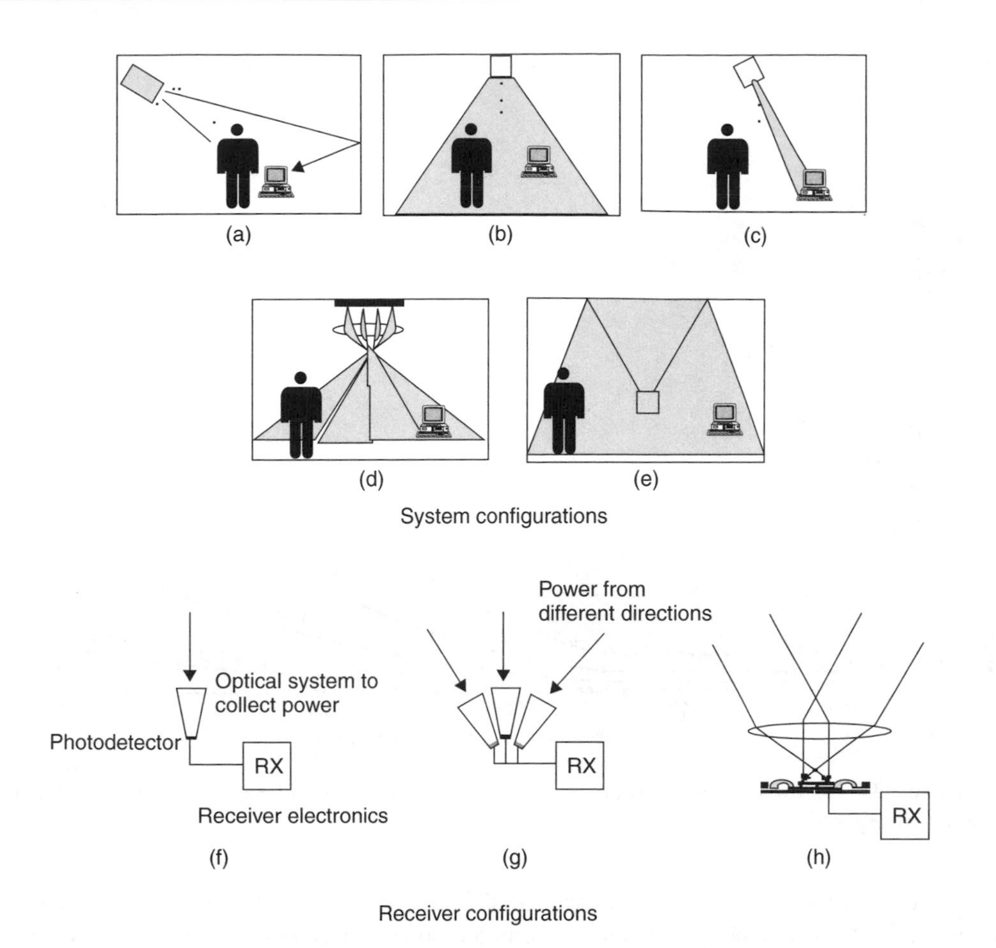

Figure 7.16 Optical wireless configurations. (a) Diffuse system, (b) Wide LOS system, (c) Narrow LOS system with tracking, (d) Narrow LOS system using multiple beams to obtain coverage, (e) Quasi-diffuse system, (f) Receiver configuration: single-channel receiver, (g) Receiver configuration: angle diversity receiver, (h) Receiver configuration: imaging diversity receiver

to a number of surfaces so that a suitable receiver may select a path from one surface only (Figure 7.16(e)).

Variants of this use structured illumination (perhaps using arrays of spots) [150–152]. There is a large amount of simulation and more limited amount of measurement data that describes these channels in some detail. In contrast to radio frequencies, simulation of indoor coverage spaces generally gives a good estimate of the channel characteristics.

7.4.2.2 System Components

Transmitter

The transmitter consists of a single, or a number of sources, and an optical element to shape the beam and also render it eyesafe if required. The main element of the transmitter is the optical source. Light emitting diodes (LED) and laser diodes are employed as the

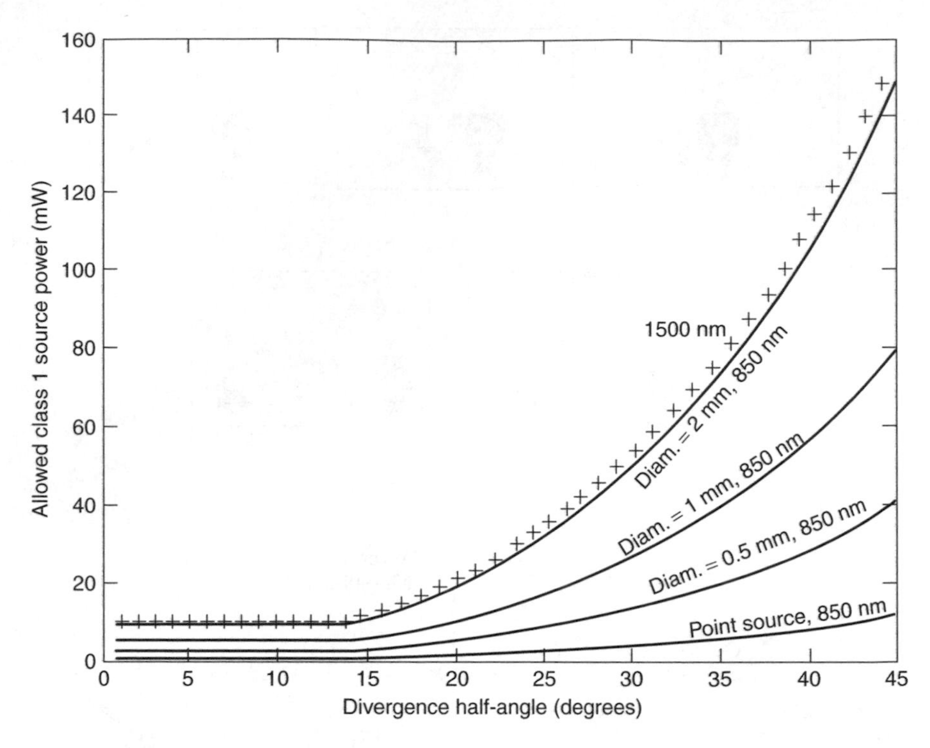

Figure 7.17 Allowed emitted power for class 1 eye safe operation of transmitter as a function of beam divergence

optical radiating element, and their transmission power is limited by eye safety regulation. Figure 7.17 shows a plot of the allowed emitted power for class 1 (the most stringent eye safety regulation) operation versus beam divergence. This is shown for an 850-nm point source, different diffuse source diameters at 850 nm, and for a 1500-nm point source. Increasing the source diameter increases the size of the image on the retina of the eye, thus reducing the possibility of thermal damage. At 1500 nm water absorption in the eye protects the retina, so the hazard is one of corneal damage, and therefore independent of source size. As might be expected more divergent sources are less hazardous as the eye, or an optical instrument such as binoculars or magnifier cannot collect all the radiation. The results in the graph are calculated using the test procedures and limits laid out in [152].

Sources of small diameter can be made diffuse using a ground glass plate, or more sophisticated engineered diffuser elements, including holographic [153] and reflective [154] examples. This latter device has been incorporated in a commercial optical link [155]. The effect of a diffuser is to increase the apparent size of the source, and the graph shows the benefit of this.

Most systems use laser diodes, due to their higher modulation bandwidth and efficiency. IR LEDs are also important optical sources being considered for establishing optical links. In addition, there is a small and growing interest in using visible LEDs that would be installed in a building to provide solid-state lighting for optical communications [154].

In such cases, multiplexing of the low bandwidth devices might be used to increase data rates.

Receiver

A typical OW receiver consists of an optical system to collect and concentrate incoming radiation, an optical filter to reject ambient illumination, and a photodetector to convert radiation to photocurrent. Further amplification, filtering and data recovery are then required (Figure 7.16(f)).

Optical Systems

Receiver optical systems can be characterized in terms of their angular Field of View (FOV) and their collection area. These are linked to the detection area by the constant radiance theorem. This states that;

$$A_{coll} \sin^2\left(\frac{FOV}{2}\right) \leq A_{\text{det}},$$

where A_{coll} is the collection area and A_{det} is the photodetector area. This is important as it limits the collection area that is available for a given FOV and photodetector. For a truly diffuse channel, the detector area sets how much power will be received. Any collection optical system changes the balance between FOV and collection area subject to the constraints above; however, if the system is receiving light from a Lambertian source such as a wall or ceiling the amount of optical power that is collected remains approximately constant for a given detector area.

Both imaging and nonimaging optics [156] can be used to collect and focus radiation onto single-element detectors. Recent designs of optical antenna [157] show good performance in compact form, although this cannot exceed that predicted by constant radiance. Various different receiver topologies have been investigated in order to circumvent these constraints. In angle diversity systems a number of single-channel receivers are combined, so that each faces in a different direction. This allows multipaths to be resolved and collection areas for each receiver to be increased [158] (Figure 7.16(g)). Imaging receivers, as developed at Oxford [159] and Berkeley [160] can also carry out this function (Figure 7.16(h)). These use a large-area pixellated detector array and an optical imaging system. Light from a narrow range of directions is collected by a single pixel, and together the array of pixels offers a large overall FOV. It also allows multipaths from different directions to be resolved as they are imaged to different pixels on the array. The array also allows the large detection area to be segmented, reducing the capacitance on each of the receiver front ends. Both of these topologies can to some extent resolve multipaths, and this may offer some means to reduce the effect of shadowing, by selecting an alternative non-shadowed path. It is also possible to use a combiner/equalizer to maximize the received signal and BER [160].

Optical Filtering

Ambient light is the most important source of interference and it may greatly deteriorate link performance [156]. Constant ambient illumination will generate a DC photocurrent, and this will normally be blocked by the AC coupling of the receiver [157]. However, the shot noise from the detection of this illumination cannot be filtered and can be large when compared with the noise from the preamplifier. Artificial illumination, particularly modern high-frequency fluorescent illumination induces electrical harmonics in the received signal, with components up to 1 MHz [161] and this can greatly affect link performance. Various studies of this have been undertaken, including [162, 163].

Optical filtering can be used to reject out of band ambient radiation and reduce the intensity reaching the detector. Various different filter types have been demonstrated; a longpass filter in combination with a silicon detector provides a natural narrowing of the bandwidth, and absorption filters can be used to reject solar and artificial illumination [164]. Bandpass interference filters can be used, although care has to be taken to allow sufficient bandwidth to allow for passband shifting with the varying angle of incidence. It is also possible to filter by incorporating appropriate layers into the photodetector. Holographic receiver front ends also allow ambient light noise to be rejected [165].

Electrical filtering can be used to reduce the effect of the illumination harmonics, but at the cost of inducing baseline wander. Work on the optimal placement of the filter cut-offs for particular modulation schemes is reported in [166].

Detector/Preamplifiers

The detector and preamplifier together are the main determining factors in the overall system performance. Both personal identification number (PIN) structures and avalanche photodiodes (APDs) have been used in single-detector systems, whilst array receivers have tended to use PIN devices. Most of the detectors are designed for optical fiber systems, where capacitance per unit area is relatively unimportant, as areas can be small, and hence commercial devices are highly capacitive. Devices for OW should be optimized for low capacitance per unit area by increasing the width of the I-region, until this effect is balanced by the increasing carrier transit time. Detectors partially optimized for this application have been demonstrated [167] but further work is required in this area.

Various approaches to mitigating the effect of input capacitance on bandwidth have been taken. Bootstrapping [168], equalization [169], and capacitance-tolerant front ends [159, 170, 171] have all been investigated.

7.4.2.3 The Optical Channel

LOS optical channels are subject to path loss, and this can be modeled using either ray-tracing or analytical techniques. The diffuse channel has both high path loss (>40 dB typically) and is subject to multipath dispersion. Both of these characteristics are dependent on the orientation of the source and receiver within the space.

There has been extensive work on predicting the characteristics of the diffuse channel, including [163, 172, 173] as well as analytical models of the channel impulse results. Most building materials are found to have a high reflectivity (0.4–0.9) and they can be approximately modeled as Lambertian reflectors. Ray-tracing techniques therefore allow generally good predictions of the channel response, even in the presence of chairs and other objects. Depending on the balance of LOS and diffuse paths within a space, channels can be modeled as Rician [174] or Rayleigh, with exponential impulse responses. Various measurements have also been made [175, 176]. Recent high-resolution data indicate that transparent 'unlimited' bandwidth diffuse channels are available in particular directions for most diffuse environments [177]. One of the major advantages of the OW channel is that there is no coherent fading, and the channel is therefore extremely stable when compared with its RF counterpart. Even though sources are coherent, the size of detectors and the scattering environment mean that any effects are removed by the spatial integration that occurs at the receiver.

7.4.2.4 Modulation Schemes

Unlike in conventional RF systems, the optical channel uses intensity modulation and direct detection. The optical power output of the transmitting source is controlled according to some characteristics of the information-bearing signal. The transmitted signal is thus always positive and its average amplitude is limited [178]. Analog and digital optical modulation is possible but, due to the intensity modulation, common modulation schemes employed in the RF domain will perform differently when applied to optical systems.

The changes in optical power produced by intensity modulation are detected by direct detection, that is, a current proportional to the incident optical power is induced in the photodetector. As pointed out in [147] two criteria should be used to evaluate the feasibility of an optical modulation system; the average optical received power required to achieve a given target BER performance and the required receiver electrical bandwidth.

Three basic modulation schemes are usually used in OW systems, namely, on-off keying (OOK), pulse-position modulation (PPM) and subcarrier modulation (SCM). An extensive account of these and other techniques can be found in [178]. Other issues to take into account when considering optical modulation schemes are their robustness to multipath propagation and, in networks, their suitability to multiple-access environments. PPM is very well suited to work in low signal-to-noise ratio scenarios, quite typical in optical channels due to blocking effects (shadowing) and ambient noise. However, multipath propagation induces intersymbol interference and PPM is particularly sensitive to these dispersive effects of the optical channel [178].

Many techniques have been considered in order to combat the deleterious effects of dispersive optical channels, among them the use of equalizers, angle diversity and spread-spectrum techniques. Different equalization approaches to chip or symbol rate have been studied for PPM-based systems, including linear and decision-feedback equalizers [178].

Spread-spectrum (SS) modulation techniques can also be used to combat multipath distortion as well as to reduce the effects of interference, in a similar fashion as they are exploited with radio systems. Direct-sequence techniques are usually used in conjunction with optical links. Since bipolar spreading sequences cannot be used to modulate an always-positive optical signal, a unipolar sequence is formed by biasing to the bipolar sequence with a fixed DC offset. This unipolar sequence preserves the correlation properties of the original sequence and it can be correlated with a bipolar sequence at the receiver. Several direct-sequence spread-spectrum (DSSS) approaches specially designed for optical systems have been proposed and studied, including sequence inversion keying modulation (SIK), complementary sequence inversion keying (CSIK), and M-ary bi-orthogonal keying modulation (MBOK) [179, 180]. Sequences with low auto- and cross-correlation sidelobes are preferred in order to minimize the degrading effects of intersymbol interference.

7.4.2.5 Optical Wireless versus Radio Communications

Over the past decade the capacity of an optical fiber link has increased by several orders of magnitude, showing almost 'Moore's Law' growth, largely due to the availability of optical spectrum. At the same time, regulation of the RF spectrum limits available bandwidth to several orders of magnitude below this. The vision of a highly connected world is likely to require unaffordable amounts of the already scarce RF spectrum. OW occupies fully unlicensed spectrum bands, and the possibility of using unregulated and unlicensed bandwidth is one of the most attractive characteristics of OW.

Unlike radio communications, the nature of the optical radiation is such that the transmitted signal is obstructed by opaque objects, and the radiation can have high directivity using sub-millimeter scale beam shaping elements. This combination of high directivity and spatial confinement gives optical channels an unmatched advantage in terms of security. Furthermore, these characteristics allow exploiting wavelength-reuse at room level, without taking special provisions for interference from and to neighboring rooms. Since the optical radiation produces no interference to electrical equipment, OW can be used in sensitive environments where conventional radio wave transmission is not allowed.

Another unique characteristic of wireless optical links is in the channel itself, and it is the fact that these links are not affected by multipath fading. This is because the dimensions of the receiver's photodetector are many orders of magnitude larger that the wavelength of the optical radiation and thus, the spatial fluctuations in signal strength due to multipath are averaged over the large detector area, which acts as an integrator. For most of the cases, and as an essential advantage, optical components are small in size, low cost and they have low power consumption. Furthermore, transceivers are relatively simple compared with their RF counterparts.

There are several drawbacks, however; since IR radiation can reach the retina and eventually cause thermal damage, the maximum power that can be transmitted is limited by eye safety regulations and extra optical elements are required to render high power sources safe.

In diffuse optical communication systems, multipath propagation caused by the dispersive optical channel will introduce pulse spreading and ISI, much as would be experienced by a radio channel, although the process is due to incoherent, rather than coherent fading. Systems above 50 Mbps or so might typically require some form of equalization.

Perhaps the major difference for the OW channel is that the detection process is usually incoherent, so that the detector responds linearly with power, rather than amplitude, as is the case with a radio receiver. Receiver sensitivity is therefore substantially lower than for radio channels, and therefore systems are more susceptible to path loss, especially in the case of diffuse systems. More complex receiver and transmitter structures can be used to reduce this and the effect of noise from ambient illumination. As the detection process is incoherent there is no inherent rejection in the detection process, so filtering mechanisms must be introduced, as mentioned earlier.

The wavelength of optical radiation makes directive channels easy to implement, and system design often leads to asymmetric channels [181]. Such directive channels are necessarily subject to blocking, which is again distinct from radio applications [149, 181] and [182].

7.4.3 Link Budget Models

In this section simple models of the RF and optical channels are developed, in order to compare the performance (A similar analysis is undertaken in [181], albeit with different emphasis.).

7.4.3.1 Radio Communications

There are many models of the path loss of a radio link, depending on environment, and on link distance. In this case, a simple set of limiting cases is considered.

If (i) the transceivers lie in each others' far field, so that Fraunhofer diffraction can be assumed, and (ii) a LOS exists between the transmitter and the receiver, the link loss can be estimated using the standard Friis' equation. In most indoor environments the antenna will be approximately isotropic, and have transmitter and receiver gains of unity. In this case the link loss L_{link} can be approximated by

$$L_{link} = \left(\frac{\lambda}{4\pi}\right)^2 \frac{1}{r^2} \quad (1)$$

where r is the link distance and λ is the wavelength of the radiation. The minimum link distance at which this occurs is the Fraunhofer distance d_f and can be estimated as

$$d_f = \frac{2D^2}{\lambda} \quad (2)$$

where D is the largest dimension of the antenna.

In a real environment, the situation is more complicated however. At distances greater than a reference distance d_{ref} from the antenna (d_{ref} is greater than the Fraunhofer distance) multiple paths from transmitter to receiver interfere and cause a path loss that varies as $r^{-\gamma}$ where γ is the path loss exponent. This gives rise to a 'dual-slope' path loss, with an r^{-2} loss at distances less than d_{ref} (but greater than d_f) and an $r^{-\gamma}$ loss beyond this.

The link loss then becomes

$$L_{link} = \left(\frac{\lambda}{4\pi}\right)^2 \frac{1}{r^2} \quad \text{for} \quad d_f \leq r \leq d_{ref} \quad (3)$$

and

$$L_{link} = \left(\frac{\lambda}{4\pi}\right)^2 \left(\frac{1}{d_{ref}{}^2}\right) \left(\frac{d_{ref}}{r}\right)^\gamma \quad \text{for } d_{ref} < r \quad (4)$$

Both the position of the break point and the slope beyond it can therefore vary widely, and the model above is at best an indication of the loss. The methods used to estimate d_{ref} and γ are detailed in each of the scenarios described later in the section.

Assuming the receiver antenna is at the standard temperature, and feeds a matched preamplifier with noise figure F then for a transmitted signal power P_t the signal to noise ratio S/N at the receiver is

$$\frac{S}{N} = \frac{P_t L_{link}}{FKTB} \quad (5)$$

where K is the Boltzmann's constant and T is the temperature in Kelvin. For a bit rate R_b average energy per bit E_b and Noise power density N_o

$$\frac{E_b R_b}{N_o} = \frac{P_t L_{link}}{FKT} \quad (6)$$

This expression allows the bit rate available R_b to be related to the range for a given required E_b/N_o. The value of E_b/N_o required for a particular BER is determined by the modulation and detection scheme used.

7.4.3.2 Optical Communications

Figure 7.18 shows the geometry of the optical links and Figure 7.19 shows an electrical model of the communications channel. Applying the Friis' formula for optical links creates results that are unrealistic in real situations, in that the diffraction limited transmitter beam that these simulate is very narrow, creating links that require very precise alignment. Two link geometries, for point and shoot links and 'hot-spot' geometry are shown in Figure 7.18.

The Intensity profile of the source is $I(r, \theta)$ where r is the link distance and θ is the angle measured from the optical axis, as shown in Figure 7.18. In this case, it is assumed

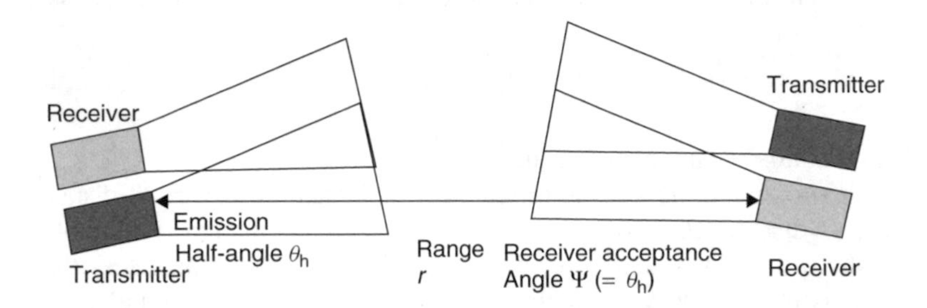

(a) Link geometry for 'point and shoot' optical link, showing 'worst case' alignment for correct operation. Transmitter and receiver are both oriented at θ_h to 'boresight' alignment.

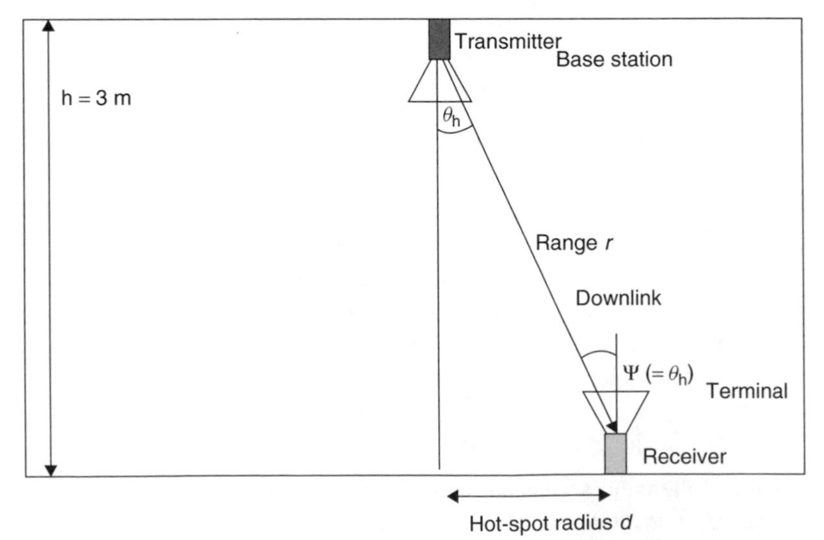

(b) Link geometry for optical hot-spot for worst case alignment

Figure 7.18 Optical communications link geometries

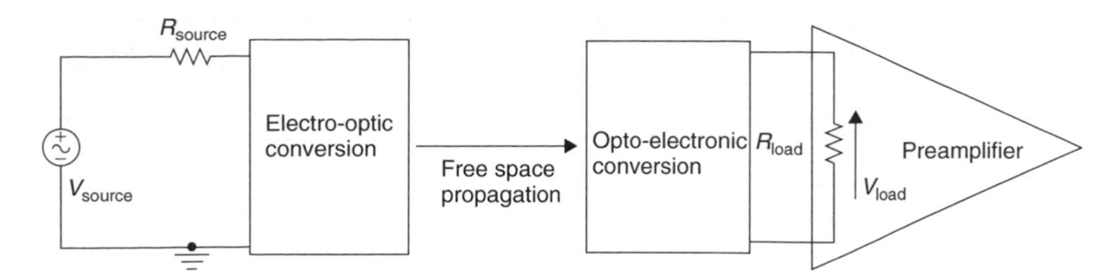

Figure 7.19 Communications channel model

that this is a constant value I_o so that $I(r, \theta) = I_o$ for θ varying from 0 to the beam half-angle θ_h. This is beneficial, compared with the more normal Lambertian pattern as the off-axis fall in intensity is reduced. Such a distribution might be achieved by using a holographic diffuser to modify the source emission profile.

Normalizing this by the power emitted by the source P_S yields;

$$I(r, \theta) = \frac{P_S}{2\pi r^2(1 - \cos\theta_h)} \tag{7}$$

This power is collected by a receiver with area A_{coll}, which in the worst case is oriented at an angle ψ from its optical axis, so the minimum power collected, P_{coll} is given by

$$P_{coll} = \frac{P_S}{2\pi r^2(1 - \cos\theta_h)} A_{coll} \cos\psi \tag{8}$$

defining the optical loss, $L_{optical}$ as

$$L_{optical} = \frac{P_{coll}}{P_S} \text{ leads to}$$

$$L_{optical} = \frac{1}{2\pi r^2(1 - \cos\theta_h)} A_{coll} \cos\psi \tag{9}$$

In both cases shown in the figure the transmitter launches power within an emission cone with half-angle θ_h and receiver optical systems have acceptance angle $\psi = \theta_h$. The beam half-angle and receiver acceptance angles are matched as this is optimum for systems with paired uplinks and downlinks as shown in the figure. Any larger receiver acceptance angle is not useful as if light is entering the receiver at this orientation, the transmitter within the same transceiver unit will then miss the receiver within the distant transceiver unit.

At the receiver the electrical power delivered to the load S is given by

$$S = i_{optical}^2 R_{load} \tag{10}$$

where $i_{optical}$ is the photocurrent. This is given by

$$i_{optical} = RP_{optical} \tag{11}$$

where R is the responsivity of the photodetector and $P_{optical}$ is the optical power incident on it.

The noise at the receiver consists of shot noise from the signal, shot noise from any DC photocurrent caused by ambient light, and amplifier noise. The noise power delivered to the load is

$$N = [2e(RP_{optical} + i_{Ambient})B + i^2_{Amplifer}B]R_{load} \quad (12)$$

where $i_{Ambient}$ is the photocurrent due to ambient illumination and $i_{Amplifier}$ is the input referred noise of the amplifier.

The ambient light current can vary over several orders of magnitude depending on the FOV and optical filtering [161]. The amplifier noise ranges from $1 - 10pA/\sqrt{Hz}$ [42] with a value of $5pA/\sqrt{Hz}$ typical for amplifiers with several GHz of bandwidth.

The overall signal-to-noise ratio is given by

$$\frac{S}{N} = \frac{(RP_{optical})^2}{[2e(RP_{optical} + i_{Ambient})B + i^2_{Amplifier}B]} \quad (13)$$

which leads to

$$\frac{E_b R_b}{N_o} = \frac{(RP_{optical})^2}{[2e(RP_{optical} + i_{Ambient}) + i^2_{Amplifier}]} \quad (14)$$

This expression allows a direct comparison with the radio link to be made. In the case of RF, the channel is subject to fading and both RF and the NLOS optical channel are subject to multipath dispersion. These expressions are therefore a simplification of the channel conditions, but are instructive as they allow a 'best-case' comparison. In the next section these expressions are used to compare the two media in typical scenarios.

7.4.4 Applications Areas for Optical Wireless

7.4.4.1 'Point and Shoot' and Narrow FOV Applications

Figure 7.20 shows a plot of $\frac{E_b R_b}{N_o}$ versus range r for both RF and OW links with different optical link half-angles. The link geometry is that shown in Figure 7.18(a).

The radio link model assumes an emitted power of −3.8 dBm at a frequency of 3.975 GHz, which is broadly representative of a UWB signal operating in the 'lower band' of the US frequency allocation. This band extends from 3.1 GHz to 4.85 GHz, and the maximum emitted power permitted is −41 dBm/MHz. The amplifier noise figure is assumed to be 6 dB.

Values of path loss γ and reference distance d_{ref} are required for the RF model. There have been numerous measurements [183–185], and these show that $\gamma \approx 2$ for LOS environments and $\gamma \approx 4$ in NLOS situations for UWB systems. As the comparison is with an optical link, the NLOS case is not relevant, so at the break point, there is no change of slope. The value of d_{ref} is therefore not required. (Measurements in [186] reinforce this assumption; a reference distance of 1 m was chosen, and measurements at this distance closely matched the free-space Friis equation. Beyond this point a value of $\gamma = 1.91$ was found, making the assumption of $\gamma \approx 2$ for all distances from the source valid.)

The optical link model assumes operation at 850 nm, using a source 2.5 mm in diameter. The optical link emits power levels allowed for class 1 emission, at each specific beam half-angle. The source aperture dimensions are large enough to be classed as diffuse, thus allowing greater than point source emission. In addition, at larger divergences greater

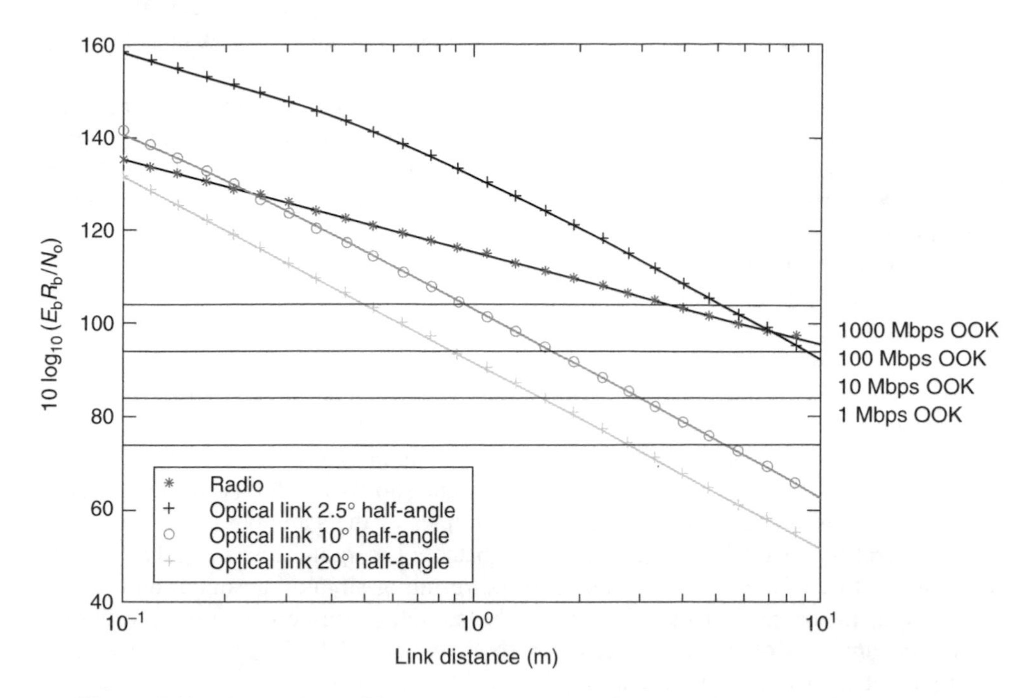

Figure 7.20 Comparison of RF and optical link performance for point and shoot links

power is allowed; for the 2.5 and 10-degree beams this is 12.2 mW, which increases to 25.8 mW for 20-degree half-angle.

The receiver collection aperture is assumed to be 5 mm in diameter. This is chosen as the maximum practical area for a small portable terminal. A detector responsivity of 0.6 A/W is assumed, and calculation suggests that a compatible receiver should be capable of Gbps performance.

The amplifier noise density is assumed to be $5pA/\sqrt{Hz}$. The current due to ambient light $i_{Ambient}$ is a difficult quantity to estimate because of widely varying conditions. A value of 125 μA is used for the 10-degree FOV and scaled appropriately for different FOV. This is equivalent to the maximum quoted in [185] when scaled for FOV and assumes coarse optical filtering. The required values of $\frac{E_b R_b}{N_o}$ for different bit rates and incoherent OOK detection with a BER of 10^{-6} are plotted on the graph also.

The optical link is superior for narrow FOV applications up to approximately 10 m, where for an FOV less than 2.5 degrees (half–angle) performance exceeds or meets that available using RF. For wider FOV the power density available at the receiver drops and the available range for a given bit rate reduces, while for the isotropic RF link this is not relevant, and performance depends only on the range. Assuming that an FOV of approximately 10 degrees (half-angle) or so is required for a 'point and shoot' application, a range of 1 m or so can be achieved with an optical link operating at 1 Gbps. It should be noted that slightly better results – perhaps a 10 % improvement in range – are obtained for an optical link operating at 1500 nm. As can be seen from Figure 7.20 the allowed power emitted from such a source is similar (though slightly lower) to that from an 850 nm source 2.5 mm in diameter for all beam half-angles. This is more than compensated by the increased responsivity at the receiver (1 A/W compared with 0.6 A/W) leading to the increase in range.

In this analysis, 850 nm is an operating wavelength as components are available at low cost. It should also be noted that this calculation does not include detailed analysis of the effects of filtering and other impairments, but this is the same for the RF link budget, where effects such as fading are ignored, so that the two are broadly comparable. In addition the results show marked differences, so that the dB is not of great importance to the overall conclusions.

From this analysis it seems that the most promising area of application, where optical links can offer a distinct channel with greater range than the equivalent RF channel is where narrow FOV LOS channels are available.

Most of the links considered here will be used to connect portable terminals to data infrastructure, so power consumption becomes a consideration. Table 7.3 shows the power consumption of several short-range RF wireless communications standards normalized by the bit rate, and that for an optical link. The data shown are for particular products, although they are broadly representative for examples of the same type.

The optical link is the most efficient, by a substantial margin. This is largely due to the base-band nature of the optical channel and the resulting simple transmitter and receiver architecture. In the case of the RF systems the complex transceiver structures dissipate substantial amounts of power. In all cases (both optical and RF) the link losses are a small part of the overall power consumption.

Within the 4G framework UWB is considered a likely solution for ad hoc networks over distances <10 m, and the example in Table 7.3 offers data rates of 114 Mbps at ranges of up to 10 m. However, its energy consumption is a factor of 3 higher than the example optical link.

At distances <1 m, where there is a LOS very HDR (1 Gbps), optical links appear competitive with UWB, offering lower power consumption and better intrinsic security. Within the IRDA the IrBurst Special interest group has identified this area, and have specified links broadly in line with the results shown here.

The IRDA is promoting 'Financial Messaging', where a secure transaction takes place between a handheld and a retail terminal, albeit at LDRs [191]. Such a concept might be extended to retailing of high-bandwidth content such as DVDs and CDs to portable players. In the future this might require several Gbps in order to achieve reasonable download times (Samsung have recently introduced a telephone with a 3 Gb disk.). Theft of content would be limited by the confined nature of the optical signal, and extremely high spatial bandwidth could be achieved within the retail environment. This is relatively

Table 7.3 Power consumption for different communications standards

Standard	Power consumption (W)	Bit rate (Mbps)	Normalized energy consumption (J/Mb)
IEEE802.11(g) [187]	1.25	54	2.31E−02
Ultra-wideband (UWB) [188]	0.75	114	6.58E−03
Bluetooth [189]	0.1	0.72	1.39E−01
Optical link [190]	0.3	150	2.00E−03

straightforward to achieve with a directed OW link used in a 'point and shoot' manner, or in a booth in which the link environment can be controlled.

7.4.4.2 Telematic Applications

Over longer distances the superior link budget of narrow FOV systems makes OW attractive for telematic applications, such as road pricing and navigation. The German government has adopted an optical communications system for its tolling system for freight vehicles [192], and an International Standards Organisation (ISO) standard (ISO CALM 204) has been defined for such systems. Several train-operating companies are also investigating FSO for communications with trains, in order to provide broadband 'to the seat', and IRDA has formed the Travel Mobility Special Interest Group (IrTM).

7.4.4.3 Hot-spots

In general, OW suffers from low receiver sensitivity when compared with RF, but has the advantage of good spatial confinement, with the ability to maintain sufficient power density for good data reception over relatively small areas.

OW 'hot-spots' that offer localized high-bandwidth connectivity have been suggested [148] and the 4G model of heterogeneous standards allows this to be included, with the optical capacity augmenting that provided by RF. Figure 7.21 shows an example of this approach. Low-bandwidth coverage is ubiquitous, and this is augmented at places where people congregate or at lobbies of buildings so that large file transfers and tasks requiring high bandwidth can be undertaken. In 'regular' environments such as open-plan offices the hot-spots may provide complete coverage, with steps taken to ensure Optical LAN coverage.

Very simple hybrid approaches combining optical and RF links have been proposed recently for short-range (indoor) communications [187, 188]. Reference [189] describes protocols that use RF signaling and reallocate the optical sources under blocking conditions.

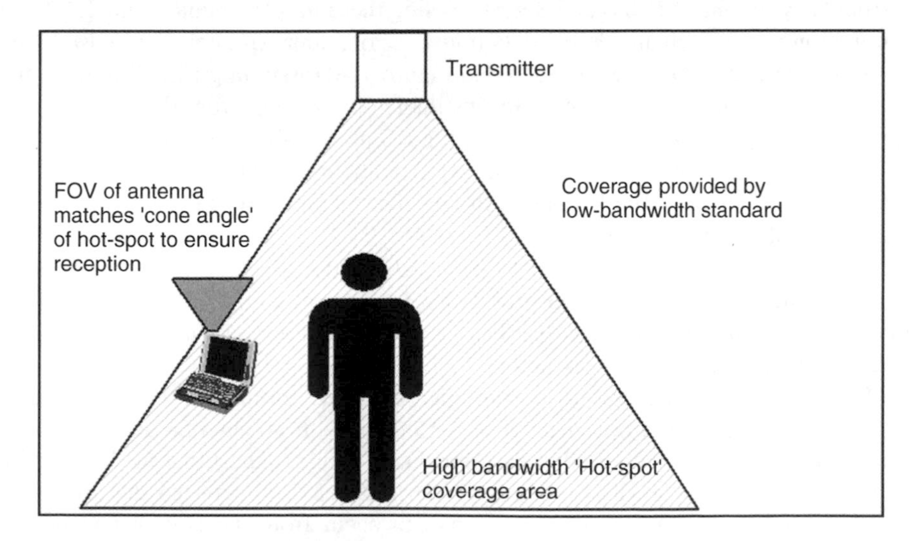

Figure 7.21 High-bandwidth 'Hot-spot'

Studies by Hou *et al* [193] have shown the advantage of this, especially in environments where the required capacity is asymmetric and the link capacity matches this. In the study a 100-Mbps optical downlink (from infrastructure to portable terminal) was combined with a 10 Mbps RF LAN. The optical link is subject to blocking, and depending on the duration of the blocking, the system performs a vertical handover to an RF LAN. Downlink traffic is four times that of the uplink, a number that is typical within the Internet, and the decision to switch between standards is taken using a fuzzy inference engine.

Under conditions of high blocking (the optical link is blocked 10 % of the time) the combined system delivers a lower delay (by a factor of 6) in packet delivery compared with the radio LAN alone, and the system is 9 times as efficient.

In this case the optical link offers increased efficiency and capacity for the wide area coverage RF LAN by augmenting capacity when it is needed. Within the framework of 4G systems the optical link can be considered to be another channel with specific distinct characteristics.

There are several routes to Gbps coverage in a hot-spot. Perhaps simplest is the use of tracking narrow FOV links, as sold by JVC[194], but this requires mechanical steering mechanisms. A potential alternative is the multiplexing of narrow LOS channels using multibeam transmitters. Further potential gains are available using imaging, multichannel receivers [159, 160, 195]. These approaches use the spatial multiplexing available because of the high gain optical antennas (lenses) and the ability to do very precise beamforming and steering when compared with RF approaches. An architecture that uses wavelength to steer narrow beams to terminals within the coverage area is presented in [196]. This offers the potential for multi-Gbps transmission to terminals using a simple passive BS. In this case there are formidable challenges in providing an uplink from terminal to BS, and this architecture is well suited to a broadcast 'hot-spot' model.

Apart from the fine control of narrow power limited channels, an alternative approach is to increase the power transmitted, and a possible method may be enabled by a switch to solid-state lighting. The efficiency of solid-state lighting using LEDs is increasing rapidly, and particularly in Japan there is an interest in using these for communications [197–199]. If illumination power is available for data transmission, approximately 30 to 40 dB more power is available than the case considered in Figure 7.20 (assuming LED lighting units of the type in [197]). This would increase the available optical range for HDRs considerably.

Figure 7.21 shows 'hot-spot' geometry with a BS placed above a receiver plane within a room. The coverage area of the hot-spot is of radius d, as shown in the diagram, and together with the height of the BS above the receiver plane h (in this case 3 m) this sets the half-angle of the hot-spot θ_h.

The optical transmitter is assumed to be a 10-W source, as would be provided by solid-state lighting, with an emission profile that is constant with angle, as was described previously. The beam half-angle is θ_h. The receiver aperture is 15 mm in diameter and all other receiver parameters are as before.

The RF channel is considered to be provided by some 'future' LAN, transmitting in the 2.4 GHz ISM band, at a power level of +15 dBm with no antenna gain.

Values for d_{ref} and γ are required; measurements in [184] suggest that for link distances of up to 10 m, $\gamma = 1.91$ at this frequency in an LOS environment, so that an estimate of $\gamma \approx 2$ is appropriate. Assuming reflections can occur from the ceiling and floor and that one may dominate, leads to two separate two-ray estimations, depending on the reflectivity of the surfaces. For the geometry shown, these yield d_{ref} in the range of 7 to 96 m. In [200], a three-ray model is proposed, and this simultaneously takes into account reflections from the ceiling and floor in order to estimate d_{ref}, but requires the BS to be mounted at a distance from the ceiling that is equal to the terminal antenna height from the floor. Assuming this value is 0.7 m d_{ref} is 30 m. In order to present 'bounds' to the break point in the figure, curves are plotted for $d_{ref} = 7$ m and $d_{ref} = 96$ m with $\gamma = 4$ beyond these distances. This assumption agrees with [184] in that it is unlikely that the effect of a break point close to 10 m would be observed in real measurements that were made up to a maximum link distance of this value.

Figure 7.22 shows $\frac{E_b R_b}{N_o}$ versus the distance the receiver is from a BS situated 3 m above the receiver plane. The geometry is that shown in Figure 7.18(b).

The RF system shows superior coverage, but the optical system would provide sufficient coverage until the next lighting unit (as these must be placed frequently enough to provide illumination).

In the case of the RF system, the challenges are to provide sufficient bandwidth within the allocated spectrum, whereas for the optical link, it is to modulate the LEDs at sufficient data rates. Research in this area is at a relatively early stage.

7.4.4.4 High-capacity 'Future' LANs

It seems likely that RF LANs will migrate to higher-frequency regions of the spectrum, as more bandwidth is required and the implementation cost falls, so it is instructive to compare high-frequency RF approaches with OW.

There has been interest in 60 GHz wireless systems over perhaps a decade or so or longer, with several demonstrations [201] and recently funded projects [202]. The

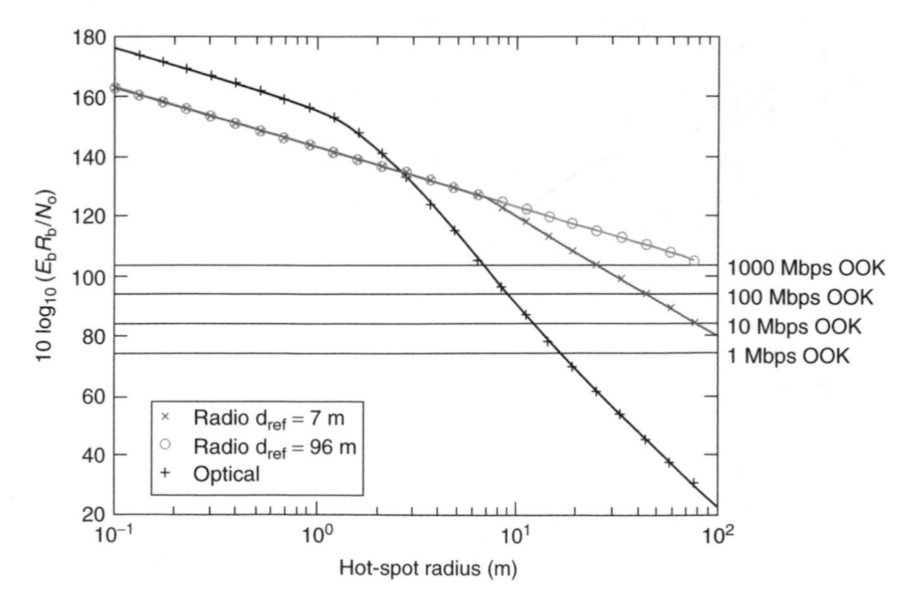

Figure 7.22 Comparison of RF and optical communications for 'Hot-spot' using solid-state lighting for communications

prospects and challenges for this medium are discussed in [203]. The availability of a large region of the spectrum is a major attraction, and the improvement in semiconductor device performance, and potential fall in infrastructure cost are seen as important enabling factors in its future widespread use. There have been a number of propagation studies that have investigated various environments and the effect of shadowing of humans and other obstacles [204, 205]. The conclusions from these are generally that 'mostly LOS' environments are required to guarantee operation [203].

One of the major problems with the acceptance of OW has been the need for geometric control of the transmission paths in LOS systems and quasi-diffuse systems. There are a number of quasi-diffuse systems [153, 206, 207] that use novel schemes to mitigate this, but the general arrangement of RF APs has always been a minor consideration in indoor systems when compared to OW. As RF LANs move to higher frequencies this distinction is unlikely to remain, with the optical and radio channels becoming 'closer' in nature.

Figure 7.23 shows a comparison between a 60-GHz RF channel with a solid-state illumination provided OW channel (emitting 10 W as previously). The BS is 3 m above the receiver plane, and the geometry is that shown in Figure 7.18(b). For the RF link, 10 mW transmitted power [203] is assumed. The transmitter gain is set to match the half-angle of the hot-spot θ_h (as shown in Figure 7.18) and the receiver gain is set to the same value (This is different from the previous 2.4 GHz example, where RF antenna gain is much more difficult to achieve.). The receiver noise figure is 6 dB.

The path loss exponent and reference distance must be determined for this configuration. Assuming either a two or three-ray model the values of d_{ref} far exceeds the range of interest, so the free-space equation holds and $\gamma = 2$ for the LOS environments considered.

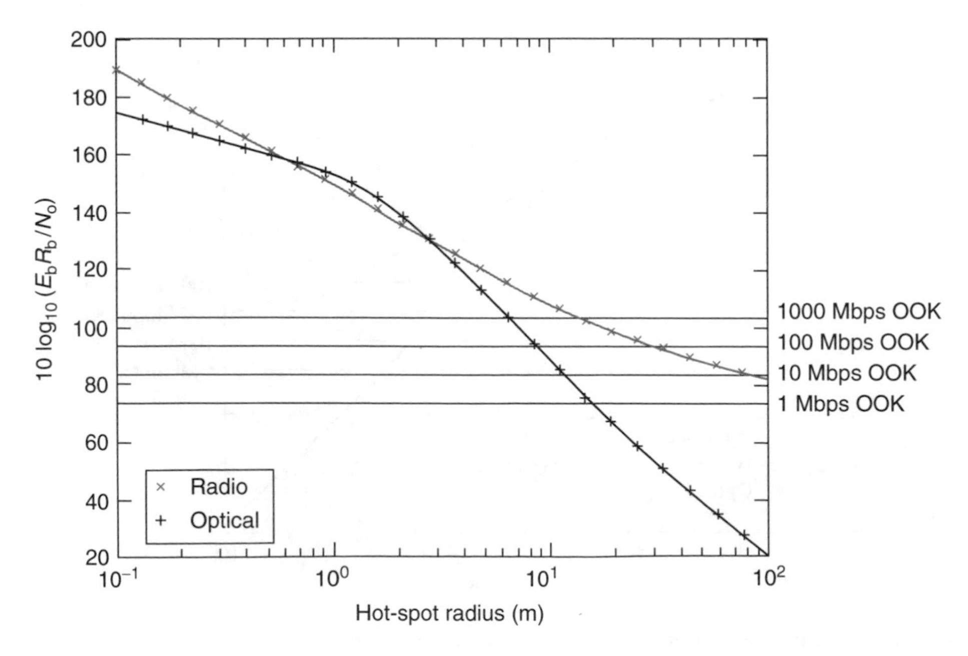

Figure 7.23 Comparison of 60 GHz and solid-state illumination hot-spot

The optical model used is the same as for the previous example, with a 10 W emitter into a half-angle that is set by the hot-spot geometry.

The optical and RF links are relatively close in performance, with the RF providing slightly improved coverage. The need to illuminate the room should mean that continuous coverage using a series of hot-spots should be available in the optical case.

Both systems are affected severely by shadowing, and require some means to provide a predominantly LOS path for a high-capacity link. It therefore seems likely that a 60-GHz LAN will be used to augment that provided by the installed 802.11 infrastructure, as is likely to be the case for optical links. Despite the inferior link budget the optical system has the virtue of simplicity, being a base-band channel. It is also not susceptible to the fading of the RF channel, and does not require complex coding and modulation.

7.4.5 Outlook for Optical Wireless

7.4.5.1 Short Term

The wide range of RF standards and the rapid increase in bit rate available makes the adoption of large-area coverage OW LANs unlikely in the short to medium term. However, within the broad range of strategies for 4G systems there is a common theme of heterogeneous wireless links, and there are situations where OW links can offer higher performance, or lower power consumption, or both than their RF counterparts. This is most likely where there are LOS or quasi LOS. Links are likely to be narrow FOV and short range, or when the geometry can be controlled, such as in optical hot-spots.

7.4.5.2 Medium Term

The use of higher frequency RF approaches to obtain bandwidth requires path management, and makes the use of OW in 'managed' situations more likely given that the alternative has a similar requirement, and OW may offer a simpler solution with low power consumption.

7.4.5.3 Long Term

The results here show that the major long-term challenge for OW is to improve the link budget to that provided by RF systems, so that obtaining LOS geometries is less critical. Coherent systems offer a potential long-term solution, albeit with formidable challenges in providing a stable low-cost geometry [208]. Vertical cavity amplifiers using modified laser structures have been demonstrated [209], although these usually operate with very small etendue, which is unsuitable for OW applications without modification. Substantial parametric gain [210] has been demonstrated over limited bandwidths (although broadband operation is possible), as well as the use of Avalanche Detectors [211]. Detector geometry and optimized devices also offer potential for increased antenna size and hence link margins [160]. The optimum solution may well be a combination of these techniques, and further work is required to compare each of these approaches and determine the best future approach.

7.4.6 Research Directions

The distinct properties of the OW channel can add to the 4G vision, with the possibility of a future terminal having a number of interfaces, both radio and optical. In order to achieve this, work in the following areas is proposed, although this is not an exhaustive list.

Link budget improvement: the major barrier to NLOS systems is the power required at the receiver, and work to improve this should be a major focus.

More comprehensive performance comparisons: between short-range optical systems and their counterpart based on conventional RF approaches; there is a need to understand the properties of both channels between the same points, so that alternative data paths can be modeled, and the performance of a network that chooses the optimum path be determined.

Network modeling: understanding how optical and radio communications might coexist.

Signal processing: examination of radio processing techniques such as Multisensor (MIMO) optical systems exploiting space and angular diversity. Space–time coding for OW channels; some work in this field has already been recently introduced by [212], where space–time codes are designed specifically for optical channels, specifically in the context of FSO communications. A MIMO channel model applied to diffuse wireless optical communications has been recently presented by [213].

Hybrid optical-RF systems: determination of the optimum method of using the alternative resources under different conditions, and the resulting performance improvement.

Visible light communications: fundamental capabilities and limitations of communication systems based on visible light (combined with illumination).

Future wireless standards offer a good opportunity for the wider adoption of OW. In particular, as 4G networks will be highly heterogeneous, OW-based air interfaces can be incorporated to terminals in addition to the conventional RF-based ones. Considerable work is still needed to fully exploit the clear advantages of the optical solutions, as well as developing low-cost subsystems and components to implement them.

7.5 Wireless Sensor Networks

WSN are emerging as a key technology, which enables a wide range of new applications and services. In this section, we describe the major technical issues that need to be addressed toward the envisioned massive deployment of these networks. We illustrate some of the key research challenges both from a networking perspective, and from the viewpoint of sensors and sensor nodes, we give an overview of the current standards and alliances that foster the use of WSNs, and look into the trends that are shaping the evolution of current standards.

The deployments of network sensors and Radio Frequency IDentity module (RFID) tags have the potential to add a whole new dimension to networking and computing. In a few years, networked sensors and actuators will outnumber traditional electronic appliances. Their importance – not just in terms of number of devices and volume of processing transactions, but also in terms of impacting our everyday life – might well eventually surpass that of human-centric devices (PCs, mobile phones, etc.). They will enable new services and applications in industrial automation, asset management, environmental monitoring, medical and transportation business, and in a variety of safety and security scenarios. A large number of objects are already equipped with sensors and tags, for example buildings, cars, and even clothes.

The need to collect and process information gathered by sensors calls for novel, efficient communication schemes, middleware, and data processing techniques. In fact, sensor networks may change one of the fundamental paradigms of data communication: the traditional 'downloading' model, where data is mainly sent from the core of the network to devices located at the network edge, is evolving into an 'uploading' model, where the main part of the traffic is 'in-bound'. Intelligent devices or smart sensors can process at least some of the data locally in order to reduce data traffic volume. By employing suitable middleware and implementing functions like data aggregation and filtering in smart sensors or at the network edge, the overhead due to the transmission of unnecessary or redundant data can be drastically reduced.

Wireless transmission is becoming a popular solution to address the problem of cost incurred by massive deployment of sensors in areas, which are hard to access. Forecasts predict that WSNs and WBANs will already represent a market of $1.5 Billion in 2007 [98]. A roadmap for the evolution and penetration of WSNs and WBANs is shown in Figure 7.24. Of course, WSNs need to be connected to the core network and back-end computing environments to enable end-to-end solutions and parallel applications. Indeed, the evolution of current ITU and IEEE standards is precisely targeting the integration of WSNs with other networks.

WSNs are short-range communication systems comprising a potentially large number of nodes (sensors and/or actuators) with a wide range of possible requirements and network topologies. We will first describe in more detail the scenarios and applications foreseen for WSNs with emphasis on a representative example. We shall next address the main characteristics and associated challenges that need to be addressed both from a node and a network perspective. Before concluding, we will consider current standardization efforts.

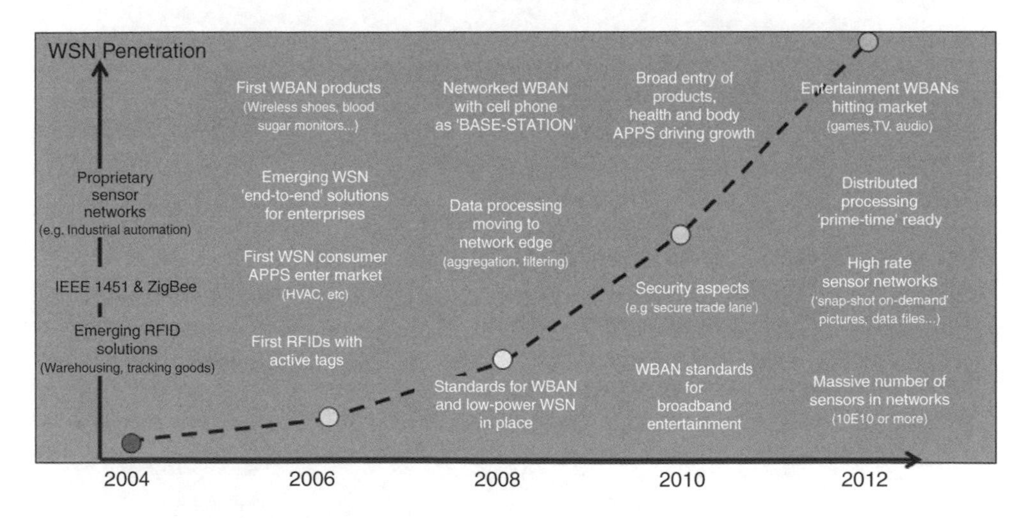

Figure 7.24 Evolution and penetration of WSNs and WBANs

7.5.1 Scenarios and Applications

WSNs can be employed in a multitude of new applications and services. In many cases, they are the key enablers for innovative solutions in the industrial, governmental, and private arenas. Basically, six different application areas can be identified as follows:

- Equipment sensors and controls;
- Remote diagnostics and service controls;
- Lighting monitors and controls;
- Automated data collection;
- Security sensors and controls;
- Heating, ventilation, air-conditioning (HVAC) sensors and controls.

Applications for WSNs fall into one (or several) of these generic areas as illustrated in Figure 7.25. The information collected by sensors is generally intended to trigger specific actions: alarms upon intrusion detection in areas of restricted access, automatic provision requests of scarce goods in commercial applications, or location-based delivery of information. A common denominator of many sensor applications is that the collected data are difficult, costly or impossible to gather by other means, and provide information that requires some type of reaction. Since the number of potential applications is probably endless, we will focus on a representative example.

Application Example: The Smart Home

The home environment is particularly well suited for WSNs as illustrated in Figure 7.26. Consider the many devices that can be controlled by sensors: lighting and heating, appliances like refrigerators or washing machines, security devices like motion detectors and cameras, and not to forget entertainment and gaming devices. By enabling sensors with wireless connectivity, the gathered information can be made accessible throughout the house, and equally important, to remote applications and services. A security-enabled residential gateway ensures that only authorized entities, that is, a cell phone, web pad, or a service provider, have access to this information and can trigger specific actions over the

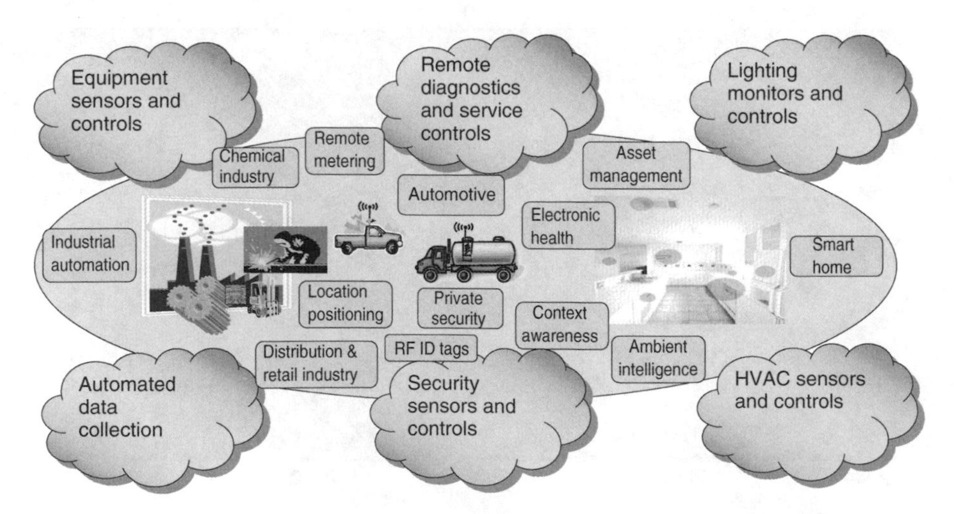

Figure 7.25 Scenarios and applications for WSNs

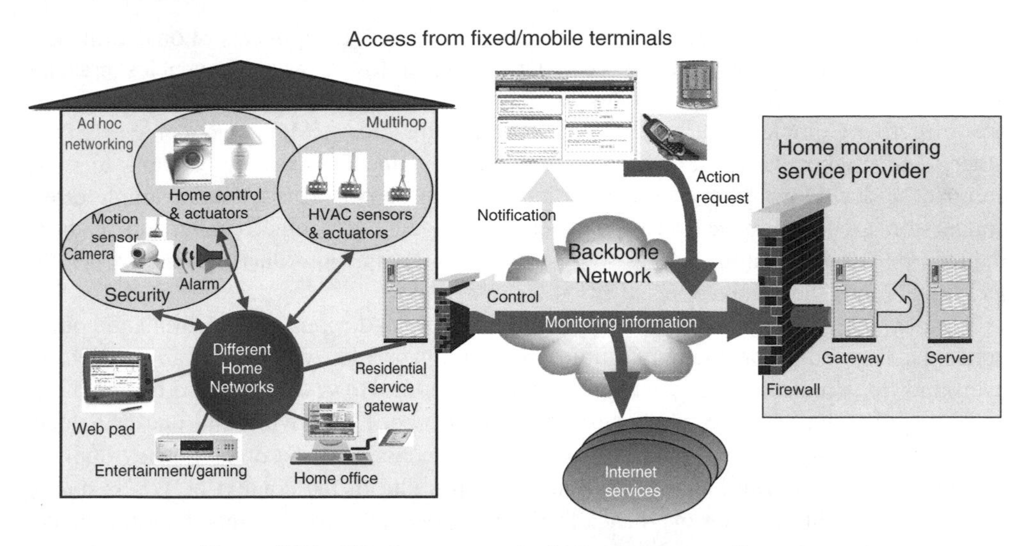

Figure 7.26 Wireless sensors in the smart home environment

network. Any data of interest might be consulted on demand or by automatic notification. Remote access to this information enables for instance, personalized security solutions that use surveillance and monitoring sensors, or shopping assistance as one might take a look at the remaining goods in the refrigerator. Certainly, the range of scenarios in which such information could be useful is very large.

7.5.2 WSN Characteristics and Challenges

WSNs differ significantly from cellular networks whose aim is to achieve high QoS standards. Except for the case of video surveillance, WSNs typically transport data packets in a sporadic or cyclic manner. The sensor network topology and the connectivity between sensors may change dynamically as a function of environment. In applications where a large number of nodes are required, the cost of each one becomes important: low complexity, self-configurability, and autonomous operation are key attributes to be fulfilled. When sensor nodes are placed in remote areas or their maintenance is prohibitively expensive, replacing batteries is often not feasible. Thus, energy efficiency becomes a crucial factor, which determines operating lifetime. With these characteristics in mind, we will describe next some of the key technical problems to be tackled.

7.5.2.1 Node Challenges

Energy efficiency is the primary objective when designing the components of wireless sensor nodes as it determines the operational lifetime of the sensor's battery. While this is facilitated in part by working at low duty cycles, it is clearly not sufficient.

Transceiver

Radio transmission and reception are the most energy-demanding operations of a sensor node. For example, a radio transmission of a 100-bit message over a distance of 100 m takes approximately 10^5 times more energy than executing a 32-bit instruction on a signal processor [214]. Moreover, the design of the RF components should aim at a high level of integration and low cost. Operating at low carrier frequencies presents advantages such as lower clock rates, smaller oscillator phase noise, and usability of a cost-effective Complementary Metal-Oxide Semiconductor (CMOS) technology. On the other hand, higher carrier frequencies allow smaller antenna structures leading to more compact nodes. Down-conversion at the receiver can be performed using a simple direct conversion architecture instead of a super heterodyne one. Direct conversion leads to reduced power dissipation, but has some drawbacks such as introducing flicker noise and DC offsets into the received signal [215].

The physical layer of a WSN should employ simple and robust modulation and coding concepts. Depending on the application, SS techniques, ultra-wideband transmission, or even traditional narrowband technologies can be chosen. However, it should be noted that narrowband techniques are more sensitive to interference than wideband ones. SS techniques can be implemented either using Direct-sequence (DSSS) or Frequency-hopping (FHSS) techniques. The latter appears more appropriate in the context of WSNs due to its reduced synchronization overhead. The use of constant-envelope modulation schemes is desirable from a power amplifier point-of-view.

Error control coding is another important design parameter to control energy consumption. Long error codes have better error correction capabilities and require less transmit power than short ones. On the other hand long codes prolong the time of a node in energy-extensive active mode, thus shortening the sleep period. Because of the limited computational capabilities of nodes, codes with simple coding and decoding schemes have to be preferred.

Finally, an important research topic is the accurate modeling of WSN channels to permit a meaningful evaluation and comparison of the different techniques.

Medium Access Control

As sensors have to share the wireless medium to communicate, a power-efficient access scheme is as important as low-power radio transceivers. Considerable efforts have been made to design medium access control (MAC) protocols for sensor networks. In the context of ad hoc networks, energy-efficient protocols such as Low-Energy Adaptive Clustering Hierarchy (LEACH) [216] and Sensor-MAC (S-MAC) (S-MAC [217] have been proposed. More recently, the adaptation of a CSMA scheme for infrastructure sensor networks led to an access mechanism that consumes very low power [218]. A common denominator in all these approaches is to optimize battery consumption by allowing the nodes to sleep as frequently as possible. Protocol overhead, packet collisions, idle listening times, and overhearing (e.g. receiving packets destined for other nodes) are frequent criteria to evaluate WSN MAC protocols. In addition, novel MAC concepts should also be investigated, such as those based on opportunistic communication.

Processing Unit

The sensor processing unit serves the purpose of controlling the sensing unit and the radio transceiver, perform power, timing and memory management, or carrying out additional tasks. Power and timing management may employ predefined sleep-mode patterns or more sophisticated workload prediction techniques. Furthermore, the processing unit may be equipped with instruction sets tailored to specific applications.

As mentioned previously, smart sensors are capable of trading off data processing and data transmission requirements, that is performing data filtering, abstraction, compression, or aggregation. The implementation of the corresponding algorithms again needs to be optimized for low energy dissipation. Additional functions that deserve attention are security features such as authentication, data integrity, and confidentiality. Considerable research is still needed to provide these security functions in WSNs, where large network size, varying topology, or ad hoc operation offer special challenges.

7.5.2.2 Network Challenges

Cost-efficient operation is also affected significantly by network-wide protocols and algorithms. Considerable efforts have been put into devising concepts, which are scalable to large network sizes [219], robust against modifications in network topology, and energy efficient. We shall next examine some of these network challenges.

Architecture
Depending on the application, network size and topology, several different architectural options arise. For ad hoc networks, there are two distinguishable classes: centralized (also called clustered) and decentralized, where only peer-to-peer communication takes place. Clearly, the network architecture needs to balance the application-specific costs of performing coordinated versus cooperative communication.

Nodes belonging to a centralized network organize themselves autonomously into interconnected clusters. Each cluster contains a cluster head (or master), one or more gateways, and several ordinary nodes (slaves) that are neither cluster-heads nor gateways. Generally, the cluster head schedules transmissions and allocates resources within the clusters. In a high-traffic environment, the performance of a centralized network is better than that of a distributed network, since resources can be controlled more efficiently and rapidly enough to avoid unacceptable levels of throughput and delay. However, a major drawback of cluster-based networks is the significant overhead related to the periodical updating of the cluster information.

In contrast to clustered networks, distributed architectures are based on a decentralized control of the network: all management functions and information on the network are evenly distributed. Hence, network management information has to be minimized to avoid overloading nodes with topological data. Overheads can be controlled by reducing the maximum number of hops through which a node keeps routing information. In decentralized networks, every node decides on its own when to access the channel. It may collect some amount of information to increase the chances of a successful transmission, such as sensing the channel to detect whether it is already being used by another station. In general, the distributed approach works well for small network loads. Moreover, for local communications, using peer-to-peer connections saves network resources tremendously. Unfortunately, due to the lack of central coordination, packet collisions and transmission delays affect network performance more severely than in a clustered network as the load increases.

Data Dissemination
Sensor data propagate through the network to reach one or several sinks. Multihop transmissions are a means to communicate at reduced energy by exploiting the lower signal attenuation of shorter hops. It also provides an opportunity to exploit data redundancy

as in cooperative relaying or to perform distributed processing by leveraging the capabilities of smart sensors. However, additional energy is required at the relay nodes for receiving, processing, and retransmitting forwarded data. Moreover, network information is necessary at each node to perform routing, which requires the exchange of routing tables between nodes and thus adds overhead to the network traffic. There is need for investigating new routing concepts that take into account the specific requirements and ramifications of WSNs. For example, the traditional node-centric approach, where routing is performed with respect to node addresses, could be replaced by a model, where routing is performed with respect to the type of data being transmitted or a given geographical area.

Connectivity in WSNs is an important issue since disconnected subnets impede data dissemination. Since the data may flow through several types of networks before reaching the final destination, inter-network connectivity is equally important. Sensor nodes with gateway functionality or dedicated gateways provide the required efficient connectivity to a backbone network, as for instance, the Internet [220]. With the increasing number of WSNs connected to an IP-based backbone network, the number of message transfers from the WSNs over the backbone network to application servers will grow further. This calls for the use of edge servers [219], which are architectural entities that are often placed at the network edge, in front of application servers, in order to reduce network-traffic load by preprocessing sensor data (e.g. filtering and aggregation). Since applications become sensitive to edge server failures, robust architectures have to be developed, for instance, by using asynchronous messaging middleware to reduce the degree of coupling between system and network components.

Many applications require location or positioning capabilities, for instance, to locate a given sensor or phenomenon, or in the context of packet routing. This motivates the development of advanced localization techniques that might use distributed signal processing, networking, and geophysical information. Deriving accurate location information is even more challenging when sensor nodes are mobile, as would typically be the case for motion sensors.

7.5.3 Standardization

There exists at present no standardization organization, which specifies all WSN aspects reaching from the physical layer, via the MAC and networking layer, up to the various sensor application profiles. The Bluetooth Special Interest Group [221] has designed a system with ad hoc connectivity for Personal Areas Networks (PAN) that has found some limited use in WSNs. It offers reasonable cost and moderate power consumption, but for most types of WSNs power consumption is still excessive, network size is limited to 8 devices per piconet, and the data rate of 1 Mbps is often more than what is required. A better match with many WSN requirements is obtained with the IEEE 802.15.4 [222] physical layer/MAC standard which defines a wireless personal area network that is operated at 868/915 MHz (Europe/North America) or 2.4 GHz and offers a data rate of 20/40 or 250 kbps. The two primary objectives of this specification are very low cost devices and low power consumption so as to extend battery lifetime to many months and even years. The number of network devices can be as large as 256 nodes. They are organized in a star topology with peer-to-peer communications enabled. The ZigBee Alliance [223] adopted the IEEE 802.15.4 physical layer and MAC, and added to it networking, security, and application functions to obtain a system specification for WSNs. Moreover, ZigBee specification addresses interoperability issues.

The IEEE 1451.5 [224] standard defines wireless communication methods and data formats for sensors and actuators, with the possibility to accommodate various existing short-range wireless communication technologies such as Bluetooth, IEEE 802.15 WPANs, and even IEEE 802.11 WLANs.

In addition to these, the Near-Field Communication (NFC) [225] forum is concerned with the implementation and standardization of NFC technology to ensure interoperability between devices and services. The NFC interface standardized in Ecma 340, ETSI TS 102 190 and ISO/IEC 18092 is based on the smart card technology at 13.56 MHz that is able to operate worldwide in a range of 10 cm achieving data rates of up to 424 kbps.

Finally, we point out in [216, 217, 226–228] some proprietary solutions that have been pioneering the field of wireless sensors networks.

7.5.4 Summary

This section has given an overview of the research issues and opportunities offered by WSNs. A considerable number of technical challenges spanning multiple disciplines need to be addressed as the development of WSNs is intensifying, driven by an abundance of new applications. Their feasibility, however, depends on optimizing sensor node components and network algorithms and protocols. In general, solutions to these problems are intimately coupled with application particularities, but generic requirements are low per-node cost and very low power consumption. Challenging research on these key issues is being pursued by numerous research groups, industry, and academia.

7.6 Acknowledgment

The following individuals contributed to this chapter: Volker Jungnickel, Eduard Jorswieck (Fraunhofer Institute for Telecommunications, Heinrich-Hertz Institut, Germany), Peter Zillmann, Denis Petrovic, Katja Schwieger, Marcus Windisch (TU Dresden, Germany), Olivier Seller, Pierre Siohan, Frédéric Lallemand, Guy Salingue, Jean-Benoît Pierrot (France Telecom R&D), Merouane Debbah, David Gesbert, Raymond Knopp (Institut Eurécom, Sophia Antipolis, France), Liesbet van der Perre (IMEC Belgium), Tony Brown, David Zhang (University of Manchester, UK), Wasim Malik, David Edwards, Christopher Stevens (University of Oxford, UK), Laurent Ouvry, Dominique Morche (LETI, France), Ian Oppermann, Ulrico Celentano (University of Oulu, Finland), Marcos Katz (Samsung, Korea), Peter Wang, Kari Kalliojarvi (Nokia Research Centre), Shlomi Arnon (Ben-Gurion University of the Negev, Israel), Mitsuji Matsumoto (Waseda, Japan), Roger Green (University of Warwick, UK), Svetla Jivkova (Bulgarian Academy of Sciences, Bulgaria), P Coronel, W. Schott, T. Zasowski (IBM Research GmbH, Zurich Research Laboratory, Switzerland), Mohammad Ghavami, Ru He (King's College London, UK), Werner Sorgel, Christiane Kuhnert, Werner Wiesbeck (IHE, Germany), Marco Hernandez (Yokohama National University, Japan), Kazimierz Siwiak (Time Derivative Inc, USA), Ben Manny (Intel Corp, USA), Lorenzo Mucchi, Simone Morosi (University of Florence, Italy), M. Ran (Holon Academic Institute of Technology, Israel).

References

[1] G. J. Foschini and M. J. Gans, "On limits of wireless communications in a fading environment when using multiple antennas," *Wireless Personal Communications*, Vol. 6, No. 3, pp. 311–335, March 1998.

[2] G. G. Rayleigh and J. M. Cioffi, "Spatio-Temporal Coding for Wireless Communications", *IEEE Trans. Comm.*, Vol. 46, No. 3, 1998.

[3] H. Bölcskei and A. J. Paulraj, "Space-frequency coded broadband OFDM systems," *Proc. IEEE WCNC*, Vol. 1, pp. 1–6, Sept. 2000.

[4] H. R. Karimi, M. Sandell and J. Salz, "Comparison between transmitter and receiver array processing to achieve interference nulling and diversity," *Proc. PIMRC*, Vol. 3, pp. 997–1001, 1999.

[5] E. Telatar, "Capacity of multi-antenna Gaussian channels," *Eur. Trans. Telecomm.*, Vol. 10, No. 6, pp. 585–595, Nov.–Dec. 1999.

[6] L. Cottatellucci and M. Debbah, "On the capacity of MIMO rice channels", *42nd Annual Conference on Communications, Control and Computing*, Monticello, USA, Oct. 2004.

[7] D. Gesbert, "Multipath: Curse or blessing? A system performance analysis of MIMO wireless systems", (Invited paper), *Proceedings of International Zurich Seminar on Communications*, Zurich, Switzerland, 2004.

[8] A. F. Molisch, M. Steinbauer, M. Toeltsch, E. Bonek, R. S. Thoma, *VTC Spring 2001*, Rhodos, Greece.

[9] M. Debbah and R. Muller, "MIMO Channel Modelling and the Principle of Maximum Entropy, Part I: Model Construction" *IEEE Trans. Inf. Theory*, April 2005.

[10] A. M. Sayeed, "Deconstructing Multi-antenna Fading Channels,", *IEEE Trans. Signal Process.*, Vol. 50, pp. 2563–2579, Oct. 2002.

[11] C. Chuah, D. Tse, J. Kahn and R. Valenzuela, "Capacity Scaling in MIMO Wireless Systems under Correlated Fading", *IEEE Trans. Inf. Theory*, Vol. 48, No. 3, pp. 637–650, March 2002.

[12] D. Gesbert, M. Shafi, D. Shiu and P. Smith, "From theory to practice: An overview of space-time coded MIMO wireless systems" , *IEEE J. Selected Areas Commun. (JSAC)*, Vol. 21, No. 3, pp. 281–302, April 2003, special issue on MIMO systems.

[13] S. T. Chung, A. Lozano and H. C. Huang, "Approaching eigenmode BLAST channel capacity using V-BLAST with rate and power feedback," *Proceedings of VTC Fall*, Atlantic City, 2001.

[14] V. Jungnickel, T. Haustein, V. Pohl and C. von Helmolt, "Link adaptation in a multi-antenna system," *Proceedings of IEEE VTC Spring*, Jeju Island, Korea, 2003.

[15] H. Boche and M. Schubert. "A general duality theory for uplink and downlink beamforming," *Proceedings of IEEE Vehicular Techn. Conf.* (VTC) fall, Vancouver, Canada, September 2002.

[16] V. Jungnickel, T. Haustein, E. Jorswieck and C. von Helmolt, "A MIMO WLAN based on linear channel inversion," *IEE Prof. Network on Ant Propag.*, Vol. 1, pp. 20/1–20/6, Dec. 2001.

[17] R. Fischer, C. Windpassinger, A. Lampe and J. Huber, *Proceedings of the 4th ITG Conference on Source and Channel Coding*, pp. 139–147, Berlin, Germany, 2002.

[18] R. Fischer, C. Windpassinger and J. B. Huber "Modulo-lattice reduction in precoding schemes," *Proceedings of ISIT*, 2003.

[19] C. Peel, B. Hochwald and L. Swindlehurst, "A vector perturbation technique for near-capacity multi-antenna multiuser communication," *Proceedings of the 41st Allerton Conference on Communication, Control, and Computing*, October 2003.

[20] G. Lebrun, T. Ying and M. Faulkner, "MIMO Transmission Over a Time-Varying TDD Channel Using SVD," *Electron Lett.*, Vol. 37, pp. 1363–1364, 2001.

[21] Qualcomm Proposal for 802.11n, *IEEE Standards Meeting*, Berlin, Germany, 2004.

[22] WIGWAM, project: http://www.wigwam-project.com/; WINNER Project: http://www.ist-winner.org/.

[23] IEEE 802.11n High Throughput Study Group, http://grouper.ieee.org/groups/802/11/Reports/tgn_update.htm.

[24] IEEE 802.11-03/940r2, *"IEEE P802.11 Wireless LANs, TGn Channel Models"*, January 9, 2004.

[25] J. C. Rault, D. Castellain and B. L. Le Floch, "The Coded Orthogonal Frequency Division Multiplex (COFDM) technique, and its application to digital radio broadcasting towards mobile receivers", *Globecom '89, IEEE*, Vol. 1, pp. 428–432, 27–30 Nov. 1989.

[26] J. Stott, *"The Effect of Phase Noise in COFDM"*, EBU technical review, Summer 1998.

[27] B. Muquet, Z. Wang, G. B. Giannakis, M. de Courville and P. Duhamel, "Cyclic-prefixed or Zero-padded Multicarrier Transmissions?" *IEEE Trans. Commun.*, Vol. 50, No. 12, pp. 2136–2148, December 2002.

[28] M. Muck, M. de Courville, M. Debbah and P. Duhamel, "*A Pseudo Random Postfix OFDM modulator and inherent channel estimation techniques*", Globecom, San Francisco, December 2003.
[29] R. W. Lowdermilk, "Design and performance of fading insensitive orthogonal frequency division multiplexing (ofdm) using polyphase filtering techniques," *Asilomar Conference*, Pacific Grove, CA, USA, November 1996.
[30] Y. P. Lin, Y.-P. Jian, C.-C. Su and S.-M. Phoong, "Windowed multicarrier systems with minimum spectral leakage," *ICASSP '04*, Montreal, May 2004.
[31] M. Pauli and P. Kuchenbecker, "On the reduction of the out-of-band radiation of OFDM-signals,", *ICC '98*, Atlanta, GA, USA, June 1998.
[32] R. Hleiss, P. Duhamel and M. Charbit, "Oversampled OFDM systems," *DSP Workshop*, Santorini, July 1997.
[33] Y. P. Lin and S.-M. Phoong, "ISI-free FIR filterbank transceivers for frequency selective channels," *IEEE Trans. Signal Process.*, Vol. 49, pp. 2702–2712, 2001.
[34] B. Hirosaki, "An Orthogonally Multiplexed QAM System Using the DFT", *IEEE Trans. Commun.*, Vol. 29, No. 7, July 1981.
[35] B. LeFloch, M. Alard and C. Berrou, "Coded orthogonal frequency division multiplex," *Proc. IEEE*, Vol. 83, pp. 982–996, June 1995.
[36] P. Siohan, C. Siclet and N. Lacaille, "Analysis and design of ofdm/oqam systems based on filterbank theory," *IEEE Trans. Signal Process.*, Vol. 50, pp. 1170–1183, May 2002.
[37] W. Kodek, A. F. Molisch and E. Bonek, "Pulse design for robust multicarrier transmission over doubly dispersive channels," *ICT '98*, Greece, June 1998.
[38] C. Siclet, and P. Siohan, "Design of bfdm/oqam systems based on biorthogonal modulated filter banks," *Globecom'00*, San Francisco, CA, USA, November 2000.
[39] H. Bölcskei, "*Orthogonal Frequency Division Multiplexing Based on Offset QAM*", in advances in Gabor theory, Birkhäuser, Boston, 2002.
[40] S. C. Cripps, *RF Power Amplifiers for Wireless Communications*. Artech House, 1999.
[41] L. Hanso, M. Münster, B. J. Choi and T. Keller, "*OFDM and MC-CDMA for Broadband Multi-User Communications, WLANs and Broadcasting*", Wiley & Sons, 2003.
[42] P. Banelli, G. Leus and G. B. Giannakis, "Bayesian estimation of Clipped Gaussian Processes with Application to OFDM," in *Proc. EUSIPCO 2002*, Toulouse, France, Vol. 1, pp. 181–184, September 2002.
[43] P. Zillmann, H. Nuszkowski and G. Fettweis, "A novel receive algorithm for clipped OFDM signals," *Proc. WPMC 2003*, Yokosuka, Japan, Vol. 3, pp. 380–384, October 2003.
[44] S. Wu and Y. Bar-Ness, "A Phase Noise Suppression Algorithm for OFDM-Based WLANs," *IEEE Commun. Lett.*, Vol. 44, pp. 535–537, May 1998.
[45] D. Petrovic, W. Rave and G. Fettweis, "Common phase error due to phase noise in OFDM – estimation and suppression," in *Proceedings of PIMRC*, Barcelona, Spain, 2004.
[46] D. Petrovic, W. Rave and G. Fettweis, "Phase noise suppression in OFDM including intercarrier interference," in *Proceedings of International OFDM Workshop (InOWo)03*, Hamburg, Germany, pp. 219–224, 2003.
[47] T. M. Ylamurto, "Frequency domain IQ imbalance correction scheme for Orthogonal Frequency Division Multiplexing (OFDM) systems," *Proceedings of IEEE WCNC*, New Orleans, USA, pp. 20–25, 16–20 March 2003.
[48] M. Windisch and G. Fettweis, "Standard-independent I/Q imbalance compensation in OFDM direct-conversion receivers," *Proceedings of the 9th International OFDM Workshop (InOWo)*, Dresden, Germany, 15–16. September 2004.
[49] http://www.iaf-bs.de/downloads/iqmod-demod_eng.pdf.
[50] D. M. Parish, F. Farzaneh and C. H. Barrat, "*Method and Apparatus for Calibrating Radio Frequency Base Stations Using Antenna Arrays*," U.S. Patent 6,037 898, Oct. 10, 1997.
[51] W. Keusgen and B. Rembold, "Konzepte zur Realisierung von MIMO-Frontends", *Frequenz, Zeitschrift für Telekommunikation*, Vol. 55, pp. 301–309, Nov./Dec. 2001.
[52] W. Eberle, J. Tubbax, B. Côme, S. Donnay, H. De Man and G. Gielen, "*OFDM-WLAN Receiver Performance Improvement using Digital Compensation Techniques,*", IEEE RAWCON, 2002.
[53] J. Craninckx and S. Donnay, "*Automatic Calibration of a Direct UpconversionTransmitter,*", IEEE RAWCON, 2003.

[54] A. Bordoux, B. Come and N. Khaled, *"Non-reciprocal Transceivers in OFDM-SDMA Systems: Impact and Mitigation,"* IEEE RAWCON, 2003.
[55] V. Jungnickel, U. Krüger, G. Istoc, T. Haustein and C. von Helmolt, "A MIMO system with reciprocal transceivers for the time-division duplex mode," *IEEE AP-S International Symposium*, Monterrey, CA, June 20–26, 2004.
[56] H. Holma, S. Haikkinen, O.-A. Lehtinen and A. Toskala, "Interference considerations for the time division duplex mode of the UMTS terrestrial radio access," *IEEE JSAC*, Vol. 18, No. 8, pp. 1386–1392, 2000.
[57] C. B. Papadias and H. Huang, "Linear space-time multiuser detection for multipath CDMA channels" *IEEE J. Sel. Areas Commun.*, Vol. 19, No. 2, pp. 254–265, 2001.
[58] T. Fong, P. Henry, K. Leung, X. Qiu and N. Shankaranarayanan, "Radio resource allocation in fixed broadband wireless networks," *IEEE Trans. Commun.*, Vol. 46, No. 6, pp. 806–818, June 1998.
[59] A. J. Goldsmith, S. A. Jafar, N. Jindal and S. Vishwanath, "Capacity limits of MIMO channels," *IEEE J. Sel. Areas Commun.*, Vol. 21, No. 5, pp. 684–702, 2003.
[60] N. Jindal, S. Vishwanath and A. Goldsmith, "On the Duality of Gaussian Multiple-Access and Broadcast Channels", *IEEE Trans. Inf. Theory*, Vol. 50, pp. 768–783, 2004.
[61] P. Viswanath and D. N. C. Tse, "Sum capacity of the vector Gaussian broadcast channel and uplink-downlink duality", *IEEE Trans. Inf. Theory*, Vol. 49, pp. 1912–1921, 2003.
[62] R. Knopp and P. A. Humblet, "Information capacity and power control in single-cell multiuser communications," *Proc. IEEE ICC*, Vol. 1, pp. 331–335, June 1995.
[63] P. Viswanath, D. Tse and R. Laroia, "Opportunistic beamforming using dumb antennas," *IEEE Trans. Inf. Theory*, Vol. 48, No. 6, pp. 1277–1294, 2002.
[64] W. Yu, W. Rhee, S. Boyd and J. M. Cioffi, "Iterative water-filling for Gaussian vector multiple-access channels," *IEEE Trans. Inf. Theory*, Vol. 50, No. 1, pp. 145–151, 2004.
[65] H. Boche and E. A. Jorswieck, "Sum capacity optimization of the MIMO gaussian MAC," *5th International Symposium on Wireless Personal Multimedia Communications*, invited paper, Vol. 1, pp. 130–134, Oct. 2002.
[66] E. A. Jorswieck and H. Boche, "Transmission strategies for the MIMO MAC with MMSE receiver: average MSE optimization and achievable individual MSE region", *IEEE Trans. Sig. Proc.*, Vol. 51, No. 11, pp. 2872–2881, 2003.
[67] D. Tse and S. Hanly, "Multiaccess fading channels: Part I: Polymatroid structure, optimal resource allocation and throughput capacities," *IEEE Trans. Inf. Theory*, Vol. 44, No. 7, pp. 2796–2815, November 1998.
[68] G. Caire, D. Tuninetti and S. Verdu, "Suboptimality of TDMA in the low-power regime," *IEEE Trans. Inf. Theory*, Vol. 50, No. 4, pp. 608–620, April 2004.
[69] R. A. Berry and E. M. Yeh, "Cross-layer wireless resource allocation," *IEEE Signal Process. Mag.*, Vol. 21, No. 5, pp. 59–68, September 2004.
[70] A. J. Goldsmith and S. B. Wicker, "Design challenges for energy-constrained ad Hoc wireless networks," *IEEE Wireless Commun.*, Vol. 9, No. 4, pp. 8–27, August 2002.
[71] I. E. Telatar and R. G. Gallager, "Combining queueing theory with information theory for multiaccess," *IEEE J. Sel. Areas Commun.*, Vol. 13, No. 6, pp. 963–969, August 1995.
[72] A. Ephremides and B. Hajek, "Information theory and communication networks: An unconsummated union," *IEEE Trans. Inf. Theory*, Vol. 44, No. 6, pp. 2416–2434, October 1998.
[73] H. Boche and M. Wiczanowski, "Stability region of arrival rates and optimal scheduling for MIMO-MAC - a cross-layer approach," *Proceedings of IEEE IZS*, Zurich, Switzerland, 2004.
[74] D. Gesbert and M. Slim Alouini, "How much feedback is multi-user diversity really worth?" in *Proceedings of IEEE International Conference on Communications (ICC)*, Paris, France, 2004.
[75] D. Gesbert and M.-S. Alouini, "How much feedback is multi-user diversity really worth?", *2004 IEEE International Conference on Communications*, Vol. 1, No. 20–24, pp. 234–238, June 2004, Paris, France.
[76] I. Toufik and R. Knopp, "Multiuser Channel Allocation Algorithms Achieving Hard Fairness," *IEEE Vehicular Technology Conference*, Los Angeles, Sept. 2004.
[77] I. Toufik and R. Knopp, "Channel allocation algorithms for multi-carrier multiple-antenna systems," *IEEE Globecom*, Dallas, Nov. 2004.
[78] D. Tse and S. Hanly, "Multiaccess fading channels: Part II delay-limited capacities," *IEEE Trans. Inf. Theory*, Vol. 44, No. 7, pp. 2796–2815, November 1998.

[79] V. Pohl, P. H. Nguyen, V. Jungnickel and C. von Helmolt, "Continuous flat fading MIMO channels: Achievable rate and the optimal length of the training and data phase," 2003, accepted for publication in *IEEE Trans. Wireless Commun.*, Vol. 4, No. 4, pp. 1889–1900, July 2005.
[80] V. Jungnickel, T. Haustein, A. Forck, S. Schiffermueller, H. Gaebler, C. von Helmolt, W. Zirwas, J. Eichinger and E. Schulz, "Real-time concepts for MIMO-OFDM," *Proceedings of CIC/IEEE Global Mobile Congress*, Shanghai, China, 11–13 Oct. 2004.
[81] M. Wouters, A. Bourdoux, S. Derore, S. Janssens and V. Derudder, "An approach for real-time prototyping of MIMO systems," *12th European Signal Processing Conference (EUSIPCO)*, Vienna, Austria, September 6–10, 2004.
[82] D. Borkowski and L. Brühl, "Hardware implementation for real-time multi-user MIMO systems," *Workshop on MIMO Implementation Issues*, IEEE RAWCON, 2004.
[83] S. Häne, D. Perels, D.S. Baum, M. Borgmann, A. Burg, N. Felber, W. Fichtner and H. Bölcskei, "Implementation Aspects of a Real-Time Multi-Terminal MIMO-OFDM Testbed," *Workshop on MIMO Implementation Issues*, IEEE Radio and Wireless Conference (RAWCON), Atlanta, GA, Sept. 2004.
[84] V. Jungnickel, T. Haustein, A. Forck, U. Krueger, V. Pohl and C. von Helmolt, "Over-the-air demonstration of spatial multiplexing at high data rates using real-time base-band processing", *Adv. Radio Sci.*, Vol. 2, pp. 135–140, 2004, (available on-line).
[85] T. Haustein, A. Forck, H. Gäbler, C. von Helmolt, V. Jungnickel and U. Krüger, "Implementation of adaptive channel inversion in a real-time MIMO system," *Proceedings of PIMRC*, Barcelona, Spain, September 5–8, 2004 (on CD-ROM).
[86] G. Brown, *"Ultrawideband: Spectrum for Free"*, Unstrung Insider (www.unstrung.com).
[87] J. Walko, "Ultra Wide Band", *The IEE Commun. Eng.*, Vol. 1, No. 6, pp. 10–13, Dec/Jan 2003/4.
[88] R. Fontana, *"A History of UWB"*, www.multispectral.com/history.html.
[89] Press Release Freescale Semiconductor Inc, http://www.freescale.com/webapp/sps/site/news_release.jsp?nodeld=093623.
[90] K. Siwiak, *"The Potential of Ultra-Wide Band Communications,"* IEEE, ICAP, 2003.
[91] K. Siwiak and D. McKeown, *"Ultra Wideband Radio Technology"*, Wiley, UK, 2004.
[92] K. Swiak, T. Babij and S. Yano, "On the relationship between multipath and wave propagation attenuation", *Electron Lett.*, Vol. 39, No. 1, pp. 142–143, 9th Jan 2003.
[93] B. Allen, "Ultra wideband wireless sensor networks," *IEE UWB Symposium*, London, UK, June 2004.
[94] T. Barrett, "History of UltraWideBand (UWB) radar & communications: Pioneers and innovators," *Progress in Electromagnetics Symposium 2000 (PIERS2000)*, Cambridge, MA, July 2000.
[95] www.uwb.org.
[96] Federal Communications Commission (FCC), *"FCC NOI: Rules Regarding Ultra-Wideband Transmission Systems,"*, ET Docket No. 98–153, Sept. 1, 1998.
[97] I.F. Akyildiz, Su. Weilian, Y. Sankarasubramaniam and E. Cayirci, "A Survey on Sensor Networks", *IEEE Commun. Mag.*, Vol. 40, No. 8, pp. 102–114, August 2002.
[98] http://www.ieee802.org/15/pub/TG4a.html.
[99] Y. Ko and N Vaidya, "Location-Aided Routing (LAR) in Mobile Ad Hoc Networks," *Proceedings of the Fifth Annual ACM/IEEE International Conference on Mobile Computing and Networking (MOBICOM '98)*, Dallas, TX, USA, August 1998.
[100] www.ubisense.net.
[101] B. Allen, M. Ghavami, A. Armogida and A. H. Aghvami, "The Holy Grail of Wire Replacement?", *IEE Commun. Eng.*, Vol. 1, No. 5, pp. 14–17, Oct/Nov 2003.
[102] Intel.com, *Industry Leaders Developing First High-Speed Personal Wireless Interconnect, Formation of Wireless USB Promoter Group Announced*, http://www.intel.com/pressroom/archive/releases/20040218corp_c.htm.
[103] M. Ghavami, L. B. Michael and R. Khono, *"Ultra Wideband Signals and Systems in Communications Systems"*, Wiley, 2004.
[104] M. Nakagawa, Z. Honggang and H. Sato, "Ubiquitous homelinks based on IEEE 1394 and ultra wideband solutions", *IEEE Commun. Mag.*, Vol. 41, No. 4, pp. 74–82, April 2003.
[105] L. Rusch, C. Prettie, D. Cheung, Q. Li and M. Ho, *Characterization of UWB Propagation from 2 to 8 GHz in a Residential Environment*. [online]. Available www.intel.com, (2003).
[106] M. Z. Win and R. A. Scholtz, "Energy capture vs. correlation resources in ultra-wide bandwidth indoor wireless communications channel," *MILCOM 97 Proc.*, Vol. 3, pp. 1277–1281, 2–5 November 1997.

[107] Y. Zhang and A. Brown, "Ultra-wide bandwidth communication channel analysis using 3-D ray tracing," *The 1st International Symposium on Wireless Communication Systems*, Mauritius, 20–22 September 2004.

[108] W. Q. Malik, D. J. Edwards and C. J. Stevens, "Experimental evaluation of Rake receiver performance in a line-of-sight ultra-wideband channel," *Proceedings of Joint UWBST & IWUWBS*. Kyoto, Japan, 18–21 May 2004.

[109] F. Argenti, T. Bianchi, E. Del Re, L. Mucchi and L. S. Ronga, "Ultra-wideband signal with polarization diversity for constant indoor QoS," in *Proceedings of 12th WWRF Meeting*. Toronto, Canada, November 2004.

[110] D. Kelly, *et al.*, "Pulson second generation timing chip: enabling UWB through precise timing," *Proceedings of UWBST*. 21–23 May, 2002.

[111] R. H. Walden, "Analog-to-Digital Converter Survey and Analysis", *IEEE J. Select. Areas Commun.*, Vol. 17, No. 4, pp. 539–550, April 1999.

[112] W. C. Black, "Time Interleaved Converter arrays", *IEEE J. Solid-State Circuits*, Vol. 15, pp. 1022–1029, Dec 1980.

[113] R. Khoini-Poorfard and D. A. Johns, "Mismatch effects in time-interleaved oversampling converters," *ISCAS '94, 1994 IEEE International Symposium on Circuits and Systems 1994*, Vol. 5, London, UK, pp. 429–432, 30 May–2 June 1994.

[114] R. C. H. van de Beek, E. A. M. Klumperink, C. S. Vaucher and B. Nauta, "Low-jitter clock multiplication: A comparison between PLLs and DLLs", *IEEE Trans. Circuits Syst. II: Express Briefs*, Vol. 49, No. 8, pp. 555–556, Aug. 2002.

[115] A. Hellemans, "French team demonstrates terahertz transistor", *IEEE Spectr.*, Vol. 41, No. 5, p. 16, 16 May 2004.

[116] Intel.com, Intel press release "*Intel's TeraHertz Transistor Architecture*": http://www.intel.com/research/spotlights/terahertzbkgdr.htm.

[117] G. Freeman, M. Meghelli, Y. Kwark, S. Zier, A. Rylyakov, J. S. Sorna, T. Tanji, O. M. Schreiber, K. Walter, J. S. Rieh, B. Jagannathan, A. Joseph and S. Subbanna, "40-Gb/s circuits built from a 120-GHz f(T) SiGe technology" *IEEE J. Solid-State Circuits*, Vol. 37, No. 9, pp. 1106–1114, September 2002.

[118] F. Furuta, K. Saitoh and K. Takagi, "Design of front-end circuit for superconductive A/D converter and demonstration of operation up to 43 Ghz," *IEEE Trans. Appl. Supercond.*, Vol. 14, No. 1, pp. 40–45, March 2004.

[119] P. Bunyk, M. Leung, J. Spargo and M. Dorojevets, "FLUX-1 RSFQ microprocessor: Physical design and test results," *IEEE Trans. Appl. Supercond.*, Vol. 13, No. 2, pp. 433–436, June 2003.

[120] W. Chen, A. V. Rylyakov, V. Patel, J. E. Lukens and K. K. Likharev, "Rapid Single Flux Quantum T-flip flop operating up to 770 GHz," *IEEE Trans. Appl. Supercond.*, Vol. 9, No. 2, pp. 3212–3215, June 1999.

[121] S. J. Tans, A. R. M. Verschueren and C. Dekker, "Room-temperature transistor based on a single carbon nanotube", *Nature*, Vol. 393, pp. 49–52, 7 May 1998.

[122] A. H. Tewfik and E. Saberinia, "High bit rate ultra-wideband OFDM," *Proceedings of Globecom*. Taipei, Taiwan, November 2002.

[123] J. Kunisch, J. Pamp, "UWB radio channel modeling considerations," *Proceedings of ICEAA*. Torino, Italy, September 2003.

[124] B. Scheers, M. Acheroy and A. Vander Vorst, "Time Domain simulation and characterisation of TEM horns using a normalised impulse response," *IEE Proc.-Microw. AP*, Vol. 147, pp. 463–468, Dec. 2000.

[125] W. Soergel, F. Pivit and W. Wiesbeck, "Comparison of frequency domain and time domain measurement procedures for ultra wideband antennas," *Proceedings of Antenna and Measurement Techniques Association, 25th Annual Meeting and Symposium*. Irvine, California, Oct. 2003.

[126] W. Q. Malik, C. J. Stevens and D. J. Edwards, "Spatio-spectral normalisation for ultra wideband antenna dispersion," in *Proceedings of the 9th IEEE HFPSC*. Manchester, UK, 6–7 September 2004.

[127] L. Zhao and A. M. Haimovich, "Performance of ultra-wideband communications in the presence of interference," *IEEE J. Select. Areas Commun.*, Vol. 20, No. 9, pp. 1684–1691, December 2002.

[128] X. Chu and R. D. Murch, "The effect of NBI on UWB time-hopping systems," in *IEEE Trans. Wireless Comm.*, Vol. 3, No. 5, pp. 1431–1436, September 2004.

[129] D. Falconer, S. L. Ariyavisitakul, A. Benyamin-Seeyar and B. Eidson, "Frequency domain equalization for single-carrier broadband wireless systems," *IEEE Commun. Mag.*, Vol. 40, No. 4, pp. 58–66, April 2002.

[130] S. Morosi and T. Bianchi, "Comparison between rake and frequency domain detectors in ultra-wideband indoor communications," in *Proceedings of 15th PIMRC*, Barcelona, Spain, September 2004.
[131] B. Allen, A. Ghorishi and M. Ghavami, "A review of pulse design for impulse radio," *IEE Ultra Wideband Workshop*, London, UK, June 2004.
[132] http://www.ieee802.org/15/pub/
[133] http://www.pulsers.net/main.shtml, 2004.
[134] IEEE, *"Draft Standard for Telecommunications and Information Exchange Between Systems – LAN/MAN Specific Requirements – Part 15.3: Wireless Medium Access Control (MAC) and Physical Layer (PHY) Specifications for High Rate Wireless Personal Area Networks (WPAN)"*, Draft P802.15.3/D17, Feb 2003.
[135] *Press Release, UWB Forum*, http://www.uwbforum.org/, May 2004.
[136] J. Ding, L. Zhao, S. R. Medidi and K. M. Sivalingam, "MAC Protocols for Ultra-Wide-Band (UWB) wireless networks: Impact of channel acquisition time," *SPIE ITCOM Conference*, Boston, MA, USA, July 2002.
[137] J. Elson, L. Girod and D. Estrin, "Fine grained time synchronization using reference broadcasts," *Proceedings Fifth Symposium on Operating Systems Design and Implementation (OSDI 2002)*, Boston USA, pp. 2–17, 2002.
[138] D. Niculescu and B. Nath, *Position and Orientation in Ad hoc Networks.* Ad hoc networks 2, Elsevier, pp. 133–151. 2004.
[139] http://www.fcc.gov/Bureaus/Engineering_Technology/News_Releases/2002/nret0203.html, FCC press rebreaklease, Feb 2002.
[140] Federal Communications Commission (FCC), *"First Report and Order in the Matter of Revision of Part 15 of the Commission's Rules Regarding Ultrawideband Transmission Systems,"* ET-Docket 98-153, FCC 02-48, released April 22, 2002.
[141] J. D. Taylor (ed.) *Introduction to Ultra wideband Radar Systems.* CRC Press, Inc., Boca Raton, Florida, USA, p. 670, 1995.
[142] EC Mandate M/329 to the European Standardisation Organisations, referenced in document RSCOM03-07, http://forum.europa.eu.int/irc/DownLoad/kxeZAgJHmuGFylKUFUC1RX_kDo04dHjCheTEj-cVifvno1 TsW42fNUnq1R3N8FpHFVXmUpoo2yLi/RSCOM03-07.pdf.
[143] CEPT/ERC Recommendation 70-03, *"Relating to the use of Short Range Devices (SRD)"*, http://www.ofcom.org.uk/consult/condocs/wireless865_868/.
[144] EC Mandate to CEPT, *"Mandate to CEPT to Harmonise Radio Spectrum use for Ultra-wideband Systems in the European Union"*, referenced in document RSCOM03-40 rev. 3.
[145] Federal Communications Commission (FCC), http://www.fcc.gov.
[146] J. D. Barry, *Wireless Infrared Communications.* Netherlands: Kluwer, 1994.
[147] J. M. Kahn and J. R. Barry, "Wireless infrared communications," *Proc. IEEE*, Vol. 85, pp. 265–298, 1997.
[148] D. J. T. Heatley, D. R. Wisely, I. Neild and P. Cochrane, "Optical wireless: The story so far," *IEEE Commun. Mag.*, Vol. 36, pp. 72–82, 1998.
[149] D. J. T. Heatley, D. R. Wisely, I. Neild and P. Cochrane, "A review of optical wireless – What is it and what does if offer?," *Br. Telecommun. Eng.*, Vol. 17, pp. 251–261, 1999.
[150] G. Yun and M. Kavehrad, *"Spot-diffusing and Fly-eye Receivers for Indoor Infrared Wireless Communications,"* presented at 25 26 June 1992, Vancouver, BC, Canada, 1992.
[151] M. Kavehrad and S. Jivkova, "Indoor broadband optical wireless communications: optical subsystems designs and their impact on channel characteristics," *IEEE Wireless Commun.*, Vol. 10, pp. 30–35, 2003.
[152] IEC 60825-1, *Safety of Laser Products Part 1*, British Standards Institution, 2001.
[153] S. H. Khoo, E. B. Zyambo, G. Faulkner, D. C. O'Brien, D. J. Edwards, M. Ghisoni, and J. Bengtsson, *"Eyesafe Optical Link using a Holographic Diffuser,"* presented at IEE Colloquium on Optical Wireless Communications, 1999.
[154] Y. Tanaka, S. Haruyama and M. Nakagawa, "Wireless optical transmissions with white colored LED for wireless home links," *Proceedings of 11th International Symposium on Personal, Indoor and Mobile Radio Communication*," London, UK, 2000.
[155] JVC, "www.jvc.com. luciole link," 2003.
[156] A. C. Boucouvalas, "Indoor ambient light noise and its effect on wireless optical links," *IEE Proc.-Optoelectron.*, Vol. 143, pp. 334–338, 1996.
[157] K. Phang and D. A. Johns, *"A 3-V CMOS Optical Preamplifier with DC Photocurrent Rejection,"* 1998.

[158] J. B. Carruthers and J. M. Kahn, "Angle diversity for nondirected wireless infrared communication," *IEEE Trans. Commun.*, Vol. 48, pp. 960–969, 2000.
[159] D. C. O'Brien, G. E. Faulkner, E. B. Zyambo, K. Jim, D. J. Edwards, P. Stavrinou, G. Parry, J. Bellon, M. J. Sibley, V. A. Lalithambika, V. M. Joyner, R. J. Samsudin, D. M. Holburn and R. J. Mears, "Integrated transceivers for optical wireless communications," *IEEE J. Select. Topics Quantum Electron.*, Vol. 11, pp. 173–183, 2005.
[160] J. M. Kahn, R. You, P. Djahani, A. G. Weisbin, B. K. Teik and A. Tang, "Imaging diversity receivers for high-speed infrared wireless communication," *IEEE Commun. Mag.*, Vol. 36, pp. 88–94, 1998.
[161] R. Narasimhan, M. D. Audeh and J. M. Kahn, "Effect of electronic-ballast fluorescent lighting on wireless infrared links," *IEE Proc.-Optoelectron.*, Vol. 143, pp. 347–354, 1996.
[162] A. J. C. Moreira, R. T. Valadas and A. M. D. Duarte, "Performance of infrared transmission systems under ambient light interference," *IEE Proc.-Optoelectron.*, Vol. 143, pp. 339–346, 1996.
[163] T. O'Farrell and M. Kiatweerasakul, "Performance of a spread spectrum infrared transmission system under ambient light interference," *Presented at Proceedings of Ninth International Symposium on Personal, Indoor, and Mobile Radio Communications (PIMRC '98)*, Boston, MA, USA, 1998.
[164] A. M. Street, P. N. Stavrinou, D. C. Obrien and D. J. Edwards, "Indoor optical wireless systems – A review," *Opt. Quantum Electron.*, Vol. 29, pp. 349–378, 1997.
[165] S. Jivkova and M. Kavehrad, "Holographic optical receiver front end for wireless infrared indoor communications," *Appl. Opt.*, Vol. 40, pp. 2828–2835, 2001.
[166] K. Samaras, D. C. O'Brien, A. M. Street and D. J. Edwards, "BER performance of NRZ-OOK and Manchester modulation in indoor wireless infrared links," *Int. J. Wireless Inf. Networks*, Vol. 5, pp. 219–233, 1998.
[167] D. C. O'Brien, G. E. Faulkner, K. Jim, E. B. Zyambo, D. J. Edwards, M. Whitehead, P. Stavrinou, G. Parry, J. Bellon, M. J. Sibley, V. A. Lalithambika, V. M. Joyner, R. J. Samsudin, D. M. Holburn and R. J. Mears, "Solid state tracking integrated optical wireless transceivers for line-of-sight optical links," *Presented at Free Space Laser Communication and Active Laser Illumination III*, San Diego, 2003.
[168] M. J. McCullagh and D. R. Wisely, "155 mbit/S optical wireless link using a bootstrapped silicon APD receiver," *Electron. Lett.*, Vol. 30, pp. 430–432, 1994.
[169] G. W. Marsh and J. M. Kahn, "Performance evaluation of experimental 50-Mb/s diffuse infrared wireless link using on-off keying with decision-feedback equalization," *IEEE Trans. Commun.*, Vol. 44, pp. 1496–1504, 1996.
[170] V. A. Lalithambika, V. M. Joyner, D. M. Holburn and R. J. Mears, "Development of a CMOS 310 Mb/s receiver for free-space optical wireless links," *Presented at Proceedings of the SPIE The International Society for Optical Engineering*, Vol. 4214, No. 17, 2001.
[171] D. M. Holburn, V. A. Lalithambika, V. M. Joyner, R. Samsudin and R. J. Mears, "Integrated CMOS transceiver for indoor optical wireless links," *Presented at Optical Wireless Communications IV*, Denver, 2001.
[172] S. H. Khoo, W. Zhang, G. E. Faulkner, D. C. O'Brien and D. J. Edwards, "Receiver angle diversity design for high-speed diffuse indoor wireless optical communications," *Presented at Optical Wireless Communications IV*, Denver, 2001.
[173] J. B. Carruthers and P. Kannan, "Iterative site-based modeling for wireless infrared channels," *IEEE Trans. Ant. Propagat.*, Vol. 50, pp. 759–765, 2002.
[174] V. Jungnickel, V. Pohl, S. Nonnig and C. von Helmolt, "A physical model of the wireless infrared communication channel," *IEEE J. Sel. Areas Commun.*, Vol. 20, pp. 631–640, 2002.
[175] J. M. Kahn, W. J. Krause and J. B. Carruthers, "Experimental characterization of non-directed indoor infrared channels," *IEEE Trans. Commun.*, Vol. 43, pp. 1613–1623, 1995.
[176] M. R. Pakravan and M. Kavehrad, *"Indoor Wireless Infrared Channel Characterization by Measurements,"* 2001.
[177] D. P. Manage, S. H. Khoo, G. E. Faulkner and D. C. O'Brien, "Novel system for the imaging of optical multipaths," *Presented at High Speed Photography and Detection*, San Diego, 2003.
[178] J. R. Barry, J. M. Kahn, W. J. Krause, E. A. Lee and D. G. Messerschmitt, "Simulation of multipath impulse-response for indoor wireless optical channels," *IEEE J. Sel. Areas Commun.*, Vol. 11, pp. 367–379, 1993.
[179] K. K. Wong and T. O'Farrell, *"Performance Analysis of M-ary Orthogonal DS System for Infrared Wireless Communications,"* 2000.

[180] K. K. Wong and T. O'Farrell, "Spread spectrum techniques for indoor wireless IR communications," *IEEE Wireless Commun.*, Vol. 10, pp. 54–63, 2003.
[181] M. Wolf and D. Kress, "Short-range wireless infrared transmission: the link budget compared to RF," *IEEE Wireless Commun.*, Vol. 10, pp. 8–14, 2003.
[182] C. C. Davis, I. I. Smolyaninov and S. D. Milner, *"Flexible Optical Wireless Links and Networks,"* 2003.
[183] Ultralab Database, "http://impulse.usc.edu/uwbloss1.html," 2005.
[184] D. Cheung and C. Prettie, *"A Path Loss Comparison Between the 5 GHz UNII Band (802.11a) and the 2.4 GHz ISM Band (802.11b),"* http://impulse.usc.edu/resources/802_11a-vs-b_report.pdf, 2002.
[185] Dai-Lu and D. Rutledge, "Investigation of indoor radio channels from 2.4 GHz to 24 GHz," *Presented at 2003 IEEE International Symposium on Antennas and Propagation: URSI North American Radio Science Meeting*. Vol. 2, 22 27 June 2003 Columbus, OH, USA, 2003.
[186] W. Ciccognani, A. Durantini and D. Cassioli, "Time domain propagation measurements of the UWB indoor channel using PN-sequence in the FCC-compliant band 3.6–6 GHz," *IEEE Trans. Antennas Propagat.*, Vol. 53, pp. 1542–1549, 2005.
[187] S. Miyamoto, Y. Hirayama and N. Morinaga, "Indoor wireless local area network system using infrared and radio communications," *Presented at Proceedings of APCC/OECC '99 5th Asia Pacific Conference on Communications/4th Optoelectronics and Communications Conference*, Vol. 1, 18 22 Oct. 1999 Beijing, China, 1999.
[188] Y. Sakurai, K. Nishimaki, S. Toguchi and M. Sakane, "A study of seamless communication method with the adequate switching between optical and RF wireless LAN," *Presented at 2003 Digest of Technical Papers. International Conference on Consumer Electronics*, 17–19 June 2003 Los Angeles, CA, USA, 2003.
[189] K. Fan, T. Komine, Y. Tanaka and M. Nakagawa, *"The Effect of Reflection on Indoor Visible-Light Communication System Utilizing White LEDs,"* 2002.
[190] DLink Corporation, *"DLINK 802.11(g) PCMIA card,"* www.dlink.com, 2005.
[191] IRDA, "www.irda.org,".
[192] EFKON, "http://www.efkon.com," 2004.
[193] Jindong-Hou and D. C. O'Brien, "Vertical handover decision-making algorithm using fuzzy logic for the integrated radio and OW system," accepted for publication in *IEEE Trans. Wireless Commun.*, Vol. 5, No. 1, pp. 176–185, Jan. 2006.
[194] JVC, "http://www.jvc.com/ds2/f_prod.htm.", 2004.
[195] P. Djahani and J. M. Kahn, "Analysis of infrared wireless links employing multibeam transmitters and imaging diversity receivers," *IEEE Trans. Commun.*, Vol. 48, pp. 2077–2088, 2000.
[196] K. Liang, H. Shi, S. J. Sheard and D. C. O'Brien, "Transparent optical wireless hubs using wavelength space division multiplexing," *Presented at SPIE Free Space Laser Communications IV*, Denver, 2004.
[197] T. Komine and M. Nakagawa, "Fundamental analysis for visible-light communication system using LED lights," *IEEE Trans. Consumer Electron.*, Vol. 50, pp. 100–107, 2004.
[198] Y. Tanaka, T. Komine, S. Haruyama and M. Nakagawa, "Indoor visible light data transmission system utilizing white LED lights," *IEICE Trans. Commun.*, Vol. E86-B, pp. 2440–2454, 2003.
[199] Visible Light Communications Consortium, "www.vlcc.net."
[200] S. Faruque, *Cellular Mobile Systems Engineering*. London: Artech House, 1996.
[201] M. Progler, "SAMBA: A mobile broadband enabler," *Presented at Proceedings of 29th European Microwave Conference*. Vol. 3, 5–7 Oct. 1999, Munich, Germany, 1999.
[202] Broadway project, "http://www.ist-broadway.org/," 2005.
[203] P. F. M. Smulders, "60 GHz radio: prospects and future directions," *Presented at Proceedings Symposium IEEE Benelux Chapter on Communications and Vehicular Technology*, Eindhoven, 2003.
[204] K. Sato and T. Manabe, "Estimation of propagation-path visibility for indoor wireless LAN systems under shadowing condition by human bodies," *Presented at VTC '98. 48th IEEE Vehicular Technology Conference. Pathway to a Global Wireless Revolution*. Vol. 3, 18–21 May 1998 Ottawa, Ontario, Canada, 1998.
[205] M. Flament and M. Unbehaun, "Impact of shadow fading in a MM-wave band wireless network," *Presented at Proceedings of 3rd International Symposium on Wireless Personal Multimedia Communications (WPCS '00)*. Vol. 1, 12 15 Nov. 2000, Bangkok, Thailand, 2000.
[206] S. T. Jovkova and M. Kavehrad, "Multispot diffusing configuration for wireless infrared access," *IEEE Trans. Commun.*, Vol. 48, pp. 970–978, 2000.

[207] A. G. Al-Ghamdi and J. M. H. Elmirghani, *"Line Strip Spot-diffusing Transmitter Configuration for Optical Wireless Systems Influenced by Background Noise and Multipath Dispersion,"* 2004.
[208] M. Jafar, D. C. O'Brien, C. J. Stevens and D. J. Edwards, *"Evaluation of Coverage Area for a Wide Line-of-Sight Indoor Optical Free-Space Communication System Employing Coherent Detection,"* In preparation, 2005.
[209] T. Kimura, S. Bjorlin, Hsu-Feng-Chou, Qi-Chen, Shaomin-Wu and J. E. Bowers, "Optically preamplified receiver at 10, 20, and 40 Gb/s using a 1550-nm vertical-cavity SOA," *IEEE Photon. Technol. Lett.*, Vol. 17, pp. 456–458, 2005.
[210] M. Idrus and R. J. Green, "Performance characterisation of a photoparametric up-converter,", *IEEE GCC*, Bahrain, paper T10P183, 5 pages, November 2004.
[211] M. J. McCullagh and D. R. Wisely, "155 Mbit/s optical wireless link using a bootstrapped silicon APD receiver," *Electron. Lett.*, Vol. 30, pp. 430–432, 1994.
[212] S. M. Haas, J. H. Shapiro and V. Tarokh, "Space-time codes for wireless optical communications," *EURASIP J. Appl. Signal Process.*, Vol. 2002, pp. 211–220, 2002.
[213] Y. A. Alqudah and M. Kavehrad, "MIMO characterization of indoor wireless optical link using a diffuse-transmission configuration," *IEEE Trans. Commun.*, Vol. 51, pp. 1554–1560, 2003.
[214] D. Niculescu, "Communication Paradigms for Sensor Networks," *IEEE Comm. Mag.*, Vol. 43, No. 3, pp. 116–122, March 2005.
[215] T.-H. Lin, "Integrated low-power communication system design for wireless sensor networks," *IEEE Commun. Mag.*, Vol. 42, No. 12, pp. 142–150, December 2004.
[216] LEACH Routing Protocol, http://nms.lcs.mit.edu/projects/leach.
[217] S-MAC Protocol, http://www.isi.edu/scadds/projects/smac/.
[218] A. El-Hoiydi and J. D. Decotignie, WiseMAC: an ultra low power MAC protocol for the downlink of infrastructure wireless sensor networks, A. El-Hoiydi and J.-D. Decotignie, in *Proc. Intl. Symp. Computers and Communications*, Vol. 1, pp. 244–251, 28 June–1 July 2004.
[219] S. Rooney, D. Bauer and P. Scotton, "Edge server software architecture for sensor applications," in *Proc. Symp. on Applications and the Internet (SAINT '05)*, pp. 64–71, 31 Jan.–4 Feb. 2005.
[220] NanoIP, http://www.cwc.oulu.fi/nanoip/.
[221] Bluetooth SIG, http://www.bluetooth.org.
[222] IEEE 802.15, http://grouper.ieee.org/groups/802/15/.
[223] ZigBee Alliance, http://www.zigbee.org.
[224] IEEE 1451.5, http://grouper.ieee.org/groups/1451/5/.
[225] NFC Forum, http://www.nfc-forum.org.
[226] Smart Dust, http://robotics.eecs.berkeley.edu/~pister/SmartDust.
[227] Pico Radio, http://bwrc.eecs.berkeley.edu/Research/Pico_Radio.
[228] WINS, http://www.janet.ucla.edu/WINS/.
[229] V. Jungnickel, V. Pohl, H. Nguyen, U. Krüger, T. Haustein and C. von Helmolt, *"High Capacity Antennas for MIMO Radio Systems,"* WPMC, Honolulu, Hawaii, 2002.

8

Reconfigurability

Edited by Panagiotis Demestichas (University of Piraeus), George Dimitrakopoulos (University of Piraeus), Klaus Mößner (CCSR, University of Surrey), Terence Dodgson (Samsung Electronics) and Didier Bourse (Motorola Labs)

8.1 Introduction

Wireless Communications comprise a multiplicity of Radio Access Technology (RAT) standards, the most commonly used being GSM (Global System for Mobile communications) [1], GPRS (Generalised Packet Radio Service) [2], UMTS (Universal Mobile Telecommunications System) [3], BRANs (Broadband Radio Access Networks) or WLANs (Wireless Local-Area Networks) [4–6] and DVB (Digital Video Broadcasting) [7]. This set of discrete technologies is currently transforming into one global infrastructure, called *Beyond the third Generation* (*B3G*) wireless-access infrastructure, aiming at offering innovative services in a cost-efficient manner. Major aspects of this convergence are the cooperative networks [8, 9] and the reconfigurability [10] concepts.

The cooperative networks concept assumes that diverse technologies, such as cellular (2.5G/3G mobile networks), BRAN/WLAN and DVB systems can be cooperating components of a heterogeneous wireless-access infrastructure. This implies that a network provider (NP) can rely on more than one RAT, for the encountered specific conditions (e.g. hot-spot situations, traffic demand alterations, etc.) at different time zones and spatial regions. At the same time, an NP may also cooperate with other NPs in order to have alternative solutions for maximising the offered quality of service (QoS) levels. Advanced management functionality is required for supporting the cooperative networks concept. The required functionality deals with the allocation of traffic to the different RATs and networks, as well as with the allocation of applications to QoS levels. Relevant research attempts have been made in the recent past [11–14].

Reconfigurability is an evolution of Software Defined Radio (SDR) [15]. It aims at bringing the full benefits of the valuable diversity within the radio eco-space, composed of a wide range of systems such as cellular, wireless local area, and broadcast. More

Technologies for the Wireless Future – Volume 2 Edited by Rahim Tafazolli

specifically, reconfigurability provides essential mechanisms to terminals and network segments, so as to enable them to adapt dynamically, transparently, and securely to the most appropriate RAT [16]. Through reconfigurability, one can envision network segments that change RAT, in a self-organised manner, in order to better handle the offered demand. In this context, reconfigurability also supports the dynamic allocation of resources (especially spectrum) to RATs [17].

This section aims at providing the basic research framework towards a successful deployment of composite reconfigurable networks. For this purpose, special attention is placed upon the concept of reconfigurability, in terms of the technology battlefields in which innovations are required. Consequently, the structure of this section is outlined in the subsequent text.

The next Section 8.2 introduces distinct categories of application scenarios aggregating technical, business, and regulatory visions. The scenarios presented are grouped in three main categories representing a common theme and corresponding to an anticipated coherent timeframe of technical availability. Such scenarios assume an end-to-end reconfigurability (E2R) mechanism available and used to identify the basic requirements for reconfigurable systems.

The next section presents the system requirements for the realisation of reconfigurability and capabilities resulting from the scenarios. For this purpose, the respective methodology is described and then the different capabilities of the reconfigurable system are identified. Each capability is further refined to extract the associated requirements.

Additionally, the section elaborates on the relevant technology roadmaps and business paths, while the last section of the chapter presents a summary and research issues needed to realise a fully reconfigurable mobile system.

8.2 Application Scenarios for Reconfigurability

This section presents a number of scenarios with the aim of highlighting the need for reconfigurability. The scenarios represent three main categories:

- Ubiquitous access;
- Pervasive services;
- Dynamic resources provisioning.

The scenarios in each category have then been reviewed in greater detail and a common scenario for each category is created by integrating the key elements of each contributed scenario into one.

Additionally, a further analysis is carried out to identify the actors involved in the scenarios and the detailed interactions of the actors so that direct links can be established with the business models. The practicalities of implementing such scenarios from both the operator viewpoint and that of the user are also discussed.

8.2.1 Methodology for Scenario Analysis

8.2.1.1 The Process

The purpose of this analysis is to define scenarios in order to illustrate E2R requirements and constraints reconfigurations and to derive scenarios that capture the key reconfigurability elements. Figure 8.1 lists the main steps of the analysis process.

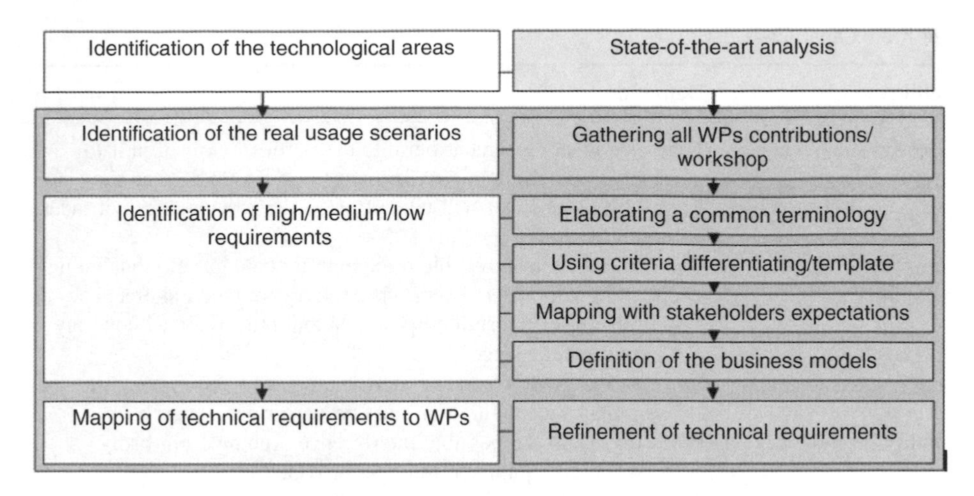

Figure 8.1 Scenarios analysis process

The first objective is to identify the main technological areas and to derive scenarios that capture the key reconfigurability elements. The next step is to clearly identify the different actors involved and the relationship between them. Scenarios will be supported by a story. This story will be used to identify the actor of the system as well as the main objects of the system (e.g. networks elements, application platform, and so on) involved in the reconfigurability process. Interactions and relations between actors will be identified. The actors involved in the process are identified and the relationships between the actors are also reviewed. After further analysis, it will lead to the definition of the business models. The defined scenarios will be analysed from various points of view to extract the corresponding requirements, business models and provide inputs for other WPs in order to refine the technical requirements.

8.2.1.2 The Actors

In order to identify who constitutes an Actor in the end-to-end scenarios, it has been proposed that a separate actor should be considered whenever the functions associated to it may be performed by an independent entity completely separated from the rest of entities in the system. Similarly, if different functions are always performed by the same entity it is proposed that those functions are grouped in one single actor.

On the basis of this principle, the actors involved in the end-to-end scenarios are identified in Table 8.1. However, before listing all the identified actors, it is interesting to provide clarification on the differentiations of some actors (such as user and subscriber, network operator and service provider) and considerations on the different approaches related to the global roaming.

User and Subscriber

The traditional situation is that the user and subscriber is the same person but it is possible, in certain cases, that the person making use of the terminal (user) is not the same one as the one who has contracted services with a service provider or network operator (subscriber). This is quite common where the Terminal User is one of many on a corporate talk plan.

Table 8.1 Capabilities list

Capability	Definition
Service level agreement	A service level agreement permits the parties involved in it to establish the minimum performance criteria for service provision and the actions that will take effect if the service does not meet these criteria
Equipment reconfiguration	The equipment must be able to change its configuration including operating parameters by means of software (such as frequency, modulation or transmitted power autonomously or without any external maintenance)
Security	In case of upgrade of one or more elements of the system, the equipment reconfiguration must be performed securely
No radio interference	Protection from the possible interference coming from badly reconfigured equipment must be provided
Download	There must exist mechanisms that allow the equipment to obtain a software module and download it in order to reconfigure to a new configuration in a new wireless network
Reconfiguration management	The end-to-end reconfiguration control management actions are assumed to be carried out by a Reconfiguration Manager. The reconfiguration process might be initiated either by the user equipment or by the network. The Reconfiguration Manager is a virtual entity that is not required but is used in order to facilitate the expression of the requirements; it will be stated by the design phase of the project if this entity has to be created or not
Service adaptation	Active services can be adapted to changes in the network status (e.g. congestion) or modifications of the equipment (e.g. reconfiguration), or alteration of the access network used by an equipment (e.g. vertical handover). Service adaptation may cause reconfiguration in equipment in an attempt to avoid major disruption on the executed service
Vertical handover	Equipments are able to move through different access networks without loosing their active connections. VHO can be equipment or network-initiated
Service provision	Service provision refers to basic telecommunication services as well as value-added services offered by operators or independent entities
System monitoring	The equipment and network must be able to monitor the current state of system operation (traffic, used spectrum, available technologies) in order to be able to estimate available resources and to facilitate the 'best' use of existing and available resources
Dynamic resource management	The network operator is able to dynamically assign its resources to the different tasks to be performed in order to make the best use of them
Spectrum transfer	The owners of the spectrum are able to transfer their spectrum to other parties through commercial agreements such as a resale or a lease. Besides, in order to make new spectrum available, move an existing technology to a different band or change the technology currently assigned to a certain band, the regulator may decide to perform a reallocation of the spectrum assigned to communication systems

In this case, the company is the subscriber who pays the bill and may dictate a policy regarding the potential reconfiguration of its terminals. Therefore, user and subscriber are separated to cover the maximum number of cases and not to limit the scope of the future business models.

Network Operator and Service Provider

The same situation happens with Network Operators (NOs) and Service Providers (SPs). NOs will continue as they traditionally have but the possibility also exists of an SP who does not own a network infrastructure and just signs agreements with one or more operators in order to serve its users. Therefore, they are considered as separate actors. This is strengthened by the fact that in the traditional model, the SP had been purely a reseller of airtime but now the role of the SP is somewhat extended. In this case, not only airtime is provided but also mobile multimedia applications, content, and other value-added services in which the SP's brand is projected potentially in competition with that of the NO. Potential brand erosion and loss of revenue could be key concerns for the Service Provider Network Operator.

Global Roaming

In order to achieve a global roaming, the potential provision of a Pilot Channel has been the subject of much discussion in the reconfigurability arena. There are some practical concerns as to how truly reconfigurable equipment would be able to communicate with the network environment on 'switch on' or after, for example, a battery failure in order to identify which network technologies are present and receive a download of the appropriate Radio Software for reconfiguration.

Various approaches have been suggested:

- The terminal could always have a default load stored in memory to which it can revert, which may vary according to the main market in which the terminal was originally sold.
- The user could be able to take the terminal to a kiosk where an appropriate load of software may be delivered by a de-facto standard radio interface such as Bluetooth.
- A Global Pilot Channel Infrastructure could be provided to which handsets revert on switch on to receive network information and provide software download (SD) information. The implementation of such an infrastructure would be a significant undertaking and could be provided by a third party either on behalf of the Government/EU or directly or indirectly by a conglomerate of the interested operators. If such a step were mandated by say, the EU, it could seriously undermine the Business Case for SDR. However, even though an Operator solution is the most likely, the Pilot Channel Provider should not be ruled out of the list of Actors.

8.2.2 Scenarios Evaluation

The aforementioned three categories of scenarios represent a significant step forward in terms of identifying where reconfigurability can play a major role in the delivery of services to the user and also in the optimisation of the network resources to achieve the best results.

In addition, this analysis has highlighted potential new actors in the Communications Business Model such as the Software provider and also raises the issue of identifying a common method to download the software over a reconfigurable device. An Over-the-air (OTA) approach, totally transparent to the user, could require a Pilot Channel available on a selected group of frequencies common in all the countries of the world. Otherwise, a kiosk model may be possible where SDs can be obtained from a kiosk, for example, at international airports, train stations, hotels, and so on. A mixed approach could be identified in the following way: a smart card or the terminal itself has in the memory several systems/standards and the activation can be generated via network input or via auto detection.

Less ambitious scenarios can also be considered. Instead of downloading a whole, new radio interface, the network operator or the user can simply upgrade the current radio interface in order to fix a bug or implement better algorithms in order to improve the network capacity. In some devices, it is not possible to reconfigure the lower layers, however, new algorithms to improve the higher layers (cell selection) can be downloaded. It is the network operator's responsibility to choose a mass software upgrade or simply to notify the users who can choose whether to download these new functionalities or not.

8.2.3 System Requirements

Requirements should describe desired characteristics of the specified system functionality. It illustrates a characteristic of the problem not a solution. Solutions are studied during the design phase. The following are the major qualities of a requirement:

- *Simple*: A requirement must have an elementary structure and state a basic need. It cannot, at a given level, be broken down into several simpler requirements. At system level, a simple requirement can be broken down into several requirements at the Component level.
- *Concise*: A requirement must be expressed in a brief and clear manner. It contains neither an explanation nor a justification.
- *Unambiguous*: A requirement must have only one possible interpretation. A requirement is said to be ambiguous if it can be semantically interpreted in several ways, implying an uncertainty with regard to the design to be elaborated.
- *Verifiable*: As much as possible, it should be possible to verify requirements: if a test cannot be associated, the validity of that requirement should be further investigated; requirements should be testable either by Inspection or Demonstration or Analysis or Test points; requirements can be refined in various (testable) sub-requirements to be testable; an effective finite procedure must make it possible to check that the system complies with the requirement.
- *Feasible*: It should be considered whether a realistic and satisfactory technical solution can be implemented or can be elaborated in the timeframe of the project.
- *Non-redundant*: A requirement must have no overlap with another requirement. In case of redundancy, requirements must be divided and refined until suppression of the overlap.
- *Non-incompatible*: No conflict with another requirement (see preceding text).
- *Classifiable*: It can be associated with an attribute belonging to a previously adopted classification system.
- *Necessary*: A requirement reflects a need.
- *Traceable*: Identifiable by a unique identifier.

8.2.3.1 Levels of Priority

Three levels of priority are defined for the requirements: mandatory, recommended, or optional. These levels are indicated by the key words SHALL, SHOULD and MAY respectively. The meaning of the key words is taken from the RFC 2119 and quoted in the following text:

- SHALL means 'that the definition is an absolute requirement of the specification'.
- SHOULD means 'that there may exist valid reasons in particular circumstances to ignore a particular item, but the full implications must be understood and carefully weighed before choosing a different course'.
- MAY means 'that an item is truly optional. One vendor may choose to include the item because a particular marketplace requires it or because the vendor feels that it enhances the product while another vendor may omit the same item'.

Table 8.1 groups the requirements identified in a list of capabilities and provides a definition of each capability.

Each of the capabilities raises sets of requirements and recommendations the E2R project developed as a complete set, see [18].

8.2.4 Roadmaps for Reconfigurability

Reconfigurability will only show its full potential if it can complement the existing legacy technologies and networks. To find the optimum way to integrate reconfigurable technologies into these systems a set of roadmaps tackling various, even nontechnical aspects are required.

These include roadmaps for:

- business models;
- regulatory models;
- technology development and availability.

8.2.4.1 Roadmap Business Models in End-to-end Reconfigurable System

The scope of this section is to expose Business Models regarding the technological roadmap and to propose a Business path coherent with the technology evolution, which will support the evolution towards reconfigurability for each identified actor in a value chain. In addition, in this section we attempt to trace a reasonable roadmap for the deployment of reconfiguration applications into equipment and these extensions into networks. Such a roadmap may serve as a guide in anticipating evolution, identifying important problem areas and if possible contribute to appropriate organisations (e.g. standard bodies, international fora) in meaningful topics. A roadmap is needed in order to put technology into perspective and for handling the complexity of difficult problems by defining an evolution path for incremental technology introduction.

Reconfigurability opens the possibility for third-party software vendors to provide high-level as well as low-level system software. It also allows different actors to trigger changes, like upgrades, to the HW/SW combination of the equipment, even after the equipment has entered the market. In such scenarios, the assignment of responsibility becomes quite crucial.

The settings and software combinations of equipment, already for non-reconfigurable technologies, are rather complex. The manufacturer installs the firmware, Operating System (OS), and basic applications while the operator may include some tailored platform software and applications. All of these installations may be correct or they may have bugs that potentially require patching. While this, in recent terminals can be done, to a certain extent rather easily, such patching will become rather problematic when system configuration software may be procured and installed even from/by third parties.

Much of the flexibility and the value that is added through reconfigurability are based on SD and controlled installation/activation. However, this relies on sufficiently secure mechanisms for download and trust into origin, download path, suitability and authenticity of the software.

For operators there are two main problem areas. In case reconfigurations cause any problems, the operator will be the main point of contact (and blame) for the user, thus failed reconfigurations can potentially harm the operator's reputation. The second problem is in the efficiency of the use of an operators' spectrum. Reconfigurations may lead to inefficiencies or misuses and consequently result in revenue loss. There may be many other potential problems, yet the shared theme of all problems identified is the need for a common scheme to assign the responsibilities for reconfiguration.

As aforementioned, there are many actors involved in reconfiguration procedures, and their interests and dealings may be rather complex; these actors, their tasks, and their relations need to be identified and the roles they play in reconfiguration processes needs to be evaluated.

A distinction of two dimensions in which actors may operate can be made: the first being the operational and the second the administrative dimension. We could identify fifteen actors for end-to-end reconfigurable systems: user, subscriber, network operator, equipment manufacturer, (value-added) service provider, content provider, software provider, service aggregator, regulator, reconfigurable equipment, reconfiguration manager, certification entity, security entity, pilot channel provider, and spectrum manager.

Focusing on some of these actors, their roles in the operational dimension include the following:

Equipment Manufacturer
Provides the reconfigurable platform, firmware and software updates/new versions.

Network Operator
Owns the spectrum as well as the infrastructure, can also act as service provider.

Software Provider
Third party providing application software, but also low-level configuration relevant software.

Service Provider
Provides the required/requested services, this may also imply the possibility that an end-user may act as service provider.

Reconfiguration Support Service Provider (e.g. Reconfiguration Manager)
Provides the basic services necessary for reconfiguration, including, for example, secure SD.

User/Subscriber
Uses the equipment and infrastructure; may request installation of new configuration of application software.

While in the administrative dimension, the same actors, see Figure 8.2 may assume different roles:

Regulator
Sets the framework for the use of reconfigurable equipment, allocates the spectrum to lease holders and governs (using policies) the usage of the spectrum and the circulation of reconfigurable equipment.

Reconfiguration Controller (e.g. Certification Entity, Security Entity, Spectrum Manager)
Verifies that intended reconfigurations will comply with the given standard or that the equipment is prevented from implementing an intended configuration. This controller also implements functions like spectrum management according to given policies and certifies the intended configurations of the reconfigurable equipment.

Equipment Manufacturer
Arranges and initiates (performs) software (firmware) updates and patch installation.

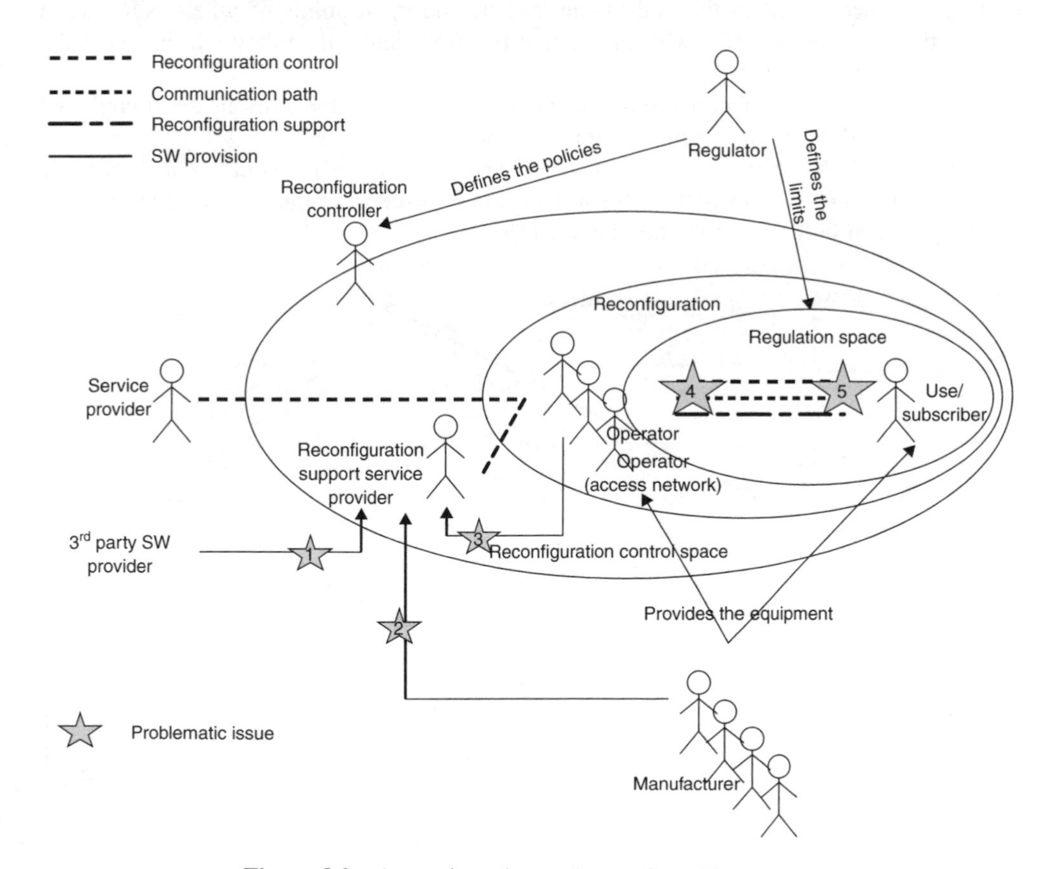

Figure 8.2 Actors in end-to-end reconfigurable systems

Software Provider
Provides third-party system, protocol and application software.

Service Provider
May request the reconfiguration of equipments to enable the provision of its services.

Reconfiguration Support Service Provider (e.g. Reconfiguration Manager)
Provides the control and security features for the reconfiguration procedure, independent of who may have initiated the reconfiguration process.

Network Operator
Provides the radio resources, mobility management (MM) and fixed capabilities to switch, route, and handle the traffic associated with the services offered to users.

User/Subscriber
May initiate, allow or decline a reconfiguration.

8.2.4.2 The Responsibility Chain Concept

This subsection looks into a complete end-to-end reconfigurable system, focusing on the administrative roles of the various actors involved (see Figure 8.3). The figure 8.3 outlines, in the context of the end-to-end system, the main points of where actors (and their activities) may interfere with the system functions and where they will have to take responsibility for the system state.

There are a number of rather sensitive areas (indicated by the stars in the figure) that may be affected during a reconfiguration procedure.

Issue 1 highlights the question of the actor who takes the responsibility for third-party software and who vouches that such software can be used to implement a radio protocol on the platform built by a specific manufacturer.

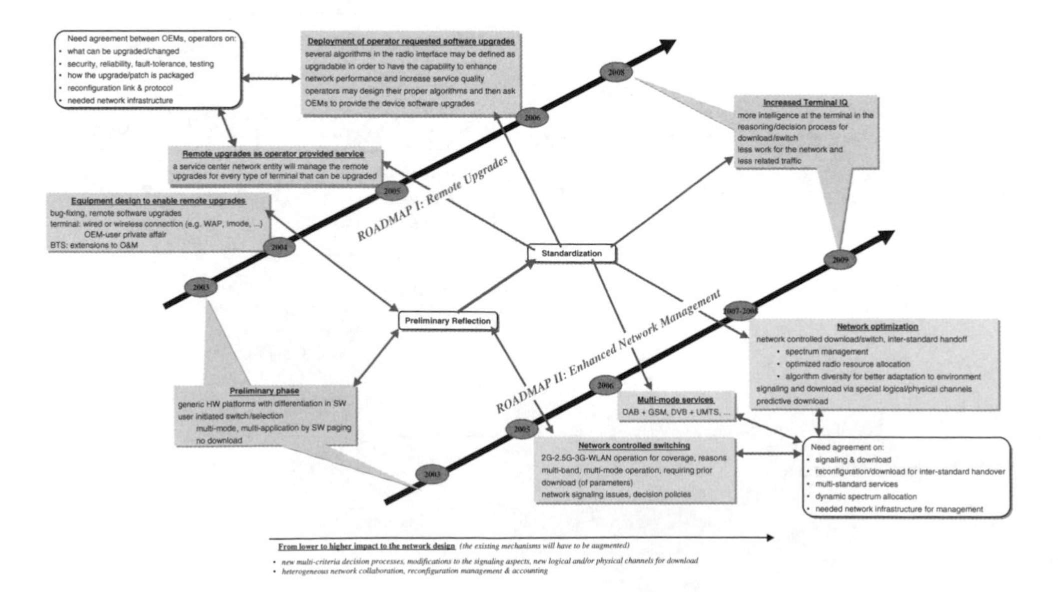

Figure 8.3 Deployment roadmaps for incremental introduction of reconfigurability applications

Issues 2 and 3 tackle the same situation but in these cases the software would be provided by the equipment manufacturer or operator, respectively, and the configurations would be used in a different administrative domain.

Issue 4 tackles the matter about permitting (reconfigured) terminals to access/use an operator's RAT.

Issue 5 deals with the problem of who can (and will) take the responsibility if a terminal is being reconfigured.

Issues 4 and 5 include the prevention of misuse of spectrum (e.g. in the Cognitive Radio Approach, when a user does not release the spectrum) as well as the spectrum control.

To approach these problems, the relationships between the actors in an end-to-end reconfigurable environment have to be defined and established.

The responsibility chain concept provides an initial overview of the different responsibilities and aims to do this definition of these relationships. The chain also needs to be connected to the value chain of mobile telecoms, with the aim of outlining possible sanctions if the assigned responsibilities are violated. The responsibility chain defines a model where the accountability for reconfigurations can be assigned to the different actors within end-to-end reconfigurable systems. Connected to the concept of value chain in the definition of the business models for end-to-end reconfigurable systems, the responsibility chain will need to identify the dynamic interactions between actors encompassing information data, control data, and money flow and will need to define a penalty scheme to penalise violations or infringements with actors rights as result of reconfiguration procedures.

8.2.4.3 Reconfigurability Roadmaps

The roadmaps treat two popular use-case scenarios. First comes the scenario for OTA software upgrades and second comes the scenario for dynamic radio mode and standard switching. These two scenarios correspond to applications of reconfiguration that appeal to operators according to previous market studies on SDR. Figure 8.3 graphically depicts the deployment roadmaps separately for each scenario.

Over the past ten years, previous work on SDR has set a solid foundation for the future evolutions represented by these two roadmaps. Technical issues have been investigated in depth by the SDR Forum. Software Defined Radio Forum (SDRF) provided feedback to 3GPP in order for the Mobile Execution Environment (MExE) to make provision for the SD required by reconfigurable radios. In this spirit, the introductory phase indicated in the roadmaps is based on this previous work. Further work in state-of-the-art survey will help to better identify the current status.

The introductory phase consists in designing the devices in such a way that part or all of their radio functionality is designed to be modifiable by means of software change. This requires, on the one hand, using generic hardware platforms and increasingly software implementations and, on the other hand, a software architecture permitting to completely or partially modify its function statically or dynamically. In addition, at the network side there is provision in the standards (e.g. 3GPP TS 22.129) for 2G to 3G seamless handoff (i.e. GSM to UMTS). This currently works under the assumption that standard switching does not need download and that core networks are compatible.

After this introductory phase, the technical focus will shift from the radio implementation to the reconfiguration aspects of SDR as well as the network involvement in the reconfiguration process. To handle these aspects, previous research work, like for instance results of EU projects, provides a good starting point. Within previous studies in the fifth Information Society Technologies (IST) FW programme, dedicated studies have been performed on downloading and SDR mechanisms for dynamic installation and testing on new SW modules in a terminal. Several projects (TRUST, CAST, PASTORAL) contributed to the elaborate proposal for the implementation of a reconfiguration control unit able to access and reconfigure SW modules at each layer of the system (Application, Protocol, Physical Layer (PHY)). These results allow to anticipate that the concept can be implemented into wireless/mobile terminals (MTs); the main limitations are linked to cost, consumption, and targeted features that have to be defined for these future systems.

Past experience shows that technologies evolve from simple towards more complex applications and on a need basis. Thanks to its simplicity, the scenario on software upgrades for bug-fixing and for performance enhancement as well as algorithm dynamic change (i.e. algorithm diversity) within a single mode of operation could be deployed first. Next, it will become simple robust schemes (e.g. based on parameter-controlled reconfiguration for multimode/multiservice operation without or with minimal network implication). These schemes will eventually permit download and reconfiguration signalling through logical/physical channels existing within the mode of operation (e.g. GSM logical channels or GSM-based wireless internet links). Alternative uplink air interfaces could be used whenever the mode of operation disposes of only a downlink (e.g. Digital Audio Broadcast (DAB)/DVB). During this period device reconfiguration mechanisms and designs will mature, a high degree of reliability of the reconfiguration processes will be attained and regulation issues will be clearer. At the same time, moving towards 4G will advance work on network interoperability and network management unification. This fact will push forward software radio applications requiring more network involvement and cooperation. An all-IP approach will certainly facilitate resolving these interoperability issues. Such an application may target things like the following:

- dynamic spectrum and network resource management;
- more intelligent air-interface selection for ‘best’ communication and service integration;
- flexible service discovery and provision;
- reconfigurable applications to support context-aware network wide reconfigurability;
- reconfigurable charging and security schemes.

Finally, progress in the domain of identification algorithms will greatly contribute in making the reconfigurable radio devices increasingly independent.

In all cases there is a need for standardisation. Standards will define the required device capabilities, the needed network infrastructure support, as well as the communication links and protocols needed for signalling and data transfers. As a conclusion it can be said that the way reconfiguration capabilities will be deployed in the future is not yet completely known.

The above roadmaps are only a plausible work hypothesis and have to be taken as such. Other interesting use-case scenarios most probably will have to be considered as well. Commercial applications will initially consider existing air interfaces. This is because work on future air interfaces (e.g. 4G) is still ongoing. However, a reconfigurable SDR approach offers the benefit of making possible future transition with a low impact on existing infrastructure.

As the deployment will be incremental; in between deployment phases experimentation is certainly needed. System prototyping will be a valuable approach in order to concretely demonstrate solutions on a smaller scale before their application on a larger scale. Such experimentation and prototyping could be part of R&D projects investigating future telecom systems.

8.2.5 Summary

Recapitulating, this section started with a presentation of some typical scenarios envisaged to occur in a reconfigurability context, outlining specific facets of reconfigurability. According to these scenarios, the basic requirements were extracted, with respect to a whole system capable of supporting reconfigurability. In addition, the commercial success of reconfigurable systems and the respective roadmaps to this success were considered.

In conclusion, the remarkable increase in the utilisation of telecommunication services has been expressed through the continuous influx and use of revolutionary applications. This unstoppable evolution of telecommunications is expected to be facilitated by the key concept of next generation's wireless systems, namely, the reconfigurability concept.

The worldwide research frameworks currently consider the requirements that need to be met in order to enable terminals and network elements to adapt transparently, securely, and efficiently to specific conditions, for example, hot-spots. In addition, it is anticipated that reconfigurability will bring advantages for all the actors of the wireless world. Roaming capabilities and applications will be offered to users; NPs will acquire more options for achieving the required QoS and capacity levels through their infrastructure and for introducing value-added services more easily. In addition, manufacturers and SPs will benefit from the flexibility offered, in order to evolve their devices and services respectively.

Consequently, special attention must be placed upon reconfigurability, in order to provide the prerequisites for the commercial vitality of newly developed wireless infrastructures and to influence users positively in using innovative applications.

8.3 Element Management, Flexible Air Interfaces and SDR

One of the aims of this section is to present a concept for a management and control system that enables elements of cooperate networks to operate in an E2R context. The main idea of this concept is a clear separation of the management and the control functions. Hardware abstraction is a research topic widely discussed in the reconfigurability community (e.g. specific RFI in the SDR Forum). The section addresses the issue of hardware abstraction in an end-to-end reconfigurable device and presents a possible approach. Some design and integration challenges for reconfigurable systems are also highlighted. The section will conclude with an overview of a verification tool for 4G/B3G systems.

A high-level view of the management and control of equipment in an E2R context is depicted in Figure 8.4.

The proposed framework consists of two main modules:

- The Configuration Management Module (CMM), which is a functional entity within the equipment (terminal, Base Station (BS)/access point (AP), or network), that manages the reconfiguration processes according to, *specified semantic*, protocols and the

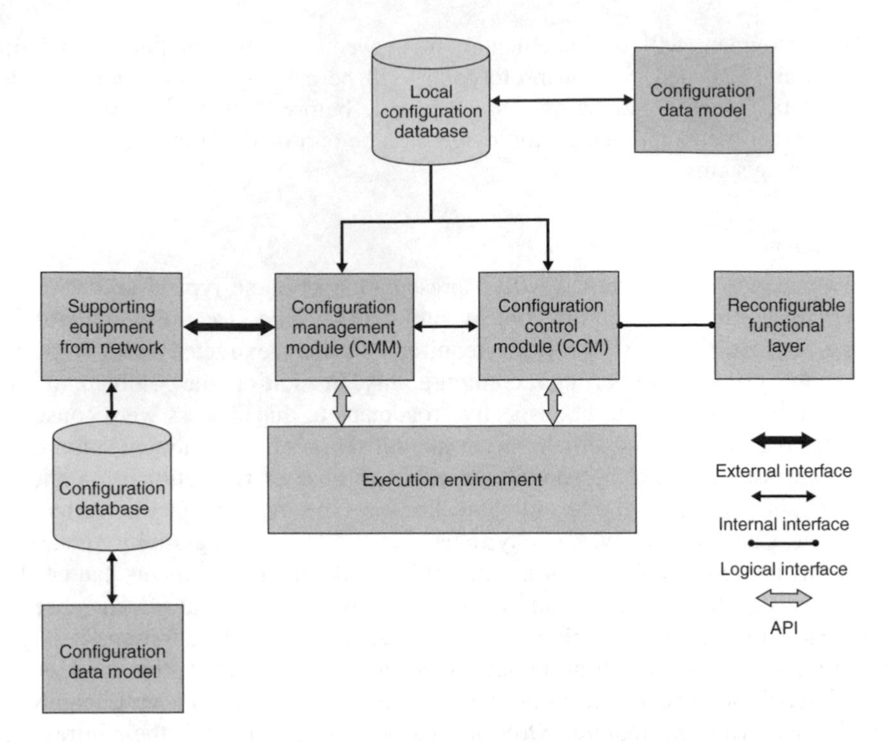

Figure 8.4 High-level view of the management and control of equipment in an E2R system

configuration data model (which may be stored in a distributed configuration database system). From the equipment perspective, the various CMMs also interact among themselves as well as with supporting equipment entities within the network, through an external (transparent) interface.

- The Configuration Control Module (CCM), which is a supporting entity responsible for the control and supervision of the reconfiguration execution. This is done using specific commands/triggers and functions of a given layer or a given execution environment. Three main layers are considered here: application, protocol stack (L2–L4), and modem (L1).

Other entities closely related to equipment management are the Execution Environment and the Reconfigurable Protocol Stack (RPS) framework (or Reconfigurable Functional Layers).

The Execution Environment is the means for providing the basic mechanisms required for dynamic, reliable, and secure change of equipment operation. The execution environment aims to offer a consistent interface to the equipment reconfiguration manager in order to apply the needed reconfiguration actions. For reconfigurable equipment, reconfigurable components need to be used. Such components are programmable processors, with reconfigurable logic, and parameterized Application-specific Integrated Circuits (ASICs) (offering software control on their parameters). The Execution Environment sits on top on this hardware platform and offers basic mechanisms enabling the exploitation of the reconfigurable hardware components.

The RPS Framework is an open protocol stack framework, which can be used to support several RATs with diversified protocols and protocol functions. This implies an architecture that supports dynamic insertion and configuration of different protocol modules in a common manner taking into account the resources and capabilities of the target devices.

The functional entities of the equipment management architecture and other internal and external entities will be described in detail in the following text.

8.3.1 Element Management

8.3.1.1 Reconfiguration Modules/Components

In the context of end-to-end reconfiguration the following scenarios have been identified:

- device management;
- multimode/multistandard;
- service adaptation;
- adaptation of device algorithms.

The following sections describe the entities (modules) of the Equipment Management required for supporting the above-mentioned scenarios.

Reconfigurable Protocol Stack Framework (RPS_FW)

In general, it is known that the wireless terminals and network equipments that are currently available in the market are complex systems that implement communication protocols, for each layer of the Open Systems Interconnection (OSI) reference model, using different platforms (i.e. Real-time Operating System (RTOS)). Moreover, it has been recognised that the OSI model enforces a strict partitioning between layers that is not always easily applicable, for instance, presentation services are required by lower layers to communicate [19]. Furthermore, most of these OSI protocols consist of a set of mandatory (core) and multiple options, with possible inconsistencies. In addition, the most critical core protocol functions have been identified, which can be used for Integrated Layer Processing architectures [20].

It is obvious from the preceding text that in most cases the same functionality is used from several layers. For this reason, it is proposed to build the common functions of each layer inside the equipment infrastructure (built-in blocks) using the host OS and the middleware of the platform. These built-in blocks and their implementation comprise the main idea of the RPS Framework (RPS_FW). However, in order to reuse and compose the built-in blocks for on-the-fly constructing and running of the stack, a software component-based architecture is recommended for developing the RPS_FW.

A software component (SWC) is a unit of composition with contractually specified interfaces and explicit context dependencies only. Stating the required interfaces and the acceptable execution environment specifies context dependencies. An SWC can be deployed independently and is subject to composition by third parties. Therefore, in addition, this RPS_FW can be extended with deploying, installing, and implementing the downloaded SWCs. These components (Resources), which are embedded inside the equipment, are used for extending protocol implementations and they finally support the particular communication system or protocol. The main goals of this framework, see

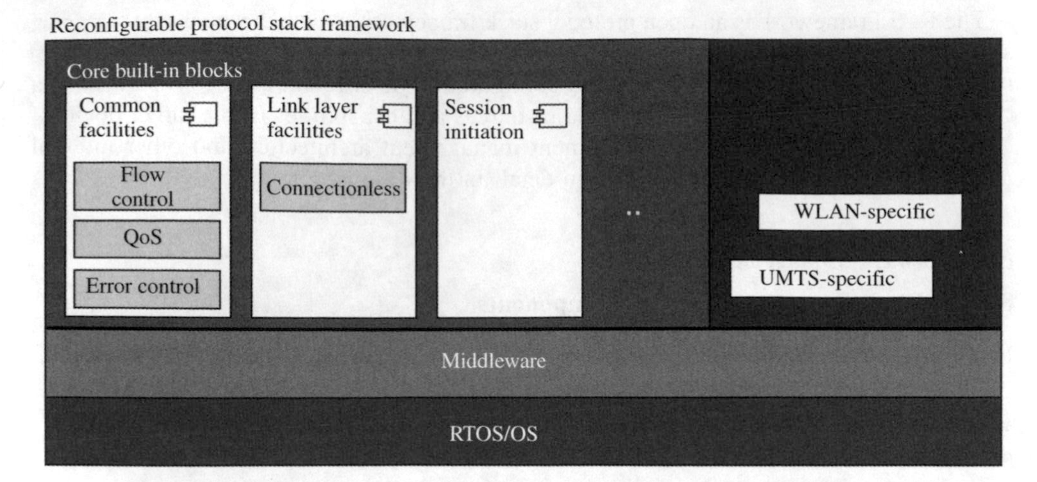

Figure 8.5 Reconfigurable protocol stack framework (RPS_FW)

Figure 8.5, are first, the ability to be extended or/and parameterized in order to support a particular functionality, and second, the ability to use shared resources.

There is tight relation between the execution environment and both the protocol stack level and the application level. The application level could be assumed to be the upper layers and specifically the layer to the user space. The following are the main requirements that the RPS_FW tries to satisfy:

- the system firmware that permits application and protocol entities to use the hardware resources (OS, Virtual Machines (VM), Middleware, low-level drivers, etc.);
- the RPS framework, which is the software that offers basic support for the implementation of flexible protocol stacks and provides the needed reconfiguration mechanisms;
- the protocol stack consisting of the logic that implements the protocol stack functionality at the different layers;
- the components and modules must be uniquely defined for each protocol function at each layer. For the construction of the reconfiguration environment, a flexible execution environment must be provided, which includes SW load modules, real-time schedules, communication maps, and interfaces between SW and HW modules;
- the operation environment that supports the SW downloading procedure. Software is formed with a component-based fashion [21].

Configuration Management Module (CMM) Entities

The CMM is responsible for managing the distributed controllers, which will initiate, coordinate, and perform the different reconfiguration functions such as monitoring and discovery, SD, mode selection and switching (multimode/multi-standard), and security. The CMM consists of the following functional entities:

- *'Interfaces with the Network Support Services' (CMM_IfNss)*: This functional entity is responsible for network-initiated reconfiguration and other services while the terminal is in on-line idle mode. This module will receive messages from the network, and it may activate other modules to start configuration or other actions in the terminal. The network-supported services can issue reconfiguration commands to the CMM. Moreover, supporting information can be exchanged between the network support services and the CMM, so as to consolidate on the best reconfiguration decisions.

- *'Monitoring and discovery' (CMM_MD)*: The role of this entity is to identify the available networks in a certain area and to monitor their status. It acquires information on the context in the environment of the device and takes into account data from multiple CCMs.
- *'Negotiation and Selection' (CMM_NS)*: This functionality is targeted for the negotiation of offers with the various available networks. It selects the most appropriate available network taking into account information such as the user and terminal profile and the offers negotiated with the networks. The user profile specifies services, QoS levels, cost levels, and so on. The terminal profile specifies the capabilities, configurations present at memory, and so on. The network offers specify the services offered, the QoS levels supported, cost information and so on. The goal of the negotiation functionality is the refinement of parameters (e.g. related to network offers). For this purpose, standard negotiation protocols need to be adopted and customised.
- *'Configuration downloads' (CMM_Dwnld)*: This functional entity provides the capability to perform downloads of the different components that may be required for the reconfiguration process. In other words, it undertakes the management of the downloading procedure. The downloaded information could be a whole protocol function in the form of components or a parameter of a component. The downloading procedure encompasses stand-alone or distributed mechanisms that are required in order to communicate with the software provider.
- *'Profiles' (CMM_Prof)*: This functional entity provides configuration profiles information on applications, user classes, equipment classes/capabilities, and configuration data models. The CMM_Prof is able to compare the profile of the current configuration and the proposed future configuration in order to recognise the absence of a particular function needed for activating the target communication system, protocol stack, or application for each level respectively.
- *'Security' (CMM_Sec)*: This module supports the security functions required during the reconfiguration process within the different layers.
- *'Decision-making and Policy (DMP) enforcement' (CMM_DMP)*: This entity communicates with the Reconfiguration Management Plane (RMP) entity. It interacts with the RMP to provide information and mechanisms for the decision of reconfiguration actions. It enables the provision of reconfiguration policies and actions throughout the network and locally in the network nodes and equipment. The interface between the RMP and the CMM_DMP mainly supports context and policy management procedures. Policy-based mechanisms and procedures are being implemented and performed by entities that are dedicated to mode selection and switching dovetail the mechanisms of RMP.
- *'Reconfiguration Installation' (CMM_Instl)*: This function is to provide, by interacting with the CMM_Dwnld and CMM_DMP functional entities the means for configuration representation and configuration deployment, which involves configuration download, validation, installation, and switching.

- *'Event handler' (CMM_Evnt)*: This entity enables the coordination of the different reconfiguration triggers, which activates scheduling and implementation procedures through the corresponding CCM according to the target reconfiguration. Events may be received from within the terminal (e.g. discovering new network: CMM_MD, or completing configuration SD: CMM_Dwnld), or externally (e.g. receiving a message via CMM_IfNss).

Configuration Control Module (CCM) Entities

The CCM initiates, coordinates, and performs the different reconfiguration functions. The CCM consists of the following functional entities:

- CCM – Application Layer (CCM_AP) that provides the interface between the CMM and the application layer.
- CCM – Protocol Stack layer (CCM_PS) that provides the interface between the CMM and the protocol stack of a protocol suite like Transport Control Protocol (TCP)/Internet Protocol (IP) (e.g. layer 3/4, namely, TCP/IP) or/and of a communication standard (e.g. layer 2/3 of UMTS/WLAN standard). This entity enables the addition of a complete new protocol function to the equipment and the parallel operation of this new protocol function with the existing functions.
- CCM – Reconfigurable Modem (CCM_RM) that provides the interface between the CMM and the PHY resources.

CCM_AP Functionality

This entity performs and activates the reconfiguration in the application layer. At the application layer, communication partners are identified, QoS is identified, user authentication and privacy are considered, and any constraints on data syntax are identified. Everything at this layer is application-specific.

Because of its nature, reconfiguration must be applied with software orientation. This orientation will enable the application layer to trigger reconfiguration to the above laying applications. Such an example could be the request from the top layer to an application in order to send data with lower bit-rate due to network congestion. The aforementioned scenario can be developed only with the addition of extra capabilities to the application layer, which is considered as the main layer inside the application level.

CCM_PS Functionality

This entity enables a complete new protocol function to be added to the equipment and operate with the existing protocol functions. For the incorporation procedure a deployment function is required, while for the instantiation a scheduling and context switching function are implemented. Consequently, the CCM for the protocol stack level consists of the following functions.

Deployment Function

This function is responsible for deploying the new components into the RPS_FW. The deployment function indicates the component's composition graph that is required for system stack construction and implementation. The deployment procedure could be triggered by the completion of the downloading procedure through the event handler. The deployment function includes some of the following procedures: configuration, releasing, installation, updating, and adapting of each component.

Scheduling Function

This function is related to the protocol stack since the actual and final outcome of this process must be the execution of the sequence task of a particular stack. The schedule consists of the ordering constraints and rules required for the perfect operation of the specific system. This procedure can be called a *communication bootstrap* for a particular protocol stack. The scheduling procedure could be triggered by the completion of the deployment procedure through the event handler.

Context Switching

In the RPS_FW, Context encompasses the operations of all layers that comprise the protocol stack. Here, context switching is considered as a function only between several protocol stack instances and not between protocol functions. The switching process could be performed using threads in the same process. Context switching to another thread in the same process is much cheaper than switching to a thread in another process. Threads can share resources easily because they live in the same address space.

CCM_RM Functionality

The CCM_RM receives the PHY-related configuration issues from the CMM. The PHY architecture has to provide the environment, which is needed to implement the functionality required by the CCM_RM.

Hardware Resources

At this level, the hardware resources appear as elements with different types of intelligence and programmability/configurability. Beside instruction programmable elements (General-purpose Processors (GPP), Digital Signal Processor (DSP)), configurable logic (Field Programmable Gate Arrays (FPGA)), special parameterized accelerators (e.g. ASICs), and also communication elements (switches, bus control logic, multi-port memory, etc.) belong to this category. The hardware components of the RF front end (oscillators, converters, filters, mixers, etc.) are not directly under the control of the CCM_RM even if they are adjustable by parameters by according interfaces (registers, I2C, SPI). In place of that, the RF front end behaves like one of the configurable elements listed in the preceding text.

The single elements have to be converted into configuration execution modules (CEM) implementing a certain level of configuration functionality, which fits to the resource interface provided by the CCM_RM. By configuration, such elements are combined into functional modules implementing a specific functionality (e.g. Down-conversion, Modulation, Decoding). This is also valid for the communication resources connecting the processing elements to guarantee the required functionality and maintain the required data throughput.

Operational Software

The operational software entity plays a substantial role in CCM_RM to manage different levels of abstraction, and the temporal scheduling of hardware and software resources that requires appropriate operational software support. As the CCM_RM is a logical element, it is a challenge for an operational software module (OSM) to manage all these processes, which are required to maintain the reconfigurability of the whole system and at the same time to guarantee the functionality of the underlying data processing system.

The execution platform of the OSM consists of the above-mentioned hardware resources, which are connected by a variety of communication elements. The OSM may incorporate an RTOS with some control interface to real-time mechanisms of other Programmable Processing Elements (PPEs) (Instruction Set Architecture (ISAs)) that requisites allocation and management of processing resources (e.g. execution time, priority, and scheduling policy) for tasks to be scheduled on them. Similarly, methods are provided for managing the associated memory resources.

Supporting of run-time reconfiguration is a critical function of the OSM. Loading and unloading of software modules on PPEs is also possible. The OSM incorporates appropriate loaders for the target PPEs while providing a common logical interface to the CCM_RM. Drivers and loaders to load and configure various reconfigurable and parameterizable hardware resources are implemented in a hardware abstraction layer. Other services for support of development and execution such as non-volatile storage, file system, logging service, diagnostics and so on may additionally be provided.

Security Architecture

Description of Entities

As shown in Figure 8.6 the security architecture consists of the following entities:

- 'Installation Manager' exists within the CMM and manages the sequencing of payload installation. It also records progress information as the installation proceeds to allow recovery or rollback after various installation errors.
- 'Recovery Manager' is invoked after an installation error has occurred, and its functions include determining the appropriate action to be taken to recover from an installation error, restoring the context of the install process at the point of the error, and instigating the recovery or rollback actions.

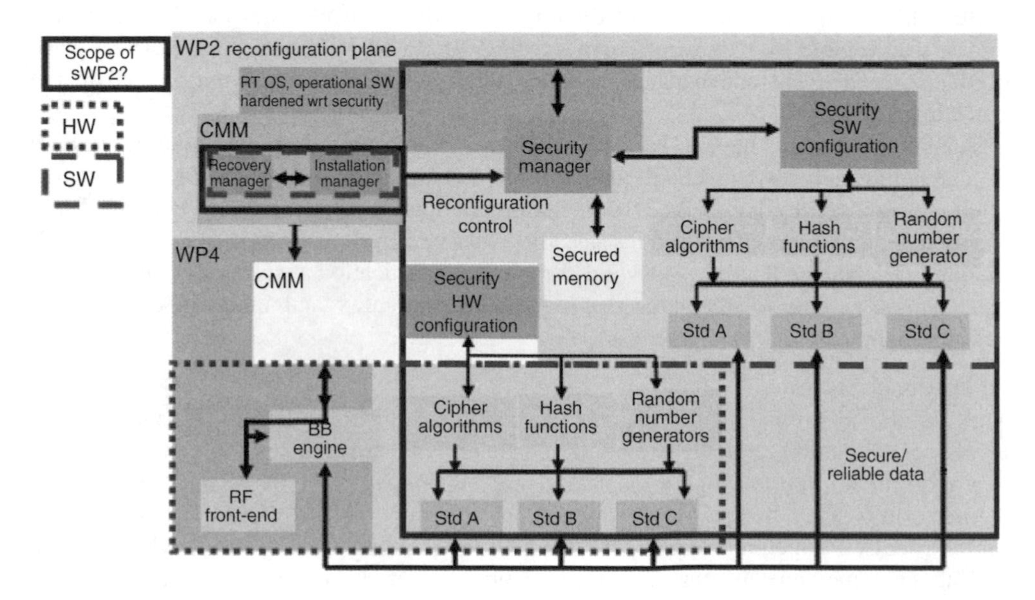

Figure 8.6 Security architecture

- 'Security Manager' communicates to the CMM (Installation recovery manager) regarding the correct data delivery, error correction, and recovery procedures.
- 'Security HW Configuration' may exist as a specialised entity in the CCM and handles all security-related hardware configuration.
- 'Security SW Configuration' is responsible for configuring algorithms, which exist in software.
- 'RTOS' will accommodate secure procedures to ensure integrity and specialised bootstrap methods.

The 'Security Manager' could also enforce flexible authorisation strategies over resources located within the terminal. The 'Security Manager' takes the security tasks from the CMM and connects to and is coordinated by the CMM. It interacts with, 'Security SW Configuration' and 'Secured Memory'.

A 'Security Manager' within the terminal provides a number of security-related tasks to ensure only authorised access to the reconfiguration management system. This Security Manager is defined as an independent process; it requires its own processor and memory space.

The Security Manager is responsible for safeguarding the access to the functions of the reconfiguration management part within the terminal and also for preventing attempts of fraudulent access from the outside world. Additionally, it provides the required security-related information to the network, stores the security information of the terminal (i.e. public keys and private keys), and the necessary external encryption keys. The Security Manager is responsible for establishing secure connections between the terminal and the network.

The Security Manager's internal architecture consists of an 'Access Manager Entity' (responsible for the establishment of secure connections between the terminal and the network and also for the processing of messages among the Configuration Manager, the network and other functional entities within the Security Manager). The Encryption and Decryption Factory (EDF) implements the security features presented by Sec_SW_CNF and encrypts both messages and reconfiguration software, before any transmission between the terminal and the network takes place. For messages and software transmissions originated within the network, the EDF decrypts the streams and passes the data to the CMM. The Security Manager in general performs the following functions:

- establishment of secure connections with the network;
- encryption and decryption of messages and data transfer;
- routing of reconfiguration messages and software to CMM;
- communication with Sec_SW_CNF for security algorithms reconfiguration;

All activities of the Security Manager are based on these basic tasks.

Security Domains

It is desirable that reconfiguration operations respect the contract between the user of the reconfigurable equipment and its operator. Depending on this contract, only a subset of reconfiguration capabilities may be granted to the user. Security mechanisms are also meant to ensure that only those reconfiguration operations that have been granted for a given user are authorised.

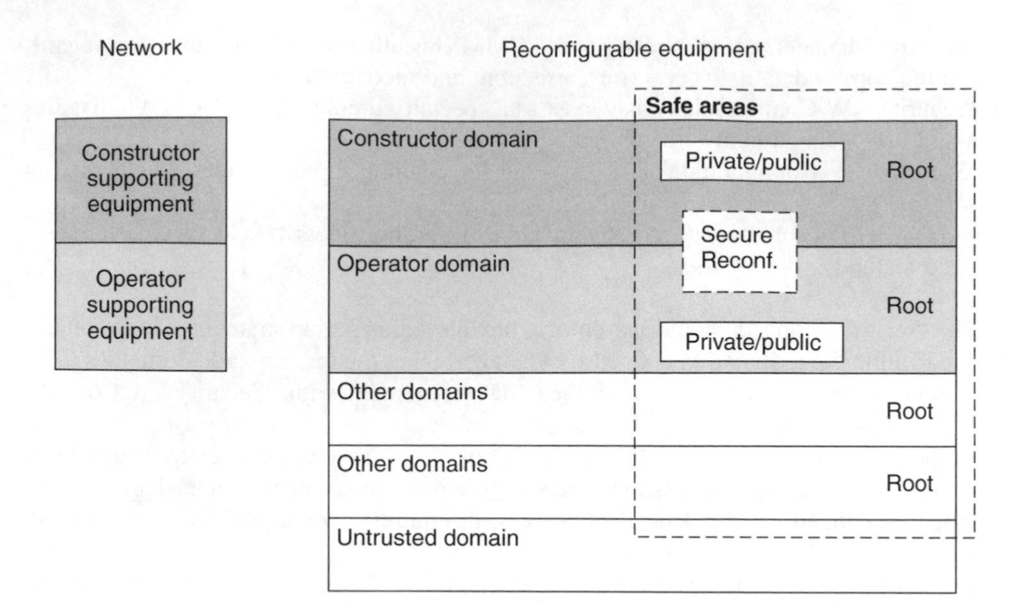

Figure 8.7 Security domains

For this purpose, it is necessary that at least two security domains are involved in reconfiguration operations: the security domain of the constructor of the reconfigurable equipment, and the one of the operator, which maintains the reconfigurable equipment (on behalf of the subscriber). Only the constructor is able to guarantee the sanity of most reconfiguration content. On the other hand, the operator should operate a fleet of devices in accordance to contracts with subscribers.

The security domains within the equipment are represented in Figure 8.7.

Each domain in the reconfigurable equipment contains a public key root that is used as the last authority of authentication chains to check signatures. Secure configuration mechanisms are assumed, which prevent any use of reconfiguration content if it is not checked against both constructor domain and operator domain.

Each domain in the reconfigurable equipment also owns a pair of private/public key on behalf of the device, which will be used for authentication of the device by the supporting equipments related to the corresponding domain.

Execution Environment Architecture

As discussed earlier, the execution environment introduces new dimension to the reconfigurable processes by supporting them for dynamic, reliable and secure exchange of their operations. It's positioning on the top of the hardware platform offering some basic mechanisms for managing and control of these processes. It can be considered as what operating system does in nowadays PCs. It will offer support to higher as well as lower layers of the system's reconfiguration.

The goal of the flexible protocol stacks is to provide an open protocol stack framework that is extensible and can be used for supporting different radio access network technologies with different protocols and protocol features. Therefore, the underlying execution environment (ExENV) is subject to several design constraints, described as follows:

- *Flexibility*: The ExENV needs to support multiple wireless standards, evolving standards, and new applications.
- *High performance*: Is required by process-intensive and latency-sensitive protocol operations and multimedia processing. For example, some real-time audio and video tasks require that the ExENV have sufficient performance to satisfy the demanding QoS constraints.
- *Power-efficient*: Means to utilise the limited amount of energy in the battery appropriately for the performance requirements. Dynamic power management comprises techniques that assign tasks to the most energy-efficient devices available and that force other unused system components into their power-down modes or even shut them off when appropriate.

A hybrid execution environment combines various hardware components (ExHW), for example, general-purpose processors (GPPs), field programmable gate arrays (FPGAs), and application-specific integrated circuits (ASICs) as well as various software environments (ExSW), for example, VM, Common Language Run-time (CLR), and OS, which can be distributed across multiple ExHW or confined to individual ExHW components.

GPPs execute applications in an ISA, which provide high flexibility. FPGAs use SRAM cells to control the functionality of logic and I/O blocks as well as routing, and can be reprogrammed in-circuit arbitrarily often by downloading bitstream of configuration data to the device, which can exploit the parallelism in algorithms better than GPPs. ASICs provide optimised solution in the power consumption, speed, and circuit area, but the development of ASICs is expensive in terms of time, manpower and cost. As a result, it is believed a heterogeneous organisation should produce an execution environment with the advantage of all the resources.

The execution environment hardware architecture is shown in Figure 8.8. In the architecture, every execution environment hardware module (ExHW) is connected to the Control Bus for exchanging control signals. The ExHW_LocalCtrl is located in each ExHW and schedules platform-independent module interaction. The bus arbitrator is to organise the traffic on the bus. The bus bridge module enables communication between share memory and the external execution environment, for example, data and commands can be passed via the bus bridge across distributed ExENVs. The ExESW can support interaction between software running on different ExHW by utilising the shared memory. For example, this could support protocol software module interaction, control command passing between CMM and CCM components or loading of configuration data or SWCs from storage (non-volatile shared memory).

An additional Data Bus can be used for simultaneous inter-module communications. The shared memory can be accessed by all ExHWs via the data bus. Therefore, an ExEHW can transfer data via the Data Bus to the share memory and concurrently the bus bridge transfers data to the external world.

Within the ExENV (which is likely to consist of a heterogeneous combination of ExHW and ExSW components), the software modules (which could be core software modules

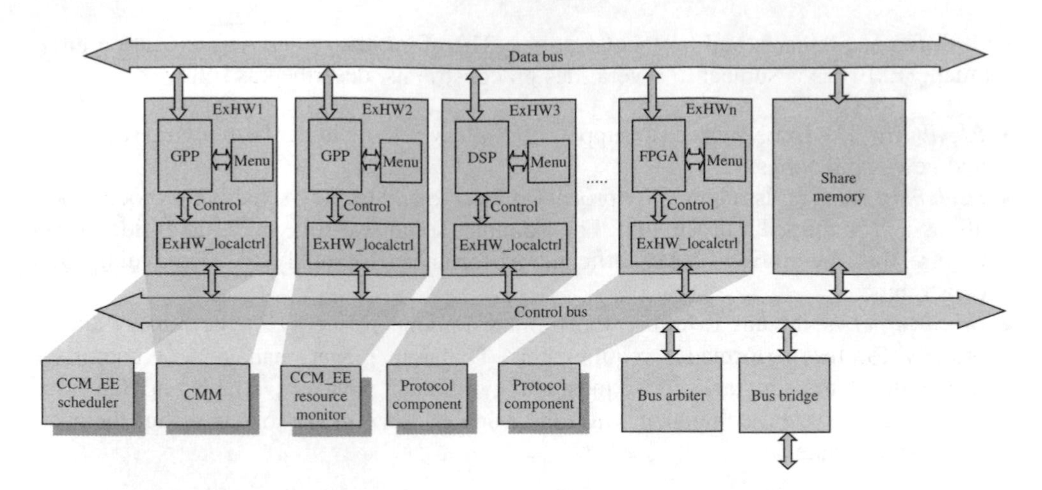

Figure 8.8 Execution environment architecture

such as CMM or CCMs or protocol stack modules) need to be mapped and run on ExSW and ExHWs and different instances of these modules can be initiated. A scheduler instance of CCM_EE schedules the reconfiguration procedure in the execution environment, for example, installation, deletion, execution, suspension, and so on. The preparation of the ExEnv includes both the preparation of the ExHW and the configuration of the necessary ExSW (OS, virtual machine, common language run-time or could even include the component support frameworks). There could be a Virtual Operating System (VOS) to abstract away from the underlying specific OS (or multiple OS in a heterogeneous ExEnv consisting of multiple ExHW), which may be COTS-based or even proprietary. The CCM_EE must also configure and control the mechanism for interaction between the ExHWs and ExSW, which may involve the use of specific Hardware Abstraction Layers (HAL) that are designed to permit ExEHW components from different vendors to be easily integrated into a heterogeneous ExEnv.

Another instance of CCM_EE is the Resource Monitor, which collects the ExHWs statistics data periodically, for example, memory, power, supply voltage, clock frequency, available FPGA area, Bus load, and so on. CMM makes the reconfiguration decision based on the monitoring data. The reconfiguration commands can be triggered by the CMM internally or by external triggers from the network interface.

8.3.1.2 Reconfiguration Procedures/Mechanisms

Internal Mechanisms

This subsection presents the need for terminal reconfiguration management, with reference to some internal procedures/mechanisms that correspond to certain functional entities of the CMM.

Extension in Protocol Functionality or/and Changes in Internal Protocol Parameters

TCP Adaptation

According to the nature of the HandOver(HO) (vertical or horizontal) and the new access network conditions, the TCP parameters that have to be modified may be different.

Potentially, all the possible TCP parameters should be modifiable:

- *srtt*: the smoothed round trip, indicating the mean time a segment lasts in the network,
- *ato*: the acknowledgement timeout,
- *rto*: the retransmission timeout,
- *cwnd*: the congestion window,
- *ssthresh*: the slow start threshold,
- *snd.wnd*: the advertised window size, that is the amount of data that can be sent,
- *rcv.wnd*: the received window size, that is the amount of data that can be received,
- *snd.nxt*: the next sequence number to be sent,
- *snd.una*: the first unacknowledged byte.

In the framework of end-to-end reconfigurability, the most common parameters to be dynamically adapted will probably be the rto, the cwnd, and the ssthresh. However, depending on the TCP stack implementation, some of the TCP parameters will be modified for the running sessions, that is the running sessions parameters will be updated in real time, and others will be modified only for the new sessions, that is only for the sessions launched after the HO completion.

Besides, in some cases, the CMM_NS may decide to give up the TCP adaptation, considering that the new TCP parameter values do not represent a significant change and will not impact the TCP sessions behaviour.

Note that this change in internal TCP parameters should be triggered by both technology and resource availability changes, thanks to a HO notification coming from a specific entity that is called here a *TMM* (*Terminal Mobility Manager*) in charge of managing the HOs.

Functional Description and Architecture of the Reconfiguration Procedure

This functional description contains the generic procedure of download, configuration, and implementation of a reconfiguration process. It defines the mechanism of delivering the reconfiguration context, storing, managing, and uploading the context to the HW platform. It also represents the implementation, alteration, and executing of the reconfiguration procedures and module control communication [22].

If a trigger for reconfiguration is passed to the CMM, the CMM will discover whether the software and policies required for an intended reconfiguration are available in the CMM_Dwnld and CMM_Prof, respectively. If policies and software are available, a further request will be dispatched to the CMM_DMP to initiate the generation of a new terminal profile. In response, the CMM_DMP will request the profiles and necessary software policies from the CMM_Prof. Upon receipt of these policies, the CMM_DMP generates the new terminal profile. The next step for the CMM_DMP is to announce the generation of the new terminal profile to the CMM and to upload it to the network for a Verification Configuration Procedure (VCP). The profile is sent to the network via the secure connection. After receiving the confirmation from the network about completion of a successful VCP, the CMM_DMP forwards a notification to the CMM that the reconfiguration procedure may proceed. The message issued by the CMM_DMP contains information about the required changes in the configuration. Using the information about required changes, the Configuration Manager generates the new CCM(s) and consequently the specified reconfigurable modules (i.e. installed and controlled by the CCM(s)). Upon completion of the Reconfigurable Module(s) installation, the CMM requests an update of the registration of the new terminal profile. Therefore, the CMM_DMP dispatches a registration request to the network. After completion of the (terminal configuration) registration, the network notifies the CMM_DMP, which then forwards the response to the CMM. At the end of any reconfiguration procedure, the CMM requests the Sec_Mng to terminate the secure connection between the terminal and the network.

Similar to the previous case, if a request for reconfiguration is submitted by the CCM and the software module to be installed is known, the CMM will issue a request whether the source code is available within the CMM_Dwnld. There are two possible answers from the CMM_Dwnld, either the software is available or it is not available. The previous sequence presented the case when the software is available, but in case the required software module is not available within the CMM_Dwnld storage, a download from the network resident software server will be required. Therefore, the CMM issues a request to the CMM_Dwnld to perform SD sequence. Furthermore, the CMM requests to the Sec_Mgr to establish a secure connection to the next available network server. After the connection is established and a confirmation is received by the CMM, the CMM passes the handle of the connection to CMM_Dwnld. The CMM_Dwnld then dispatches a SD request via secure connection, to the network. The request carries a number of parameters specifying the software type and location (i.e. URL, IP address, FTP site, etc.). After completion of the download negotiation, the actual download path leads from the software store to the CMM_Dwnld via the Sec_Mgr. Once the last packet is received from the network, the CMM_Dwnld notifies the Configuration Manager about accomplishment of the SD. A similar sequence is followed for the download of the policies, in which case however, the download will be initiated by the CMM_Prof. After completing the download of the policies, the CMM_Prof will also inform the CMM that all policies are available and the reconfiguration procedure may proceed. Once all policies and software modules required are available, the procedure follows the same sequence as the one described above.

This case describes one of the core parts of a reconfiguration procedure, with a particular focus on the mechanisms of the terminal profile handler and the CCM. The sequence depicted assumes that the preliminary signalling has been completed and the CMM now requires the actual generation of a new terminal profile and the implementation of the subsequent steps of the reconfiguration procedure. Configuration and software policies are then requested from the CMM_Prof and interpreted by the CMM_DMP, which in turn generates and compiles the source code for the new profile. In case the compilation failed, this process needs to be repeated or the reconfiguration procedure abandoned. After successful completion, the new profile then becomes uploaded to the network where its validity and correctness will be evaluated during VCP. Assuming this evaluation proves the validity of the file, the network issues the permission (to the terminal) to implement the new configuration. If the validation failed, the terminal profile generation process must either be repeated or the reconfiguration sequence abandoned.

During the validation process, the CMM_DMP still retains the old terminal profile and awaits the acknowledgement about the validation from the network. After that, the CMM_DMP's comparisons block extracts the differences between the old and the new tag-file and forwards it to the CMM together with a confirmation about the successfully performed (or, if necessary, failed) VCP. The next step in the reconfiguration procedure is then the implementation of the required CCMs (i.e. they are instantiated by the CCM).

The scenario, when configuration of the software radio platform takes place during system initialisation (i.e. boot-time), offers a slightly different case for the definition of the reconfiguration terminal. A boot sequence provides distinctive differences to the afore-described reconfiguration procedures. In the beginning, after switching on the terminal, the initialisation of the different reconfiguration management modules (i.e. their processes) is taking place. Each of these processes initialises its main thread and passes a handle to the CMM. After this initialisation procedure is finished, the CMM triggers a configuration sequence by dispatching requests to the CMM_Dwnld and the CMM_Prof to check the availability of software and policies, respectively. The procedure is required to ensure that the policies and the software are in the storage places (although in this case this is rather a formality, because when the terminal is in the process of turning off it stores its last state, including the policies and software in CMM_Prof and CMM_Dwnld, respectively), this however, is not the case when the terminal is in a restricted (manufactures) state (i.e. the manufacture may need to include the initial policies for the terminal to be able to start the very first time). The configuration procedure continues with the activation of the profile, which, in this particular case, does not require a VCP. The rational for omitting the VCP is that the intended configuration has already been approved by the network in the course of the previous reconfiguration procedure. However, this is not the case when the terminal changes the RAT, then the complete reconfiguration of the terminal is needed so as a follow-up the VCP will be required. The Configuration Manager then creates the required number of CCMs and starts their initialisation. Once a required CCM is ready and functional, it confirms its availability to the Configuration Manager, which then announces that the terminal is readily configured.

In this particular scenario, when the terminal boots up to a previously known configuration, there are some parts of the reconfiguration sequence that can be omitted (i.e. due to the availability of policies and software, and the already validated profile).

This reconfiguration procedure, compared with the previously described scenarios, relies on the condition that the terminal has its main modules already installed but there is no trace of any previous configuration (i.e. no validated profile). It may be assumed that this case will only occur in the manufacturer stage, when the required minimum radio modules are to be installed. For this reason, the first active reconfiguration procedure of the terminal has a minimum or no radio stack available. A possible way to install modules would be through wired connections and direct installation of the software modules and policies (in particular, the profile and software policies) in CMM_Dwnld and CMM_Prof, respectively. After this, the terminal needs to be reset and to reboot. Using the (then) already installed software and policies, the CMM will install the minimum possible terminal configuration, which suffices to provide a basic configuration of the reconfigurable part and enable future reconfigurations. The scenario starts the reconfiguration process with a mutual authentication procedure between the terminal Sec_Mng and the vendor's software server (the PC, sim card etc.). After the completion of the authentication, the CMM receives a request to install an initial set of configuration software. The CMM responds once the system is ready to pursue the procedure and to install the required software; the CMM also dispatches a notification to both CMM_Dwnld and CMM_Prof to inform them about the procedure. In addition to the notification, similar to the previously described download cases, the CMM starts to request the availability or download of software and policies from the CMM_Dwnld and CMM_Prof, respectively. These two procedures can be performed concurrently and may last until the required data is completely received and stored. The sequences may be repeated, like in the previous cases, until both entities have to respond to the CMM that software (policies) download is completed. The difference in the sequence is that the course of actions finishes as soon as the downloads are completed, that is, no reconfiguration procedure is required or followed. Then the reconfiguration can be performed following the boot procedure. Reconfigurable terminals require this initial installation of a radio configuration; otherwise the terminal will not be able to connect to an air interface.

Software Updates CMM Installation

After the downloading process, the installation process takes place (Figure 8.9). The deployment is related to the installation process since the downloaded code must be mounted in the RPS_FW in order for the instantiation process that follows the installation to complete in the appropriate manner. The installation information must be preserved inside a file including information about where the components must be located and what the interfaces between components are to allow linking with the existing components. For this reason, the downloaded code must reside in local memory, at least temporarily.

Scheduling of Protocol Stack Reconfiguration

Using the built-in components residing inside the RPS_FW, a communication bootstrap must be performed. The communication bootstrap is the procedure that uses the schedule of the protocol tasks for running the particular stack of the protocol suite. The stack is running since the components have been composed and instantiated, as in Figure 8.10.

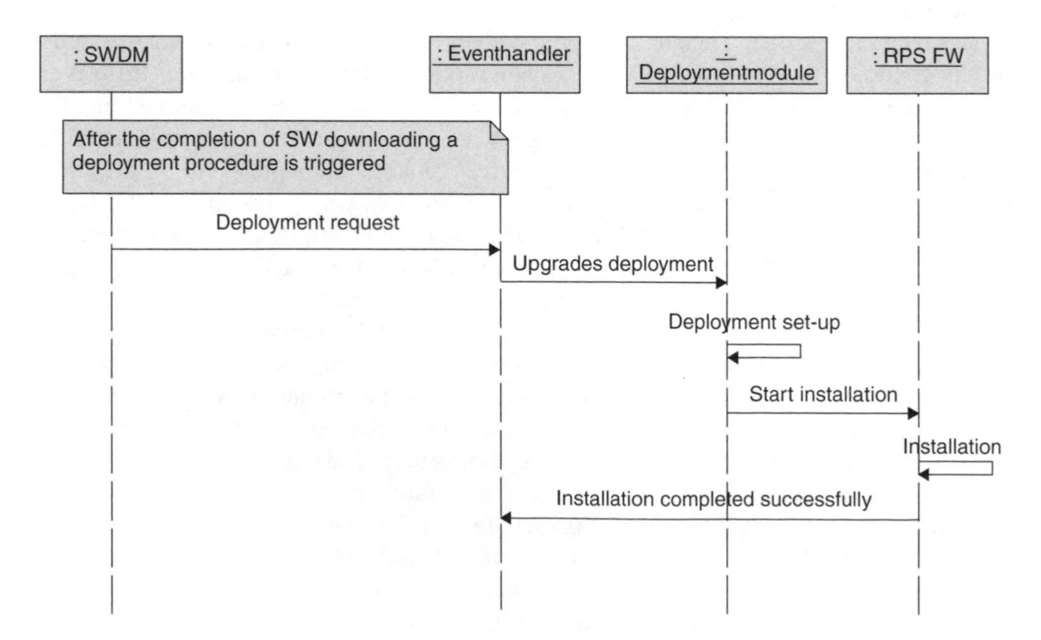

Figure 8.9 Software updates installation

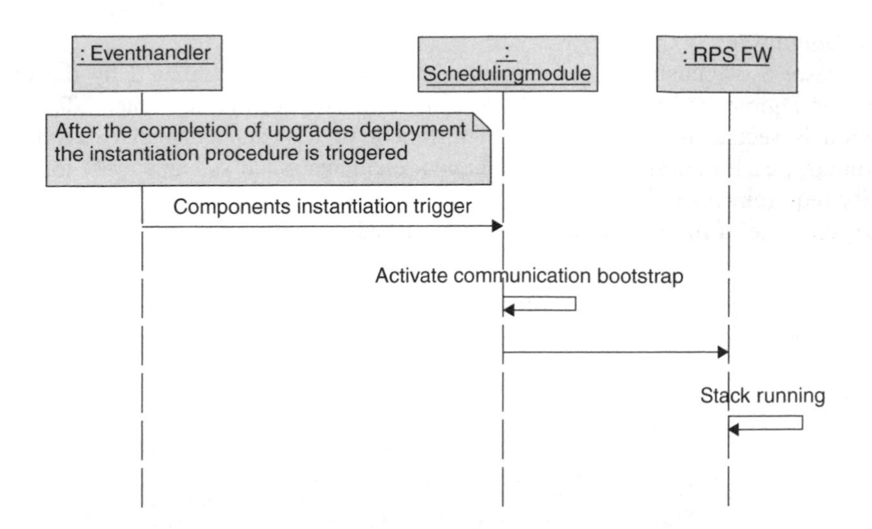

Figure 8.10 Running the schedule of protocol suite tasks

External Mechanisms

This subsection presents the need for terminal reconfiguration management, with reference to some external procedures/mechanisms that correspond to certain functional entities of the CMM.

Monitoring and Discovery

A terminal, in particular, should be constantly monitoring the environment. This procedure is an essential entity foreseen for the CMM, as in the terminal centric scenario, the need for reconfiguration is imposed by the terminal, and thus by the CMM of the reconfigurable terminal. This need would become apparent, when the terminal 'comes to realise' that its operation parameters could be better, if it would use an alternate RAT instead.

So, the purpose of the 'Monitoring and Discovery' procedure is the identification (within the monitoring range of the terminal) of an alternate RAT (other than the one operating so far), with better offers, in terms of better circumstances of coverage, QoS, and so on.

Therefore, the CMM should include a functional entity that implements this procedure. This entity (CMM_MD) should be capable of interacting with external to the whole management module entities, such as the network support functions, through the appropriate interfaces. Furthermore, it should also be capable of communicating and exchanging monitoring with the rest of the internal entities, through the CCMs.

Negotiation and Selection

After the discovery of alternative RAT choices, the terminal should be capable of interacting with these RATs, in order to negotiate in terms of quality and cost factors. The negotiation phase is of course followed by the selection phase, during which the terminal selects the best reconfiguration pattern.

The above necessitate the existence of a 'Negotiation and selection' functional entity within the CMM, in order to realise this procedure. The CMM_NS functional entity should interact with the network support functions, in order to exchange the necessary negotiation information, as well as with the rest of the, internal to the equipment, entities.

Secure Diagnostic

In this case, a diagnostic on the reconfigurable equipment is initiated by the operator supporting equipment, and driven by the constructor supporting equipment. Though this operation is secure, no preliminary establishment of a secure channel is assumed. On the contrary, each transaction includes its own elements (such as signatures) for insuring security requirements.

The sequence of message exchanges is described in Figure 8.11 below.

'G':

The operator supporting equipment finds out that some diagnostic driven by the constructor of the reconfigurable equipment should be performed. A list of possible diagnostics is

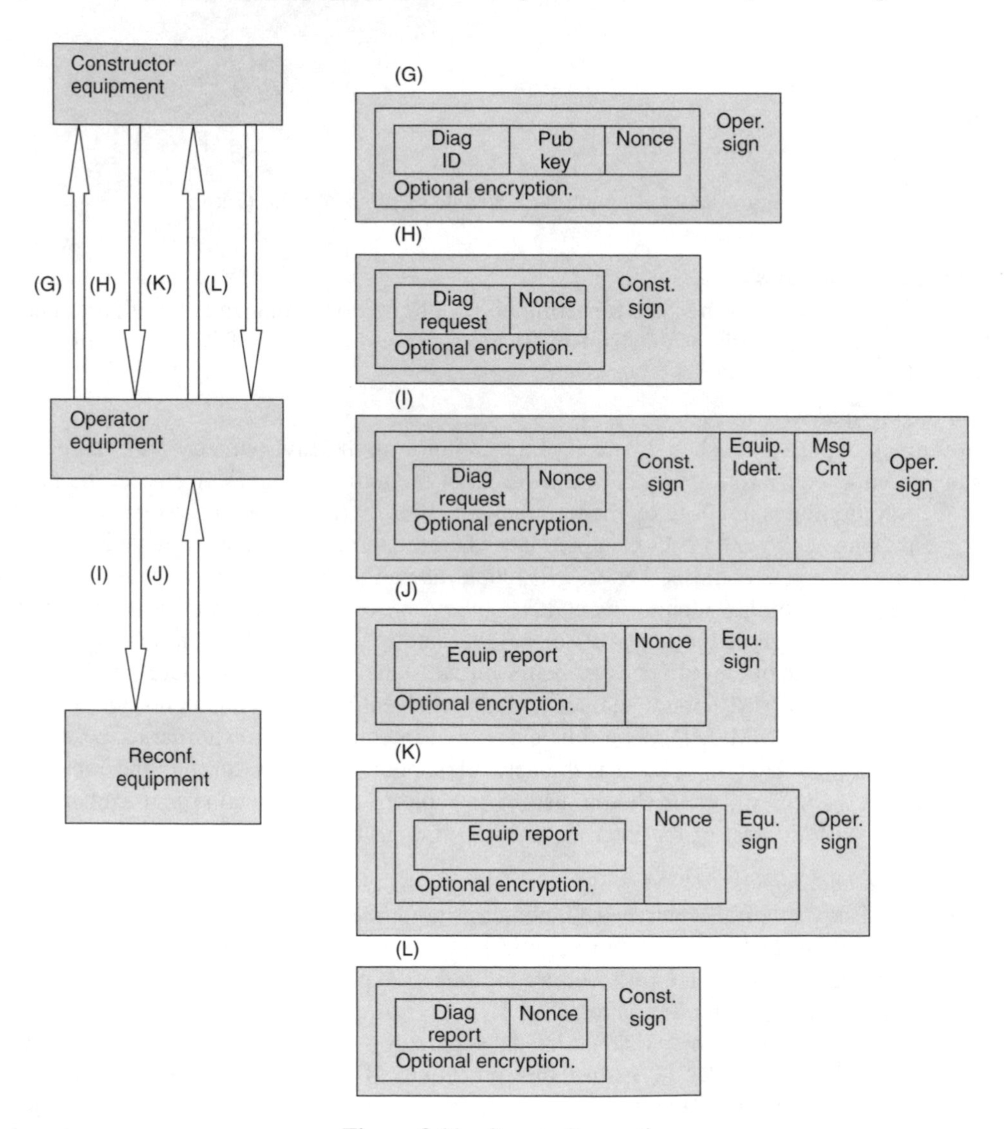

Figure 8.11 Secure diagnostic

already established between this operator and the constructor of the reconfigurable equipment. The operator supporting equipment builds a message containing a diagnostic ID, which is a pointer into this list.

It appends to this diagnostic ID the public key for this reconfigurable equipment, which pertains to its constructor domain. It also appends a nonce value (as an anti replay measure), and it may encrypt the result by using the public key of the constructor equipment. Finally, it signs the total by using its own private key.

'H':
Upon receiving the message 'G', the constructor supporting equipment checks the operator signature (against its own certificates and trusted entities). If necessary, it decrypts the message and retrieves its content. Then the constructor supporting equipment builds a detailed diagnostic request specifying to the recipient reconfigurable equipment which measurements and test it should perform. It appends the nonce as just received from the operator, and it signs the resulting message by using its own private key. The resulting message is forwarded to the operator equipment.

'I':
Upon receiving H, the operator equipment checks the constructor signature, and if OK, it appends a message counter and an equipment identifier, and then it signs the resulting message, in order to obtain message 'I' which is forwarded to the reconfigurable equipment.

'J':
Upon receiving message 'I', the reconfigurable equipment does the following:

- it checks the operator signature against its operator domain;
- (if OK) it checks that the received equipment identifier matches its own identifier;
- (if OK) it checks that the received message counter is greater than the value received on the last similar message;
- (if OK) it checks the constructor signature according to its own constructor domain;
- (if OK) it retrieves the detailed diagnostic request that was built by the constructor supporting equipment;
- it performs any measurements and test which are required, and it builds a corresponding report;
- optionally, it encrypts this report by using its private key pertaining to its constructor domain;
- it appends the received nonce;
- it signs the resulting message by using its private key pertaining to the constructor domain, so that it obtains message J which is forwarded to the operator supporting equipment.

'K':
Upon receiving message 'J', the operator equipment checks that the nonce matches its initial nonce selection, and if OK, it appends its signature in order to obtain message K. This message is forwarded to the constructor supporting equipment.

'L':
Once the constructor equipment receives message 'K', it performs the following operations:

- check of the operator signature;
- check of the reconfigurable equipment signature, with respect to the public key received in the initial message G;
- check that the nonce value matches what was received in message 'K';
- elaborate a final diagnostic report intended for the operator supporting equipment;
- add the nonce;
- optionally encrypt the report plus nonce;
- append its signature;
- forward the message to the operator supporting equipment.

Eventually, the operator supporting equipment receiving message 'L' will check constructor signature, nonce correspondence, and will read and exploit the final report.

Software Downloading
Terminal Initiated
- As in Figure 8.12, the Event Handler receives the download event from the corresponding CCM and then it sends a trigger for downloading to CMM_Dwnld.
- The CMM_DMP chooses the appropriate SW running using a set of decision algorithms and applies them on the information retrieved from the CMM_Prof and RCM.
- The Decision Component makes the decision on the SW Version (i.e. module, component) based on the list of software available and the current SW Profile.

Network-initiated
The diagram in Figure 8.13 depicts the network-initiated trigger for downloading reconfiguration procedure.

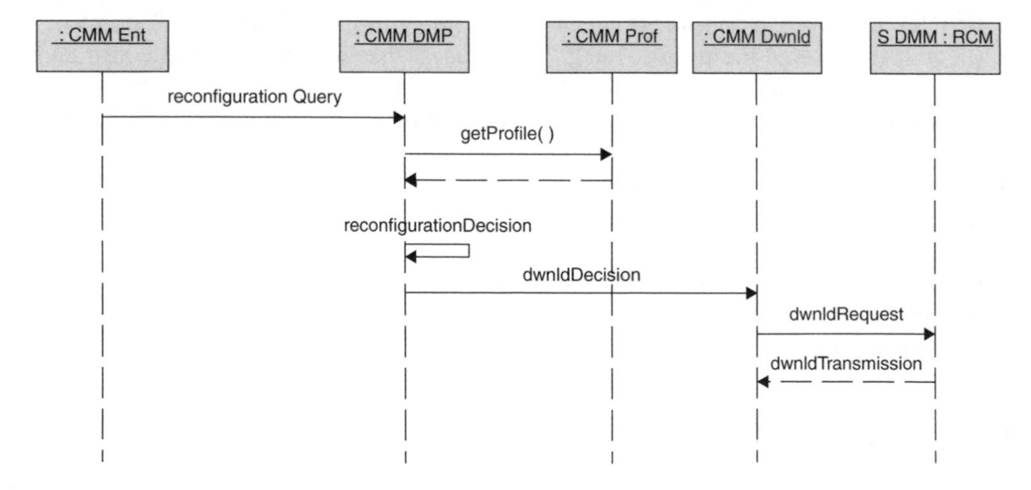

Figure 8.12 Terminal-initiated SW download management

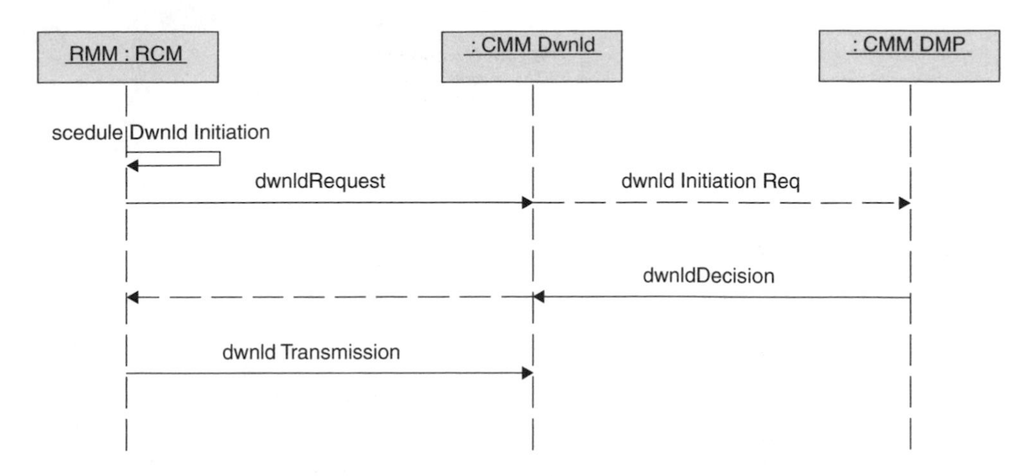

Figure 8.13 Network-initiated trigger for downloading reconfiguration procedure

Relationships between Entities (Modules/Components) and Procedures

Relations between Procedures and Modules

The entities depicted in Figure 8.4 are intended to support mainly the following procedures:

- Reconfiguration decision-selection;
- SW Downloading;
- SW Deployment;
- Tasks scheduling and instantiation.

These procedures have been described in the above subsection, however, in the following text, there is a more detailed description of the role of each module regarding the reconfiguration procedures.

The following sections describe the relationships between the various modules and the procedures of the multimode/standard switching scenarios. The first subsection gives an overview of the internal relations and the subsequent subsections provide sequence diagrams that illustrate the internal relationships as well as the external interactions in the respective scenarios.

Overview of Module Relationships

The multimode/standard switching distinguishes different entities inside the local CMM on the terminal:

- Monitoring and Discovery (CMM_MD);
- Negotiation and Selection (CMM_NS);
- Reconfiguration Implementation (CMM_DMP);
- Configuration Download (CMM_Dwnld);
- Profile information database (CMM_Prof).

Figure 8.14 depicts the relationships between the CMM modules and the different CCMs.

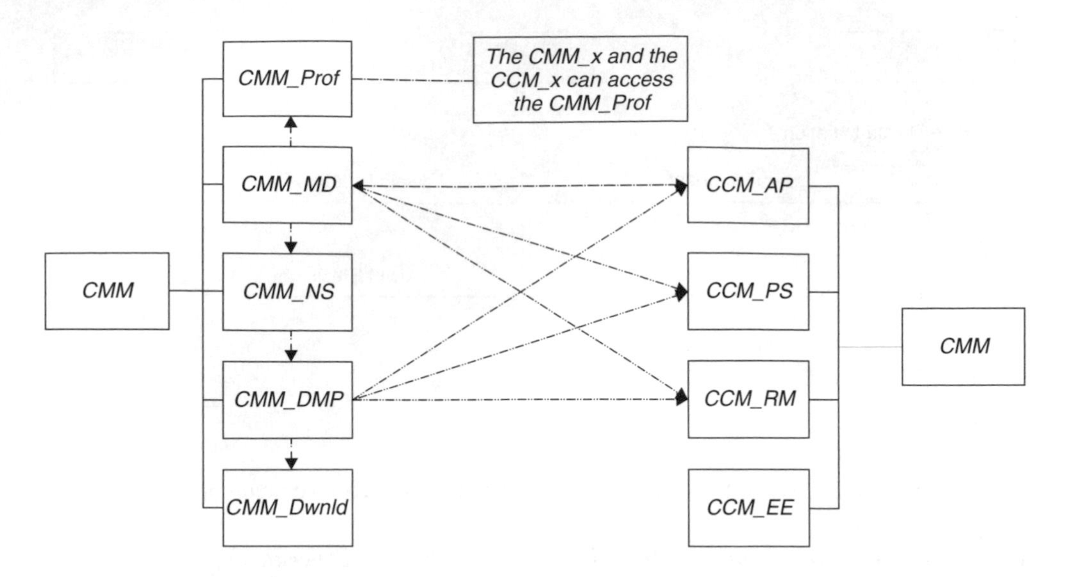

Figure 8.14 Overview on the internal relations between the modules

CMM_MD and CCM_AP

The application layer configuration controller (CCM_AP) informs the monitoring and discovery entity of the CMM (CMM_MD) about changes in the applications requirements and user preferences. Upon this information, the CMM_MD can search for a technology that is suitable to the changed requirements. A switch to the new RAT is made.

The CMM_MD on the other hand, provides the CCM_AP with information on the current technology capabilities, that is available RATs and their service levels.

CMM_MD and CCM_PS

The CMM_MD interacts with the protocol layer configuration controller (CCM_PS) to query information on the current protocol stack configuration and to change the protocol stack configuration.

CMM_MD and CCM_RM

The CMM_MD requests the modem reconfiguration controller (CCM_RM) to scan for additional access technologies and to monitor the available access technologies. The CCM_RM provides information on the available access technologies to the CMM_MD.

CMM_MD and CMM_NS

When the CMM_MD has received information on the currently available access technologies, it instructs the negotiation and selection entity of the CMM (CMM_NS) to negotiate and decide which technology shall be used in the future.

CMM_NS and CMM_DMP

After the CMM_NS has selected the RAT to which the terminal shall switch, it issues the implementation of this reconfiguration to the reconfiguration implementation entity of the CMM (CMM_DMP).

CMM_DMP and CCM_x
During the implementation of a new configuration, the CMM_DMP interacts with the different CCMs in order to reconfigure and install the software for the new configuration. Furthermore, the CMM_DMP is responsible for initiating the HO from one radio chain to another in terminals with multiple radio chains.

CMM_DMP and CMM_Dwnld
In case the necessary software modules for the reconfiguration are not available in the terminal, the CMM_DMP requests the download of these software modules at the CMM_Dwnld.

CMM_x, CCM_x, and CMM_Prof
All the CMM and CCM modules can request profile information from the CMM_Prof as well as they can store/change information in the CMM_Prof. There are number of support functions the different methods of the Management scheme have to implement:

Change of Profile Information
Changes of the user preferences, for example, the required service level, the connectivity, and the costs are passed from the CCM_AP to the CMM_MD. The CMM_MD stores the changes in the profile and decides whether these changes require a transfer of profile data to the network or not. Furthermore, it decides whether a change of the RAT should be considered. This would mean starting monitoring and discovery actions as well as initiating the negotiation and selection process.

Profile Transfer
A terminal and the network exchange profile information during initialisation and whenever a change in the shared parts of the profile occurs. The profile can be transferred on request or automatically in both directions when the sending side feels the transfer makes sense.

Monitoring and Discovery
During the monitoring and discovery process, single radio chain terminals suspend their current RAT temporarily. They inform the RCM on the network side when they start the monitoring and discovery activity and as soon as they have finished it. These notifications allow the RCM to react appropriately, for example, buffer data packets.

Negotiate and Select
When the CMM_MD considers that a change of the RAT might be useful, it triggers the CMM_NS to select an appropriate RAT. Therefore, the CMM_NS uses the information stored in the profile database as well as it interacts with the RCM on the network side to include the current dynamic capabilities of the network side into the decision. Profile information may be exchanged between the network and the terminal to update the information stored on both sides before the decision is made.

Reconfigure Radio Chain
The reconfiguration of a radio chain and the HO (seamless or non-seamless) from one RAT to another involves multiple interactions between the CMM_DMP and the various CCM modules.

If software modules required for the reconfiguration are not yet available on the system, the CMM_DMP interacts with the CMM_Dwnld to download the required modules from the network.

The terminal informs the RCM before starting the reconfiguration and after finishing it. These notifications allow the RCM to react on the RAT switch, for example, routing all ongoing communication to the new connection and buffering data packets during an interruption of the connection.

Single radio chain terminals cannot handover seamlessly between two RATs. Multi-radio chain terminals stay connected to the old RAT until the configuration of the new RAT is completed. They seamlessly hand over from the old RAT to a new RAT.

Many steps in the configuration of a radio chain are optional, depending on the actual configuration. Especially, changes in the protocol stack layer and the application layer are only needed during mode switch (e.g. UMTS to WLAN), but not during a handover between two APs of the same technology (e.g. switch between two UMTS Base Stations (BS)).

Reconfiguration of TCP Parameters

The TCP adaptation process takes place after detection of a vertical or horizontal HO. First, it is supposed that the TMM informs the CMM about the nature of the HO, and the new access network conditions, via the CMM_IfNss. Then, the CMM_IfNss delivers the information enclosed in the HO notification (HO type, network conditions and access technology information) to the CMM_NS. This entity can then consult the CMM_Prof in order to associate new TCP parameters values to the new access network conditions and characteristics, and/or the user profile. Then, the CMM_NS decides if the TCP parameters change is necessary. Note that this decision will be taken according to configuration parameters set by the user or the mobile device administrator.

If the TCP parameters modification is decided by the CMM_NS, then it generates a request to the CMM_DMP, that is the entity in charge of pushing the TCP parameters change. At last, the CMM_DMP forwards the new TCP parameters values to the CCM_PS in order to make this change operational.

After the TCP parameters modification completion, acknowledgement messages can be sent back to the CMM_DMP and the CMM_NS if required, but this kind of notification, in most cases, is not required.

Definition of Relationship between Reconfiguration Module Controller and CMM

The network node (terminal/BS) is configured to a standard (or agreed transmission scheme) 'x' (e.g. WLAN) and needs to be reconfigured to a completely different standard 'y' (e.g. UMTS).

Assuming a scenario in which a reconfiguration procedure, requiring the complete terminal reconfiguration from an access standard x to a standard y, is triggered, and the required downloads of policy and software and the activation of the profile (including the VCP) have been accomplished, the network dispatches a message about the successful completion of the VCP. The CMM_DMP receives this notification and informs the CMM about the permitted reconfiguration and also forwards information about the changes to be implemented in the reconfiguration part. The CMM uses the information about the changes necessary to generate a set of new CCMs, which, in turn, install the reconfigurable modules. Every CCM then performs an initialisation of both state machines and its parameters; it then allocates memory and the I/O parameter types of the reconfigurable module. After this initialisation, the CCM issues a request to the CMM_Dwnld to provide handles to the required software modules. The CMM_Dwnld responds to the requesting CCM, by forwarding the required handles. After this exchange, the CCM implements the modules and sends a notification to the CMM. Once the installation is complete, the result is passed to the CMM.

This sequence has to be repeated for every CCM; the responses and outcomes are collected by the CMM. If the installations of the reconfigurable modules are completed, the CMM sends requests to the modules (i.e. to their CCM) to create the connection points between the modules. Confirmations about the establishment of the ports between the modules are to be forwarded from the CCM before the CMM can send a test signal to the CCMs. The test signal (or sequence of test signals) ensures that every module individually and also the complete module structure can be tested. After receiving a positive response from every CCM confirming the functionality of every connection point and the new radio configuration as a whole, the CMM issues a connect message to all CCMs. The 'old' reconfigurable modules continue to function until the 'new' implementation is operable. During the transition period, the first (if there are more then one) module in the chain of the 'old' chain implementation starts to buffer incoming information (i.e. to prevent possible loss of data) and simultaneously the signal is passed through the 'new' module chain. The transition period constitutes a second function test for the new radio implementation. After a series of performance checks of the new configuration implementation, the CMM issues a request to all CCMs to finally hand over from standard x to the new standard y. The hand over is implemented in a sequential way, whereby the buffered data in the first (old) module is sent to the first CCM of the 'new' module chain, and at the same time the old CCM continues buffering the incoming signal. This CCM then processes the buffered data from the first old module through the new modules and then forwards the (meanwhile) buffered data. This is repeated until the last 'new' module is connected within the module chain. Once the new configuration is in place and every message delivered from the reconfigurable modules to CCMs is within the set parameter limits, notification messages from the CCMs are sent to the CMM to finally confirm the reconfiguration. As a further step, the CMM requests the old CCMs to destroy their modules. If the destruction of all old reconfigurable modules is confirmed, the CMM destroys the old CCMs. Finally, the CMM requests, from the CMM_DMP, the update of the status of the new terminal profile and the registration of the new terminal configuration to the network. The registration of the new configuration and the acknowledgement sent by the network concludes this complete terminal reconfiguration sequence.

Definition of Relationship between Security Manager and CMM

If the Security Manager receives a request, from the CMM, to establish a secure connection between the terminal and the network, it performs a mutual authentication procedure with the network part. The next step then is to respond to the CMM; the response carries an indication that authentication has taken place and a connection may then be established. In case the request was sent by the network, an identical process has to be performed (i.e. in the opposite direction).

If any message is transmitted between CMM_Dwnld, CMM_Prof, CMM_DMP, CMM and the network, it becomes processed and packed into a secure frame. After encryption, it becomes transmitted to the network. If a packet arrives from the network, the Access Manager authenticates and forwards it to the 'E&D Factory' where the necessary decryption is performed; after this, the message is passed to the Appropriate Functional Entity (AFE). The secure connection remains open until its termination by the Configuration Manager. The same sequence of events is repeated if the network requires establishing a secure connection with the terminal for reconfiguration request.

8.3.2 Flexible Air Interfaces

This chapter introduces a reference model for a multimode protocol stack of a flexible, dynamic reconfigurable air interface for future wireless networks. This future wireless network has the vision of a ubiquitous radio system concept providing wireless access from short-range to wide-area, with one single adaptive system for all envisaged radio environments. It will efficiently adapt to multiple scenarios by using different modes on a common technology basis. The generic protocol stack enables an efficient realisation of reconfigurable protocol software as part of a completely reconfigurable wireless communication system. Following a bottom-up approach, this chapter considers parameterizable modules of basic protocol functions corresponding to the Data Link Layer (DLL) of the International Standards Organisation (ISO)/OSI reference model. System specific aspects of the protocol software are realised through adequate parameterization of the modules. Further functionality and behaviour can be added through the insertion of system specific modules or inheritance. The subsequent sections will elaborate on how such a generic protocol stack can be constructed in a general way, followed by the more specific example of a 'Generic Link Layer (GLL)'. Additionally, the last section throws light on PHY-related aspects of a flexible air interface in introducing a multi-antenna–based approach for adaptive data transmission.

8.3.2.1 Multimode Reference Model Based on Modes Convergence

A Multimode Protocol Architecture

Figure 8.15 illustrates the architecture of a multimode protocol stack for a flexible air interface [23]. The layer-by-layer separation into specific and generic parts enables a protocol stack for multiple modes in an efficient way: The separation is the result of a design process that is referred to as cross-stack optimisation, which means the identification and grouping of common (generic) functions. The generic parts of a layer, shown in Figure 8.15, can be identified on different levels. The generic parts are reused in the different modes of the protocol stack. All generic parts together can be regarded as the generic protocol stack [24–26]. The composition of a layer out of generic and specific parts is exemplarily depicted in Figure 8.15.

The composition and (re-)configuration of the layer is performed by the (N)-Layer Modes Convergence Manager [(N)-MCM]. The protocol modules of generic functions are exemplarily introduced: Some of them are reused in a layer and/or additional functions are taken from the toolbox of common protocol functions as part of the generic protocol stack. The Radio Resource Control (RRC) on the control-plane and the RLC on the user plane are generic to the layers located above. A mode specific protocol stack has an individual management plane. Radio Resource Management (RRM), the Connection Management (CM), and the MM are located in the RRC layer. The cross-stack

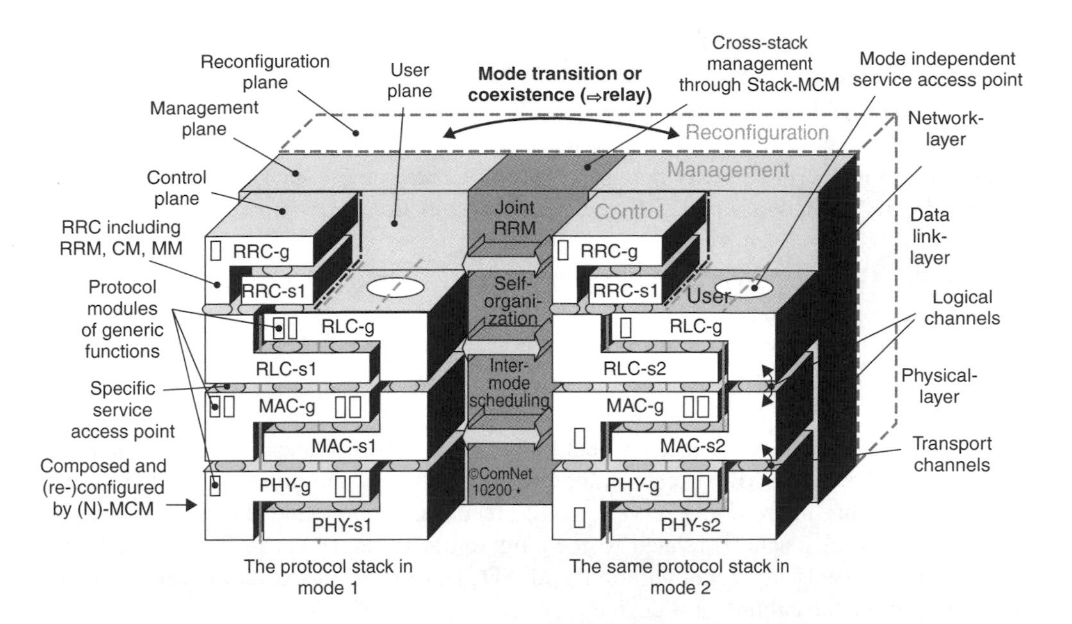

Figure 8.15 The multimode protocol architecture, facilitating transition between modes (inter-mode HO) and coexistence of modes (in relay stations connecting different modes) by way of the cross-stack management supported by the modes convergence manager of a layer or stack

management of different modes completes the reference model for multimode protocols in connecting the management-planes of the device's modes with the help of the Stack Mode Convergence Manager (Stack-MCM). Stack-MCM and (N)-MCM exchange in a hierarchical order data between two modes. The transition between modes and the coexistence of several modes are performed by the Stack-MCM. The (N)-MCM enables composing a layer out of different parts, which is dealt with in the previous section. The split between user and control-plane is limited to the network layer, as known from H/2, and the DLL is used for both, signalling and user-data transfer.

The (N)-MCM is the intermediator between the generic and specific parts of the multimode protocol stack's layers: All SAPs, that is interfaces, touched during the mode transition of a layer are administrated by the (N)-MCM. In the classical view of protocols as state machines, the (N)-MCM transfers all state variables of the protocol layer between two modes with the help of the Stack-MCM. This could, for instance, imply the state transfer of being connected from one mode to the other together with a data transfer of received but unconfirmed data frames.

Concrete, the (N)-MCM manages a single layer and has the following tasks and responsibilities that are dealt over later in this section:

- Layer composition and reconfiguration, considering all interfaces related to the transition between two modes;
- Protocol convergence:
  - horizontally – between 'generic' and 'specific' parts;

— vertically – mapping of higher-layer user-data flows for RLC as known, for instance, from Asynchronous Transfer Mode (ATM).
- Data preservation and context transfer.

Furthermore, the convergence between mode-specific protocol stacks is realised through the Stack-MCM implying implicitly the following functions:

- Joint RRM (Radio resource coordination) between different modes;
- Inter-mode scheduling;
- Self-organisation (frequency allocation of adjacent relays and APs, user-data flow routing).

The reconfiguration plane located behind the management plane, is not considered in this section. The Stack-MCM realises the reconfiguration of the protocol stack from one mode to the other in providing services to the reconfiguration plane. The reconfiguration plane contains all functions related to reconfiguration management [27] as for instance, the security aspects of reconfiguration and SD as well as the communication of the reconfiguration capabilities of a device.

Composition of a Layer from Specific and Generic Parts
Figure 8.16 shows the general structure of a protocol layer conforming to the reference model in [23]. It is assumed that the functionality inside the layer is always composed of a generic (common to all modes) part and mode specific parts, which jointly provide the modes' services of the layer via Service Access Points (SAPs). Through this, layers can be configured for one mode at a time. Modes can also coexist temporarily or permanently. The specific SAPs of a layer are defined via the currently used mode or set of modes (as a result of the actual configuration of a protocol layer (N) for a mode X, it makes available mode X's (N)-service number Y via a mode X specific (N)-SAP for the service

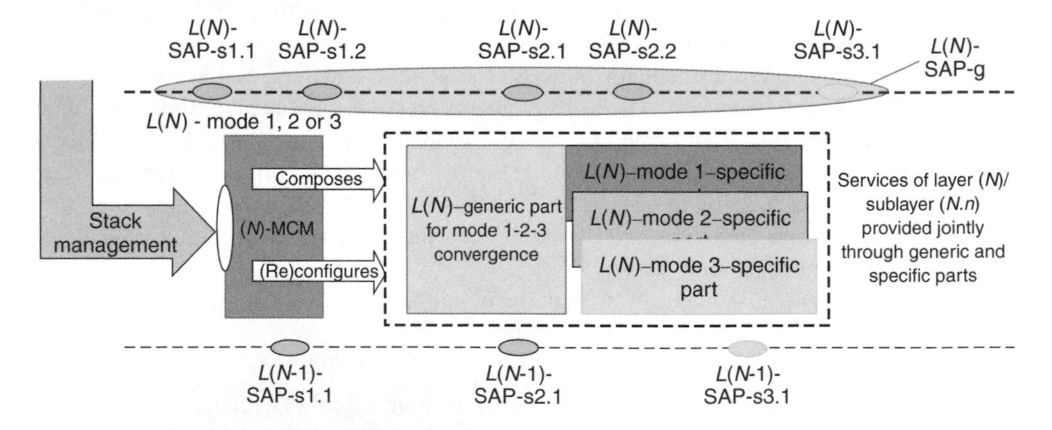

Figure 8.16 Composition of a layer (N) from generic and specific functions. The composition and (re-)configuration is handled by the (N)-MCM. The (N)-MCM is controlled by a layer-external stack management entity, namely, the Stack-MCM

Y (Notation used: (N)-SAP-sX.Y)). This does not preclude the possibility that SAPs of different modes can be accessed by higher-layer entities in a common way as visualised by L(N)-SAP-g.

The composition and (re-)configuration of the layer is taken care of by a layer-internal instance, the (N)-MCM, which resides in the management plane. The (N)-MCM enables an (N)-layer to provide multiple modes and makes functionality of one mode or common to several modes available. An instance of the MCM serves as a reconfiguration handler in each layer of the air-interface protocols.

Figure 8.17 depicts two exemplary cases of an (N)-layer: In Figure 8.17(a), a completely generic layer configured (if necessary) for mode 1 by the (N)-MCM is shown. The common, generic part is thereby composed by the (N)-MCM. In contrast Figure 8.17(b) illustrates a layer, where the different integrated modes exhibit no commonalities. Here, the layer cannot provide generic services. The role of the MCM is restricted to support the Stack-MCM by choosing the demanded mode-specific part.

'Mode Transition' versus 'Mode Coexistence'

The Stack-MCM administers the protocol stacks of several modes under consideration of the environment, that is receivable modes at the location of a device and requirements of the user and its applications. The reconfiguration of the protocol stack from one mode to the other is referred to as mode transition. The mode transition may be done on several levels depending on the architectural constraints of the protocol layers and the general treatment of generic parts: In case of a unique permanent existing DLL [28] of functions which are reconfigured (re-parameterized, restructured, or extended) the mode transition is limited to the layers below – the Medium Access Control (MAC) and PHY. The same stands for the case of a protocol stack of two modes used for relaying as defined below: The termination of an end-to-end retransmission protocol above the relay limits the considered protocol layers during transition from one relaying mode to another. Thus, mode transition may be done (i) on the MAC level or (ii) on the RRC level depending on the termination of the generic parts of the DLL.

Temporal or permanent modes of a protocol stack existing in parallel are referred to as mode coexistence. Mode coexistence can be reasoned through: (i) a relaying function in case of two simultaneous existing modes or (ii) the simultaneous connection to multiple modes for other reasons (a dedicated mode for broadcasts, applications of the user with different QoS requirements, cost preferences from the user, etc.) or (iii) short-term coexistence for inter-mode HO, and (iv) seamless mode HO.

Functions of the (N)-layer Modes Convergence Manager [(N)-MCM]

Protocol Convergence

The convergence of multimode protocol stacks has two dimensions: First the convergence between two adjacent layers, in the following text referred to as vertical convergence as it is known from the user plane of H/2 protocol stack. Second the convergence between layers located in the different modes of the protocol stack, which have the same functions: In the following text referred to as horizontal convergence. The generic protocol stack, managed by the (N)-MCM as explained above, enables both the horizontal as well as vertical protocol convergence.

From the perspective of higher-layer protocols the multimode protocol stack is transparent on the user- as well as on the control-plane, that is generic parts terminate the stack

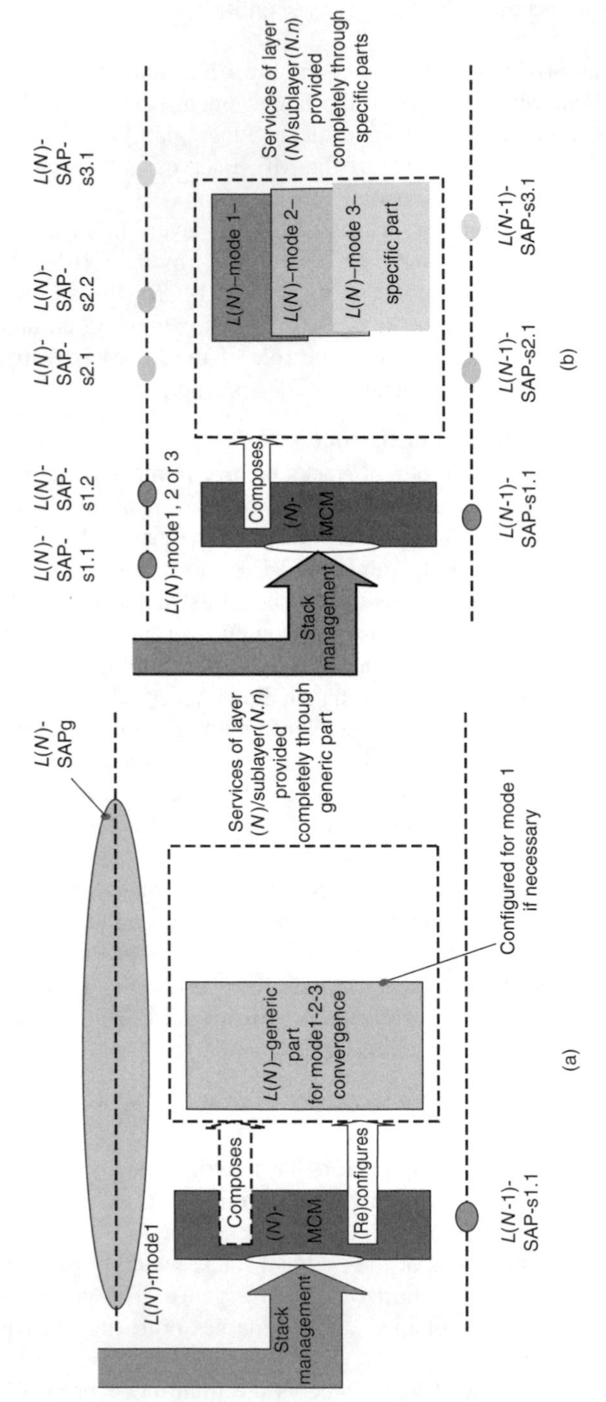

Figure 8.17 (a) and (b) Exemplary composition of a layer (N). Two extremes are depicted

to the layers above, as depicted in Figure 8.17. The vertical conversion of the (N)-MCM implies the adaptation of the multimode protocol stack working packet data protocols to the specific mode. This may be for instance, the conversion of an IP datagram in compressing the IP-Header.

Layer Composition and Reconfiguration
The separated approach of generic and specific parts requires an administration when taking the transition between modes into account: The common generic parts of the old mode need to be adopted for being reused in the new mode of the protocol stack. It is assumed that the generic parts of a layer exist permanently and are to be reconfigured and/or recomposed by the (N)-MCM corresponding to the characteristics of the targeted new mode. This assumption may imply a module-based composition concept of the generic parts as introduced in [29, 30]. The composition and configuration of the layer out of generic and specific parts are done by the (N)-MCM.

Data Preservation and Context Transfer
The communication between two modes and the mapping between generic and specific parts of a mode is done by the (N)-MCM (inside layer) and the Stack-MCM (transfer between modes). The transition between two modes can be optimised in using the data from the old mode to the new mode. The ability of a user-plane protocol to reuse status information in the generic part and protocol data after transition to another mode requires an extension of the protocol into the control-plane though it performs only user-plane tasks. Depending on the status of the related protocol parts, the data transfer is referred to as 'data preservation' or 'context transfer'. If the generic part is reconfigured and recomposed the data needs to be preserved, that is adopted, to the new mode. In the case of a deletion of the old specific/generic part the data transfer is named *context transfer*, which implies preservation for the new mode.

Functions of the Stack Modes Convergence Protocol (Stack-MCM)
Joint Radio Resource Management
The functions of the user and the control-plane are administrated by the RRM, which may be coordinated centralised or decentralised, and the RRM decisions are executed by the RRC of the corresponding modes. The RRM may assign multiple modes to one specific data flow. The RRC provides status information about the mode-specific protocol stack in a generic structure to the RRM of the multimode protocol. In case of a (semi-) centralised coordination of the radio resource allocation, this generic information structure about the status of the different modes of the protocol stack can be transmitted to enable an adequate decision.

The RRM of a single multimode device may also support the coordination across neighbouring operating devices as, for instance, the coordination across BS.

Inter-mode Scheduling
The Stack-MCM as intermediator between modes performs scheduling among different modes, as illustrated in Figure 8.16. In contrast the scheduling inside a mode across logical/transport channels: It is done in MAC-g or RLC-g of the specific mode's protocol stack. The inter-mode scheduling considers the dynamic scheduling of different user-data flows over multiple modes. The scheduling strategy may for instance, be based on the modes' interference situation, which requires a provision of necessary information directly from the PHY if the decision is done in the MAC independent from RRC/RRM. This

information about the quality of the radio link is again provided in a generic information structure.

Self-organisation

The envisaged communication system is able to autonomously decide about its radio resource allocations in taking the environment into account. This implies for instance, the adequate selection of frequencies used for transmission or the routing of user-data packets. The radio resource is selected under consideration of interference avoidance with other radio systems. Further, the optimised spectrum utilisation coordinated with neighbouring radio systems of the same technology is taken into account, which may also be related to efficient multi-hop relaying, depending on the selected deployment scenario. The self-organisation comprises scenarios of breaking down and installation of additional devices in an operating communication system. The Stack-MCM has to support the addressed functionalities in activating for instance, different modes to provide information about the interference situation or the role of devices (if it is acting as relay or AP) in reception range.

8.3.2.2 Definition of a Generic and Reconfigurable Link Layer

In mobile communications, the radio link is, in general, the bottleneck of the end-to-end path and it is costly or even impossible to increase its capacity. Therefore, it is required to utilise the available radio resources in the most efficient way. This role is performed by sophisticated radio PHYs and radio link layers, which are optimised to the RAT in use. In Figure 8.31 (left side) a simplified protocol stack is depicted, in which a correspondent node communicates with an MT. The end-to-end connection is established with for example, the IP. The radio link layer (here called *GLL*) and the radio PHY enable data transmission over the radio link.

8.3.3 SDR

Future mobile communications systems have been studied for a long time, leading to the emergence of various cheap and efficient communications devices: GSM/GPRS, UMTS, 802.11, DVB-T, Bluetooth, and so on. Integration of the various, associated, protocols into a global framework, enabling a transparent seamless diagonal HO, is a key objective to E2R. During the last decade, reconfigurability has been evangelised by techno-addicts advocating that technology could achieve radio reconfigurability. Joe Mitola mentions (about Software Radio): ‘A software radio is a radio whose channel modulation waveforms are defined in software’. Such an approach can be described as techno-oriented because it refers to digital conversion and processing power enhancement.

In its beginning, Software Radio used to be compared to the PC concept: a standard platform composed of cheap boards provided by multiple companies. This paradigm has been observed with the emergence of the Personal Digital Assistant (PDA) and its ensuing success – ensuring a market corresponding to the need of the ‘easy to use’ and ‘easy to transport’ ‘computer’. The PDA concept has introduced a different way of using a computer without a keyboard and mouse, using instead a stylus. The success of the PDA rests on the use of a common hardware and software platform. As soon as a platform has been stable and made ‘open’ to software developers, an increased number of software programs are created – enabling value-added services. Despite these first concepts on what SDR might be, the PC paradigm is yet to be fully exploited in application to the mobile phone market, which is now facing a new phase of its development.

The convergence of mobile phones with PDA-type technology and the emergence of Smart-phones begin to pave the way for SDRs or terminals. Mobile phones have become, and are becoming increasingly a way for a user to display their own identity and many companies target various market segments in terms of the products themselves (for example, producing a line of phones for the Business User, another for those who are fashion conscious and perhaps phones aimed towards the youth end of the market, incorporating the latest computer games). Mobile phone platforms are becoming more flexible as they evolve to cater to the increased expectations of their users. The ability to change the mobile device's cover exists, along with downloading ringtones, games, and so on. However, although the emergence of an all-in-one concept is being witnessed, the underlying communications technology is currently not reconfigurable.

Within the past few years extensive research on reconfigurability has been conducted. A very strong heritage in reconfigurability was gained through former EU-IST FP5 projects such as DRIVE, OVERDRIVE, TRUST, SCOUT, CREDO, and MOBIVAS, where the expertise in the functions offered to user terminals, applications and services, was capitalised. Each of these projects concentrated on a variety of different technical aspects such as terminals, value-added service provision, enabling technologies, applications, reconfigurable devices, network provisions, security, and proof of concept of reconfigurability.

Wireless World Research Forum (WWRF) has defined SDR Reference Models and has analysed issues and problems with respect to SDR Architectures in a recent White Paper, titled 'Reconfigurable SDR Equipment and Supporting Networks' [31], that has been produced within the WWRF WG3 SDR Group [32, 33]. Several existing SDR systems and supporting network reference models and architectural approaches were reviewed. The section also pointed out some of the main research areas on SDR for the next decade. The work presented in this section is complementary to the previous work mentioned above. In this direction it addresses an important relative research topic, hardware abstraction. This issue is currently widely discussed in the reconfigurability community (e.g. specific RFI in the SDR Forum) as physical implementation will be proprietary to the manufacturer. In this direction, this section discusses on Hardware Abstraction in an End-to-end Reconfigurable Device and presents a possible hardware abstraction approach.

A further issue with respect to ongoing and future work on the exploration and development of SDR and of new radio interfaces in general, is performance validation. This is understood not as the testing of new interface features by the company or group that develops them, but as third-party verification. This is an issue of special importance for telecommunications operators, who additionally have to test interoperability between equipment and systems coming from different manufacturers. From this background, a need can be seen for the implementation of B3G verification tools. Section 8.3.3.3 describes a test bed targeting the evaluation of new air interfaces and interoperability testing.

8.3.3.1 Hardware Abstraction in an End-to-end Reconfigurable Device

Hardware Abstraction Reasoning

In the past, the configuration control for the PHY in a wireless terminal has been limited to changes requested by the Radio Resource Controller (RRC) or a similar entity defined in the supported standard. These changes would be limited to a subset of the standard, as defined by the terminal's class mark, and would include services such as switching between data services and speech calls or a change of speech codec. Each of these different configurations would be known at design time and would be created and tested in DSP/CPU software (sometimes in assembler code but more recently in C) and then implemented in an embedded ROM. A ROM solution was used rather than RAM or FLASH to save silicon area and allow operation at the high clock speeds required in a data processing application. Typically, the only mechanism available for modifying the PHY to implement functionality above and beyond that conceived at design time was a small amount of patch RAM in the program space of the DSP. This was combined with some form of patch vectoring and used to fix bugs in the ROM code.

More recently research projects have been looking at methods for reconfiguring the PHY to implement functionality not conceived of at design time. Earlier projects such as TRUST, SCOUT, and CAST have directly addressed the problem of Configuration Control. A common theme exists, that is all projects have used object-modelling techniques to encapsulate functionality and all have three primary components, as shown in Table 8.2.

Both SCOUT and CAST address heterogeneous architectures and recognise the requirement to support communication between the processing elements. Only SCOUT appears to have addressed the issue of timing deadlines. ADRIATIC addresses this problem indirectly by working on the reconfiguration times of the processing elements. None of these projects directly address the problem of verifying that a new configuration will always operate correctly. MuMoR reduces this problem by constraining the configuration space to a number of RATs.

Hardware Abstraction Possible Approach

The PHY architecture of a reconfigurable device consists of a set of reconfigurable functional elements (RF front end, communication, and digital processing). They appear as (re)Configurable Execution Modules (CEMs) implementing a specific functionality (e.g. Down-conversion, Modulation, Decoding, Rake). A specific entity called *CCM* reconfigures such modules and also manages the communication resources between to guarantee the required functionality and maintain the required data throughput.

Configuration Control Module (CCM)

The role of the CCM is to supply an abstract configuration interface to the signal processing system in a wireless terminal, BS or AP.

Table 8.2 Comparison of configuration components in FP5 projects with E2R

TRUST	SCOUT	CAST	E2R
BPC (Baseband Processing Cell)	Proxy	Java Class	CEM (Configurable Execution Module)
TMM (Terminal Management Module)	TMM	RSC (Reconfigurable Resource Controller)	CMM (Configuration Management Module)
RMM (Reconfigurable Baseband Management Module)	RMM	PLC (Physical Layer Controller)	CCM (Configuration Control Module)

The CCM is located in the System Abstraction Layer (SAL) or the high-level Hardware Abstraction Layer (HAL). This layer implements a set of interfaces that allow high-level entities to configure resources, that is create and link signal processing functions (e.g. modulation, demodulation, channel coding, source coding, Radio Frequency (RF) transceiver, AFE, etc.). By supplying a platform-independent interface, called *Service API*, a system can reconfigure either end of the wireless link without a detailed understanding of the underlying implementation.

The configuration of the wireless terminal includes downloading of software (ISA objects, FPGA code, etc.) and parameters into resources to implement different functions as well as the configuration of the communication fabric to implement the required data flow and control flow structures.

Control and supervision of the reconfiguration process takes place in cooperation with the CMM and/or by autonomous processes implemented by dedicated CCM functions and optional modules like the adaptation module.

The CCM will be implemented by means of software modules, built on top of the underlying operational software, that is a certain OS, using its extensions like device driver to gain access to the real hardware interfaces or additional OS extensions to implement diverse CCM functionalities.

Configurable Execution Modules (CEMs)

The CEMs may support different levels of reconfiguration (configuration capabilities):

- Dedicated signal processing entities, programmable IPs blocks;
- Algorithm-specific accelerators;
- Programmable logic (e.g. FPGA, CPLD);
- Application tailored DSPs, AIPS;
- General-purpose processors, DSPs.

A manufacturer will exercise the choice and selection of suitable CEMs, implementing certain PHY functionality. This manufacturer dependency requires an implementation independent view of the PHY hardware in order to support certain reconfigurations and use cases.

Hardware Abstraction Layer Concept

Hardware abstraction will be used on various levels to support the CCM and the RMP on top. In particular, the exchange of reconfiguration capability parameters between certain entities (e.g. network entities, operator) has to be addressed.

A hardware abstraction layer concept might be used, which introduces hardware abstraction on certain levels (See Figure 8.18):

- The CCM can be seen as the high-level abstraction of the reconfiguration capabilities of the PHY hardware. Abstraction has to be done on the function rather than the implementation level. This level implements a kind of Service API.
- The Generic Resource Interface, abstracts the different classes of Configurable Execution Modules (CEMs) for the CCM.
- The Physical Interface abstracts the real implementation of a CEM.

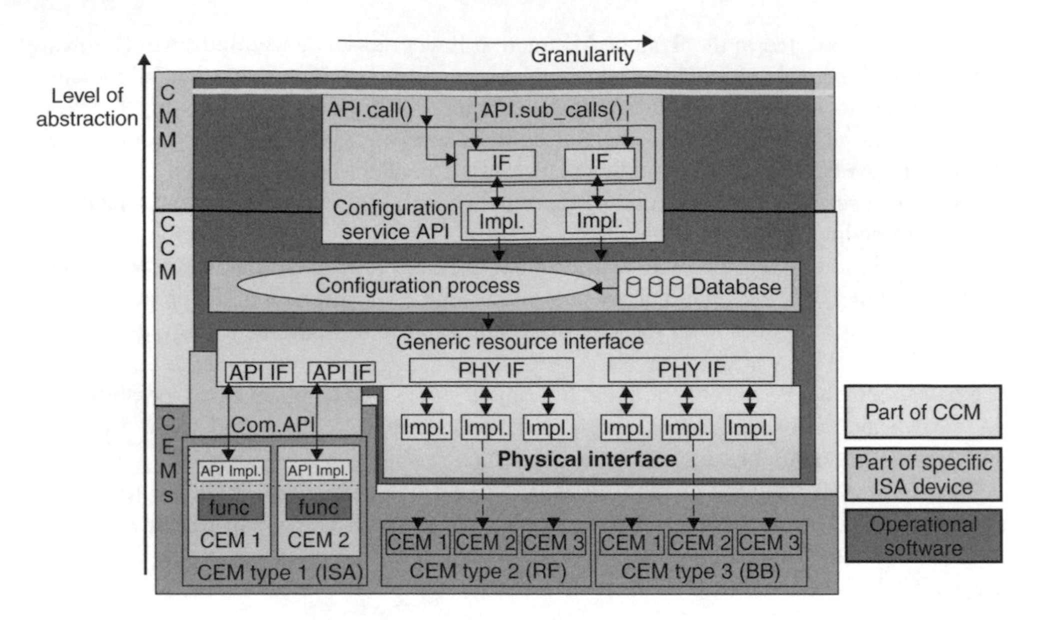

Figure 8.18 Abstract functional view

- Intelligent CEMs have at least a basic OS running on top. Their physical interface will be replaced by a low-level communication API, which enables communication (i.e. exchange of configuration data, program code).

Configuration Service API

The Configuration Service API is the mentioned interface towards the upper layer between the CCM and the platform-independent CMM. The Configuration Service API is implemented in the CCM, and is used by CMM. This is a platform-independent entity and provides services for the configuration. Different levels of configuration granularity can be envisaged as follows:

- *Coarse grained*: 'configure WLAN chain with param1 = x & param2 = y'
- *Mid grained*: 'configure channel estimation = algo1'
- *Fine grained*: 'configure FIR coefficient1 = 0.654'

Typically, the upper-level management entity (i.e. the CMM) is using the coarse-grained layer, but optimisation of the radio link or certain adaptation functions may require the direct access to the mid-grain or even the fine-grain configuration layer, which could be done either by the CMM or additional adaption or radio link modules.

Inside the Configuration Service API, the coarse-grain level will use functions of the mid-grain level and the mid-grain level those of the fine-grained. Thus, the architecture of the underlying hardware is hidden, but assessable in an abstracted common way.

Generic Resource Interface

The intention for the definition of the Generic Resource Interface is to have a common single interface to the lower layer. Abstract HW models will encapsulate front-end, digital baseband resource respective functions, and communication resources. The CCM has to interact with the resources available, which are executing the PHY processing.

The physical implementation will be proprietary and might be organised in several different ways. Functional partitioning will lead into several possible allocations of functionality into multifarious CEMs. In the described approach for the way of reconfiguration, the CEMs can be grouped into two different interface classes:

- *ACTIVE*: ISA devices that have their own firmware respective OS, and therefore an API for handling accesses. This is named *the Low-level Communication API*.
- *PASSIVE*: All other devices without their own firmware respective OS. They do not provide an API, and therefore need an active handling. This is referred to as a *low-level Physical Interface*.

The Generic Resource Interface (i.e. the upper-level interface) is therefore subdivided in the different properties of generally different access-types of CEMs as shown in Figure 8.6.

Physical Interface
The Physical Interface is part of the low-level HAL for devices without their own firmware or OS. They need an active handling of each configuration, which is demanded. Some examples are as given in the following text:

- *Analogue circuits*: for example, AD-converter (configure required number of bits)
- *Accelerator*: for example, upload Microcode
- *Programmable ASIC*: for example, Set certain register values and read status registers

The Physical Interface provides mapping of Generic Resource function calls to the proprietary configuration procedures of dedicated CEMs, that is the real physical addresses contained in the database are utilised for the reconfiguration.

Low-level Communication API
The low-level Communication API provides common services for configuration of ISA devices running under certain OSs. They are implemented on certain ISA devices. ISA devices have their own firmware respective OS, and therefore an API for handling the reconfiguration requests. The configuration through the Communication API includes downloading of software modules and parameters.

A UML (Universal Modelling Language) description of the described hardware abstraction is shown as a summary in Figure 8.19.

The previous sections have indicated an approach towards Hardware Abstraction in an end-to-end reconfigurable device. This approach will lead to a clear understanding of the requirements for each relevant abstraction layer within the reconfigurable PHY and combine different classes of reconfigurable devices. This includes the exchange of capability parameters for the PHY. The capability parameters are used by the user, service supplier, and the operator to decide on the most suitable terminal/base station configuration for the given context.

The abstraction layer will translate an implementation-specific configuration, proprietary to a certain manufacturer and specific to a certain implementation, into a functional

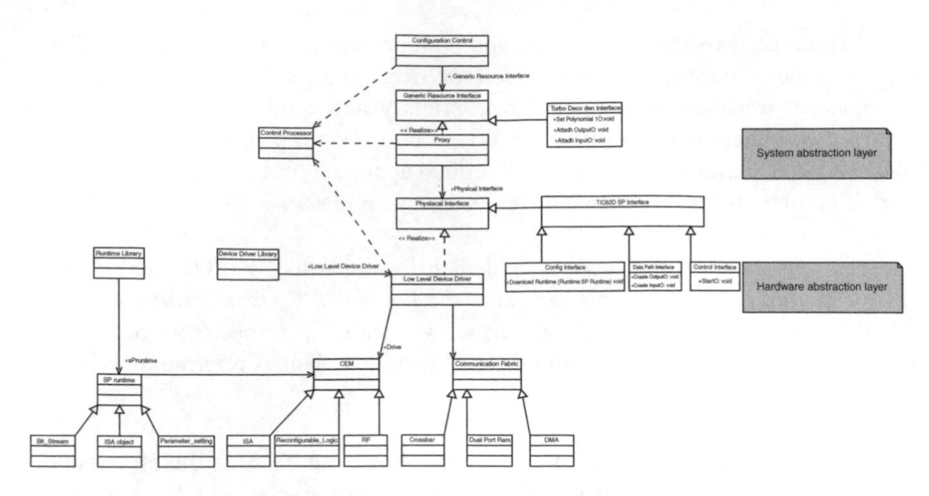

Figure 8.19 UML structure of the hardware abstraction

description. These translations and descriptions have to be generic and flexible enough, in order to serve different requirements. At certain levels, harmonisation between manufacturers is required.

Additional certain CEMs will be evaluated and modelled in an abstract way, in order to be integrated into the system architecture.

8.3.3.2 Design Exploration and Integration Challenges for Reconfigurable Architectures

Most wireless PHYs are characterised by several signal-processing blocks that are similar in their structure and functionality but differ in their implementation. The digital baseband processing complexity is dominated by channel estimation, detection, and decoding in the presence of interference. Interconnect is a limited design resource needed for achieving parallelism in these architectures [34–36]. In recent research, algorithms and architectures for these three major blocks [37–47] have been proposed for both base stations and mobile handsets in cellular and indoor wireless LAN systems. The proposed fixed-function ASIC-like implementations simultaneously achieve high performance and area-power efficiency by exploiting special algorithmic and architectural structures, and it is important to preserve these features when developing reconfigurable systems.

With the rapid development and adoption of advanced wireless systems, it is important to reduce the time to develop hardware research prototypes of these new algorithms [48, 49]. It is well known that VLSI implementations consume minimal power but typically require long design cycles. On the other hand, DSPs are completely programmable but are in general, unable to meet real-time deadlines and power budgets for high data rate mobile communications. A new class of application-specific processors is an area of current research [50]. These architectures are programmable and may also be reconfigurable to allow for application-specific instructions [51, 52]. FPGAs present an excellent platform for prototyping and evaluation of these architectures, particularly the class of heterogeneous FPGAs that contain both programmable fabric and embedded processor

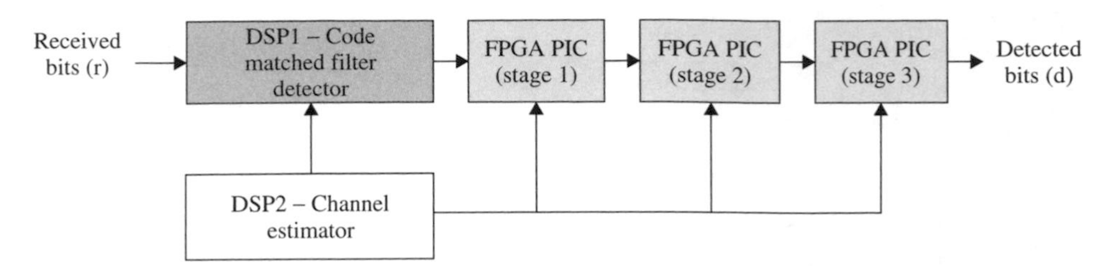

Figure 8.20 Parallel interference cancellation detector example on a heterogeneous DSP–FPGA system. Interconnect among modules may reduce system performance

cores. However, there are many new challenges in interconnect design to effectively integrate the embedded host and programmable fabric co-processor functions in these FPGAs, see Figure 8.20.

Processor Challenges for 4G Systems

The ongoing definition of 4G wireless systems underscores several urgent challenges for system realisation and design. Higher data rates and complex antenna and modulation schemes will require efficient algorithms and flexible chip architectures for baseband processing. Not only will area, time, and power consumption be the main design constraints, but also system reconfigurability and design reuse will be needed for fast creation of new systems. This flexibility will be a key for Multiple Input – Multiple Output (MIMO) antenna-processing blocks that may be configured with a varying number of antennas.

As data rates have increased, a single DSP processor is not capable to keep up with the signal processing requirements of advanced algorithms. Multiple DSP processors do not have the power efficiency to exploit the instruction and data parallelism in the algorithms. A solution to this DSP limitation has focused on custom ASIC co-processor designs clustered around a programmable DSP host. However, the multiple co-processor solution requires extensive design and verification time and with the spiralling costs of ASIC fabrication, these systems are increasing in cost and have little ability for reuse or modification. The class of application-specific instruction set processors has recently emerged to fill this middle ground between DSP and ASIC solutions. For example, stream processors, as shown in Figure 8.21, contain multiple SIMD-like clusters with specialised memory systems. Our recent research on chip equalisers for High-speed Downlink Packet Access (HSDPA) and 1X-EV-DO [53] systems and LDPC decoding has shown the flexibility of these ASIP solutions, which will lead to System on Chip (SoC) processors. From design exploration of parallelism and flexibility requirements, these future SoC processors will be an efficient integration of DSP, ASIP, and ASIC structures.

For high data rate 4G systems operating at greater than 100 Mbps, there are several research trends that are emerging that will require significant advances in VLSI architectures to provide performance with limited area, time, and power constraints. For many MIMO Orthogonal Frequency Division Multiplexing (OFDM) systems, effective coding will be important. In [53], an iterative turbo-like system integrating LDPC decoding with the receiver is presented. However, for outdoor highly mobile environments, equalisation will be needed even in OFDM systems [54, 55]. Furthermore, parallel interference cancellation can also be used to enhance these MC-CDMA systems [56]. With the focus

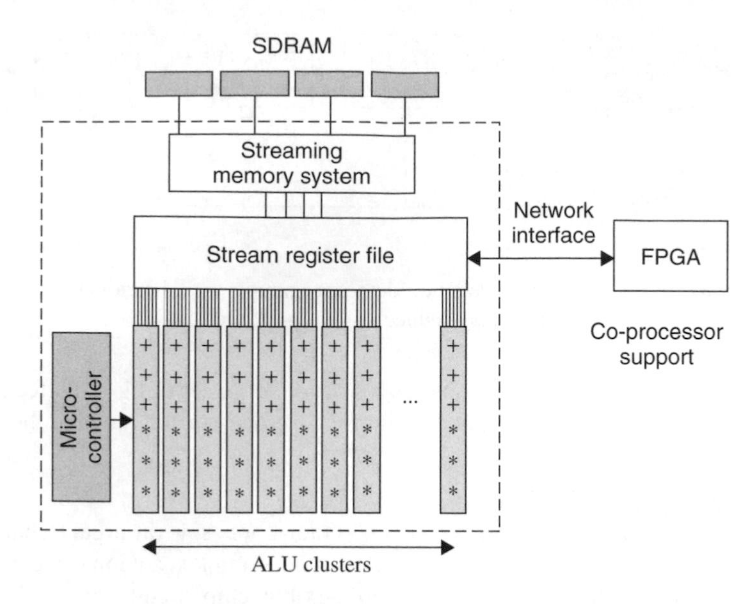

Figure 8.21 Programmable multi-cluster stream processor architecture with FPGA co-processor highlighting interconnect challenges between functional units and register files

in 4G on adapting OFDM, there can also be benefits from applying techniques from WLAN applications, and also from low-power Ultra-wideband (UWB) OFDM architectures [57].

In addition to the flexibility needed for adaptation as 4G standards continue to evolve, there is growing interest in system reconfigurability and design modularisation and reuse. However, system reconfigurability typically introduces overhead (lower data rate and higher power) that may not be acceptable in many applications. The commercial aspects of the military XG and JTRS hardware research are leading to cognitive radio and also are studied in the European E2R project and the WWRF WG-6 on Reconfigurability. These current system initiatives highlight research challenges in increasing data rate and lowering power consumption. In order to develop low-cost, low-power systems, there is much research to be done in creating a low-power architecture with an efficient hardware abstraction layer (HAL). This HAL would allow for efficient hardware and software partitioning and for the use of appropriate modules to create an adaptable 4G system. Current wireless system architectural blocks do not have the design standardisation and modularisation to allow for efficient mix and match for reconfigurable (or cognitive) radio systems. Research in the use of design and architecture simulation tools, such as Xilinx System Generator, are leading to more efficient design reuse strategies.

Reconfigurability, system design, and reuse will be important for 4G wireless systems. The mixture of DSP, ASIP, and ASIC structures will require efficient interconnection interfaces for high performance system design. The design of these ports, buffers, queues, and the resulting HAL will need to be optimised for overall system performance, see Figure 8.21. Simulation environments will need to be enhanced to collect system statistics for design exploration of processor utilisation and power efficiency.

8.3.3.3 4G/Beyond 3G Verification Tools

SDR Test Bed for New Air-interface Evaluation and Interoperability Testing

Ongoing and future work on the exploration and development of new radio interfaces raises the issue of performance validation, understood not as the testing of new interface features by the company or group which develops them, but as third-party verification. This is an issue of special importance for telecommunications operators, who additionally have to test interoperability between equipment and systems coming from different manufacturers. From this background, a need can be seen for the implementation of B3G verification tools.

Broadly speaking, new interfaces cannot be in-depth tested with standard commercial equipment, because such equipment is usually adapted to the evaluation of those systems or elements for which there is a market, that is backed by standards. Besides, in the fluid environment of new mobile communication systems research, it is becoming increasingly important to perform third-party testing at intermediate development levels: quite often what has to be measured is not a complete element, like a transmitter or receiver, but a hardware, software or even middleware block, such as a resource demanding signal processing algorithm.

A possible answer to this evaluation request are test beds that simultaneously fulfil the following requirements. Their interfaces are open at all levels, to foster voluntary testing requests. They are as flexible as possible: modular in their construction and expansion capabilities, standard in equipment practice and with externally programmable signal processing sections. They can be configured and run remotely through Internet connections. Another interesting requirement is that of being simple to replicate, to enable low scale trials that require the interplay of several network elements. Such is the case, for example, of trials that involve testing MAC procedures or ad hoc networking algorithms.

8.4 Network Architecture and Support Services

Reconfigurable technologies enable efficient access management for terminals, in addition to flexibility in operating the radio network. Reconfigurability is expected to realise the main requirements for adaptability in 4G systems, where a significant advance is the expectation of improved capabilities. Such capabilities include support for smaller cells, self-planning dynamic topologies, full integration with IP, more flexible use of the spectrum and other resources, as well as user location-dependent usage. In this section, an End-to-end Reconfiguration system architecture is presented, detailing the RMP that implements protocols and mechanisms supporting efficient context, policy and profile management, service provisioning, SD, secure reconfiguration, and dynamic network planning. Specific mechanisms such as one-to-many SD facilitators are also introduced.

The work presented in this section has been partly carried out within the IST End-to-end Reconfigurability (E2R) [10] project, motivated by results from the SDR Forum [15], in addition to those of the IST TRUST [58] and IST SCOUT [16] projects. In particular, under the perspective of TRUST and SCOUT, SDR is not restricted to terminal but also takes network aspects into account. An overview of SDR from a technical, market and regulatory perspective is given in [59, 60].

Today's state-of-the-art multi-band terminals usually consist of several dedicated transceivers (i.e. hardware components restricted to a single RAT), and a general-purpose processor that runs higher level functions such as applications, graphics/sound, and so on. The usage of various Time Division Multiple-Access (TDMA) and Code Division Multiple-Access (CDMA) technologies in different regions of the world leads to a fragmentation of the global market for mobile devices, thus making it difficult for equipment manufacturers to leverage the economy of scale expected from the growing global market for their products. SDR terminals, with generic transceivers that can be reconfigured to different radio standards through software, promise to allow such economy of scale thus reducing manufacturing costs for terminals as a common platform for all kinds of radio access technologies.

The development of configurable hardware devices and fast DSPs for digital baseband processing has already begun (see [61], for example). However, the full exploitation of the possible benefits of the SDR concept requires far more than sheer processing power in terminals. In order to allow for maximum flexibility, terminals should be able to download new radio software via the mobile network. Through this downloaded software, the terminal can be reconfigured to new radio access technologies that have not been considered by the manufacturer. Moreover, bug fixes to software can be delivered quickly, thus minimising the effects of known bugs on the operation of the terminal and the network. Unfortunately, such SD requires a sophisticated network infrastructure to be performed efficiently and securely; furthermore, there will be a gradual evolution towards SDR terminals from the present current devices.

Key requirements for network infrastructure should be identified through use cases [62]. Terminal reconfiguration affects a wide range of network entities, protocol layers, and operational schemes. Among the different management aspects for reconfiguration, we consider the following important cases that cover the essential network support required for terminal reconfiguration: SD Management, Profile Management, CM, and Programmable Network Element Management. Related work has already been carried out in IST-FP5 projects (e.g. TRUST, SCOUT, and MOBIVAS [63]), which served as a basis for the research of E2R.

The delivery of seamless mobile multimedia services, enabled by the interoperation of heterogeneous radio access networks, poses new requirements and invokes new challenges in radio network planning. Inevitably, the concise definition, development, and integration of novel methods and mechanisms for network planning and deployment in a reconfigurability context are vital. Operators should coordinate their heterogeneous wireless-access systems, a process that could materialise through agreements to absorb traffic from other networks with predefined coupling structures. This would assist in the handling of changing circumstances (e.g. hot-spots, traffic demand alterations, etc.) or service management requests.

The support of advanced services requires the engagement of additional management intelligence and associated modules, for the realisation of a Management Plane spanning all layers. Moreover, the support of application and service adaptation, in conjunction to network reconfiguration, is very important.

Taking SD as an example, scenarios can be classified into Terminal-initiated SDs, Mass-upgrade SDs, or Delayed/Rejected Software Upgrades. These three key classes of SDs require functional components to be defined in networks, such as reconfiguration managers at different hierarchy levels.

Network planning is considered an important research working area, offering challenges for detailed RRM, including parameter fine-tuning. The essential inputs to network planning consist of propagation modelling, traffic modelling, traffic forecasting and RRM performance modelling. A number of IST projects have considered network planning for particular RATs, in the sphere of geographical deployments and multimedia services offering [64–67]. In the reconfiguration environment, the optimisation of radio resource deployment through cost-effective RAT selection, with different coupling structures and traffic splitting policies [68], is presently under investigation within the IST E2R project:

- System architectures supporting reconfigurability;
- Reconfiguration services for context, policy and profile management;
- Enabling procedures – for example, download mechanisms, etc.
- Security;
- Flexible network elements.

8.4.1 Approaches and Research Ideas

Reconfigurable radio can be used together with traditional network infrastructures, which do not provide any support for network reconfiguration. However, the full benefits of SDR only become apparent if the utilised network infrastructure takes into account the specifics of a particular terminal, and provides support for it.

In order to sufficiently support SDR, network infrastructure has to be carefully reconsidered. In particular, the following issues have to be taken into account in the design of network elements for future mobile networks:

- The distribution pattern of terminal and user-related information (profiles) is different compared to conventional networks.
- SD support has to be integrated into the network infrastructure.
- Integrated usage of different radio technologies available in the same area should be allowed.
- Focus on fixed radio standards is likely to be reduced in favour of greater flexibility.
- Greater openness of network infrastructure must allow device sharing, RAT sharing and resource sharing amongst NOs.
- Full usage of reconfigurable network elements to support optimum working points for tuning of RRM parameters.

Technical solutions for these issues attain to aspects of reconfiguration platforms, stretching over reconfigurable network entities and terminals. The interfaces between the entities participating in reconfiguration procedures have to be considered, and their required functionalities have to be identified. Above all, an environment has to be defined to enable the deployment, download, installation, and initialisation of network services for reconfiguration. The objectives here are as follows:

- The definition and outlining of requirements and strategies of end-to-end reconfiguration policies within the reconfigurability management framework. Also, the specification of RMP concepts.

- The definition of discovery mechanisms for available reconfiguration services and downloadable reconfiguration software.
- Specification of reconfiguration and download services, as well as their respective triggering interfaces and mechanisms.
- The definition of management procedures and Application Programming Interfaces (APIs) for reconfiguration negotiation and initiation, between adaptable end-user services and the reconfigurability management framework.
- The design of reconfiguration platforms, over reconfigurable entities and their respective interfaces.

8.4.1.1 System Architectures Supporting Reconfigurability

The network architecture supporting end-to-end reconfiguration [69] aims to define functional entities for deployment in physical infrastructures, along with inter-working with legacy control, management, and user-data planes. This architecture is based on the RMP, a management functionality conceived for the flexibility to reconfigure all the layers of the protocol stack.

The overall E2R system architecture provides the logical functions relating to layering principles for protocol modelling. These principles separate protocol structure into three layered subsystems, namely, Transport, Radio Network, and System Network layers (see [60] for more detail). A layer encompasses elements of similar functionality. Protocols within each of the layers operate across multiple interfaces; they are inter-working entities that belong together because of the nature of the mechanisms they provide. These logical elements also extend across multiple network nodes, thus they comprise a set of protocols contributing to the distributed execution of a common system-wide function.

The RMP architecture facilitates inter-operator negotiations, involving the exchange of information that is required for terminal reconfiguration and advanced RRM and to provide mechanisms for the dynamic planning and management of heterogeneous, coupled, and multi-standard radio access networks. In the MT, RSFs access reconfigurable layers through the Local Reconfiguration Manager. From the Application layer, through the Radio Network layers and down to modem baseband and HW, RSFs in the MT control the reconfiguration process of the complete communication chain. They also process all reconfiguration process related information for reconfigurable terminals, originating in any of the layers.

Following the design principles of IP-based networks, the functions related to wireless connectivity are logically and physically separated. The concept of a common IP-based core network, serving multiple heterogeneous radio access networks, requires the encapsulation of access-specific functions in order to allow for the definition of an abstract set of functions that commonly apply for all access networks.

On this rationale, Radio RSFs have been defined to fulfil the requirements identified in the introduction section, spanning the RMP down to the composite RAN, in order to habilitate faster environment scanning when managing the context of the heterogeneous wireless access to support seamless services.

The terminal communicates with the network for assistance when a reconfiguration trigger provokes a state change in the terminal, and launches the reconfiguration process. The Radio RSFs assist the terminal by interacting with Radio Network layer elements of available RANs, implementing inter-working functions where needed [62].

Inter-working refers to interactions that occur in the access stratum between entities of different systems – these entities may reside in domains belonging to different operators. To assist inter-working, the terminal reconfiguration process benefits from the rapid detection, identification and selection of the most suitable RAN [70]. In deploying these functions in the RAN, the reconfiguration process also benefits from locality, thereby performing acceptably in time-constrained scenarios such as vertical HOs in delay and jitter-sensitive sessions (i.e. Conversational and Streaming in UMTS); furthermore, interactions between control-plane functions are kept within the radio access system domain. Traffic management for spectrally efficient downloads is enhanced, because interactions are closer to radio resource controlling servers. Moreover, the accuracy of time-sensitive information, such as radio channel conditions, is improved by shortening paths in retrieving context information. Locality consequently reduces traffic in the network as well.

Radio RSF for BS also fall within the access stratum. RSF for BS implement Dynamic Network Planning and Management (DNPM) mechanisms that inter-work with Operation System Functions local to Radio Network Subsystems OAM plane.

8.4.1.2 Reconfiguration Services for Context, Policy, and Profile Management

The main task of the RMP is to provide layer abstractions to applications and services on the one hand, and to terminal equipment and network devices on the other. The RMP also comprises RSF responsible for coordination of the reconfiguration process and for the provision of the required resources [71].

The RMP aims to provide the basis for integrated plane management and layer management support functions. Traditional plane management embraces configuration control, resource management, performance management, fault management, access and security management, and accounting management [62]. Traditional layer management necessitates the existence of interfaces to all protocol layers, both in the control and in the user plane. Layer management handles Operation and Management (O&A) functions on a per-layer basis.

The additional functional entities required to cater to reconfigurable environments, specify a new plane (e.g. RMP) that stretches across both network elements and terminal equipment. The modular entities describing the RMP functional architecture and comprising the core reconfiguration services, are shown in Figure 8.22, and are described as follows:

Plane Management Functions

- **Context Management:** This functional entity monitors, retrieves, processes, and transforms contextual information. Contextual information includes profile information as well as resource-specific information, regarding the reconfiguration progress, the current operational mode, state information, congestion indications, and so on. Contextual information affects the service provision phase, and provides inputs to policy decisions and reconfiguration strategies.
- **Profile Management:** The profile management entity handles profile definition and provision, managing and combining the different profiles. This profile information originates in different parts of the system, and includes user profiles, network profiles, application/service/content profiles, terminal profiles (i.e. so-called *Reconfiguration Classmarks*), charging profiles, security profiles, and so on. The collection of profile

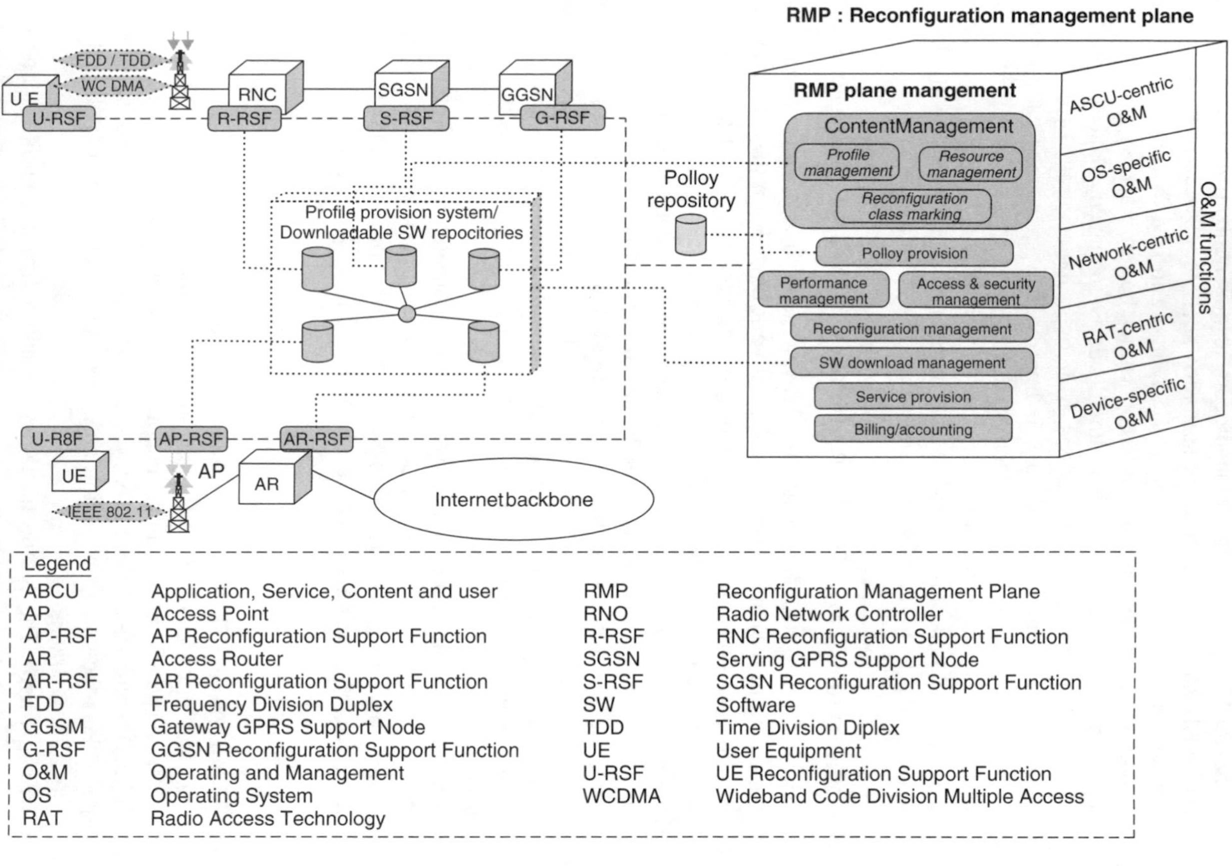

Figure 8.22 RMP architecture

repositories in a reconfigurable beyond 3G system should be viewed as a composite Profile Provision System (PPS). The PPS should apply to an n-tier platform capable of disseminating profile management policies to an n-layered architecture. This multitier architecture can be constructed on the basis of topological and/or semantic considerations. Segmentation and distribution of profile data representation via profile staging, and a two dimensional (topology-based) multi-tier (semantic-oriented) hierarchical organisation of profile managers, should offer performance and flexibility benefits.

- **Reconfiguration Classmarking:** This entity keeps track of the different network nodes and their states regarding reconfiguration (e.g. the protocol versions that are installed). Each terminal is assigned a Reconfiguration Classmark, which specifies its level of dynamism and capabilities regarding reconfiguration (e.g. enhanced MExE classmark). The calculated value of the classmark depends on the type of reconfiguration requested and negotiated, on the type of software to be downloaded, on business incentives, and on individual or operational chains of stakeholders involved in the reconfiguration process.
- **Policy Provision:** This functional entity is the main decision-making entity for reconfiguration, comprising the entry point for reconfiguration-related system policies. Furthermore, it exploits contextual information and redefines policy rules and reconfiguration strategies. This functional entity makes up to date decisions about the feasibility of reconfigurations, as well as the respective actions required to be triggered. Additionally, the Policy Provision function caters to inter-domain issues, interacts with Policy Enforcement Points, and facilitates the mechanics for end-to-end reconfiguration differentiation.
- **Reconfiguration Management:** The reconfiguration management function initiates network-originated and coordinates device-initiated configuration commands, by communicating with peer CCMs in the terminal equipment. In order to supervise end-to-end reconfiguration, it incorporates the necessary signalling logic, including trading and negotiation services. In the case of scheduled SD, the Reconfiguration Control function hands-over control of the residual reconfiguration steps to the SD Management function.
- **Software Download Management:** This function is responsible for identifying, locating, and triggering the suitable protocol for software for download, as well as for controlling the steps prior to, during, and after the download. The target software is fetched from the appropriate repository under the control of a RSF.
- **Service Provision:** This entity is responsible for interaction between the RMP and the application/service. It accepts and processes reconfiguration requests for the network, in order to provide the necessary environment for an application and service to execute. Additionally, it provides feedback to the application about the feasibility of the request, and can also initiate a reconfiguration command on behalf of the application; for example, it can initiate network configuration changes or selection of different settings by the users, or it can initiate mobility-related actions. Furthermore, the Service Provision function can trigger service adaptation actions based on network or device capability modifications, or based on updated policy conditions. Finally, roaming issues for service provisioning are also tackled by the Service Provision functional entity.

Layer Management Functions

In order to accomplish end-to-end reconfiguration, traditional layer management functions must be enhanced to collaborate with the RMP plane management functions. For example, functions for O&M may be exploited for the service provision stage, and thus should be adapted on the basis of input related to the definition and enforcement of reconfiguration policies. End-to-end differentiation of reconfiguration services should also take into account the outcome of reconfiguration functions for O&M, such as monitoring the reports and capabilities of network elements.

O&M functions can be classified into five categories: Application, Service, Content, and User-centric (ASCU) functions; OS-specific; Network-centric; RAT-centric; Device-specific. The provision of customer care information is a typical example of an ASCU-centric O&M function. Logging is an important feature, offering the history of reconfiguration actions (e.g. recent OTA upgrades), statistical information on the latest faults, and alarms reported to the user, and so on. The other four categories of O&M functions are described as follows:

- *OS-specific O&M functions*: These coordinate the auditing, testing, and validation procedures at the Reconfigurable User Equipment (UE).
- *Network-centric O&M functions*: Addresses the impact of mobility and QoS on the SD process, also dynamic network planning and its impact on traffic split, which comprise important O&M functions for reconfigurable network elements.
- *RAT-centric O&M functions*: Manages RAT-specific issues for a single RAT, or guarantee efficient collaboration of multiple RATs. The Composite Radio Environment Management function handles stability, conflict resolution, and certification issues, and ensures proper collaboration between network infrastructure manufacturers and terminal providers in the reconfiguration process. The Radio Element Management functional entity cooperates with the Performance Management RMP plane management entity. Analysis of RAT-specific performance data is an example of performance management, which may in turn affect real-time reconfiguration. The Function Partitioning and Reallocation entity coordinates coupling issues, as well as the distribution of functional entities for multi-RAT environments owned by a single administrative authority. Finally, the Inter-working function verifies the correct operation of control-plane functionality between radio-elements owned by different operators, as well as in network-sharing scenarios.
- *Device-specific O&M functions*: These include, for example, functions for UE Management. Although security hazards exist, remote equipment diagnosis assists in the remote identification of equipment faults. Coordination with HAL configuration modules can also be accomplished through device-specific RMP O&M functions.

Radio Reconfiguration Supporting Functions

The work on the definition of a general reconfiguration procedure started in the TRUST project [58], in which high-level system schematic descriptions were provided. The following phases were identified as taking part of the reconfiguration process:

- *1st phase*: Available Mode Lookup;
- *2nd phase*: Negotiation;
- *3rd phase*: Decision-making;

- *4th phase*: SD;
- *5th phase*: Location Update.

In addition, a number of reconfiguration management strategies were identified for the negotiation and decision-making stages. To develop all of these phases, a modular approach was used in TRUST [72].

The SCOUT project continued the detailing of each stage. Work was focused on the definition of the functions needed to address each step, the network elements involved, and the signalling flows between them. Figure 8.23 illustrates the functional architecture of the Radio RSF, instantiated in the Proxy Reconfiguration Manager (PRM) located in the RAN; this is further grained in the C-plane [Software Download and Reconfiguration Controller (SDRC)] and User-plane (SPRE).

Inter-working covers the necessary requirements needed for providing interoperability between different RATs. A high degree of integration between different RATs should be supported, improving network efficiency for important functionalities such as inter-system vertical HOs. The SCOUT project analysed Tight and Very Tight Coupling approaches, in terms of their impacts on the reconfiguration process of a terminal, as well as the consequent distribution of functions in network elements [73].

Figure 8.24 is a composite illustration showing the proposed message exchange process between the Radio RSF entities involved in the different steps of the reconfiguration process.

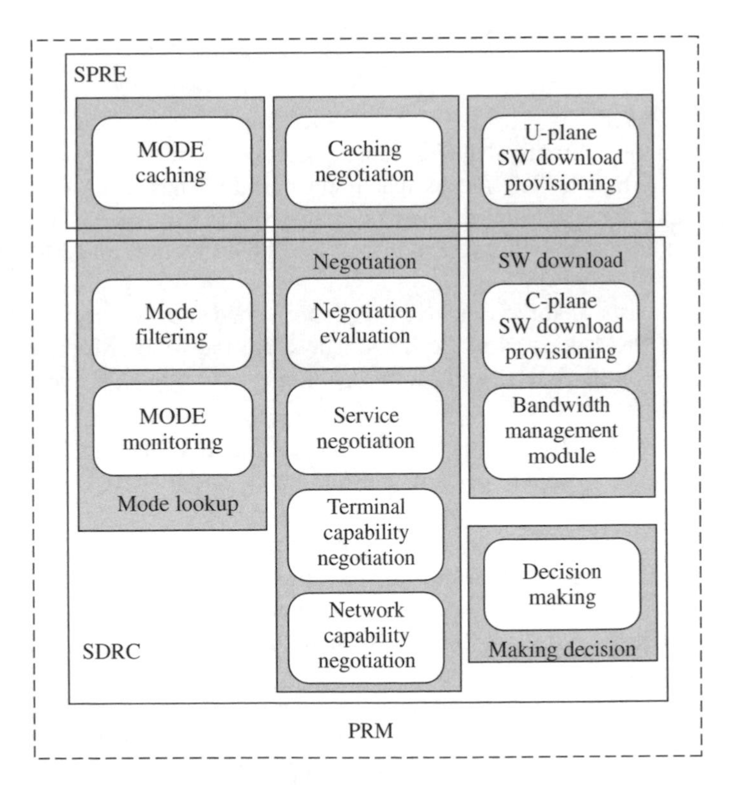

Figure 8.23 Radio RSF architecture

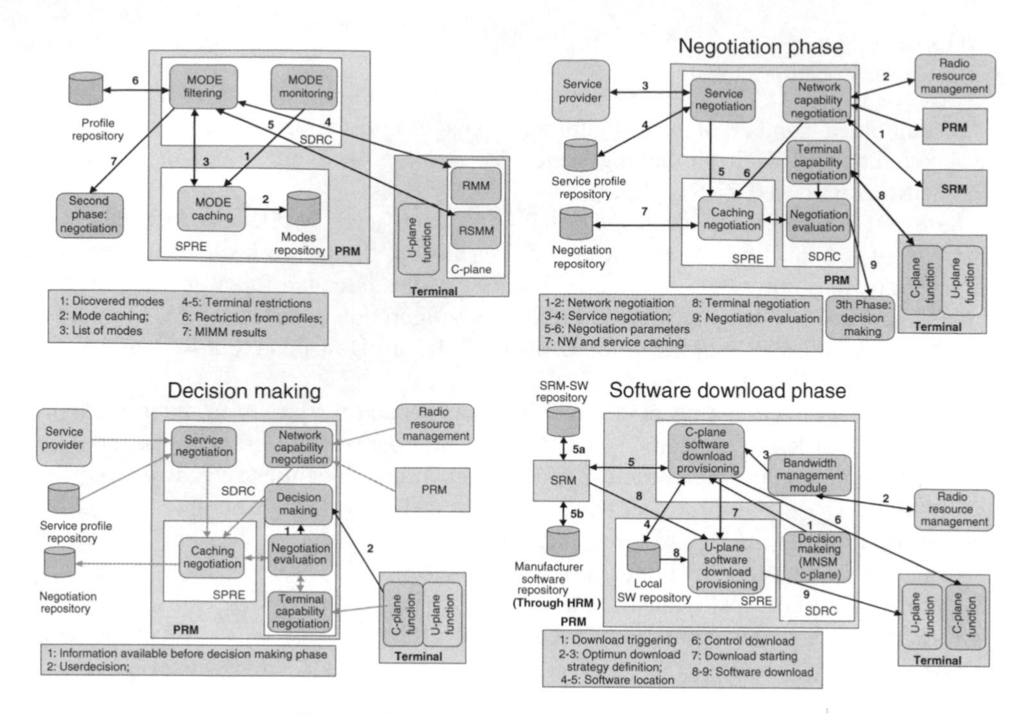

Figure 8.24 Reconfiguration process phases

Available Mode Lookup

The identification of alternative modes can be network assisted or not, depending on the status of the terminal. This means that if the terminal has insufficient resources to autonomously scan the spectrum, the network can provide this kind of information. The module specified to address this task is the Mode Identification and Monitoring Module (MIMM).

The MIMM consists of three functional blocks mapped into the user (U-) and control (C-) planes of the PRM, located in the RAN. These are the Mode Monitoring and Mode Filtering functions in the SDRC, and Mode Caching function in SPRE as defined below.

- *Mode Monitoring*: This function is able to discover, identify, and monitor existing alternative modes. It is also able to perform measurements, and it provides a list of available modes to the Mode Filtering process.
- *Mode Filtering*: When the reconfiguration process is ongoing, this function interacts with the Mode Caching function in order to retrieve the list of alternative modes, and then filters them according to the criteria specified both by the user and the terminal constraints. Mode Filtering provides a list of modes with the minimum requirements to address to the Negotiation phase.
- *Mode Caching*: This function manages the information obtained from Mode Monitoring, and stores it in a Repository. This function interacts with the database where profiles are stored. The role of this repository is to share, with all the terminals, a list of available modes.

Figure 8.24 illustrates the locations of the functionalities in the PRM and their role in the 'Available Mode Lookup' mechanism in detail.

Negotiation

The Available Mode Lookup phase has the goal of providing a list of available RATs to the Negotiation phase, addressing mainly user- and terminal-specific criteria. The Negotiation phase then executes a filter based on Service, Network, and Terminal Capability Negotiations (TCNs). Networks should report information that could be reused for future reconfiguration processes.

Service Negotiation (SN) has the aim of retrieving service capabilities related to services offered by a particular Service Provider, while the Network Capability Negotiation (NCN) has the main goal of identifying the QoS that will be offered to the user by a particular RAT.

A further filter to the RAT mode list is provided on the basis of Service and Network negotiation, and results are cached in the network near the terminal (i.e. in the PRM).

Finally, if the reconfiguration is ongoing, the terminal interacts with the network and negotiates the Terminal capability, matching the status of the terminal resources with the requirements of the RATs acquired in the previous phases – this is in order to find the best solution for a possible switching mode. Results of this evaluation are sent to the Making Decision phase. The module specified to address the Negotiation phase is the Mode Negotiation and Switching Module (MNMS). It is composed of the following functionalities:

- *Network Capability Negotiation Function (NCN)*: Gathers information about the network resources available for new modes (i.e. for QoS reasons).
- *Service Negotiation Function (SN)*: Retrieves the service requirements needed to support the service provisioning.
- *Terminal Capability Negotiation Function (TCN)*: Interacts with the terminal C-plane in order to gather information on dynamic terminal parameters (i.e. remaining power, available memory, etc.)
- *Cache Negotiation Function (CN)*: Results from the NCN and SN functions are stored in Negotiation Repositories by the Cache Negotiation Function.
- *Negotiation Evaluation (NE)*: Collects information from the Negotiation Repository and TCN function, and matches this information to find the best solution for a possible mode switch.

Decision-making

Information collected in the previous phase is input to the 'Decision-making' algorithm, which aims to understand whether the reconfiguration is feasible or not.

- The Decision-making function is located in the MNSM module, is mapped in the C-plane of the PRM and interacts with the caching negotiation module and the terminal.

Software Download

This phase consists of the requested software module's download to the device. The module to support this process is the SDM (SD Module) located in the U-plane. The C-plane also participates in the download, providing control functions and signalling.

After the definition of the optimum download strategy, the Software Download Management (SDM) is in charge of downloading the software from a certain entity in the network, for example, a software cache situated within the PRM or the Home Reconfiguration Manager (HRM). It is composed of the following functions:

- *Bandwidth Management Module (BMM)*: Depending on several important variables, this calculates the optimum download strategy. It interacts with the RRM in the network.
- *SW Download Provisioning (SDP) function*: The SDM also interacts with a local software Repository. The SDP function is able to download the software from the network to the terminal.

Location Update
This phase has the goal of supporting terminal mobility during the reconfiguration process. This task is supported by the LUM (Location Update Module). In micro-mobility situations, the functionalities supporting terminal mobility should be located in the PRM (C-plane), while in macro-mobility situations the management of mobility should be done within the core network.
Base Station RSF
RSF for BSs implement BS Service Profile Management, describing the Hardware, Software, and Functional (e.g. Air-interface) capabilities; SW Download Management of additional/new software modules required for a specific configuration; Load/Traffic Management to control the allocation of resources to a specific standard; and Performance/Load Monitoring to observe the hardware and software resources within a BS.

8.4.1.3 Enabling Procedures

One-to-many Software Download Protocol for End-to-end Mass-upgrade Downloads
Mass upgrades are a particularly important area of research at present. To provide software, firmware, or other information to MTs for mass upgrades, OTA downloads ultimately offer a panacea, as is the case for single-user downloads [74]. Through OTA downloads, no prerequisites are needed to create channels for the SD as it is assumed that wireless channels, or an automated means to create them, already exists ubiquitously. Hence no effort from the user is needed to perform the SD, and it is ensured for security purposes that the download can be enforced if the terminal is to be allowed to use the resources of the system.

One-to-many Downloads at the Network Layer
At the network layer, there are three basic methods for one-to-many SDs: n-times unicast (where n is the number of receivers of the SD), which we will refer to as n-unicast, reliable multicast, and reliable broadcast. Practically, mass upgrades over the Internet are usually performed using n-unicast connections to receivers, because receivers are likely to initiate the upgrade at different points in time. This is very inefficient, as the same information is sent repeatedly, particularly on links close to a source. Future one-to-many SDs should be achieved using reliable multicast for improved efficiency, and a range of reliable multicast protocols have been envisaged for this purpose [75, 76] (see [77] for some further examples). However, many reliable multicast solutions suffer from operational difficulties, which cannot easily be overcome.

Given the challenges introduced by the use of reliable multicast, TCP n-unicast would be advisable if there were only a small number of receivers of the download. TCP is far superior to reliable multicast in congestion control and flow control performance, and it is also able to offer a separate data rate to each receiver of the download (i.e. it is multi-rate), whereas many reliable multicast protocols only offer the download at the rate that the worst performing receiver in the group can sustain (they are single-rate). Alternatively, given a very high proportion of receivers of the download in the network, multicast routing and state requirements in the source and intermediate nodes might be excessive and unnecessary, so broadcasting of packets might be a better option.

For a range of reasons in mobile communication scenarios, it is unlikely that all receivers would be able to, or wish to, receive the download starting from the same point in time. Hence a mechanism should be incorporated allowing receivers to join the download after it has been initiated. Furthermore, at some point in the download receivers might forsake participation, due to leaving a radio resource coverage area or the loss of terminal power, for example. For these reasons amongst others, the number of download receivers and hence the optimum form of one-to-many download methods might change during the download, and consequently, a switching scheme between one-to-many download methods would be beneficial [78].

Unified Protocol

From a high-level perspective, there are two basic approaches for the implementation of dynamic switching between one-to-many download methods [78]. Bridging schemes may be applied between existing protocols, which are relatively simple to achieve but exhibit bad performances. However, a unified protocol, merging elements of n-unicast and multicast/broadcast techniques, would provide far better computational and operational efficiency than bridging. A unified protocol at receiver and source sides is depicted in Figures 8.25 and 8.26 respectively.

To trigger the switching process, we define four threshold values:

Tu = Threshold for switch from multicast to n-unicast

Tm1 = Threshold for switch from n-unicast to multicast

Tm2 = Threshold for switch from broadcast to multicast

Tb = Threshold for switch from multicast to broadcast

where Tu < Tm1 < Tm2 < Tb. We have introduced a form of hysteresis in return switches to improve the stability, for example, switches from multicast to n-unicast occur at a lower value of the considered metric than associated switches from n-unicast to multicast. This averts unnecessary switching as a result of small fluctuations in the metric.

8.4.1.4 Security

In using a reference model for the reconfiguration system, threats and security objectives related to reconfiguration need to be thoroughly understood. To achieve the overall objective of providing a reliable service that fulfils the expectations of all involved stakeholders, the following points have to be addressed:

- the provision of a secure SD and execution environment to protect against malicious (or non-operational) software;

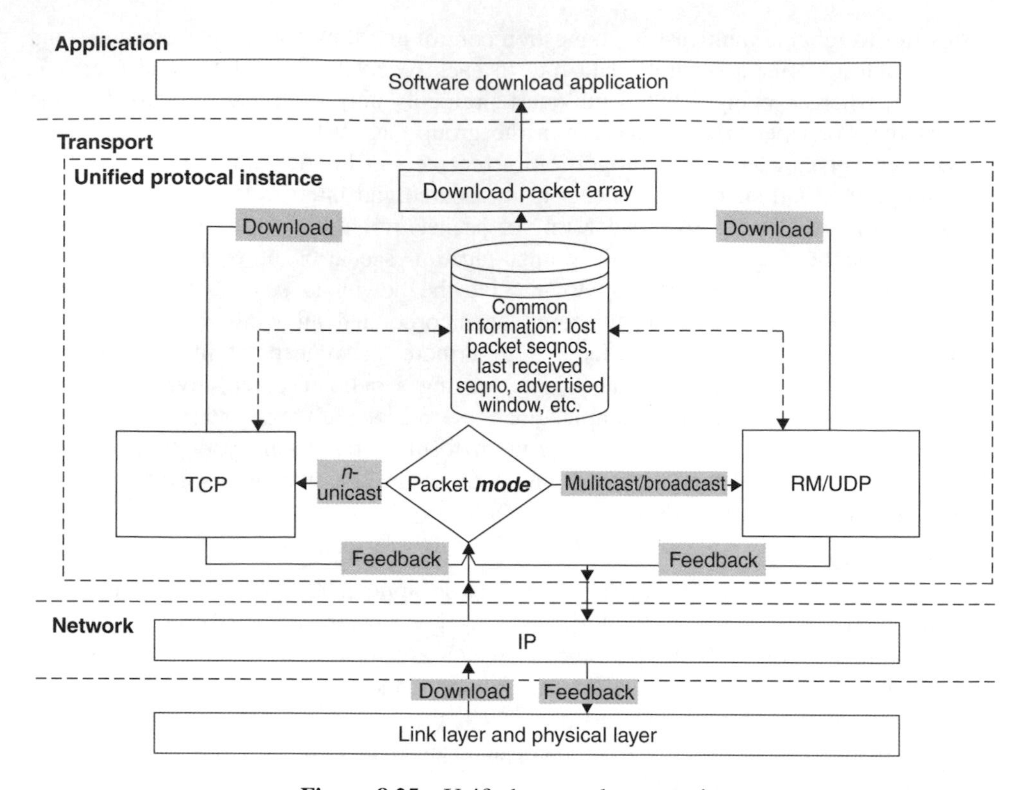

Figure 8.25 Unified protocol at a receiver

- specification of a secure reconfiguration process to ensure that the configuration matches the preferences and expectations of the end-user, service provider, and NOs;
- the control of radio emissions to ensure that, irrespective of the possibility to reconfigure a terminal's radio part, specified conformance parameters cannot be invalidated.

Suitable splits of responsibility between the control and the reconfiguration processes, as well as different approaches regarding the responsibility and technical means to ensure conformance must all be investigated. The most suitable chosen approach depends on the required flexibility, end-user, and/or service provider preferences, as well as the economic/business environment.

For reconfiguration control, the following scenarios need to be investigated:

- *Centralised Reconfiguration Control*: A single, central entity is responsible for deciding on and performing the reconfiguration. This means that the entity decides on which software (configuration) must be downloaded and installed on a reconfigurable device, and on when to do the reconfiguration. The reconfiguration control can lie in a network-based entity, or be controlled directly by the end-user. Candidate stakeholders controlling the reconfiguration must be identified; in particular, possible stakeholders are the user's (communication) service provider, the network operator (in case this is different from the service provider), or an independent reconfiguration service provider

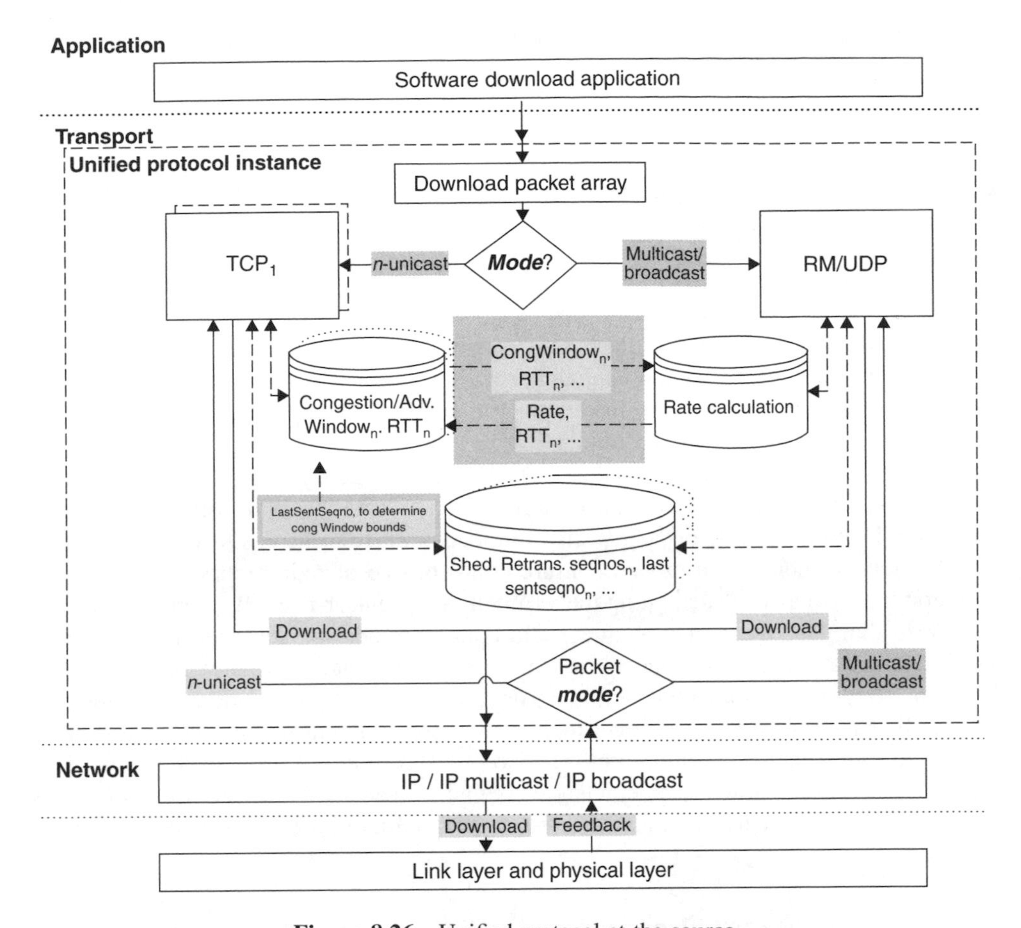

Figure 8.26 Unified protocol at the source

(which must in any case be certified and approved by the service provider). Some form of distributed control for the reconfiguration can be achieved by using a central reconfiguration decision authority (RDA), which acts as an arbiter or mediator to solve possible conflicts of interest between the involved administrative domains. While the final decision is still taken by a central entity, several stakeholders can signal status information and proposals to perform a reconfiguration, and therefore can influence the reconfiguration.

- *Decentralised Reconfiguration Control*: This is the more challenging approach where control of the reconfiguration is distributed between administrative domains. This means that possible conflicts of interest have to be solved directly by the involved administrative domains and the reconfigured device, without relying on a central entity to take the final decision. In particular, the split between a currently used visited network and the user's service provider needs to be investigated. While the current network has detailed information about local radio link properties and current network conditions, it is the user's service provider who is responsible to ensure reliable provision of subscribed services.

Conformance for radio emissions can be ensured on the basis of the following approaches:

- Reliance on a trusted provider for radio software (a manufacturer or independent third party);
- Network-based configuration validation;
- Terminal and/or network-based radio-emission monitoring to identify and deactivate incorrectly configured radio equipment.

While using a network-based server to validate an intended radio configuration has been discussed in literature, it has been left open as to which properties can actually be verified, as well as whether the properties that can be verified are sufficient to ensure conformance of radio emissions and/or standard compliance. However, network-based tests and checks can be used to ensure reliability; thereby reconfiguration attempts can be prevented that would be rejected later on by the terminal in any case.

In TRUST, an approach has been developed to monitor radio emissions, either terminal-based or network-based, and to detect and deactivate terminals with a rogue radio configuration. From the regulatory perspective, it needs to be clarified as to which properties are sufficient to monitor, in order to ensure conformance of radio emissions acceptably. Furthermore, instead of deactivating the terminal, a more user-friendly approach has been developed that deactivates the 'Rogue' radio configuration, and activates a fixed failure-mode configuration instead, that allows the terminal to contact a network-based repair function. The terminal may contact this repair function to obtain a working configuration, without being required to be brought to a service point or returned to the manufacturer. This fixed failure-mode configuration can also be used to ensure that, irrespective of the currently used configuration, emergency calls can be set up. However, since this issue deals with service deactivation, non-authorised terminal modifications should be carefully analyzed.

8.4.1.5 Flexible Network Elements

The Dynamic Network Planning and Flexible Network Management (DNPM) [79] concept investigates network-planning issues for coupled networks, based on reconfigurable network elements such as the BS and Radio Network Controller (RNC). It aims to deliver the principles of network reconfiguration, and to show the advantages of reconfigurable networks for capacity enhancement. In order to deliver reasonable results, both analytical and simulation approaches should be carried out by means of task splitting amongst work packages and activities. The demanded dynamic network simulations, taking into account multi-standard radio network elements, must be performed, and recommendations for network planning must be derived. Automatic network planning is another use-case for reconfigurable, multi-standard network elements (e.g. autonomous selection of carrier frequencies, etc). Mechanisms and signalling for such 'self-tuning' RAN's must be developed.

Compared to conventional situations, additional capabilities that arise from composite radio infrastructures include the abilities for alternate RATs to be activated in network segments and for available resources (spectrum to be assigned to RATs taking into consideration varying temporal and spatial spectrum needs in reconfigurable communications). There are several key tasks that are covered by network planning:

The modelling task is a logistic task for fulfilling network reconfiguration, in order to design the system signalling supporting reconfigurability. This task delivers time-variant and site-variant traffic intensities, with detailed models for Busy Hours (BH) in different service types, as well as combinations of models for different traffic splitting policies. The reason to model time-variant and site-variant traffic types and user behaviours is to test the performance of advanced network management schemes, including network reconfiguration working in an outer-loop and joint RRM and spectrum-sharing schemes being modelled as an inner-loop.

Theoretical work to test the requirements of the radio interface, can be compared to trials from snap-shot simulation in typical scenarios; different network configurations can be tested, where the network configurations include: Feasibility of setting radio interfaces; Location of BSs of available air interfaces; Propagation parameters; Antenna patterns; Coupling structures and depth among sub-networks; Policy of JRRM; Statistical values of required spectrum in different scenarios with available RATs. In the optimisation campaign, algorithms like Greedy, Taboo Search, and Simulated Annealing can be considered.

During network performance evaluation procedure, the following policies from reconfiguration feasibility viewpoints are taken into consideration: Joint admission control and adaptive radio multi-homing policy (e.g. whether alternative access is allowed, whether simultaneous connections are allowed); maximum delay, mean delay for sub-streams if traffic has been split.

The output of a problem that handles such network-planning instances includes the network elements (BS) that will be deployed and their locations, the selection and configuration of multiple RATs in certain elements of the network, the partitioning of traffic to various RATs and networks (e.g. from UMTS to 4G or vice versa, from UMTS to BRAN, etc.) as well as the resources per BS/RAT.

New functions required to exploit the benefits flexible network management are as follows:

- *Real-time network reconfiguration*: The inter-working between performance management and the radio-element manager can be reused for the purpose of observing, triggering, and evaluating the changes of the radio network. However, the existing architecture and signalling only support the changes of the selected implementation parameters.
- *Optimum radio environment setting*: Different levels of radio entities will be supported. The end-to-end reconfiguration aims at optimum reconfiguration with the involvement of reconfigurable MT and the radio network. The network management architecture is designed to optimally schedule the network reconfiguration and terminal reconfiguration.
- *Function partitioning and reallocation*: A very important character of the reconfigurable radio-elements is the changing of functions and changing of associations. A typical example can be given on the basis of the hotel Base Transceiver Station (BTS)/RNC concept. The signal from one RF set can be processed by one node B or another from time to time, which allows the enhancement of the system reliability, availability, and room for the design of advanced RRM.
- *Flexible network management*: The statistical behaviour of the user traffic shows the features of the traffic as a temporal-spatial variant with short coherence time. It requires

the database management to be efficient targeting at the optimum selection of the effective parameters, which triggers the network reconfiguration. For instance, a local jitter/variance of the traffic can be amended by a local implementation parameter change, whereas, statistical thorough traffic change with confidence of long-term analysis will result in a big change of the radio network, for example, RF reconfiguration and function reallocation.

- *Distributed intelligence*: From the nature of the hierarchy architecture defined by 3GPP, indirect inter-working between radio-elements in different levels becomes inefficient, for example, between node B and the extension points, between the neighbouring node Bs, and so on, are envisaged to be included in the emerging reconfigurable system. In addition, the operators might operate the same radio network (network-sharing scenario) or allow coupling the network architecture with each other. In this case, strategies implicating the signalling design must be developed, which supports efficient reconfiguration. The possibilities of establishing common databases being accessed by different operators are under investigation, which targets at a shared network infrastructure.
- *Network reliability*: Sudden changes of the radio network resulting in degradation of the availability should be avoided, whereas, the loss of availability is measured by the dramatic drop of ongoing calls or the high probability of the high rate of call blocking for the admission control functions. Therefore, steps ensuring the reliable network reconfiguration and the levels of network reliability should be reached. Decision-making process enhances the network reliability based on accurate traffic estimation.

8.4.1.6 One-to-many Software Download: Unified Protocol Concept

The performance advantages of using unified one-to-many SD protocol, able to dynamically switch between one-to-many download methods, are wide ranging. For example, consider a switch from n-unicast to multicast. Given a bridging scheme between protocols, it is likely that congestion control in multicast mode would have to start from a transmission rate of zero; taking time to converge and causing considerable packet loss in the process (see Figure 8.27, [78]). A unified protocol would be able to share information more easily between one-to-many download instances, hence could inherit/transform congestion control measurements into the target mode and in many cases would maintain the transmission rate throughout a switch.

Further advantages of using a unified protocol are compelling: drastically reduced memory consumption (see Figure 8.28, [78]), improved processing efficiency across the transport layer, and the flawless continuation of the download throughout the switching process amongst many others.

Realisation of the Unified Protocol Concept

Much work is being undertaken to extend the logical implications of the unified protocol architecture for dynamic switching, as well as to consider its incorporation into a range of systems in more depth. We hope to achieve an actual implementation of a unified switching protocol in the medium–long term, but in the short term we are improving some of the finer details of the protocol's architecture.

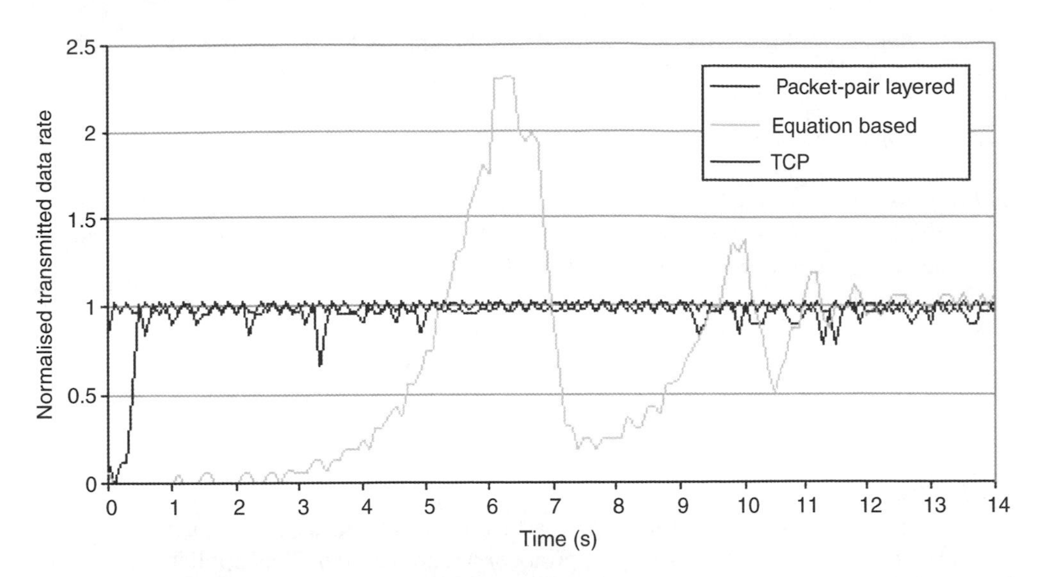

Figure 8.27 Plot of convergence to available bottleneck data rate of layered and equation-based multicast, compared to TCP

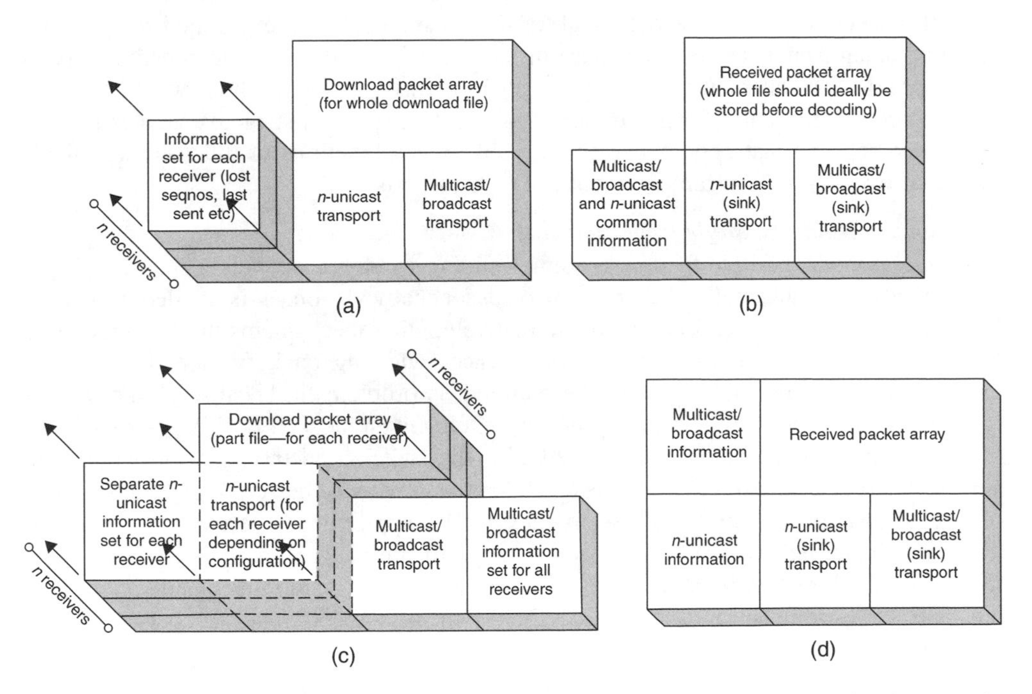

Figure 8.28 Volumetric depiction of memory usage for the unified protocol approach (a) at the source, (b) at a receiver; alternatively, for a form of bridging between protocols (c) at the source, (d) at a receiver

Particular areas of research interest for the unified protocol concept include the following:

- *Packet mode marking*: n-unicast or multicast/broadcast mode must be marked on each packet. How is this achieved?
- *Progression of congestion window edges immediately after switching from multicast to n-unicast*: This is problematic because of the lack of in-flight positive feedback immediately after a switch to many-unicast;
- *Further refinement of parity transmission approaches*: To further optimise dynamic switching (particularly between n-unicast and multicast) and late receiver joining, in addition to providing improved recovery from wireless losses;
- *Incorporation of layered multicast into the dynamic switching protocol*: Requires in-depth consideration of the benefits.

Packet Mode Marking

The mode to which they apply must be marked on packets in order for them to travel to the correct part of the unified transport protocol upon reception. This applies for both the source and receivers, hence download and feedback packets must be marked.

For reliability and congestion control purposes, there are two approaches used in the unified protocol: TCP-like unicast and reliable multicast. Only one bit is therefore required to mark packets in order to select which of these parts packets are destined for. This bit could be taken from common free space in the transport headers for the constituent parts of the protocol, or alternatively two ports could be used, both of which would apply to the protocol. Clearly, the former option is preferable, so an area of research is to look for a way of using that approach with some likely combinations of unicast and reliable multicast over User Datagram Protocol (UDP) transports.

Switching to N-unicast Mode from Multicast Mode

In n-unicast mode, positive feedback (using a sliding-window mechanism similar to TCP) is assumed. In reliable multicast/broadcast modes, negative feedback is applied in order to limit the risk of a feedback implosion; alternatively, other mechanisms in which feedback traffic load is sparse might also be used. Hence switching back from multicast to n-unicast mode, should such a switch be required, is problematic because of the lack of in-flight positive feedback at the time of the switch. Given no such positive feedback, the sliding-window edges would not proceed after the switch, and transmission would halt abruptly. The solution suggested by us in [78] simply involves the automated transmission of packets, at the correct transmission rate, for an approximate period of one round-trip time. However, we are working on a range of alternative approaches to maintain transmission after the switch, using an interim measure. The important considerations here are that congestion control performance should not be impeded and acceptable fairness and reactivity must be maintained.

Parity Transmissions

We are working on a range of analyses and simulations of parity coding performance in general, particularly with a view to coding parity over the whole file as one, instead of coding in blocks. Coding over the file as one offers several performance advantages. Namely, each received parity packet is able to correct any original data packet from

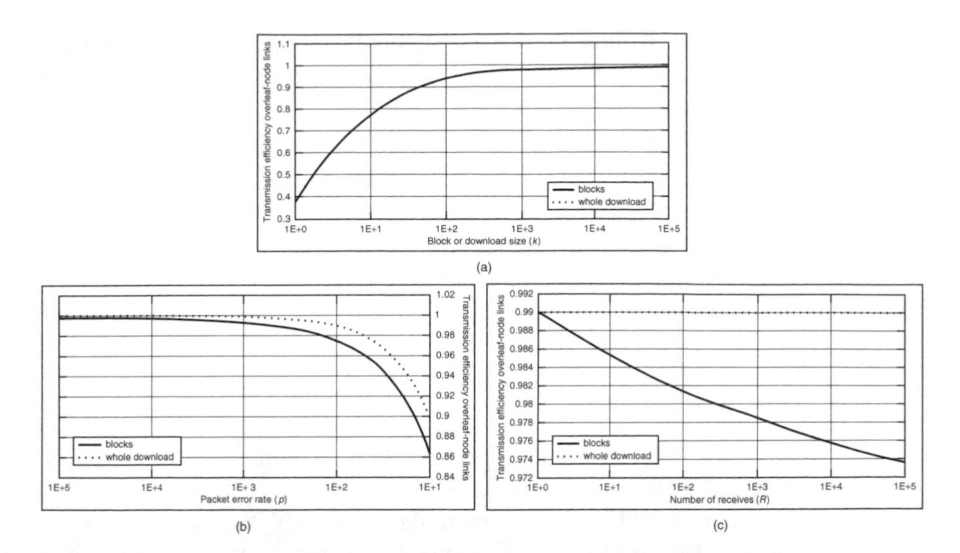

Figure 8.29 Transmission efficiency over leaf-node links for the 'whole download' and 'block' parity coding approaches. $k = 1000$, $p = 0.01$, and $R = 1\,000$, although these are varied in turn in the plots

the whole file, instead of just a subsection (block) of the file; furthermore, a scheme can be employed allowing receivers to automatically leave the download once they have received enough packets to reconstruct the download. In recent papers, we have quantified these benefits both through analysis and simulation. Figure 8.29 shows the transmission efficiency performance gains that can be achieved by such an approach. We plan to significantly expand on this work in the context of our unified protocol.

This work is relevant because, further to the above advantages, parity transmission if coded over the whole file as one assists our unified dynamic switching transport protocol in the following ways:

Late Receiver Joining
Parity transmission helps recover the download at late-joining receivers. This is important in accounting for the temporary lack of availability of some receivers at the start of a download, where such a requirement is particularly relevant in large one-to-many downloads to MTs.

Switching between N-unicast and Multicast
If a download was in n-unicast mode for a period of time before a switch to multicast, receivers would have obtained different numbers of packets of the download at the time of the switch. This is because a different transmission rate to each receiver is allowed in n-unicast mode.

Given conventional retransmissions, after a switch from multi-rate n-unicast to single-rate multicast, the source would have the option of either continuing transmitting from min{lastSentSeqnon}, max{lastSentSeqnon}, or a point in between. Continuing from min{lastSentSeqnon} is the simplest choice, but this would erode many of the advantages of the prior n-unicast mode through data repetition to many receivers. Continuing from max{lastSentSeqnon} avoids repetition, but requires the later retransmission of sequence numbers min{lastSentSeqnon} through to max{lastSentSeqnon}. In a subsequent retransmissions pass, these packets could probabilistically be a cause of inefficiency to those receivers not requiring them.

Alternatively, the transmission of parity packets in the subsequent pass ensures optimum transmission efficiency, as each transmitted parity packet is useful to every receiver.

Parity Transmission Orders
In our unified protocol [78], and in the related work we are performing, we suggest possible approaches for parity transmission ordering (see Figure 8.30). Further work is being undertaken to analyse and simulate such approaches. Particular areas of interest include parity processing loads at receivers for the transmission orders, and the overall transmission efficiency implied by these respective approaches.

Layered Multicast
We are also working on schemes for incorporating layered multicast into the switching protocol concept. Clearly, the use of broadcast in conjunction with layered multicast is not an option, because it would require flooding the network with the information in all layers of the multicast. Hence in such a scenario the dynamic switching scheme would only operate between n-unicast and multicast. Furthermore, such are the performance gains of using layered multicast in preference to equation-based multicast congestion control, that layered multicast may indeed be sufficient as a stand-alone solution in place of dynamic switching, should sufficient technologies exist to achieve layered multicast in an acceptable fashion. In the light of this, the benefits of using layered multicast for the scheme require further investigation.

Figure 8.31 shows a basic transmission order for original data and parity packets using layered multicast [78]. Given the use of layered multicast, we particularly plan to look at performances of various distributions of parity and transmission rates amongst layers. It is

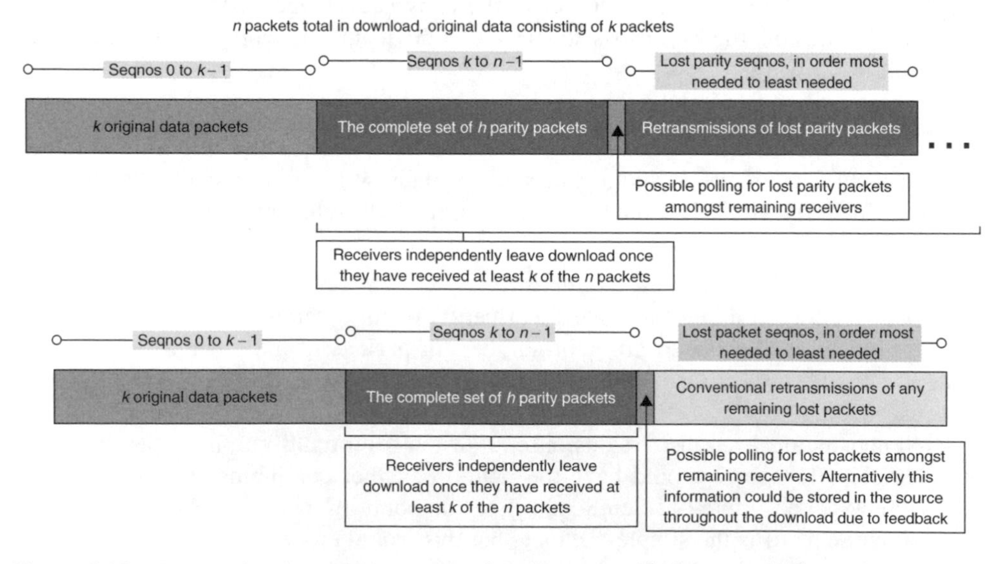

Figure 8.30 Two parity transmission approaches for our unified protocol (these also apply for reliable multicast downloads in general)

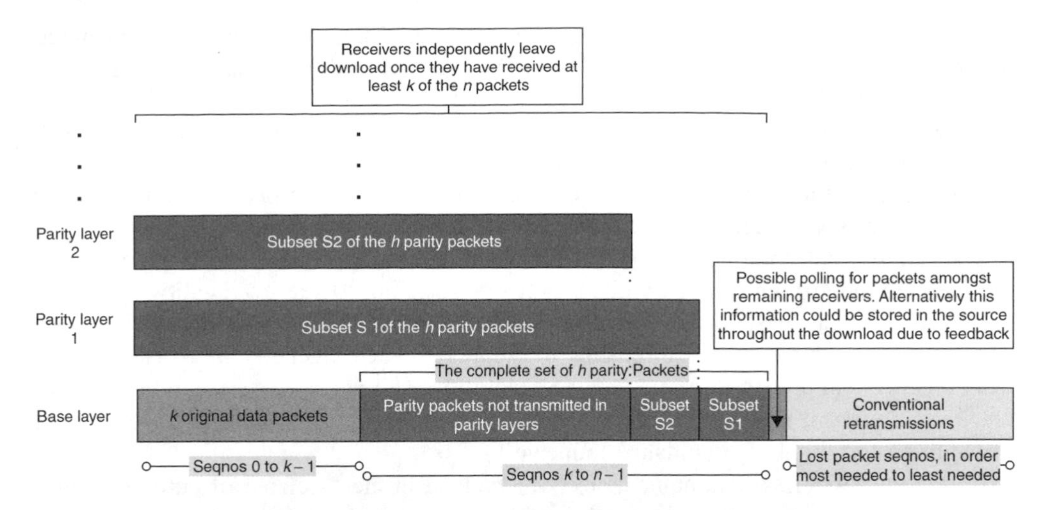

Figure 8.31 Packet transmission order for the layered multicast

unclear how different our conclusions would be from other established layered multicast mechanisms – this is to be found upon deeper consideration of the implications of our requirements.

8.4.2 Summary

This section has presented many areas of research undertaken by WWRF members that fall into the realm of reconfigurable networks. Since reconfigurability is envisaged to prevail in the world of telecommunications in the coming years; it is mandatory to embrace the overall system's characteristics from the network point of view in order to support reconfigurability at all levels. Hence starting from appropriate system architecture, this section presented the key issues that need to be addressed; these issues are topics of ongoing research particularly within the IST-E2R project. In addition to this, it has presented a novel concept for an end-to-end download transport protocol, which potentially offers many performance advantages over currently assumed one-to-many download techniques. This protocol offers an approach for dynamic switching between one-to-many download methods such as many-unicast, reliable multicast, and reliable broadcast. Moreover, the architecture for this approach has been introduced, highlighting the way in which elements of reliable unicast and reliable multicast transports can be merged into a single one-to-many download transport layer. As a proof of concept, we presented achievable gains due to congestion control performance and protocol memory usage that can be achieved through the use of a unified transport protocol. It has also shown the benefits of some of the facilitating technologies for the unified protocol. Furthermore, it has discussed some of the implementation complexities that should ideally be overcome to smoothly incorporate the protocol into a range of systems.

8.5 Cognitive Radio, Spectrum and Radio Resource Management

The innovative communications services and applications imposed by ever-increasing customer demands render necessary the development of efficient radio resource exploitation mechanisms and intelligent network-planning methods. The whole process of planning and managing a reconfigurable network must be reconsidered, in order to live up to the expectations created by the migration towards a new era of communications. This section deals with such issues, emphasising the role of new engineering technologies in reconfigurable networks. For this purpose, it presents the respective RRM, that is Spectrum Management and Joint RRM, and Network Planning technical approaches, aiming at their application in next generation wireless-access infrastructures.

The future of telecommunications is anticipated to be an evolution and convergence of mobile communication systems with IP networks, leading to the availability of a great variety of innovative services over a multitude of RATs. To achieve this vision, it is mandatory to embrace the requirements for support of heterogeneity in wireless-access technologies, comprising different services, mobility patterns, device capabilities, and so on. Furthermore, it is equally important to promote research in networking technology, through the provision of a guidance framework.

Present-day wireless communications, which stand at the forefront of current technological advances, comprise a multiplicity of RAT standards. Of these, the most commonly used are the Global System for Mobile communications (GSM) [1], GPRS [2], the UMTS [3], BRANs, various types of WLANs [4], [5], [6], and DVB [7]. Furthermore, during the period of this work, new wireless standards like Worldwide Interoperability for Microwave Access (WiMAX) have emerged. Moreover, the complete set of wireless technologies is currently being transformed into one global infrastructure vision, called the *Beyond 3rd Generation* (B3G) wireless-access infrastructure. This is aimed at offering innovative services, based on user demands, in a cost-efficient manner. Major contributing concepts towards this convergence are cooperative networks [8, 9] and reconfigurability [10].

The cooperative networks concept assumes that diverse technologies, such as cellular 2.5G/3G, BRAN/WLAN, and DVB systems, can be joint components of a heterogeneous wireless-access infrastructure. This allows an NP to rely on more than one RAT, dependent on the encountered specific conditions (e.g. hot-spot requirements, traffic demand alterations, etc.) at different times and in different areas. The NP may also cooperate with other NPs in order to make alternative solutions available for maximisation of QoS levels offered to users. Advanced management functionality is required to support the cooperative networks concept, and much associated research has been done in the recent past [11–14]. This envisaged functionality deals with the reallocation of traffic to different RATs and networks, as well as the mapping of applications to QoS levels.

The move towards the Reconfigurability concept was initiated as an evolution of SDR [15]. It aims to provide essential mechanisms to terminals and networks, so as to enable them to adapt dynamically, transparently, and securely to the most appropriate RAT dependent on the current situation [16]. Through reconfigurability, we envisage network segments being able to change RAT in a self-organised manner, allowing them to better handle offered demand. In this context, reconfigurability also allows for the dynamic allocation of resources (such as spectrum) to RATs [17]; consequently, many new possibilities are invoked with respect to the efficient management of the whole reconfiguration process. This management involves the available resources, mostly spectrum, as well as the joint management of demand deriving from the cooperative RATs. The configuration process and the respective management needed are presented in detail in Section 8.5.2.

In accordance with the above observations, this section has been produced within the WWRF WG6 Reconfigurability group, and is aimed at providing the basic principle that must be adhered to in order to make cooperating reconfigurable networks commercially successful. These principles lie in the effective management of the available resources, that is (i) more efficient utilisation of available spectrum, (ii) management of radio resources belonging to different RATs with fixed spectrum allocation, and (iii) an intelligent network-planning process.

This section is structured as follows. The first part considers the general characteristics of RRM, providing an analysis of RRM and requirements for the effective management of resources, along with associated technical considerations. A summary of some envisaged RRM solutions is also provided. The next part is dedicated to Spectrum Management. As spectrum today is a scarce resource, it is necessary to use it efficiently; cooperation amongst networks will assist the efficient use of radio spectrum in future communication systems. After that the section presents Joint RRM (JRRM), consisting of a feasibility study for JRRM, a functional overview of the proposed JRRM scheme, and some important JRRM-related research topics, along with a novel scheme for managing resources of different RATs, namely, Hierarchical Radio Resource Management (HRRM). Section 8.5.1.2 considers Dynamic Network Planning, which is an absolutely necessary task for network designers. Finally, the section discusses enabling technologies that can be utilised to facilitate the provision of Flexible Spectrum Management and JRRM, in a reconfigurability context.

8.5.1 RRM in a Reconfigurability Context

This section presents the basic principles of RRM in a reconfigurability context. Its requirements, technical aspects, and solution methods are presented.

8.5.1.1 Analysis of RRM

The E2R has a strong impact on all aspects of the system, ranging from the terminal, to the air interface, up to the network side. Future network architectures must be flexible enough to support scalability as well as reconfigurable network elements in order to provide the best possible resource management solutions in hand with cost-effective network deployment. The ultimate aim is to increase spectrum efficiency through the use of more Flexible Spectrum Allocation (FSA) and RRM schemes, although suitable load balancing mechanisms are also desirable to maximise system capacity, to optimise QoS provision, and to increase spectrum efficiency. Once in place, mobile users will benefit from this by being able to access required services when and where needed, at an affordable cost.

From an engineering point of view, the best possible solution can only be achieved when elements of the radio network are properly configured and suitable RRM approaches/algorithms are applied. In other words, the efficient management of the whole reconfiguration decision process is necessary in order to exploit the advantages provided by reconfigurability.

For this purpose, future mobile radio networks must meet the challenge of providing higher QoS through supporting increased mobility and throughput of multimedia services, even considering scarcity of spectrum resources. Although the size of frequency spectrum physically limits the capacity of radio networks, effective solutions to increase spectrum efficiency can optimise usage of available capacity.

Through inspecting the needs of relevant participants in a mobile communication system, that is the Terminal, User, Service and Network, effective solutions can be used to define the communication configuration between the Terminal and the Network, depending on the requirements of services demanded by the users. In other words, it is necessary to identify proper communication mechanisms among communications apparatuses, based on the characteristics of users and their services. This raises further questions about how to manage traffic in heterogeneous networks in an efficient way.

MTs are characterised by a wide range of capabilities, influenced by the RAT processing capability, for example, multi-modality, processing power, display size, and so on. Fast-developing SDR technologies enable MTs to be adaptable to different RATs, through SD procedures. Reconfigurability is an extension to SDR aimed at optimising communication, the mechanism between the terminal and network, comprising protocols defined across all OSI (Open System Interaction) layers.

From the user perspective, services are expected, which not only depend on a traditional single traffic type, but also multiple traffic types from one or more radio networks. Multimedia applications are becoming popular and are already beginning to demand wireless transport. They cover information types from voice, control data to audio, video or any combination thereof. Supporting multimedia traffic with data rates of even up to several Mbps (Megabits per second) is an important aspect of future radio communication systems. Mobile users might use the MT to access regular voice services in high mobility environments; however, they might use a laptop access to data services with higher data rates in low-mobility environments. For instance, the data rates offered by the emerging UMTS Frequency Division Duplex (FDD) system can be as high as 384 kbps (kilobits per second) for wide-area coverage, and several Mbps for local-area coverage, even using a single frequency band. With a multimode terminal, the users are not restricted to a single RAT. They are able to access a set of RATs alternately or simultaneously. The radio network is responsible for selecting the most appropriate accessible RAT, or a combination of RATs, taking into account user service and network information.

In conclusion, RRM is a complex process, but is necessary in the deployment of 4G networks. It consists of dynamically managing resources like the spectrum, as well as allocating traffic dynamically to the RATs participating in a heterogeneous wireless- access infrastructure. Consequently, RRM can be seen as a superset of Spectrum Management and Joint RRM (JRRM). This is the starting point for the decoupling of these ideas presented herein.

8.5.1.2 Spectrum Management Overview

The configuration of resources is recently a pertinent area of research in the plethora of telecommunications technologies. Current spectrum allocation approaches, fixed by nature, do not allow for the allocation of frequency bands to different RATs dynamically. However, the coexistence and cooperation of diverse technologies, which form part of a heterogeneous infrastructure, has brought about the possibility of flexibly managing the spectrum in a dynamic manner. No longer are fixed frequency bands guaranteed to apply to specific RATs, but conversely, through intelligent management mechanisms, bands can be allocated to RATs dynamically in a way such that the capacity of each RAT is maximised and interference is minimised.

In the context of this section, the work on spectrum management involves a brief presentation of the technical requirements for flexible spectrum management, some current research activities, and the vision of spectrum for the coming years, as well as related regulatory issues.

8.5.1.3 Joint Radio Resource Management (JRRM) Overview

In this section, we specially propose and study some mechanisms controlling the communication schemes between MTs and radio networks, which consist in a number of RATs coexisting together and cooperating with each other, but with fixed spectrum allocation. Such mechanisms are defined as the JRRM for the heterogeneous networks. The multimode capabilities are enabled by reconfigurability.

To design the wireless systems, we encounter typical problems, such as the signal attenuation, terminal noise, fast fading due to multi-path phenomenon, shadowing, Multiple-Access Interference (MAI) and other typical system related features, for example, the mutual relation between interference strength and duration period given by link adaptation. All these problems prevent us from using radio resource efficiently. On the other hand, the radio resource is not only, by definition, the radio spectrum, but also realised in the real radio network as, access rights for the individual mobile user, time period, a mobile user being active, channelisation codes, transmission power, connection mode, and so on, which require the management functions being designed in different timescales. Furthermore, radio resources from different radio networks can be managed jointly in order to solve the encountered problems more effectively. JRRM is therefore generalised as a set of network controlling mechanisms that support intelligent admission of calls and sessions, distribution of traffic, power and the variances of them, thereby aiming at an optimised usage of radio resource and maximised system capacity. These mechanisms work simultaneously over multiple RATs with the necessary support of reconfigurable/multimode terminals.

Moreover, in the context of JRRM, we propose the novel solution of HRRM. As the specifications for HRRM differ slightly being dependent on the exact usage scenario, we essentially give a generalisation of it. HRRM could either be implemented in entirety, or as a general concept that might be used as a facilitator for radio-efficiency, allowing sub-parts of RATs to be automatically dynamically partitioned. HRRM is a very basic idea that takes advantage of the fact that common core blocks of functionality are often present in RATs, which allows for the specification of complete RATs through a basic form of building blocks. These building blocks in essence constitute elements in a class-hierarchy, which are inherited sequentially to generate the complete RAT.

8.5.2 Spectrum Management

8.5.2.1 Introduction to Spectrum Management in a Reconfigurability Context

Along with the advent of the 4G era in telecommunications technology, new techniques must be developed for the intelligent management of spectrum among the RATs forming a heterogeneous infrastructure. The term reconfiguration applies not only to the selection among a set of available RATs in the service area, but also to the appropriate configuration of the resources utilised, that is careful selection of the operating frequency band. Such issues imply that there is a need for flexible managing of the spectrum and are dealt with in this section.

Furthermore, there is a tight relationship between spectrum management and cognitive radio. Flexible spectrum management is needed for wireless devices that operate in either the licensed band or the unlicensed band, or both, as Illustrated in Figure 8.32. Cognitive radio will provide the technical means for determining, in real time, the best band and best frequency to provide the services desired by the user at any time. In this section, even though separate subsections are provided for technical and regulatory aspects of spectrum management, the two are closely intertwined. For example, the European R&TTE Directive requires that the Declaration of Conformity by the manufacturer be a statement of conformity with: harmonised standard, and/or consultation with a Notified Body (the manufacturer remains responsible for the Declaration of Conformity).

Although many administrations outside Europe do not use procedures similar to the Manufacturer's Declaration of Conformity, the regulatory agencies (RAs) in these administrations typically do rely on technical specifications/standards/ITU Recommendations as part of their certification procedures. Thus, the technical, regulatory, and standardisation aspects of spectrum management are closely intertwined.

8.5.2.2 Technical Aspects of Spectrum Management

Technical Spectrum Management Requirements

This part aims at identifying the required PHY functions that radio equipments (UE and Radio Network Access Point (RNAP)) have to be enabled with so that multi-band

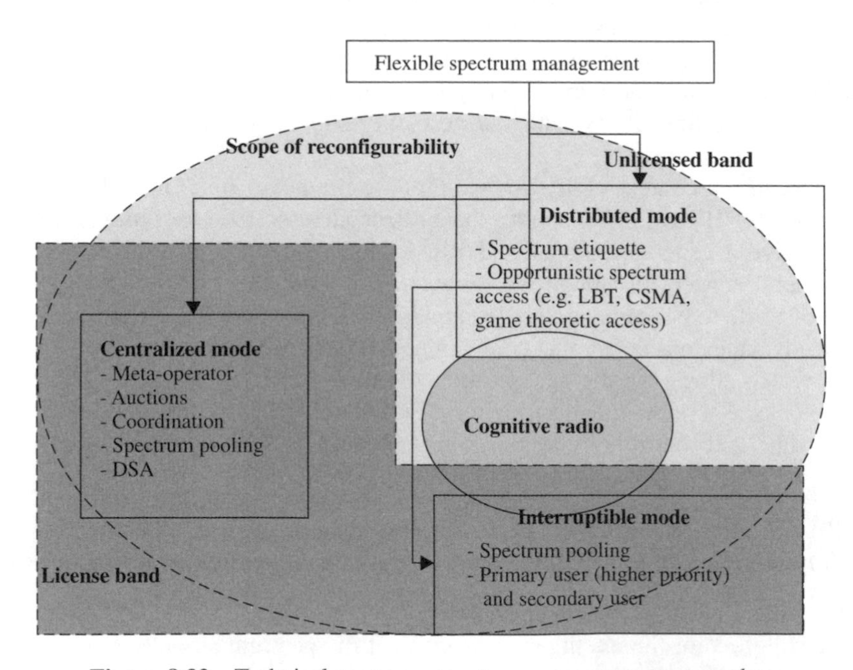

Figure 8.32 Technical scope on spectrum management approaches

capabilities are supported. The notion underlying the multi-band capabilities is discussed hereafter respectively from the frequency range, spectral resolution, spectral coexistence, and switching standpoints.

Current Spectrum Management Research

Spectrum Management Research in the United States

There is considerable research in the United States on spectrum management technologies including cognitive radio. This section provides a brief summary of several major research programs in the United States that are focused on cognitive radio and spectrum and RRM. This brief summary does not purport to cover all research in the United States in these areas, but it does give a sample of such ongoing research.

Some of the research on cognitive radio and related topics has been fostered by the Federal Communications Commission concepts on cognitive radio and interference temperature as described in the Federal Communication Commission (FCC) Spectrum Policy Task Force Report (Federal Communications Commission, 'Spectrum Policy Task Force Report', ET Docket 02–135, November 2002. Available at: http://www.fcc.gov/sptf/). This report noted that: 'Preliminary data and general observations indicate that portions of the radio spectrum are not in use for significant periods of time'.

The FCC report notes, however, that more information is needed in order to quantify and characterise spectrum usage more accurately so that the FCC can adopt spectrum policies that take advantage of the 'holes' in the spectrum usage.

Subsequent to the issuance of the Spectrum Policy Task Force Report, the FCC has issued other related documents that include the following:

- Facilitating Opportunities for Flexible, Efficient, and Reliable Spectrum Use Employing Cognitive Radio Technologies, Notice of Proposed Rule Making, ET Docket 03–108.
- Establishment of an Interference Temperature Metric to Quantify and Manage Interference and to Expand Available Unlicensed Operation in Certain Fixed, Mobile and Satellite Frequency Bands, Notice of Proposed Rule Making, ET Docket 03–237.
- Unlicensed Operation in the TV Bands, Notice of Proposed Rule Making, ET Docket 04–186.

These documents and the comments and reply comments may be found at:
http://wireless.fcc.gov/spectrum/proceeding.htm?pagenum=2

The FCC is pushing for a new paradigm for spectrum management – cognitive radio, policy-based radio, reconfigurable radios and networks, and revised RRM will be keystones of this new paradigm.

Defense Advanced Research Projects Agency

The Defense Advanced Research Projects Agency (DARPA)is developing a new generation of spectrum access technology under the NeXt Generation (XG) communications program [80–86]. Although this program is oriented towards military communications, it is applicable to advanced spectrum management for any band and any communications services. This multi-usage is analogous to the development of data communications protocols which were originated by DARPA and used in the DARPA-net, but which have evolved to what is now know as the World Wide Web. The motivation for this new spectrum access technology is the same for the commercial communications community as it is for the military communications community.

Apparent Spectrum Scarcity: The current method of allotting spectrum provides each new service with its own fixed block of spectrum. Since the amount of usable spectrum is finite, as more services are added there will come a time when the spectrum is no longer available for allotment. We are nearing such a time, especially due to a recent dramatic increase in spectrum-based services and devices.

However, as noted by the FCC, there are large portions of allotted spectrum that are unused. This is true both spatially and temporally. Thus, there are portions of assigned spectrum that are used only in certain geographical areas and there are some portions of assigned spectrum that are used only for brief periods of time. Studies have shown that even a straightforward reuse of such 'wasted' spectrum can provide an order of magnitude improvement in available capacity. Thus, the issue is not that spectrum is scarce – the issue is that we do not currently have the technology to effectively manage access to it in a manner that would satisfy the concerns of current licensed spectrum users.

The DARPA XG Program is pursuing an approach wherein static allotment of spectrum is complemented by the opportunistic use of unused spectrum on an 'instant-by-instant' basis in a manner that limits interference to primary users. This approach is called *opportunistic spectrum access* spectrum management and the basic parts of this approach are as follows:

- Sense the spectrum in which you want to transmit.
- Look for spectrum holes in time and frequency.
- Transmit so that you do not interfere with licensees.
- There are a number of research challenges to this adaptive spectrum management including the following:
  - Wide-band sensing.
  - Opportunity identification.
  - Network aspects of spectrum coordination when using adaptive spectrum management.
  - Need for a new regulatory policy framework.
  - Traceability so that sources can be identified in the event that interference does occur.
  - Verification and accreditation.

More information on the DARPA XG Program for spectrum access management may be found in general briefings available at the DARPA web site (http://www.darpa.mil/ato/programs/XG/) as well as in the following two documents that are also available at the DARPA web site:

The XG Vision – Request for Comments, v2.0, Prepared by BBN Technologies

The XG Architectural Framework – Request for Comments, v1.0, Prepared by BBN Technologies

National Science Foundation

The National Science Foundation (NSF) has a research programme titled 'Programmable Wireless Networking' (NeTS-ProWiN) [87]. This NSF research programme is addressing issues that result from the fact that wireless systems today are characterised by wasteful static spectrum allocations, fixed radio functions, and limited network and systems coordination. This has led to a proliferation of standards that provide similar functions – wireless

LAN standards (e.g. Wi-Fi/802.11, Bluetooth) and cellular standards (e.g. 3G, 4G, CDMA, and GSM) – which in turn has encouraged stovepipe architectures and services and has discouraged innovation and growth. Emerging programmable wireless systems can overcome these constraints as well as address urgent issues such as the increasing interference in unlicensed frequency bands and low overall spectrum utilisation.

The research sponsored by NSF under the Programmable Wireless Networking Program is addressing these issues by supporting the creation of innovative wireless networking systems based on programmable radios. The objectives of this NSF research program are to:

- capitalise on advances in processing capabilities and radio technology and on new developments in spectrum policy;
- improve connectivity and make more effective use of shared spectrum resources;
- enhance the wireless networks community by intermixing the networking, radio, and policy communities, integrating education with research through focused activities, and diversifying participation.

Programmable radio systems offer the opportunity to use dynamic spectrum management techniques to help lower interference, adapt to time-varying local situations, provide greater QoS, deploy networks and create services rapidly, enhance interoperability, and in general, enable innovative and open network architectures through flexible and dynamic connectivity. The specific research areas under this NSF Programme are as follows:

- dynamic spectrum management architectures (including sensor-based architectures) and technologies; includes investigation of issues such as techniques to implement policies, security and robustness, QoS, and enforcement;
- topology discovery, optimisation, and network self-configuration technology;
- techniques for the interaction among routing, topology, and administrative/network management including the development of the policy and security framework;
- flexible radios for networking research.

Software Defined Radio Forum

The SDR Forum has an embryonic program for investigating technical, regulatory, and market aspects of cognitive radio and spectrum efficiency. The technical aspects include the development of security requirements. Some spectral occupancy measurements have been made under the auspices of the SDR Forum. The results of these measurements will be useful in determining the parameters to be used in designing policy-based cognitive radio systems and identifying which portions of the band should be considered for the initial application of such radio systems.

Spectrum Management Research in Europe

The current scheme of allocating spectrum in a fixed manner provides the advantage that the equipment always knows about the operating frequency and system parameters of each RAT. However, communication traffic load is time and space dependent, hence introducing flexibility in the spectrum allocation will allow spectrum to be allocated in a manner that better matches the operators' and users' temporal usage/provision needs within a geographical area, thus optimising the use of spectrum.

The traditional way for allocation of spectrum was (is) to allocate a fixed amount of spectrum in a particular communication system to each operator. A more dynamic way of reallocating spectrum according to traffic needs would be required. To this purpose, a comprehensive set of new Advanced Spectrum Management concepts are introduced using SDR technologies as an enabler.

The goal of FSA needs an approach that identifies and develops new methodologies for the spectrum management in order to increase the spectrum utilisation efficiency and find efficient solutions to solve the problem of 'scarcity of spectrum'. Spectrum as such is not really a scarce resource; yet it is not used in an optimised manner. And optimisation of its usage would be much easier if a 'Full FSA' approach would be implemented.

Spectrum Brokerage (SB) is a part of this approach. Here, spectrum is looked upon like economic goods like stocks or real estate. Automatic spectrum agents act as brokers between suppliers and spectrum purchasers. Spectrum purchasers give purchase orders with maximum prices and deadlines just like when buying stocks. Rates will develop very dynamically depending strongly on the location and the time of the 'spectrum transaction'. Even buying spectrum from users, which have already allocated spectrum, is imaginable. After the identification and development of SB methods, mechanisms and algorithms to enable automatic SB processes need be developed. Then suitable user demand models and accounting and billing methods need to be investigated and included in the approach.

Within the FSA approach optimisation processes for optimum spectrum management need to be defined, the goal being to propose a generic methodology for allocating/deallocating spectrum in a free/unlicensed spectrum band (this includes the management of inter-system interference, the sharing rules, the operator's and regulator's role, and so on). Steps towards full FSA may include enabling dynamic spectrum allocation (DSA) and efficient spectrum sharing between operators.

Another approach aiming to increase spectrum efficiency through higher flexibility is based on the principle of Spectrum Pooling (SP). This approach takes advantage of the cognitive radio approach, first introduced by Mitola. The ability of such a radio to detect free sub-bands within a certain frequency band would enable the equipment to exploit unused frequencies without a disadvantage for the license owner of the spectrum and to enable dynamic spectrum sharing (DSS) and DSA. Classical SP can be applied for increasing spectrum utilisation without changes in existing networks. Obviously, spectrum utilisation efficiency may be further increased if the network provides signalling information to a spectrum pooling system.

Within the SP approach, a subset of algorithms/mechanisms follows the Highest Priority Intelligent Air Interface. The underlying idea is to look upon SP the other way around: Instead of designing a system for the rental users, we develop an intelligent system for spectrum owners, in which the spectrum owners applications (e.g. military, search and rescue, firefighters, police) are given access to the next available spectrum resource in any frequency region with highest priority. Formerly reserved spectra for these applications may then be released for civilian use.

The main assumption for future spectrum management schemes is that the allocation itself will be increasingly flexible and that the relevant timeframes between reallocations also can be adapted in a rather flexible and demand driven way. The present spectrum management research in the EU is being conducted in parallel by a number of research organisations and collaborative projects.

Spectrum Management Vision

Recommended Policy Principles to Guide Future Spectrum Regulations

Policies for the licensed band and the unlicensed band are different. For the licensed band, spectrum management needs to agree on the following principles:

Principle 1: Allowed RAT is limited in an internationally agreed set of technologies (e.g. UMTS band only can be accessed by International Mobile Telecommunications (IMT) 2000 technology family), that is not completely technology neutral.

Principle 2: Using the auctioned, traded or coordinated spectrum to harmonise among those RATs in Principle 1, not to harm them.

Principle 3: Secondary users should not disturb the primary user under the specified 'Interference temperature' criteria.

Principle 4: RAT using this band should be backward compatible or terminal can be reconfigurable or legacy terminal can be handed over.

Arguments for Principle 1:

Technology neutrality is currently an intensively debated issue in European Conference of Postal and Telecommunications (CEPT)/European Communication Committee (ECC) and in ITU-R. The request for technology neutrality is based allegedly on corresponding treaties of the WTO but might also be initially caused by requests from the US government to improve market chances in the European market during the licensing process for UMTS in the year 2000 [88, 89].

Technology Neutrality

The term *Technology Neutrality* is used for some time as a slogan. The term *technology* in this document is defined as 'RAT'. Under this context, Neutrality is understood as the absence of regulatory rules or conditions to be followed by decision makers.

It is proposed to adopt the definition of 'Technology Neutrality' as the freedom of an operator to use whatever radio standard he wants in a certain spectrum. Technology Neutrality in the field of radio telecommunication therefore means in fact 'Standards Neutrality'.

Technology Harmonisation

The term *Harmonisation* is generally understood as a process to increase commonalities and to decrease differences with the aim to improve interoperability and compatibility (between standards). A primary goal of Reconfigurability is to realise technology Harmonisation.

From the spectrum usage viewpoint, harmonised spectrum is a prerequisite for economics of scale and eases global roaming.

Harmonisation versus Technology Neutrality

Harmonisation and Technology Neutrality can be considered as antithetical tendencies. While Harmonisation aims at minimisation or even avoidance of differences between standards, Technology Neutrality aims to permit different standards and their usage, and to leave it to the market preferences and other influences whether alignment will occur.

Harmonisation aims at maximum interoperability and compatibility for mobile communication, while Technology Neutrality consciously risks the usage of different incompatible standards resulting in non-interoperability and missing roaming capabilities between different systems in adjacent or even same areas and/or spectrum bands.

However, Harmonisation and Technology Neutrality to a large extent do also coexist – as we can learn from the concept of IMT-2000 radio technologies. IMT-2000 radio technologies are harmonised at a certain level, but they remain partly incompatible technologies, which would require multimode terminals for roaming between family members. Identification of harmonised spectrum for IMT-2000 means basically that certain standards out of a limited set of standards can be used in this spectrum. Such a decision will be, on the other hand, in favour of the reduction of the complexity of reconfigurable terminals. The comparison between Harmonisation and Technology Neutrality is shown in Figure 8.33.

Short- and long-term history show strong trends towards the increase of harmonisation. The roadmap of the technology evolution of Reconfigurability is to nowadays harmonise the existing internationally recognised and standardised RATs, as Harmonisation is considered as one means to fulfil the WTO requirement to avoid Technical Barriers to Trade [90]. In the future, it needs to evolve itself as a part of new standard supporting backward compatibility.

8.5.2.3 Timeframe for Implementation

The roadmap for the rollout of DSA relies on both the complexity of the technical implementation and the evolution of the regulation policy.

From the technical standpoint, given the achievable reconfigurable performance of the current enabling technologies with associated constraints (power consumption, compact size), the following short-term scenarios for the operation of DSA can be envisaged:

- *RNAP side (Figure 8.34)*: This is the case of unconstrained equipment – like BSs, broadcast Tx/Rx, or embedded devices (e.g. in a car or in a car). These equipments can support operations with a large frequency range combined with a quite good spectrum resolution.

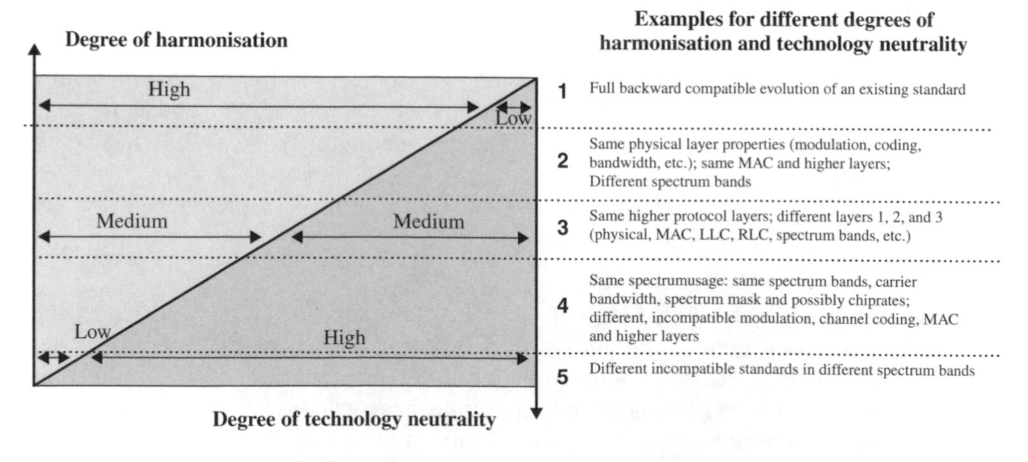

Figure 8.33 Harmonisation versus Technology Neutrality

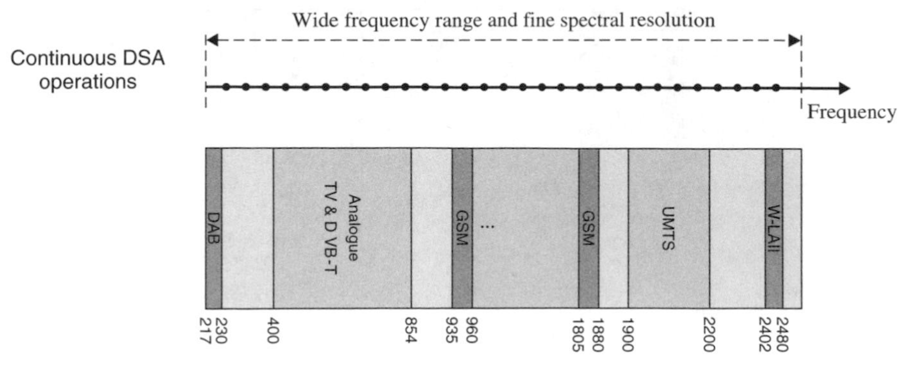

Figure 8.34 RNAP implementation scenario

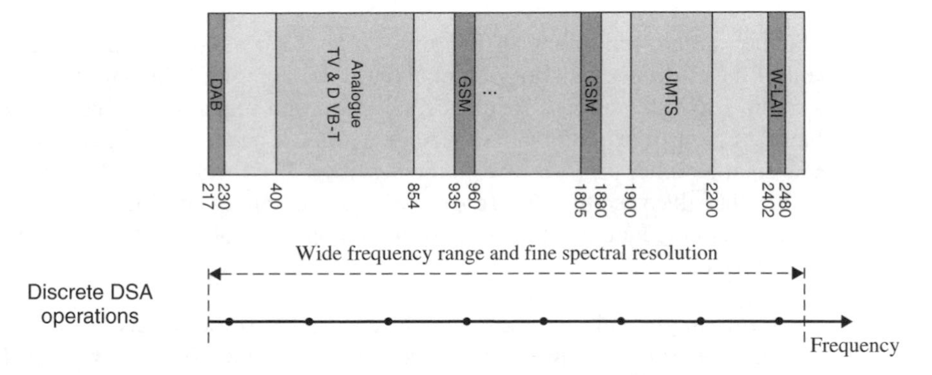

Figure 8.35 Scenario 1 for mobile handset implementation

- *Mobile handset side*: At the opposite of RNAP side, the constraints associated to the handsets only enable the support of either:

Scenario 1 (Figure 8.35): Alternatively, handset capabilities could allow for a narrower frequency range but with a higher degree in spectrum resolution for a similar amount of complexity to the previous point.

Scenario 2 (Figure 8.36): For handsets, the current capabilities enable the support of a wide frequency range by using a limited number of predefined RF carriers.

8.5.2.4 Regulatory Aspects of Spectrum Management

General Aspects

There is general interest in cognitive radio, reconfigurable radio, and SDR by a number of RAs. Regulators are primarily concerned with the security of radio software so that these types of devices cannot operate outside the spectral parameters for which they are certified. They are less concerned about the other types of operational software (e.g. OSs) unless download of these other types of software could either inadvertently or intentionally cause improper operation of the radio software.

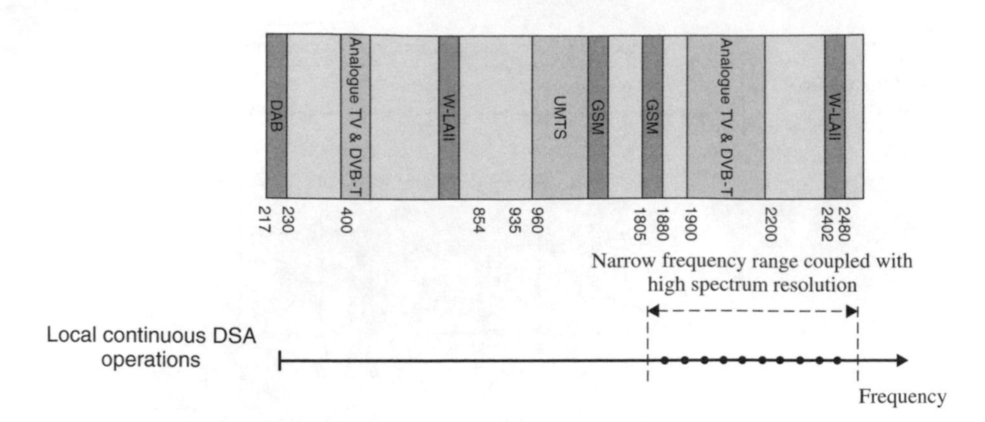

Figure 8.36 Scenario 2 for mobile handset implementation

In the application arena, the failure of the application such as a game may result in disappointment of the customer with no additional harm done. In the radio software arena, failure of the radio software, such as frequency selection and modulation, may result in a customer having a non-functional radio device or, worse, a device that functions inappropriately and disturbs other users or other radio services. The mechanisms to download may be the same, but the specifics and scope can vary to satisfy the imposed criteria. The following regulatory requirements apply to one component of operational software, namely, radio software:

- Regulators require, in general, that the device must provide some means of indicating its current 'type approval' or 'conformance acceptance', often a physical label attached to the device. However, the specific need with cognitive radio, reconfigurable radio, and SDR is that the indications must be associated with the radio configuration that is downloaded because the radio characteristic of the device is changed after a download and reconfiguration. Consequently, the currently accepted practice of a physical label becomes untenable. Therefore, download and device management may need to include how an electronic variant (proposed in some regulatory jurisdictions) of this labelling might be accommodated and how the device could provide information as to its current version/variants.
- Regulators require that the radio must not be able to operate with an unapproved configuration. Therefore, a security mechanism must be built into the radio to prevent malicious or accidental reconfiguration.
- Regulators may require a 'stronger' authentication for radio SD. Verification of a downloaded module may need different techniques than other downloads.
- To support the interest of regulators, the reconfigurable device should support a feature whereby it is possible to carry out a post-download audit to ensure that the radio software executing in the device is an approved software load for that device.

The requirements for ensuring proper operation of reconfigurable wireless devices are common to all RAs worldwide. However, various RAs have somewhat differing approaches as to how these requirements should be satisfied. Some of the differing perspectives of different RAs are provided in the following subsections. Also, in different regulatory jurisdictions, the actors who are responsible for carrying out the solution may not be the same.

Existing Regulatory Environment

General Issues

The design and roll out of DSA schemes requires the support of some regulation moves in both the spectrum management policies. This section gives an overview of the main regulation effort carried out in this area through the review of different worldwide initiatives. The retained new policies will directly impact the reconfigurability requirements for DSA.

After all the spectrum auctions, the new distribution of spectrum for commercial use, and the comprehension that the limited spectrum available is not at all efficiently used, it is necessary to re-evaluate the spectrum management policies that govern all wireless industries ranging from communication and broadcast to satellite communications. Globally, most countries have applied central planning approach to spectrum management so far, but there are many alternatives with varying degrees of flexibility and market-based incentives.

Traditionally, agencies of national bureaus have taken over the central planning for frequency allocations; however, this approach contrasts significantly with the global trend towards market-based policies in other aspects of telecommunications. There are a number of issues that have to be addressed to achieve a timelier scheme for allocation of spectrum.

Some spectrum regulation authorities investigate new approaches to allow new spectrum allocation practices. The investigations include at least the following topics: spectrum-sharing extent, flexibility extent, and the temporary rights versus permanent rights. They also investigate to determine who gets the spectrum in terms of distribution new licenses, auctioning licences and license renewals, and government held spectrum.

All these topics impact the spectrum-sharing coordination process for DSA. The previous topics and some additional ones have been more developed in Section 8.4 of the OverDRiVE Project deliverable D06 [91].

One the most proactive regulating agencies in the field of flexible spectrum management in Europe is the Radio Communications Agency (RA) in the United Kingdom. The RA [92] (UK regulating authority), had launched a consultation in 1998 to discuss the possible introduction of spectrum trading in UK. In March 2002, Professor Martin Cave had recommended spectrum trading [93]. In October 2002, the UK government had largely assented to Martin Cave' recommendations. At the same time (July 2002), RA has published a consultation document (namely, *Implementing spectrum trading*), and has received many answers from all spectrum-concerned actors, mainly assenting to spectrum trading. OFCOM (Office of Communications – Since 29 December, 2003, Ofcom will be the new regulator for the TV, radio, telecommunications and other communications industries. One of the regulators Ofcom will replace is the RA. Ofcom will be taking over their responsibility for licensing wireless transmission equipment. Ofcom is independent of the Government. Besides Ofcom's aims at looking for the most efficient way to use spectrum and thinking about ways of allowing companies within a particular industry to regulate themselves rather than being regulated by Ofcom has suggested a proposal for the introduction of spectrum trading in the United Kingdom. This proposal is currently under consultation [94] before the final decision. This aims at seeking comments on the proposal from all concerned parties. Some kind of roadmap for this introduction of spectrum trading for different systems (in term of license classes) has been proposed by OFCOM.

Similarly, some active discussions on the relaxing of some important constraints on current spectrum use are discussed by the Federal Communications Commission (FCC) [95] in the United States to enable more unlicensed practices (namely, *open spectrum*). In December 2002, the FCC has released a Notice of Inquiry (NOI) [96] regarding the feasibility to allow unlicensed devices to operate temporally and spatially in the TV broadcasting bands when the spectrum is not used. These sharing schemes will be supported by the introduction of new classes of radio systems: the cognitive radio [97].

Finally, from the worldwide perspective, a resolution of World Radio Conference (WRC) in 2003 has mentioned the possibility of using 5 GHz band to allow new spectrum allocation techniques by the use of smart radios. The term 'cognitive' function for dynamic frequency selection (DFS) has been used.

The subsections below provide a brief summary of some of these regulatory activities.

Europe

As noted by Babb *et al.* [97] there are three types of rules:

- *Regulatory*: For a mobile phone to carry the European Community's CE mark it must conform to the radio parameters (RF, interference, etc.) and safety regulations in force throughout the European Union.
- Type approval.
- *Operator approval*: Assurance that the phone functions in the desired manner when connected to a particular operator's network.

The Telecommunication Conformity Assessment and Market (TCAM) Surveillance Committee in Europe has developed a questionnaire regarding SDR and is evaluating the results of that questionnaire. It is not anticipated that security requirements for operational SD will be substantially different from those of other administrations.

Faroughi-Esfahani *et al.* [98] provide a summary of the European perspective of reconfigurable systems, including a view of the regulatory changes between the present and the future.

Currently, there is no common view by the member states of the European Union on the regulatory requirements of SDR under the R&TTE Directive. There are some views within the European region that regulatory requirements will be dependent upon the type of marketplace for SDR and cognitive radios. Horizontal markets in which third parties can provide software for use in an SDR or cognitive radio device is of concern because regulators have interest in establishing clear responsibilities in case of non-compliance. There are some views within the European Union member states that standardisation for SDR (and presumable cognitive radio) seems to be essential for the following:

- security mechanisms;
- open interfaces.

The ECC Project Team 8 is in the process of developing a report on a more flexible regulatory structure. The scope of the report, which is to be completed by the end of 2005, includes the following:

- Conduct a study on the overall direction of harmonisation policy, bearing in mind that harmonising measures should be technology neutral, flexible, and include review stages
- Investigate ways and possibilities of establishing a more flexible regulatory structure for spectrum management to better enable the introduction of new radio technologies and adapt to the changing market demand.
- The introduction of increased opportunities to share, including sharing on the basis of geographical area(s), time, and service should be studied.
- It is considered that the project team should also study the methods and conditions under which parts of spectrum may be designated to technology- and service- independent allocations, with a view to facilitating easier spectrum access.

United States

The FCC's First Report and Order [99] on SDR stated that industry standards organisations are still investigating security issues, and further stated, in paragraph 32:

> We continue to believe that the best approach is to rely on a general requirement that manufacturers take adequate steps to prevent unauthorized changes to the software that drives their equipment.

The FCC also asserted that it may need to specify more detailed security requirements at a later date as SDR technology develops. Thus, it is clear that in defining the operational SD problem, it is necessary to define the timeframe being addressed. The FCC, in December 2003, issued a Notice of Proposed Rule Making and Order for Facilitating Opportunities for Flexible, Efficient, and Reliable Spectrum Use Employing Cognitive Radio Technologies [100] which, inter alia, asked for comments on the following:

> We, therefore, believe it is time to revisit the SDR rules to determine if changes are needed concerning whether the SDR rules should be permissive or mandatory, the types of security features that an SDR must incorporate, and the approval process for SDRs that are contained in modular transmitters.

In particular, the FCC is investigating the following statement from the preceding NPRM [99]

> Required SDRs to incorporate security features to ensure that only software that is part of an approved hardware/software combination can be loaded into an SDR. The exact methods are left to the manufacturer.

The FCC is investigating whether to require that an SDR-capable device be declared as such or whether this is left as an option to the manufacturer asking for a license for the device.

Japan

A Japanese perspective of SDR regulatory issues is provided by Suzuki [101], which identifies the following regulatory issues related to SD security that would also apply to cognitive radio:

- security system for granting certification and preventing illegal modification of SDR equipment;

- test methods that permit hardware and software to be tested separately;
- configuration control of modification history.

The last two items are particularly significant, and considerable ongoing research in Japan is particularly related to the second item [102–106].

8.5.2.5 Standards, Technical Specifications, ITU Recommendations Needed for Spectrum Management

There is a need for a common understanding of terminology, concepts, visions, and a roadmap for cognitive radio and spectrum and RRM. Work on this type of high-level guidance document should be undertaken by WWRF in collaboration with other international organisations in the near future.

There is also a longer-term need for the International Telecommunication Union to develop a recommendation on the regulatory aspects of cognitive radio devices and reconfigurable networks. The ITU Radiocommunication Sector (ITU-R) is currently developing two reports on SDR; one is being developed in ITU-R Working Party 8F (IMT-2000 and Systems Beyond IMT-2000) and Working Party 8A (Mobile Telecommunications). These documents will provide useful technical information on SDR systems. However, in the long term, a recommendation that covers issues such as the global circulation of reconfigurable terminals and networks need to be addressed by the ITU.

8.5.3 Joint Radio Resource Management

With reconfigurable terminals becoming available, a mobile network operator is in the position of offering a multi-system access (GERAN, UTRAN, WLAN, etc.) to ensure that users are 'always best connected'. In such a role as an 'integrated operator' he can offer a seamless personalised access to mobile multimedia applications charged together over one bill of the same operator.

In such a scenario, the operator has full control of all single RATs involved and will be motivated to use JRRM techniques for the optimised usage of the radio resources of the different RATs.

JRRM concepts and algorithms have to meet requirements from the viewpoint of overall network performance, the individual user's point of view, and the operator's point of view. JRRM solutions should integrate all these aspects.

Overall Network Performance

From a technical perspective, it is the general goal of JRRM to optimise the overall performance of the multi-RAT network. Users should be served on the basis of the QoS needs of their applications and subscriptions. Radio resources should be distributed throughout the network so that all users in the network should be as satisfied as possible and are 'always best connected'.

User's Preferences

The developed concepts should at least consider the possibility for the user to choose the RAT. While it could be advantageous to make the usage of RATs seamless, there might be users that are aware of the technology and have a favourite RAT.

Operator's Strategy

Operators might prefer to generate traffic on a certain RAT independent of the aim to maximise user performance. Motivations for these policy-based JRRM concept could be the return on a certain investment or the independence of a certain manufacturer or a certain technology. For example, the operator might want to determine if UTRAN or WLAN should preferably be used independent of performance aspects.

Adaptive Radio Multi-homing Concept

The Internet Engineering Task Force (IETF) proposed IP Multi-homing concept [107, 108] and this concept is contributed by research projects like IST MIND [109, 110]. This framework manages IP traffic being routed through different radio access networks to the same mobile node. At the IP layer, the multi-homing algorithm allows to route the traffic for each individual stream through a specific interface according to the type of the traffic. An extension to the conventional multi-homing concept is to run simultaneous connections on the radio-frame level, which we call with reference to reconfigurable terminals as the *Adaptive Radio Multi-homing* approach.

Definition 1: Adaptive Radio Multi-homing (ARMH), as in Figure 8.37, is an overall management framework extended from IP Multi-homing concept. It provides multiple radio accesses for multimode/multiband terminals in order to allow the terminal maintain simultaneous links with radio networks. It selects the most proper JRRM function based on the identified information from the cooperating sub-networks, terminals, user, and services. In order to support the selected JRRM functions, proper traffic classification, calibration, and inter-working between the service application server and Radio Resource Controller (RRCR), and the configuration of transmission format as well as MAC protocols are managed by ARMH.

The system capacity gain obtained from the JRRM is in principle the enlargement of the number of operational servers from the queuing model viewpoint, which therefore results

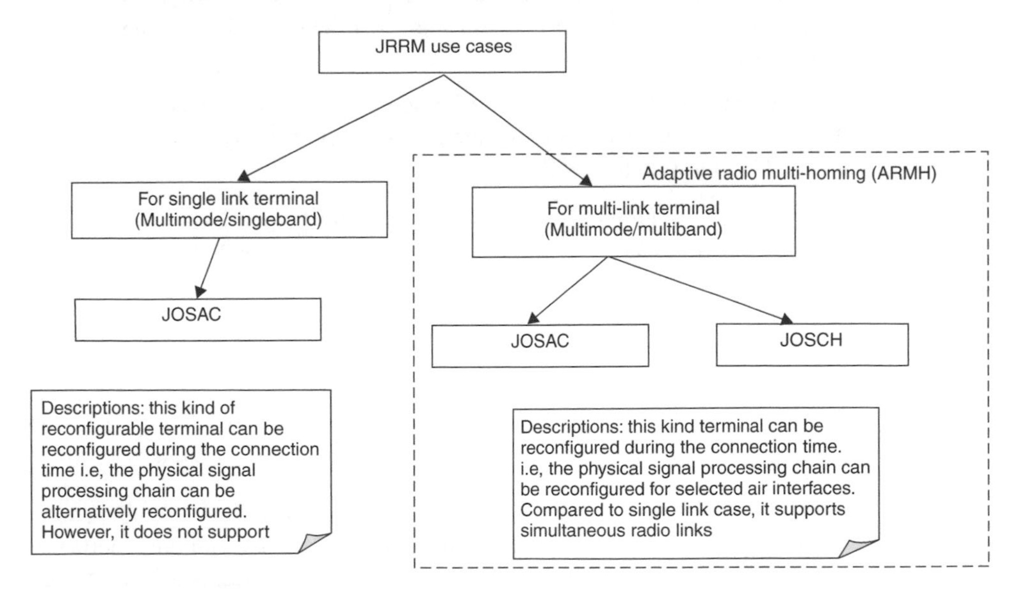

Figure 8.37 Illustration of use cases of JRRM and radio multi-homing

in a higher trunking gain. On the other hand, by alternatively allocating the resource to call units among the inter-working radio networks or frequency layers, the load balancing effect among the radio networks are realised. In a typical soft blocking sensitive radio network, such effect is very significant.

Besides the capacity gain from the network operation point of view, the advantage of having concurrently parallel streams is manifold: If one bearer service has a high availability in the network (low data rate bearer services result in high coverage, for example, a 16 kbits/s service is available in 99 % of the cases), this link would be used for transferring important information to the terminal. On the other hand, a low data rate service cannot fulfil the requirements for multimedia traffic resulting in high data rate demands. If traffic is intelligently split into rudimentary and optional information streams, a higher QoS to the user is provided. Whenever possible, the user combines both streams for yielding a higher QoS, and due to the higher availability of a lower data rate service in UMTS, a minimal QoS to the user can be fulfilled.

Traffic Splitting over Heterogeneous Network

Joint Session Admission Control (JOSAC) takes the neighbour RAT system load into account. The traffic stream is routed through the cooperating systems according to the restrictions and advantages of each system. From the service point of view, different levels of service calibration can be identified to meet the user's satisfaction, for example, for video transmission, the video traffic can be split into base layer and enhancement layer, where the base layer consists of the most important lower frequency information. The user would not be satisfied if one only receives the enhancement layer information. On the other hand, from the user mobility point of view, the connection of UMTS is not restricted by the location of the user, but by the WLAN.

Generally, three key points are covered by radio multi-homing:

Traffic prioritisation: As Figure 8.38 depicts, the incoming traffic is split over two (more) sub-streams. The important information goes through a reliable RAT, whereas, the rest goes through other RATs.

Traffic scheduling: Segments of users' traffic presented by packets are sent at a suitable period through optimally selected RATs associated with optimum physical mode, that is the Modulation and Coding Scheme (MSC).

Synchronisation: Packets belonging to the sub-stream are multiplexed back to the original traffic stream in the receiver based on proposed synchronisation schemes (in MT and RNC).

Buffer management: The jitters and average delay parameter are controlled by buffer size and synchronisation approaches. The static terminal and user profiles stored in network side will be retrieved by RNC to determine the calculation power and buffer size of the terminal and to evaluate the user preference and cost. The synchronisation methods are used mainly to compensate average delay, whereas buffers are used to compensate jitters.

As Figure 8.38 shows, suppose a reconfigurable terminal is demanding a scalable video service from the remote server through tight coupled sub-networks. Suppose both sub-networks are controlled by one RNC. In order to establish simultaneous sub-streams belonging to the same video context, the following procedure should be fulfilled:

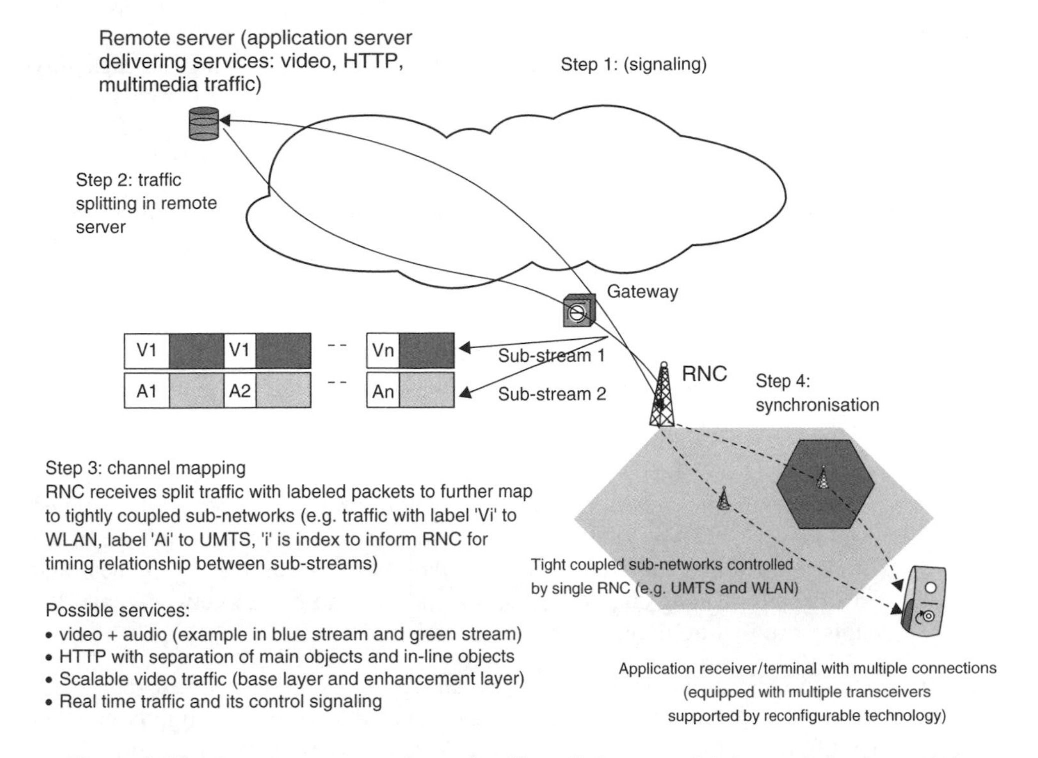

Figure 8.38 Overview on procedure of traffic splitting over tightly coupled sub-networks

Step 1. Signalling and initialisation step. RNC receives an application from MT with multiple radio accesses/addresses. After estimating the available radio resource in a controlled sub-network; RNC will apply to the remote server for traffic splitting indicating average rate in each sub-link.

Step 2. Traffic is split according to RNC's application (Step 1) and sent to RNC. Sub-streams are labelled differently.

Step 3. RNC receives split traffic with labelled packets to further map to tightly coupled sub-networks (e.g. traffic with label 'Vi' to WLAN, label 'Ai' to UMTS, 'i' is index to inform RNC for timing relationship between sub-streams). The possible services that could be applied are as follows:
- Video and audio sub-streams
- HTTP with separation of main objects and in-line Objects
- Scalable Video Traffic (Base Layer and Enhancement Layer)
- Real-time traffic and its control signalling

Step 4. The synchronisation mechanism in RNC remedies delays generated by sub-radio-networks due to:
- Different TTI value for bearer services
- Automatic Repeat Request (ARQ) actions due to different connection qualities
- Different Processing Power of different BTS (especially in the shared processing component case, which is valid for functional partitioning scheme).

Principles of Deploying JRRM

The ARMH approach basically requires reconfigurable terminals obtaining the capability of multiple radio access, which allows incoming traffic being split over different air interfaces with simultaneous connections carrying prioritised sub-streams. To apply JRRM, a number of principles embedded in the radio resource controller are listed as follows:

Principle 5: In a highly loaded system, the radio-controlling entity must be prepared to apply for available resources, from another spectrum pool, or from another operator.

This principle implies that the spectrum pool could be shared by the involving operator or RAT. The physical limitation given by inter-RAT spectrum sharing is rather an empirical input to the inter-operator sharing approach. On the basis of that the spatial orthogonality, opportunity-based orthogonality and the necessity of inter-access point synchronisation can be obtained.

Principle 6: In coupled radio networks, traffic allocation through the very tight coupled sub-network must be jointly designed. The allocation must consider the traffic pattern including the QoS detailed to the objects level, the deployment and overlapping among sub-networks, the current load of sub-networks, and the noise rise feature upon a new call arrival, and so on.

Thanks to the ARMH approach, the resource allocation over the coupled radio sub-networks balances the system load. Assuming RAT1 and RAT2 are very tightly coupled and they have different noise rise characteristics. Without losing generality, name RAT1 suffers more from interference, that is a typical soft blocking system; whereas, RAT2 is only limited by its physically available radio units, for example, a synchronised TDMA system. By considering the water filling principle, RAT2 is the candidate sub-network to receive more traffic.

Principle 7: System capacity is enhanced thanks to both the resource and traffic scalability. As can be explained by a more reliable network and resource management, in this case, the outage probability is significantly reduced.

This principle can be proven by many facts. Take the power controlled CDMA system as an example. By pouring a big trunk required traffic, for example, video traffic, into a single frequency layer will result in a higher average power requirement and higher interference to the existing sessions in the system compared to the case when the traffic is split over two frequency layers. The same principle is applied to hard blocking system, which can be modelled by a Markov finite state machine. In fact, admission of a high throughput required traffic unit would occupy more servers compared to normal traffic, for example, voice. Without the traffic split mechanism, the higher call blocking/dropping rate occurs.

8.5.3.1 Research Topics

Here we list a number of heated research topics in the field of JRRM for reconfigurable systems:

- Advanced transmission technique
- Cross layer optimisation

- Advanced mobility technique
- Roaming and MM
- Optimum service provisioning

For instance, the cross layer optimisation technology aims at jointly finding the optimum end-to-end communication rates, routing, power allocation, and transmission scheduling for wireless communication links that take place simultaneously in the radio network. The resulting transmission format, and the time schedule for the individual communication link as well as the parameter setting for higher-layer protocols will significantly increase the spectrum efficiency. In general, reconfigurability needs to have an optimum solution to maximise the diversity gain, antenna array gain, and data scheduling gain. The multidimensional problem complicates such an optimisation paradigm.

8.5.3.2 Hierarchical Radio Resource Management

In this subsection, we outline the basics of a novel scheme for the efficient management of radio resources, which is an adapted generalisation of what was first conceived by us in [111]. This scheme is intended to either work as a stand-alone solution, or for elements of it to be used to facilitate the other approaches to radio resource and spectrum management mentioned in this section.

We envisage a scenario where a target frequency range is unregulated or pooled, or where networks/NOs are fully cooperating in the resource allocation process in order to serve the common task of greatly improving spectrum efficiency and QoS in the vicinity as a whole. We envisage most if not all base stations being reconfigurable to a number of RATs; furthermore, we note that it would be useful, although not a prerequisite, for MT reconfiguration capabilities to also exist. Taken in entirety, such a scenario allows for optimum RRM methods to be used, in conjunction with dynamic radio resource adaptations as users move in space or their resource requirements alter.

On the basis of this scenario, we introduce our approach to hierarchical dynamic RRM. In our approach, through the division of RATs into constituent parts and the treatment of these parts in a hierarchical manner, radio resource allocations can be dealt with autonomously and efficiently by a hierarchy of Agents operating in networks. This is strongly in line with the functions of RATs, which can be considered as elements that can be hierarchically inherited in sequence, to constitute the complete RAT [112].

Use Cases

There are a number of possible use cases for this scheme, of which we pinpoint some simple examples here. This list of use cases is not exhaustive – many further possibilities exist.

Resource Reallocation

There are two mobile networks or two parts of a single mobile network, one configured to provide FDMA/TDMA access and one configured to provide FDMA/CDMA (i.e. several frequency bands of CDMA spread-spectrum access). All FDMA/TDMA resources are already spent, but some terminals supporting only FDMA/TDMA still wish to join a network.

Fortuitously, spare capacity remains in the CDMA parts of the network(s). To make use of this, one of the CDMA bands is made available, possibly requiring some simple reorganisation of the allocated CDMA resources to achieve this. This CDMA band is then reclassified as FDMA/TDMA (and the associated CDMA base stations are reconfigured accordingly) in order to allow the GSM-like terminals to join the network.

Demanding Resource Requirements
Everyone is very happy, and the TDMA and CDMA parts are coexisting with reasonably sufficient spare resources. However, banking group Money Inc. came along, and decided that a complex (high bandwidth) video telephony session was needed between their board members on the move, also involving high-speed file transfers. Given the revenue to be made, NOs have a prewritten agreement that their network(s) should be adapted to support such an eventuality. They reallocate some bands from TDMA to CDMA, and reconfigure base stations accordingly. These bands are then switched back to TDMA once the meeting is finished.

Automated QoS Maximisation
There are four groups of users in the vicinity: software downloaders, videophone users, audiophone users using CDMA technology, and audiophone users only able to support a simple GSM-like radio interface. The former three groups of users would all appreciate their QoS being improved, through faster downloading, higher resolution/frame rate video calling, and an advanced high bit-rate clearer audio calling, respectively. The latter GSM users do not require any improvement, as they are already using the highest-quality codec that their system can support. There is considerable overcapacity in the FDMA/TDMA network at that time.

Given the overcapacity, FDMA/TDMA resources are freed and reconfigured to become CDMA resources. The first three sets of users are then a lot happier, and the fourth set experiences no change in perceived QoS.

Coexisting Technologies
Mobile NOs already commonly provide coexisting 2G and 3G networks. Our scheme could allow a form of load balancing between these 2G and 3G parts. It is not even a prerequisite for BS to be reconfigurable to achieve this, as Agents at a higher level in the network could be responsible for deciding upon radio resource allocations to the 2G and 3G parts.

8.5.4 Network Planning for Reconfigurable Networks

8.5.4.1 Requirements for Dynamic Network Planning for Reconfigurable Networks

The ever-growing demand for high-speed access to all kinds of telecommunication systems has rendered necessary the reconsideration of traditional network-planning methods. Taking into account that the advent of composite reconfigurable networks has become an inseparable part of almost every communications conference and journal, dynamic network planning is essential in order to handle the alternations that take place in frequent time periods, with respect to the demand pattern in a specific geographical area.

With the support of reconfigurability and flexible network reconfiguration, the cost for network deployment evaluated by the Capital of Expenditure (CAPEX) and Operational Expenditure (OPEX) can be reduced, while significantly enhancing the efficiency of user traffic handling. During the network-planning phase, the following mechanisms are included:

- setting feasible radio interfaces;
- determining allowed location of BS;
- setting antenna patterns;
- creating coupling structure among sub-networks;
- configuring policy of Joint Radio Resource Management (JRRM);
- planning statistic values of required spectrum in different scenarios with available Radio Access Technologies (RATs)

Furthermore, taking into consideration that reconfigurability does include utilisation of the current sites and/or deploying the necessary reconfigurable sites, the goal of dynamic network planning is to reduce the cost for network deployment, through the selection of the appropriate RATs for operation at different time and space regions.

Consequently, the network-planning topic is considered as a subset of a more general framework, namely, DNPM. It consists of the planning phase and the management phase. During the initial planning phase, feasibility of setting radio interfaces; location of BSs; antenna patterns; coupling structure among sub-networks, as Base Station (BS), APs and RNC, and so on; policy of Joint RRM (JRRM); statistic values of required spectrum in different scenarios with available Radio Access Technologies (RATs) are developed. In the management phase, radio network elements are subject to be reconfigured, which is triggered by the management entities, for example, the network element manager, so that self-tuning of a radio network targeting at optimum parameter settings can be carried out. From our immediate intuition, the capacity gain will result in a reduction of the number of BSs to be deployed in the radio network.

8.5.5 Cognitive Radio

8.5.5.1 System Functional Requirements

Figure 8.39 is an adaptation of a figure that was part of a presentation at the Plenary Session of WWRF 10 in NYC. The figure depicts the technological functionalities that are needed to accomplish DFS using cognitive radio, which can be referred to as policy-based radio. Clearly, a system approach is needed to accomplish the goals of DFS.

The functionality requirements in Block 1 of Figure 8.39 are critical technology drivers. DFS requires the ability for a real-time, wide-band sensing of the spectral environment. This is the process of sampling the channel in order to determine occupancy. It should be noted that there is no agreed definition of when a channel is occupied; several factors are involved including receiver sensitivity, the sampling time and the sampling interval, thresholds for discriminating wide-band noise from signals, and so on. It is noted in Block 1 that along with this sensing capability is the need for policy agility, which is the ability to change the policies controlling the behaviour of the radio to be changed dynamically. Policy agility allows adaptation to policies changing over time and geography. Such policy changes could be downloaded from the Internet in a machine-readable format. The timescale for such changes of course is much different from the frequency agility timescale needed for DFS.

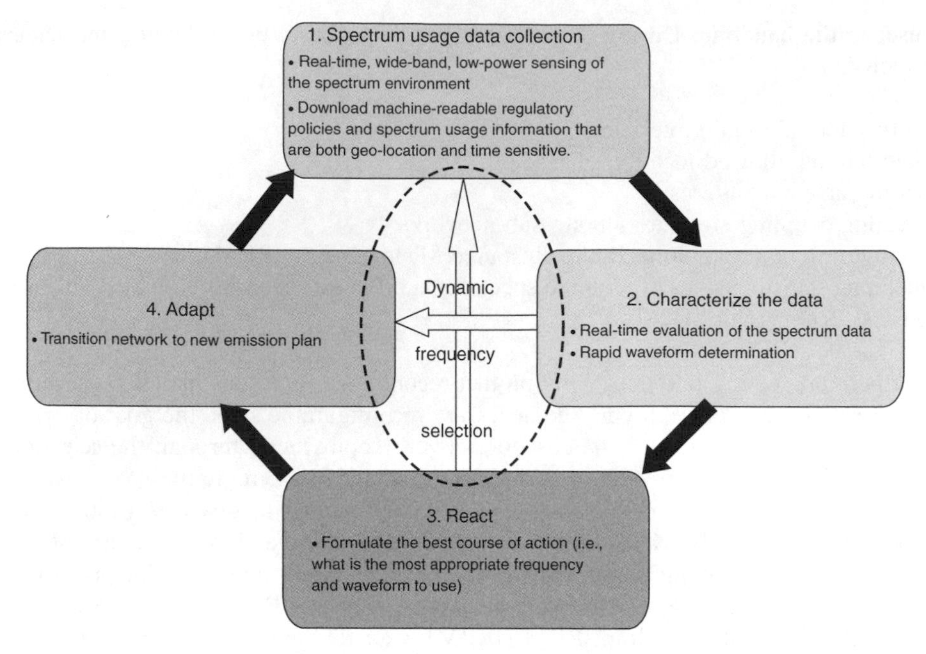

Figure 8.39 Key technology components of policy-based cognitive radio needed for dynamic frequency selection

The functionality requirements in Block 2 of Figure 8.39 include an analysis of the data to determine if the particular channel is an opportunity for usage. This identification process includes the characterisation of the data and uses this information to determine if the channel can be used by another communications service or system. The identification process also includes communication with some subset of its neighbours because what may appear to be a clear channel at one end of the link may not be a clear channel at the other end of the link. For some mobile wireless subsystems, this communication may require a narrow-band pilot channel.

Block 3 of Figure 8.39 is the synthesis of the specific dynamic waveform and frequency that are appropriate for use at this time and this location. This leads to the need for the network to adapt (Block 4).

8.5.5.2 Physical Layer Functions

So far we have discussed the different functions (PHY and network) required to support DSA multi-band operations. This part reviews how these functions can actually be designed and implemented with the current technologies. In particular, the state of the art of the current SDR technologies enabling the support of the PHY functions is reviewed. In the meanwhile, the possible emerging and promising enabling technologies that could overcome the limitations of the current enabling technologies is reviewed.

A summary of current and emerging (breakthrough) technologies for the support of RF reconfigurable modules at Analogue to Digital Conversion (ADC), the antenna, RF front end and digital processing can be found respectively from Tables 8.3 to 8.6.

Table 8.3 Current and emerging enabling technologies for ADC

	Current enabling technologies	Enabling technologies breakthrough
ADC	• Silicon based – best achievable Nyquist sampling rate performance is around 10 Gsamples/s	• Superconductivity-based • Optically-based

Table 8.4 Current and emerging enabling technologies for antennas

	Current enabling technologies	Enabling technologies breakthrough
Miniaturisation	Planar Inverted F Antenna (PIFA)	• Lengthening of current paths–based antennas (e.g. fractal antennas) • Capacitive or self charging–based antennas (e.g. PIFA antenna)
Multi-band	Associations of several resonators with the introduction of slits	• Association of several resonators with the use of particular antenna geometry (e.g. fractal antennas) • Radio wavelength adaptation with commuting or variable components (e.g. MEMS or diode PIN based)
Wide-band	Planar antenna-based	• Dielectric resonator • Association of several resonators frequency shifted • Independent frequency-based antennas (e.g. b-conical or spiral antennas) • MEMS – Micro Electro Mechanical Systems

8.5.6 Summary

Recapitulating the research approach, it must be noted that reconfigurable networks will become commercially successful, only if:

- new applications are introduced and massively adopted (if offered at acceptable costs);
- the required QoS and capacity levels are achieved in a cost-effective manner;
- appropriate spectrum management techniques for dynamic frequency sharing are developed including the ability to perform the following functions in real time:
  - — real time, wide-band, low-power sensing of the spectrum environment;
  - — real-time characterisation of JRRM spectrum data;
  - — real-time formulation of the best course of action (i.e. dynamic frequency and waveform selection);
  - — real-time adaptation in the network layers;
- needed regulatory changes are adopted on a global basis.

Table 8.5 Current and emerging enabling technologies for some RF front-end modules

	Current enabling technologies	Enabling technologies breakthrough
Filter	• Filter rank to cover a wide range of frequencies • Tunable filters electronically controlled based on varactor	• Superconductivity-based tenable band pass filters – analogue based • MEMS switches–based tunable band pass filters – analogue-based
Amplifier	• Specific to each band – analogue silicon RF technology based (below 2 GHz) • Software switching between amplifiers dedicated to different bands	• Silicon–germanium transistor based (would enable operations up to 40 GHz) • Ultra linear amplifier
Oscillator	• Fixed frequency oscillator • Phase locked oscillator (VCO) – analogue-based: Narrow-band VCO (High Q), Wide-band VCO (free running). • Phase locked oscillator (VCO) – digital-based: Numerical control oscillator (NCO)	• MEMS-based wide-band VCO • Precise balanced quadrature oscillator to enable direct down-conversion architectures • MMIC-based multi-port junctions
Mixer	• Analog mixer – MMIC based. Several mixers are usually needed to cover large bands (by switching) • Digital mixer – ASIC, FPGA, DSP, DDS-based	• MMIC-based multi-port junctions

This Section 8.5 outlined the working framework for the design, development and implementation of the functionalities needed for (i) the efficient management of spectrum in an end-to-end reconfigurability context, (ii) the joint management of the RATs participating in a wireless-access heterogeneous infrastructure, and (iii) the network-planning strategies necessary for the design phase of next generation wireless networks.

For this purpose, in the beginning, it gave an overview of the main aspects that were dealt with.

After that it introduced RRM as an overall strategies superset, divided in the following subsets:

- Spectrum Management, that is efficient spectrum allocation among RATs that were originally selected in order to increase the overall spectrum efficiency.
- Joint RRM, that is load balancing and traffic splitting among RATs with fixed amount of spectrum granted to each RAT and in certain cases only, redistribution of demand to RATs.

Section 8.5.2 dealt with Spectrum Management. More specifically, it presented the basic requirements that arise from the increased user demands in terms of spectrum and the

Table 8.6 Current and emerging enabling technologies for digital processing

	Current enabling technologies	Enabling technologies breakthrough
Configurable	• ASIC	
Reconfigurable	• DSP • FPGA	• Reconfigurable computing machine (algorithm and operational levels) • General-Purpose Processors (GPP) using portable software code • ASIP-based approach
Dynamically reconfigurable		• Technologies convergence (hybrid DSP/ASIC/FPGA architectures) • Enhanced FPGA • FPAA • Reconfigurable computing machine (algorithm and operational levels) – only ACM provides dynamical reconfigurability • General-Purpose Processors (GPP) using portable software code • ASIP-based approach

necessity for DSA in an end-to-end reconfigurability context. Then it outlined the basic technical approaches in flexible spectrum management and finally it gave an overview of the respective regulatory issues that should be taken into account.

Section 8.5.3 presented the key characteristics of JRRM in reconfigurable networks, with respect to the challenges that must be met by such a joint optimisation approach and the functionality to be developed, in order to handle specific traffic conditions in heterogeneous networks with a given spectrum. This chapter also introduced the novel scheme of HRRM. The scheme was presented in a generic fashion so as not to limit its scope. The logic behind the use of a hierarchy in the radio resource allocation process was particularly discussed, and a basic description of the implementation architecture for the scheme was given.

In addition, Section 8.5.4 dealt with network-planning issues, which are now mandatory to be reconsidered and further evaluated compared to the classical network-planning methods. More specifically, this section presented the way for utilising existing network infrastructures in terms of reconfigurable networks with many RATs, among which a selection of the most appropriate is essential.

Furthermore, Section 8.5.5 presented the technologies that are today considered for successfully encompassing all the previously mentioned features in next generation terminals and network segments.

8.6 Acknowledgement

The following individuals contributed to this chapter: Vera Stavroulaki, George Dimitrakopoulos Flora Malamateniou, Apostolos Katidiotis, Kostas Tsagkaris (University of Piraeus, Greece), Karim El Khazen, David Grandblaise, Christophe Beaujean, Soodesh

Buljore, Pierre Roux, Guillame Vivier (Motorola Labs, France), Stephen Hope (France Telecom, UK), Jörg Brakensiek, Dominik Lenz, Ulf Lucking, Bernd Steinke (Nokia, Germany), Hamid Aghvami, Nikolas Olaziregi, Oliver Holland, Qi Fan (King's College London, UK), Nancy Alonistioti, Fotis Foukalas, Kostas Kafounis, Alexandros Kaloxylos, Makis Stamatelatos, Zachos Boufidis (University of Athens, Greece), Jim Hoffmeyer (Western Telecom Consultants, Inc, USA), Markus Dillinger, Eiman Mohyeldin, Jijun Luo, Egon Schulz, Rainer Falk, Christian Menzel (Siemens, Germany), Ramon Agusti, Oriol Sallent (Universitat Politechnica de Catalunya Barcelona, Spain), Miguel Alvarez, Raquel Garcia – Perez, Luis M. Campoy, Jose Emilio Vila, W. Warzansky (Telefonica, Spain), Byron Alex Bakaimis (Samsung Electronics, UK), Lars Berlemann, Jelena Mirkovic (RWTH Aachen University, Germany), Alexis Bisiaux (Mitsubishi, France), Alexandre de Baynast, Michael C., Joseph, R. Cavallaro, Predrag Radosavljevic (Rice University, USA), Antoine Delautre, JE. Goubard (Thales Land & Joint Systems, France), Tim Farnham, Shi Zhong, Craig Dolwin (Toshiba, UK), Peter Dornbush, Michael Fahrmair (University of Munich, Germany), Stoytcho Gultchev (CCSR, University of Surrey, UK), Mirsad Halimic, Shailen Patel (Panasonic, UK), Syed Naveen (I2R, Singapore), Jorg Vogler, Gerald Pfeiffer, Fernando Berzosa, Thomas Wiebke (Panasonic, Germany), Christian Prehofer, Qing Wei (DoCoMo, Germany), Jacques Pulou, Peter Stuckmann, Pascal Cordier, Delphine Lugara (FTRD, France), Gianluca Ravasio (Siemens IT, Italy).

References

[1] M. Mouly, M.-B. Pautet, *"The GSM System for Mobile Communications*", published by the authors, Palaiseau, France, 1992.

[2] R. Kalden, I. Meirick, M. Meyer, "Wireless Internet Access Based on GPRS", *IEEE Personal Communications*, Vol. 7, No. 2, pp. 8–18, April 2000.

[3] 3rd Generation Partnership Project (3GPP) Web site, www.3gpp.org, 2005.

[4] Institute of Electrical and Electronics Engineers (IEEE) 802 standards, Web site, www.ieee802.org, 2004.

[5] J. Khun-Jush, P. Schramm, G. Malmgren, J. Torsner, "HiperLAN2: Broadband Wireless Communications at 5 GHz", *IEEE Communications Magazine*, Vol. 40, No. 6, pp. 130–136, June 2002.

[6] U. Varshney, "The Status and Future of 802.11-based WLANs", *IEEE Computer*, Vol. 36, No. 6, pp. 102–105, June 2003.

[7] Digital Video Broadcasting (DVB), Web site, www.dvb.org, January 2002.

[8] P. Demestichas, L. Papadopoulou, V. Stavroulaki, M. Theologou, G. Vivier, G. Martinez, F. Galliano, "Wireless Beyond 3G: Managing Services and Network Resources", *IEEE Computer*, Vol. 35, No. 8, pp. 80–82, August 2002.

[9] D. Kouis, P. Demestichas, V. Stavroulaki, G. Koundourakis, N. Koutsouris, L. Papadopoulou, N. Mitrou, "A system for enhanced network management towards jointly exploiting WLANs and other wireless network infrastructures", accepted for publication in the IEE Proceedings in Communications Journal.

[10] End to End Reconfigurability (E2R), IST-2003-507995 E2R, http://www.e2r.motlabs.com, 2005.

[11] P. Demestichas, N. Koutsouris, G. Koundourakis, K. Tsagkaris, A. Oikonomou, V. Stavroulaki, L. Papadopoulou, M. Theologou, G. Vivier, K. El-Khazen, "Management of Networks and Services in a Composite Radio Context", *IEEE Wireless Communication Magazine*, Vol. 10, No. 4, pp. 44–51, August 2003.

[12] P. Demestichas, V. Stavroulaki, "Issues in Introducing Resource Brokerage Functionality in B3G, Composite Radio, Environments", *IEEE Wireless Communications Magazine*, Vol. 11, No. 10, pp. 32–40, October 2004.

[13] P. Demestichas, G. Vivier, K. El-Khazen, M. Theologou, "Evolution in Wireless Systems Management Concepts: From Composite Radio to Reconfigurability", *IEEE Communications Magazine*, Vol. 42, No. 5, pp. 90–98, May 2004.

[14] P. Demestichas, V. Stavroulaki, L. Papadopoulou, A. Vasilakos, M. Theologou, "Service Configuration and Distribution in Composite Radio Environments", *IEEE Transactions on Systems, Man and Cybernetics Journal*, Vol. 33, No. 4, pp. 69–81, November 2003.
[15] Software Defined Radio Forum, www.sdrforum.org, 2005.
[16] IST Project SCOUT (Smart user-centric communications environment), www.ist-scout.org, 2002.
[17] L. Paul, D. Grandblaise, R. Tönjes, K. Moessner, M. Breveglieri, D. Bourse, R. Tafazolli, "Dynamic Spectrum Allocation in Composite Reconfigurable Wireless Networks", *IEEE Communications Magazine*, Vol. 42, No. 5, pp. 72–81, May 2004.
[18] E2R, WP2, "D2.1: State-of-the-Art, Scenario Analysis and Requirements Definition for Equipment Reconfiguration Management", www.e2rmotlabs.com, June 2004.
[19] Software Defined Radio Forum, "Specification for PIM and PSM for SW Radio Components", Draft Revised Submission, 2004-01-01, www.sdrforum.org, 2005.
[20] D. Clark, D. Tennenhouse, "Architectural considerations for a new generation of protocols", Proc. ACM SIGCOMM, Philadelphia, USA, September 1990.
[21] S. Clemens, *"Component Software - Beyond Object-Oriented Programming*", Addison-Wesley/ACM Press, 1998.
[22] S. Gultchev, K. Moessner, R. Tafazolli, "Reconfiguration mechanisms and processes in RMA controlled soft-radios signalling", SDR Technical Conference, Orlando, FL, 17–19 November, 2003.
[23] L. Berlemann, R. Pabst, B. Walke, "A Reconfigurable Multi-Mode Protocol Reference Model Facilitating Modes Convergence", EW '05, Nicosia Cyprus, April 2005, submitted.
[24] M. Siebert, "Design of a Generic Protocol Stack for an Adaptive Terminal", Proc. First Karlsruhe Workshop on Software Radios, Karlsruhe, Germany, March 2000, pp. 31–34.
[25] M. Siebert, B. Walke, "Design of Generic and Adaptive Protocol Software (DGAPS)", Proc. Third Generation Wireless and Beyond (3Gwireless "01), San Francisco, CA, June 2001.
[26] M. Siebert, M. Steppler, "Software Engineering in the Face of 3/4G Mobile Communication Systems", Proc. 10th Aachen Symposium on Signal Theory, ISBN 3-8007-2610-6, Aachen, Germany, September 2001, pp. 89–94.
[27] N. Alonistioti, A. Glentis, F. Foukalas, A. Kaloxylos, "Reconfiguration Management Plane for the Support of Policy-based Network Reconfiguration", Personal Indoor and Mobile Radio Communications (PIMRC) Symposium, Barcelona, Spain, 5–8 September, 2004.
[28] S. Joachim, A. Schieder, "Generic Link Layer", Wireless World Research Forum, Tempe, Phoenix, 7–8 March, 2002.
[29] L. Berlemann, A. Cassaigne, B. Walke, "Generic Protocol Functions for Design and Simulative Performance Evaluation of the Link-Layer for Re-configurable Wireless Systems", WPMC '04, Abano Terme, Italy, September 2004.
[30] L. Berlemann, A. Cassaigne, R. Pabst, B. Walke, "Modular Link Layer Functions of a Generic Protocol Stack for Future Wireless Networks", Software Defined Radio Technical Conference 2004, Phoenix, USA, November 2004.
[31] "Reconfigurable SDR Equipment and Supporting Networks Reference Models and Architectures", WWRF Working Group 3 – White paper, 2002.
[32] S. Mann, M. A. Beach, P. A. Warr, J. P McGeehan, "Increasing the Talk-time of Mobile Radios with Efficient Linear Transmitter Architectures", *IEE Electronics and Communication Engineering Journal*, Vol. 13, No. 2, pp. 65–76, April 2001.
[33] P. A. Warr, M. A. Beach, J. P. McGeehan, "Gain-element Transfer Response Control for Octave-band Feedforward Amplifiers", *IEE Electronics Letters*, Vol. 37, No. 3, pp. 146–147, February 2001.
[34] S. Rajagopal, J. R. Cavallaro, S. Rixner, "Design Space Exploration for Real-Time Embedded Stream Processors", *IEEE Micro*, Vol. 24, July–August 2004.
[35] M. Sgroi, M. Sheets, A. Mihal, K. Keutzer, S. Malik, J. Rabaey, A. Sangiovanni-Vincentelli, "Addressing the System-on-a-Chip Interconnect Woes through Communication-based Design", Proceedings of the Design Automation Conference, Las Vegas, NV, pp. 667–672, June 2001.
[36] I. Verbauwhede, P. Schaumont, C. Piguet, B. Kienhuis, "Architectures and Design Techniques for Energy Efficient Embedded DSP and Multimedia Processing", Proceedings of the Design Automation and Test in Europe Conference, Vol. 2, pp. 988–993, February 2004.
[37] S. Das, S. Rajagopal, C. Sengupta, J. R. Cavallaro, "Arithmetic Acceleration Techniques for Wireless Communication Receivers", 33rd Asilomar Conference on Signals, Systems, and Computers, Pacific Grove, CA, pp. 1469–1474, October 1999.

[38] S. Rajagopal, B. A. Jones, J. R. Cavallaro, "Task Partitioning Base-station Receiver Algorithms on Multiple DSPs and FPGAs", International Conference on Signal Processing, Applications and Technology, Dallas, TX, August 2000.

[39] C. Sengupta, J. R. Cavallaro, B. Aazhang, "On Multipath Channel Estimation for CDMA Using Multiple Sensors", *IEEE Transactions on Communications*, Vol. 49, pp. 543–553, March 2001.

[40] B. Aazhang, J. R. Cavallaro, "Multi-tier Wireless Communications", *Kluwer Wireless Personal Communications*, Special Issue on Future Strategy for the New Millennium Wireless World, Vol. 17, pp. 323–330, June 2001.

[41] S. Rajagopal, J. R. Cavallaro, "A Bit-streaming Pipelined Multiuser Detector for Wireless Communications", IEEE International Symposium on Circuits and Systems (ISCAS), Vol. 4, Sydney, Australia, pp. 128–131, May 2001.

[42] K. Chadha, J. R. Cavallaro, "A Dynamically Reconfigurable Viterbi Decoder Architecture", Proc. Asilomar Conference on Signals, Systems and Computers, Pacific Grove, CA, pp. 66–71, November 2001.

[43] J. R. Cavallaro, M. Vaya, "VITURBO: A Reconfigurable Architecture for Viterbi and Turbo Decoding", IEEE International Conference on Acoustics, Speech, and Signal Processing (ICASSP), Vol. II, Hong Kong, China, pp. 497–500, April 2003.

[44] G. Xu, S. Rajagopal, J. R. Cavallaro, B. Aazhang, "VLSI Implementation of the Multistage Detector for Next Generation Wideband CDMA Receivers", *Journal of VLSI Signal Processing*, Vol. 30, pp. 21–33, March 2002.

[45] S. Rajagopal, S. Bhashyam, J. R. Cavallaro, B. Aazhang, "Real-time Algorithms and Architectures for Multiuser Channel Estimation and Detection in Wireless Base-station Receivers", *IEEE Transactions on Wireless Communications*, Vol. 1, pp. 468–479, July 2002.

[46] B. A. Jones, "Rapid Prototyping of Wireless Communications Systems", Master's thesis, Department of Electrical and Computer Engineering, Rice University, Houston, TX, January 2002.

[47] S. Rajagopal, S. Bhashyam, J. R. Cavallaro, B. Aazhang, "Efficient VLSI Architectures for Baseband Signal Processing in Wireless Base-station Receivers", 12th IEEE International Conference on Application-specific Systems, Architectures and Processors (ASAP), Boston, MA, USA, pp. 173–184, July 2000.

[48] A. Bria, F. Gessler, O. Queseth, R. Stridh, M. Unbehaun, J. Wu, J. Zander, "4th-Generation Wireless Infrastructures: Scenarios and Research Challenges", *IEEE Personal Communications*, Vol. 8, pp. 25–31, December 2001.

[49] U. Varshney, R. Jain, "Issues in Emerging 4G Wireless Networks", *IEEE Computer*, Vol. 34, pp. 94–96, June 2001.

[50] H. Corporaal, "A Different Approach to High Performance Computing", Pro. Fourth International Conference on High Performance Computing, 1997, Bangalore, India, December 1997.

[51] S. Srikanteswara, J. Neel, J. H. Reed, P. Athanas, "Designing Soft Radios for High Data Rate Systems and Integrated Global Services", Proc. Asilomar Conference on Signals, Systems and Computers, Pacific Grove, CA, November 2001.

[52] I. P. Seskar, N. B. Mandayam, "A Software Radio Architecture for Linear Multiuser Detection", *IEEE Journal on Selected Areas in Communications*, Vol. 17, pp. 814–823, May 1999.

[53] B. Lu, G. Yue, X. Wang, "Performance Analysis and Design Optimization of LDPC-Coded MIMO OFDM Systems", *IEEE Transactions on Signal Processing*, Vol. 52, No. 2, pp. 348–361, February 2004.

[54] W. Bai, L. Tang, Z. Bu, "An Equalization Method for OFDM in Time Varying Multipath Channels", 2004 IEEE 60th Vehicular Technology Conference, 2004, VTC 2004-Fall, September 2004, Paper 0303_16.

[55] M. B. Breinholt, M. D. Zoltowski, T. A. Thomas, "Space-Time Equalization and Interference Cancellation for MIMO OFDM", Proc. Thirty-Sixth Asilomar Conference on Signals, Systems and Computers, 2002, Vol. 2, pp. 1688–1693 3–6 November, 2002.

[56] S. Iraji, J. Lilleberg, "Interference Cancellation for Space-Time Block-Coded MC-CDMA Systems over Multipath Fading Channels", 2003 IEEE 58th Vehicular Technology Conference, 2003, VTC 2003-Fall, Vol. 2, pp. 1104–1108, 6–9 October, 2003.

[57] A. Batra, J. Balakrishnan, and A. Dabak, "Multi-band OFDM: A New Approach for UWB", ISCAS '04. Proc. 2004 International Symposium on Circuits and Systems, Vol. 5, pp. 365–368, 23–26 May, 2004.

[58] Transparently Reconfigurable Ubiquitous Terminal, IST-1999-12070 TRUST, http://www.ist-trust.org.

[59] W. Tuttlebee (Editor), *Software Defined Radio. Origins, Drivers and International Perspectives*, Wiley Series in Software Radio, John Wiley & Sons, 2002.

[60] D. Bourse, "WWRF WG3 White Paper: 'Reference model for end-to-end reconfigurable mobile communication systems'", WWRF6 Meeting, London, UK, 25–26 June, 2002.
[61] SandBlaster™ DSP, Sandbridge Technologies, Inc., http://www.sandbridgetech.com.
[62] M. Dillinger, K. Madani, N. Alonistioti, *"Software Defined Radio: Architectures, Systems and Functions"*, John Wiley & Sons, ISBN: 0-470-85164-3, April 2003.
[63] IST Project MOBIVAS, http://mobivas.cnl.di.uoa.gr.
[64] IST Project CREDO, http://credo.nal.motlabs.com.
[65] IST Project Nr. 2000–28088, Models and Simulations for Network Planning and Control of UMTS (Momentum), Web site: http://momentum.zib.de.
[66] IST Project Nr. 18863, System for management of quality of service in 3G networks, (SOQUET).
[67] IST Project Nr. 20275, Simulation of Enhanced UMTS Access and Core Networks (SEACORN).
[68] L. Jijun, R. Mukerjee, M. Dillinger, E. Mohyeldin, E. Schulz, "Investigation on Radio Resource Scheduling in WLAN Coupled with 3G Cellular Network", *IEEE Communication Magazine*, Vol. 41, pp. 108–115, June 2003.
[69] M. Dillinger, G. Ravasio, N. Olaziregi, N. Alonistioti, K. Kawamura, R. Garcia, T. Wiebke, "Network architecture supporting the End to End Reconfiguration (E2R)", WWRF8bis Meeting, Beijing, China, February 2004.
[70] N. Olaziregi, H. Aghvami, "A Novel Approach for Reconfigurable Systems at RAN Level", ICT '2003 Proceedings, Tahiti, Papeete, French Polynesia, February 2003.
[71] Z. Boufidis, M. Stamatelatos, N. Alonistioti, A. Delautre, M. Dillinger, "Actors, Management Plane, and Policy Provision Challenges for End-to-End Reconfiguration", E2R Workshop, Barcelona, Spain, September 2004.
[72] N. Olaziregi, C. Niedermeier, R. Schmid, D. Bourse, T. Farnham, R. Haines, F. Berzosa, "Overall System Architecture for Reconfigurable Terminals", IST Summit 2001, Sitges, Spain, September 2001.
[73] S. Micocci, N. Olaziregi, G. Ravasio, J. Bachmann, F. Berzosa, "Architecture of IP based Network Elements Supporting Reconfigurable Terminals", Proc. IST Summit 2003, Aveiro, Portugal, June 2003.
[74] W. H. W. Tuttlebee, "Software-Defined Radio: Facets of a Developing Technology", *IEEE Personal Communications Magazine*, Vol. 6, No. 2, pp. 38–44, April 1999.
[75] Multicast Dissemination Protocol version 2 (MDPv2), homepage: http://cs.itd.nrl.navy.mil/5522/mdp/mdp_index.html.
[76] L. Rizzo, L. Vicisano, "RMDP: An FEC-based Reliable Multicast Protocol for Wireless Environments", *ACM Mobile Computing and Communications Review*, Vol. 2, No. 2, April 1998.
[77] K. Obraczka, "Multicast Transport Protocols: A Survey and Taxonomy", *IEEE Communications Magazine*, Vol. 36, No. 1, pp. 94–102, January 1998.
[78] O. Holland, A. H. Aghvami, "Dynamic Switching Between One-to-Many Download Methods in 'All-IP' Cellular Networks", *IEEE Transactions on Mobile Computing*, Vol. 5, No. 3, pp. 274–287, March 2006.
[79] M. Dillinger, E. Mohyeldin, P. Demestichas, G. Dimitrakopoulos, J. Luo, N. Olaziregi, N. Alonistioti, "Dynamic Network Planning and Functional Components supporting End to End Reconfiguration (E2R)", WWRF10 Meeting, New York, NY, October 2003.
[80] "The XG Vision Request for Comments, V 2.0"; Defense Advanced Research Agency, available at: http://www.darpa.mil/ato/programs/xg/rfc_vision.pdf.
[81] "The XG Architectural Framework Request for Comments V1.0"; Defense Advance Research Agency, available at: http://www.darpa.mil/ato/programs/xg/rfc_af.pdf.
[82] P. Marshall, "XG – Next Generation Communications", WWRF 10 Opening Plenary, New York City, 27 October, 2004.
[83] P. Marshall, "XG Communications Program Briefing", ITU Study Group 8; September 2004; available at: http://www.itu.int/ITU-R/study-groups/rsg8/sem-tech-innov/docs/.
[84] P. Marshall, "What's Beyond Software Defined Radios? Policy Defined Radios", Panel Session at SDR Forum Technical Conference, Phoenix, Arizona, USA, November 2004.
[85] R. Krishnan, "Towards Policy Defined Cognitive Radios", NSF Workshop on Programmable Wireless Networking Information Meeting, 5 February, 2004; available at: www.cra.org/Activities/workshops/nsf.wireless/BBN_Krishnan2.pdf.

[86] P. Marshall, "XG Communications Program Overview", XG Industry Workshop, 30 June, 2004.
[87] J. Evans, "Programmable Wireless Networking Details and Logistics", National Science Foundation Programmable Wireless Networking Informational Meeting, 5 February 2004; available at: http://www.cra.org/Activities/workshops/nsf.wireless/Evans_ProWiN_Openin2.pdf.
[88] Testimony Before the Subcommittee on Trade of the House Committee on Ways and Means Hearing on Trade Relations with Europe and the New Transatlantic Economic Partnership, Kevin Kelly, July 28, 1998, http://waysandmeans.house.gov/legacy/trade/105cong/7-28-98/7-28kell.htm.
[89] United States Welcomes EC Statement of Support for ITU Process on Setting New Mobile Telecommunications Standards, http://www.fcc.gov/Speeches/Kennard/Statements/stwek905.html.
[90] "Agreement on Technical Barriers to Trade", WTO, http://www.wto.org/english/docs_e/legal_e/17-tbt.doc.
[91] P. Leaves "Dynamic Spectrum Allocation and Coexistence – Functional Specification and Algorithm", March 2003, IST-2001-35125/OverDRiVE/WP1/D06.
[92] United Kingdom Radiocommunications Agency, http://www.radio.gov.uk/.
[93] M. Cave, "Review of Radio Spectrum Management", UK Department of Trade and Industry, 2002, Available at: http://www.spectrumreview.radio.gov.uk/.
[94] "Spectrum Trading Consultation", November 2003, OFCOM.
[95] U.S. Federal Communications Commission, "Spectrum Policy Task Force Report", ET Docket No. 02-135, 2002.
[96] U.S. Federal Communications Commission, "Notice of Inquiry: Additional Spectrum for Unlicensed Devices Below 900 MHz and in the 3 GHz Band", ET Docket No. 02-380, December 11, 2002.
[97] D. Babb, C. Bishop, T. Dodgson, "Security Issues for Downloaded Code in Mobile Phones", *Electronics and Communications Engineering Journal*, Vol. 14, pp. 219–227, 2002.
[98] J. Faroughi-Esfahani, R. Falk, N. Drew, P. Bender, "A Regulatory View on Security Requirements for Reconfigurable Radio", SDR Forum Technical Conference, San Diego, CA, Paper No. SW3-03, 2002.
[99] FCC, First Report and Order, Authorization and Use of Software Defined Radios, ET Docket No. 00-47, September 2001.
[100] FCC, Notice of Proposed Rule Making and Order, Facilitating Opportunities for Flexible, Efficient, and Reliable Spectrum Use Employing Cognitive Radio Technologies, ET Docket No. 03-108; Authorization and Use of Software Defined Radios, ET Docket No. 00-47, December 2003.
[101] Y. Suzuki, "Interoperability and Regulatory Issues around Software Defined Radio (SDR) Implementation", *IEICE Transactions on Communications*, Vol. E85-B, No. 12, pp. 2564–2572, December 2002.
[102] H. Harada, Research and Development on Regulatory Issue of SDR, Yokosuka Radio Communications Research Center, Communications Research Laboratory (CRL), Independent Administrative Institute, Japan, 17 September, 2003, presentation to SDR Forum.
[103] H. Harada, M. Kuroda, H. Morikawa, H. Wakana, F. Adachi, "The Overview of the New Generation Mobile Communication System and the Role of Software Defined Radio Technology", *IEICE Transactions on Communications*, Vol. E86-B, No. 12, pp. 3374–3384, December 2003.
[104] Y. Suzuki, K. Oda, R. Hidaka, H. Harada, T. Hamai, T. Yokoi, "Technical Regulation Conformity Evaluation System for Software Defined Radio", *IEICE Transactions on Communications*, Vol. E86-B, No. 12, pp. 3392–3400, December 2003a.
[105] Y. Suzuki, H. Harada, K. Uehara, T. Fujii, Y. Yokoyama, K. Oda, R. Hidaka, "Adaptability Check during Software Installation in Software Defined Radio", *IEICE Transactions on Communications*, Vol. E86-B, No. 12, pp. 3401–3407, December 2003b.
[106] Y. Suzuki, T. Yokoi, Y. Iki, E. Kawaguchi, N. Nakajima, K. Oda, R. Hidaka, "Development of Experimental Prototype System for SDR Certification Simulation", *IEICE Transactions on Communications*, Vol. E86-B, No. 12, pp. 3408–3416, December 2003c.
[107] J. Abley, B. Black, V. Gill, RFC 3582: Goals for IPv6 Site-Multihoming Architectures, IETF, August 2000.
[108] T. Bates, Y. Rekhter, RFC 2260: Scalable Support for Multi-homed Multi-provider Connectivity, IETF, January, 1998.
[109] M. Andrej, T. Suihko, M. West, "Aspects of Multi-Homing in IP Access Networks", IST Mobile Communications Summit, Thessaloniki, Greece, pp. 115–119, June 2002.

[110] IST-2000-28584 MIND Deliverable D2.2, "MIND Protocols and Mechanisms Specification, Simulation and Validation", November 2002.
[111] H. Oliver, Qi Fan, A. Hamid Aghvami, "A Dynamic Hierarchical Radio Resource Allocation Scheme for Mobile Ad-Hoc Networks," IEEE PIMRC 2004, Barcelona, Spain, September 2004.
[112] J. Mitola, *"Software Radio Architecture Object-Oriented Approaches to Wireless Systems Engineering"*, Wiley-Interscience, ISBN 0-471-38492-5, 2000.

9

Self-organization in Communication Networks

Edited by Amardeo Sarma (NEC), Christian Bettstetter (DoCoMo) and Sudhir Dixit (Nokia)

9.1 Introduction and Motivation

Mark Weiser's vision of ubiquitous computing [1] is about to become a reality. Wireless technologies continue to penetrate rapidly, connecting not only mobile phones and computers, but also a myriad of small devices, sensors, and everyday items. This trend creates new applications but, adds to the spatial-temporal complexity and dynamics of the network. It thus increases the burden on network administrators and users. An important question is how this complexity can be reasonably managed without requiring users to become technical experts in the field, and the network owners to spend time and resources in managing networks.

One promising approach is to increase the degree of self-organization, that is, to design and develop networks that minimize human intervention but organize themselves in an automated manner. In fact, the world presents many examples where self-organization provides efficient solutions to non-trivial problems: birds organize themselves to fly in well-structured flocks, water and dust particles in the air form cloud streets of beautiful shapes, and insects can emit light flashes in perfect synchrony [2]. By 'self-organization', we intuitively mean that an organizational structure is achieved in a distributed manner. The individual entities (e.g. animals, human beings) exchange information locally and there is no need for an external control instance. But self-organization is more than that; it is about the application of simple, high-level behaviour rules in the individual entities that lead to sophisticated functionality of the overall system.

Although many phenomena of self-organization can be observed in nature, sociology, and ecology, their application and transfer to technological systems is still in the fledgling stages. Some examples can be found in the areas of autonomous mobile robots,

Technologies for the Wireless Future – Volume 2 Edited by Rahim Tafazolli

genetic algorithms, and pattern recognition. The goal of the Special Interest Group on Self-organization in Wireless World Systems (SIG3) within the Wireless World Research Forum (WWRF) is to address the topic of self-organization in wireless communication and computing. We envision a world of ubiquitous wireless communication in which many functions are self-organized and thus require the least manual intervention by users and administrators. The specific goals for this vision are as follows [3]:

- understand the needs and requirements for self-organization in wireless systems, to allow systems to operate the way they are expected to do;
- develop solutions (concepts, tools, and techniques) for self-organization, where humans are part of the overall adaptive self-organization system;
- enable new business models, for example, for micro operators.

To achieve these goals, it will be essential to understand the technical and user requirements of self-organized systems. Solutions will include concepts, tools, and techniques for self-organization. The technical work should encompass the network with its control, transport, and management functionality. All levels or layers, such as physical, network, middleware, and application, are potentially affected. Often pure self-organized solutions are not enough, but in many cases, the system must meet some constraints; it should behave as the user expects, and the user should retain some high-level control in order to correct misbehaviour or prevent security attacks from outside.

9.2 Self-organization in Today's Internet

Today's communication and computer networks already have several functions that contribute to a higher level of self-organization. Especially the Internet and its applications give us many examples where user-friendly self-configuration and distributed operation can be used to reduce operational costs, enhance usability, and create completely new services. This section gives some specific examples in this area. We address the following topics: (1) self-configuration, (2) peer-to-peer networking, and (3) open-content Web sites.

9.2.1 Self-configuration in the Internet

The goal of self-configuration is to disburden the users and administrators from configuration tasks, such as entering IP addresses of servers and configuring routers. While various network technologies, such as Dynamic Host Configuration Protocol (DHCP) [4], already perform some self-configuration, the Internet Protocol Version six (IPv6) promotes this paradigm even further [5]. The ultimate goal is to achieve full network self-configuration and 'plug-and-play' functionality for end-users. To achieve this goal, several protocols have been defined, which can be classified into host, router, and service configuration technologies. In the following sections, we give some specific examples in each of these areas.

9.2.1.1 Host Auto-configuration

IPv6 contains several mechanisms for auto-configuration of hosts. Specifically, these mechanisms allow hosts to obtain an IP address and other information that enables them to access the network and search for advanced and more specific configuration data. These auto-configuration techniques can be classified into stateless and stateful methods.

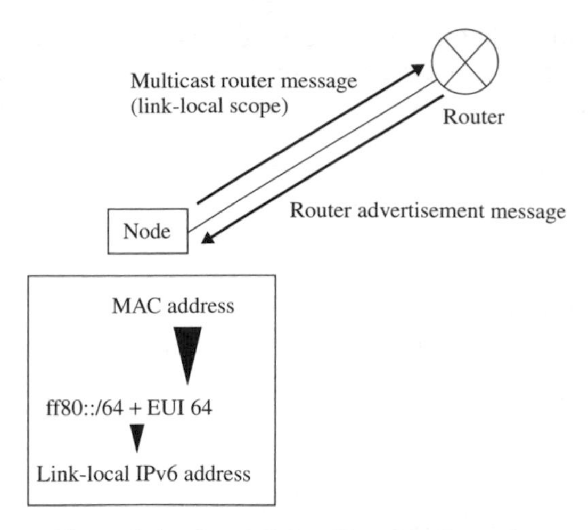

Figure 9.1 Stateless IPv6 auto-configuration

As illustrated in Figure 9.1, stateless auto-configuration is performed as follows [6]; when a node is activated, it generates a link-local address by adding the EUI-64 node address to the link-local unicast prefix (FE80::/64). The EUI-64 address is computed by converting the MAC address of the interface, thus creating a unique identifier for the interface. A duplicate address detection (DAD) algorithm should be performed to confirm the uniqueness of the address. Every interface joins the multicast group (FF02::1) in order to receive all multicast messages on the link. The node sends a solicitation message to the multicast router address (FF02::2). In case a router is available, it will reply with a router advertisement message. This message indicates whether stateless auto-configuration is allowed or stateful auto-configuration is offered or is even mandatory.

Stateful auto-configuration is performed using the DHCP [7]. As shown in Figure 9.2, the host has to discover the address of a DHCP server. Therefore, the node sends a DHCP

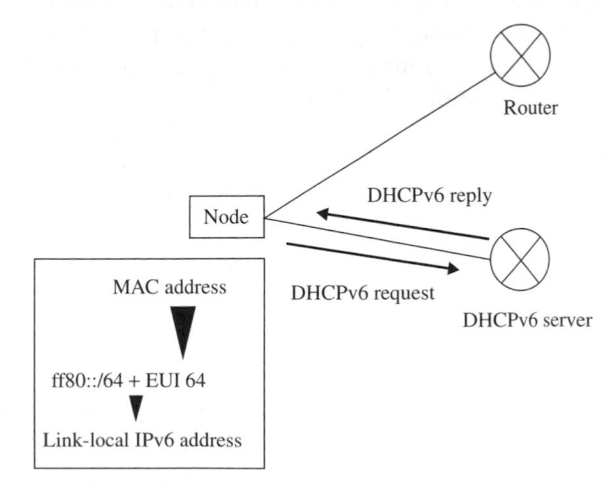

Figure 9.2 Stateful IPv6 auto-configuration via DHCPv6

solicitation message to the link-local multicast address (FF02::1:2). The packet arrives at the DHCP server responsible, either directly or is relayed to another server in case a server is responsible for several sub-networks. The server replies with a server advertisement before a request-reply message exchange delivers all required information (e.g. link-local, site-local and global IP address as well as Domain Name Server (DNS) address, time offset, maximum hop limit, etc.) to the node. DHCP allows for straightforward reconfiguration of addresses in case of topology changes and avoids duplicated addresses without additional mechanisms. However, like all centralized approaches, the DHCP server constitutes a single point of failure, which requires server duplication.

The optimal method to perform the address assignment is the combination of stateless and stateful mechanisms. While stateless auto-configuration provides basic access information, such as the IP address, stateful auto-configuration is able to support enhanced service information.

9.2.1.2 Router Auto-configuration

Router Prefix Delegation

The Automatic Prefix Delegation Protocol [8] enables routers to exchange IP address prefixes among each other. The protocol is based on a query–response mechanism between a delegating router and a requesting router. The protocol is based on two main message types: prefix request and prefix delegation. The requesting router can ask for three different types of actions to be done in a prefix request message: prefix delegation, prefix refresh, and prefix return. A delegated prefix is always associated with a limited lifetime. The protocol, although being very simple and lightweight, has latency problems in large networks as well as possible collisions with other delegating routers.

Another protocol capable of router prefix delegation is OSPFv3 for IPv6 router auto-configuration [9], which is an improved version of the Open Shortest Path First (OSPF) routing protocol. It defines a router auto-configuration method by the addition of three new LSA (Link State Advertisement) messages that can be used to choose the SLA (Service Level Assignment) prefix assignment to routers and to check its uniqueness. Since OSPF is a very well-known routing protocol, a good router auto-configuration mechanism can be achieved by adding three new SLA messages. However, it is only useful for the SLA IPv6 prefix auto-configuration, while the TLA (Top Level Assignment) part must be delegated in another way. Although router auto-configuration via IPv6 can be employed to define the SLA, another protocol is required in order to assign the TLA.

Finally, the Unique Identifier Allocation Protocol (UIAP) [10] enables an application to validate and defend the uniqueness of an identifier presented by an application within a domain. It is a general-purpose protocol that can be applied to any identifier that should be tested for uniqueness. The protocol can be very useful to check for duplications in IPv6 prefixes before assigning them to a router.

IPv6 Router Renumbering

The idea behind the router renumbering protocol [11] is to enable routers to perform prefix maintenance operations, such as adding or changing IP prefixes. Such operations might be needed because of topology changes in the network. However, the operation in large scale networks has not been tested so far.

Zero router Automatic ReConfiguration Protocol (ARCP)

The development of the Automatic ReConfiguration Protocol (ARCP) [12] is taking place in the Zero router Internet Engineering Task Force (IETF) working group. ARCP is established between an ARCP client, an ARCP server, and a DHCP server. The ARCP client is a router that wants to be configured. Either initial configuration after the router is activated in the network or there is reconfiguration due to changes in the network topology. The ARCP server is the node that processes the requests from the ARCP clients. The DHCP server must assign an IP address to a new router in the network and inform the router about the ARCP server address.

9.2.1.3 Service Auto-configuration

In order to allow the discovery and configuration of services as offered by the network, the following protocols have been proposed. These services are often high-level services, such as printing service or local proxy settings.

Service Location Protocol

As shown in Figure 9.3, the architecture of the Service Location Protocol (SLP) [13] consists of three main components:

- user agents (UAs) that perform service discovery on behalf of the user or application;
- service agents (SAs) that advertise the location and characteristics of services on behalf of services;
- directory agents (DAs) that collect service addresses and information received from SAs in their database and respond to service requests from UAs.

When a new service connects to a network, the SA contacts the DA to advertise its existence (service registration). When a user needs a certain service, the UA queries the available services in the network from the DA (service request). Before a client (UA or

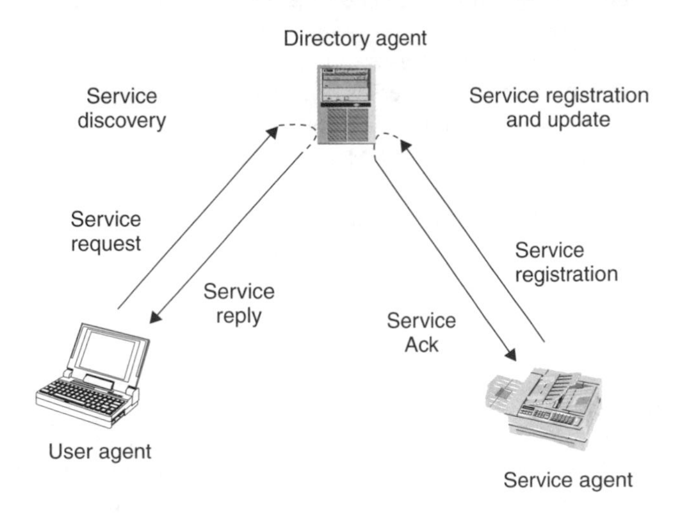

Figure 9.3 Service location protocol (SLP)

SA) is able to contact the DA, it must discover the existence of the DA. It either queries the DHCP server which knows the address of the SLP DA, sends a request to an SLP multicast group address, or receives a DA advertisement message that is periodically sent out from the DA.

SLP has two operational modes, depending on whether a DA is present or not. If a DA exists on the network, it will collect all service information advertised by SAs. UAs will send their request to the DA and receive the desired service information. If there is no DA, UAs repeatedly send out their requests to the SLP multicast address. All SAs listen for these multicast requests and, if they advertise the requested service, they will send unicast responses to the UA. Furthermore, SAs multicast an announcement of their existence periodically, so that UAs can learn about the existence of new services.

Service Discovery Service

The Service Discovery Service (SDS) [14] is a scalable, fault-tolerant, and secure information directory providing client access to all available services on the network. SDS can store many types of services that are available for execution, services running a specific host, available service platforms, and passive data. Furthermore, SDS supports both passive and active discovery. SDS uses XML service descriptions as message payload, which provides a great flexibility. Furthermore, the use of XML allows validation of the service descriptions on a per-tag granularity.

Universal-plug-and-play

Further service discovery protocols, such as Universal-Plug-and-Play (UPnP) [15], address only small network installations or are scoped to a link-local application.

9.2.1.4 Zeroconf and the Rendezvous Protocol

The 'zero configuration networking' (Zeroconf) working group in the IETF defines protocols for network auto-configuration that do not need dedicated servers. The goal is that these networks become interoperable with any other network, independent of whether the other network is auto-configured or not. The objectives and main working areas are:

- allocate addresses without DHCP servers [16];
- translate between names and IP addresses without DNS servers;
- find services (e.g. printing) without a directory server;
- allocate IP multicast addresses without MADCAP servers [17].

The rendezvous protocol is a commercial implementation of the Zeroconf mechanism made available by Apple in 2002. It comprises the following:

- *Address allocation*: When a new computer is added to a network, the device randomly selects an IP address and probes the network if this address is already in use. In case of address conflicts, the process is repeated until the device acquires a unique address and is ready to exchange IP traffic.
- *Name-to-address resolution and service discovery*: Rendezvous uses a variant of DNS called *Multicast DNS-Service Discovery* (*mDNS-SD*). A device advertises its service, including type of service, name of the service, IP and port address, and so on. Each device on the network receives the notification and stores the information, which can

be queried by applications. To share a service, a device must create a unique name for each of its services and advertise them to other nodes. Therefore, every host acts as a DNS server.

9.2.2 Peer-to-peer Networking

Peer-to-peer networking is a communication service in which each participant has the same capabilities and every member can initiate a connection to another member. This is why the members are often referred to as peers (one among equals) or servants (pointing out that they can act as both SERVer and cliENT at the same time). The resulting peer-to-peer (P2P) network is no longer managed by one central instance but by the peers themselves. Each peer has to follow some basic rules to build up or maintain the network topology. Communication is achieved either by message flooding or by means of a structured architecture. P2P protocols establish overlay networks that are mostly based on User Datagram Protocol (UDP), Transport Control Protocol (TCP) or HyperText Transport (or Transfer) Protocol (HTTP/) connections, but the overlay connections do not reflect their physical counterparts. To be able to enter the virtual network, a new peer has to know at least one IP address of a node already participating in the overlay network. Address caching or making use of a bootstrap server is the common way to solve this problem.

In some backbone networks, more than 60 % of the network traffic results from peer-to-peer (P2P) applications and protocols [18]. File sharing is still the major application area, but voice-over-IP and redundant storage are advancing new areas.

Figure 9.4 provides a summary of the characteristics of different kinds of peer-to-peer networks and compares them to classical client–server networks.

9.2.2.1 Unstructured P2P Networks

Let us first take a look at unstructured P2P networks. As the term indicates, the nodes form a network with a random arrangement, that is, the connections between the nodes in

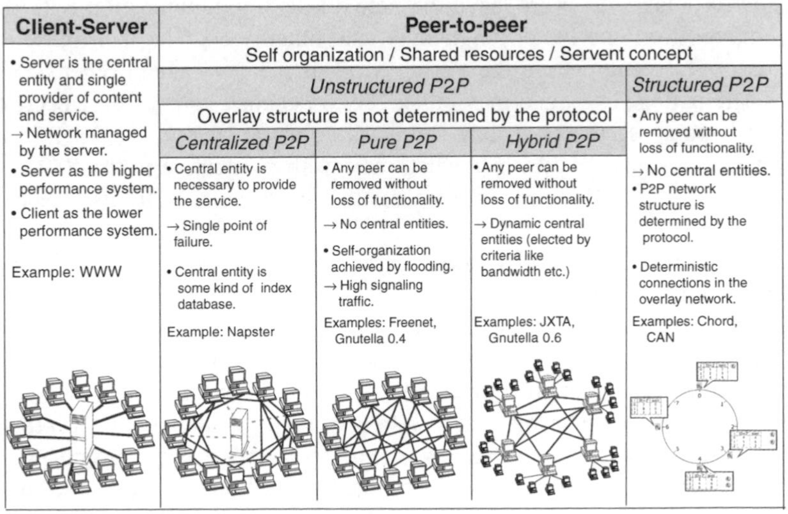

Client-Server	**Peer-to-peer**			
	Self organization / Shared resources / Servent concept			
	Unstructured P2P			*Structured P2P*
	Overlay structure is not determined by the protocol			
	Centralized P2P	*Pure P2P*	*Hybrid P2P*	
• Server is the central entity and single provider of content and service. → Network managed by the server. • Server as the higher performance system. • Client as the lower performance system. Example: WWW	• Central entity is necessary to provide the service. → Single point of failure. • Central entity is some kind of index database. Example: Napster	• Any peer can be removed without loss of functionality. → No central entities. • Self-organization achieved by flooding. → High signaling traffic. Examples: Freenet, Gnutella 0.4	• Any peer can be removed without loss of functionality. → Dynamic central entities (elected by criteria like bandwidth etc.) Examples: JXTA, Gnutella 0.6	• Any peer can be removed without loss of functionality. → No central entities. • P2P network structure is determined by the protocol. • Deterministic connections in the overlay network. Examples: Chord, CAN

Figure 9.4 Peer-to-peer networking

the overlay topology are not predetermined. Furthermore, the network does not put any effort in the management or distribution of the shared content. It is like a deterministic chaos; simple but important basic rules exist, but the complete network structure cannot be determined. We can find several parallels in nature. For example, ants build huge hills, but each individual ant needs to follow only a few, basic instincts. Unstructured P2P networks come in different variants that are explained in the following sections.

Centralized P2P Networks

The first P2P file sharing application was introduced by Napster in 2000 – with it a real P2P boom started. However, Napster is still client–server based; a central database maintains an index of all files that are shared by the peers currently logged on. Each peer that logs into the network first has to register all its content at the central database. The database can then be queried by the peers to learn about the ports and IP addresses of peers sharing the requested content. Hence, Napster can be classified as a centralized unstructured P2P network, where a central entity is necessary to provide the service. There is some ongoing discussion if this network architecture should be counted among P2P networks, as it is highly client–server based and no real self-organization is involved. However, we argue that only querying the central index database makes use of the central instance, whereas content is exchanged single-handed.

Pure Unstructured P2P Networks

The main disadvantage of Napster's central architecture is its single point of failure. For this reason, pure unstructured P2P networks have been developed. An example is Gnutella 0.4. Such networks work without any central entity, and all peers provide the same functionality. This makes the network highly fault resistant since any peer can be removed without loss of functionality.

Each peer maintains a list of other peers – it stays in contact with its so-called 'neighbours'. For example, Gnutella 0.4 peers simply select the first peers they come in contact with. Other protocols prefer peers in geographical proximity or ones that can be contacted with the lowest delays. As peers leave the network and communication with them fails, they are removed from the list of neighbours, and other peers take their place in the list.

A peer can only participate in the P2P network if it knows about other peers that are currently participating. The node may either rely on cached addresses of peers that were active in a previous session or it may contact a bootstrap server. This server is a well-known host with a stable IP address, which may itself participate in the P2P network, or which simply caches the IP addresses of peers that used the bootstrap server to enter the network.

Self-organization is achieved by flooding the messages through the network. A peer forwards every incoming message to all neighbours, except to the neighbour the message was received from. Each message contains a unique identifier and time-to-live value to avoid message loops and infinite propagation. Because of the limited time-to-live, it is not guaranteed that a peer finds the desired content in the network. The small-world experiment [19], however, showed that members of large social networks (in this case, the population of the United States) would be connected to each other through short chains of roughly six acquaintances. Affirmed by the Small-World Project, we can estimate that social searches and thus searches for content can reach their targets in a median of five to seven steps; also the actual success depends strongly on individual incentives. In addition, the more interesting the content is for users, the more it will be shared by many participants, resulting in an even higher success probability.

Hybrid Unstructured P2P Networks

A main disadvantage of Gnutella 0.4 is that all request messages are flooded through the network. This results in a large overhead and high network load. Gnutella 0.6 tries to reduce network traffic by introducing hierarchies through the differentiation between ultra-peers and leaf-nodes. Hence, Gnutella 0.6 is a mixture of pure and centralized P2P networking approaches and is thus classified as a hybrid unstructured P2P network.

In general, a leaf-node maintains only one connection to one ultra-peer (also called *super-peers*). An ultra-peer, in turn, is connected to a few other ultra-peers and many leaf-nodes. Each ultra-peer undertakes the task of the central entity for the leaf-nodes it is connected to. Hence, a Napster-like network is maintained between them. The ultra-peer indexes the shared content of all of its leaf-nodes and answers queries directly if one of its peers hosts the requested content. Additionally, the ultra-peers establish Gnutella 0.4–like connections among each other. In case the requested content is not hosted by any of their leaf-nodes, the ultra-peers broadcast the request in the ultra-peer layer. The ultra-peers thus shield their leaf-nodes from the signalling traffic and also provide better query routing functionalities. The network decides whether nodes connecting to the network should act as ultra-peers or leaf-nodes. Available bandwidth, processing power, and storage capacity as well as the need for a new ultra-peer are taken into consideration for this decision.

9.2.2.2 Structured P2P Networks

Pure and hybrid unstructured P2P networks have two main disadvantages. Firstly, flooding generates a high signalling overhead in the network. Secondly, there is no guarantee that content would be found.

Hence, a new generation of P2P protocols, the structured P2P networks, has been developed. Here, each peer obtains a specific identifier (ID) that is created, for example, by hashing the IP address and port number. Neighbours are no longer chosen by a random process, but by arranging the peers in the 'identifier space'. The CHORD protocol, for example, arranges its nodes in a one-dimensional ring-shaped identifier space. In addition to peer IDs, content IDs are also generated with the same hash function. The content is then copied to the peer whose ID is closest to the content ID. As this generates a lot of traffic, in the majority of protocols, only a description of the content, together with a link to the peer where it is stored, is moved to the responsible peer. The main advantage of this structure is that a more efficient search is possible. As the placement of peers and content is predetermined, lookups benefit from knowing the location where the queried content should be located in the network. The query messages can be forwarded on an almost direct path without the need for flooding the message through the network. Arriving at its deterministic destination, a lookup can be resolved at any rate; if the content or a link is available, it is returned to the initiator of the query.

The structured P2P approach also has disadvantages. High churn rates, that is, nodes joining and leaving the network frequently, stress the stability of structure and consistency of data. Additionally a permanent shifting of content is necessary, resulting in high maintenance traffic and incorrect lookups. Therefore, it is questionable if structured P2P protocols should be applied in environments with low average online time.

9.2.3 Open-content Web Sites

Recently several Web sites appeared in the Internet that can be edited by anyone, thus forming a dynamic open platform for knowledge exchange. A popular example is the online encyclopaedia Wikipedia (Figure 9.5, [20]).

In such open-content Web sites, the usual process for content creation – that is, having expert authors writing the content and reviewing the content–is replaced by a self-organized, user-centric approach. In Wikipedia, for example, anyone in the world can change, add, and edit encyclopaedic articles. The changes have immediate effect on the Web. On the one hand, this approach lowers the hurdle to add and correct content, and it has the potential to collect a huge amount of content. On the other hand, the approach opens the Web site to intentionally or unintentionally false or imprecise information.

9.2.3.1 Wiki

Most open-content Web sites are based on so-called 'Wiki' software. The first Wiki tool was created by W. Cunningham in 1995. The name has been derived from the Hawaiian word 'wikiwiki', meaning 'quick', with the idea of putting content fast on the Web. More than 100 different implementations are available today, most of them being open source.

The Wiki software uses databases or files for storing the content and provides a user interface to search, display, and edit the content. In many systems, there is an additional

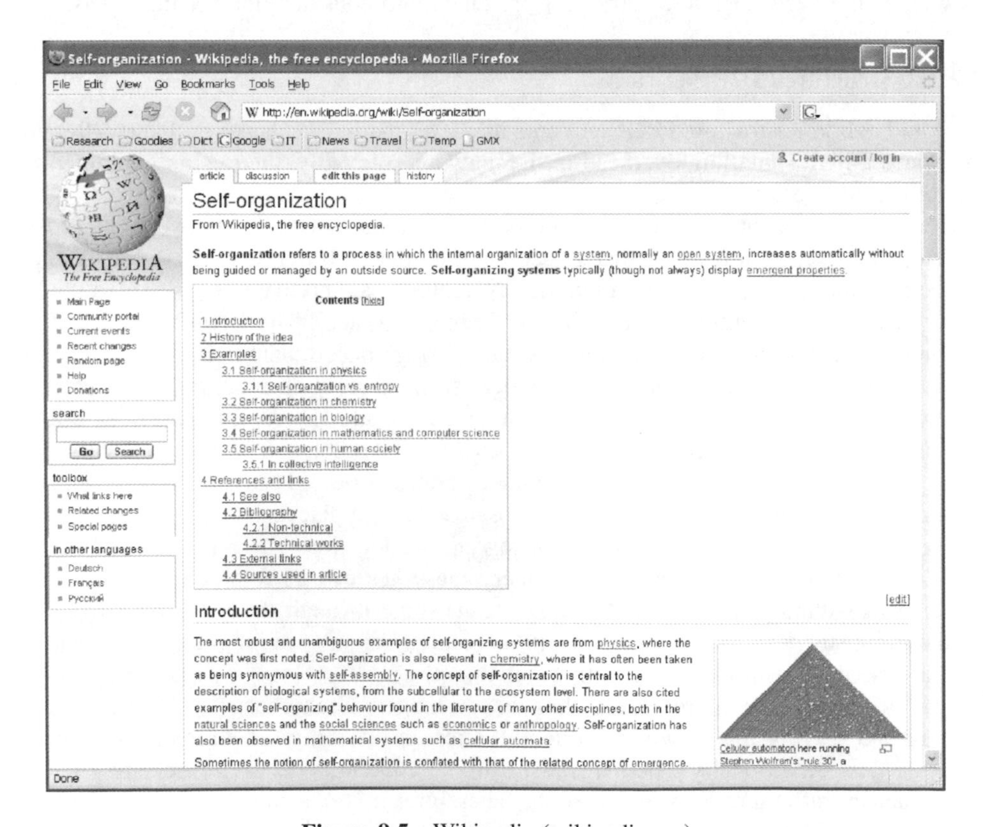

Figure 9.5 Wikipedia (wikipedia.org)

'talk' functionality to discuss the content. Changes to the content can be traced, and older versions can be displayed to facilitate better discussion and improvement of the content. It is assumed that the quality of the content improves with a large number of readers and reviews, updates, and changes. The process of improving the content is to a great extent a self-organized process between users, often also described as an 'anarchic', 'democratic', or 'socially Darwinian evolutionary' process. In contrast to content management tools, like TYPO3 [21], Wikis do not have a multilevel access right policy, but anyone can change, add, and edit the content. In contrast to chats, newsgroups, and fora, there is no fixed structure, like threads, and also no time ordering.

9.2.3.2 Wikipedia

Wikipedia is a free Internet encyclopaedia that anyone can edit. Started in 2001, currently it has more than 1.6 million articles online, around 600 000 of them in English (June 2005). Thousands of changes and new articles are created every hour.

Although no expert editing team is present and no review process is available, the quality is interestingly highly ranked. Several professional evaluations have shown that Wikipedia is easily on par with expensively edited digital encyclopaedias, like Brockhaus Premium or Microsoft Encarta [22]. From a scientific point of view, Wikipedia ranks best in this comparison. One major advantage is its timeliness. The content is updated regularly and reflects current political and social events and thus represents an extremely contemporary knowledge database. Critics who are editors of traditional encyclopaedias state that there is no authority behind the information, as anyone could have entered it and no expert (authority) has checked the content. Further critics lead in the direction that the content is not always precise and concise enough.

To maintain Wikipedia as a valuable encyclopaedia, some basic rules and policies have been set up. The main policy, which cannot be changed even by voting, is the policy to maintain a neutral point of view. If there are contradicting points of view, all of them should be mentioned rather than discriminating one. Another basic rule is that no advertisement is allowed. Wikipedia is also not the place for opinions and original (unpublished) research. To protect Wikipedia against vandalism, volunteer administrators can lock articles from being edited, delete articles, and block individual users temporarily from the system. In case of strong or continued breaking of rules, the founder of Wikipedia, Jimmy Wales, can block some users permanently.

9.3 Self-organization in Ad Hoc and Sensor Networks

In wireless ad hoc and sensor networks, the mobile devices communicate with each other without the need for any fixed network infrastructure. As shown in Figure 9.6, such networks are based on the 'multi-hop principle', that is, each device may act as a relay for data of other devices. There has been a huge effort in the research community in developing protocols and understanding the theory of such completely decentralized, purely wireless networks. Among other things, a set of routing protocols has been standardized within the IETF [23], distributed algorithms for network organization and auto-configuration have been proposed, and new understanding of theoretical aspects (e.g. capacity and connectivity) has been obtained. In this section, we explain three selected

Figure 9.6 Ad hoc networking

concepts for self-organization in ad hoc and sensor networks: (1) cooperation and fairness, (2) distributed topology control (TC), and (3) address self-configuration.

9.3.1 Cooperation and Fairness

Cooperation among nodes in ad hoc networks does not come by nature. Since nodes tend to save their battery power, they may act selfishly in the sense that they do not forward data from other nodes. Countermeasures are needed to stimulate nodes to 'play fair', either by punishing unfair nodes or by providing an incentive to increase participation. These measures are known as cooperation approaches, and may be classified into detection-based and motivation-based approaches.

Detection-based cooperation approaches focus on the detection of potential misbehaviour in the network with the aim of punishing the originator for misbehaving. In contrast, motivation-based approaches do not aim to detect the misbehaving party itself. Instead, they aim at preventing misbehaviour by stimulating the involved parties to act in a 'protocol conform' manner to benefit the community. It is only when the parties are aware that a punishing mechanism is in place that this may prevent selfish behaviour to some degree.

9.3.1.1 Detection-based Approaches

The detection-based model proposed by Marti, Giuli, Lai, and Baker [24] mitigates the effects of misbehaviour in the data-forwarding phase. Using the promiscuous mode, misbehaviour, such as packet dropping of a node, is detected. With this information, routing packets through these nodes can be avoided.

Yang, Meng, and Lu [25] present a network layer security solution to observe selfish neighbour nodes. This serves to make nodes aware that they may be excluded from the community if they misbehave. The problem of wrong accusation is handled by k-voting, where k denotes the threshold for the minimum number of ingoing accusations from observing neighbour nodes. An additional feature is the collaborative registration of each node by its direct neighbours for a particular lifetime. Only registered nodes can send traffic.

The CONFIDANT protocol by Buchegger and Boudec [26] aims at detecting and isolating misbehaving nodes, which makes it difficult for a node not to co-operate. Each node consists of a monitor, a trust manager, a reputation system, and a path manager. The monitor observes the misbehaviour of nodes and informs the reputation system. The reputation system is responsible for judging the actual rating of a node on the basis of observations sent by the monitor. The trust manager sends and receives alarms to and from other nodes about ratings. The trust manager is also responsible for assessing the truthfulness of the alarms on the basis of the trustworthiness of the sender of the alarm. The path manager decides whether to deny or provide malicious nodes of any service. This is a complete architecture for a community-based rating system. However, it does not discuss in detail how to measure trustworthiness of accusation or how to rate originators.

Paul and Westhoff [27] propose an accusation scheme on the basis of majority voting for detecting attacks on the Dynamic Source Routing (DSR) protocol. The source node can subsequently advertise these ratings along with adequate proofs to trusted nodes. Such ratings are used by the knowledgeable nodes to deny any future service to the attacker. This causes the malicious nodes to be aware that their actions are being observed and rated, which, in turn, helps in reducing mischief in the system.

9.3.1.2 Motivation-based Approaches

Two protocols are known from literature that use digital signatures to increase participation and support charging and billing in multi-hop ad hoc networks. Both the Simple Cheat-Proof protocol (Sprite) [28] by Yang, Meng, and Lu and the Secured Incentive based Charging Protocol (SICP) [29] by Lamparter, Paul, and Westhoff present technical solutions for a similar business model. Each node that forwards foreign data receives credits (a negative charge) and gets some monetary reward, whereas the sender, in the latter case, and in [29] also the final destination, have to pay (are charged) by an ISP that is responsible for the (temporary) connection of the ad hoc cloud to the fixed network. In cases where the sender and the final destination belong to the same ad hoc cloud, a decentral traffic forwarding remains possible. The traffic does not have to be diverted over the access point to the fixed network. This business model becomes viable because of a technical realization that cannot be forged, which is necessary to convince the participators of the ad hoc community that they will indeed get the rewards if they co-operate. The forwarding of data by power restricted devices that have limited battery power, causing users to be potentially selfish and save their battery power, seems to be more likely in the presence of protocols like Sprite or SICP.

A third charging and rewarding protocol for packet forwarding has been proposed by Salem, Buttyan, Hubaux, and Jakobsson [30]. They suggest forwarding all traffic in a multi-hop manner over an access point to the fixed network, even when the sender and the receiver belong to the same ad hoc cloud. Owing to this restriction for traffic forwarding, the scalability requirements for the security associations are relaxed for this approach. Only the access point (and not each mobile node) needs to establish security associations with each mobile node. As a result, choosing a symmetric MAC scheme for authentication is appropriate.

9.3.1.3 Effect of Cooperation Approaches

Although the need for protocol extensions dealing with cooperation issues for ad hoc routing protocols is doubtlessly accepted, it is still a question of belief whether such extensions really intensify participation within the network and benefit the community. This is applicable to both detection-based and motivation-based approaches. Motivation-based approaches, which not only intensify cooperation, but also show how to integrate charging and billing into an available Authentication, Authorization, and Accounting (AAA) architecture in the fixed network, are very promising.

It should be noted that cooperation and fairness in ad hoc and sensor networks have to take fairness issues at the MAC layer into account.

9.3.2 Distributed Topology Control

TC can be regarded as the art of coordinating the transmit powers of the nodes such that certain network-wide properties (e.g. maintaining connectivity, reducing node energy consumption) are achieved. Given the fully distributed nature of ad hoc networks, TC should be implemented without the intervention of any centralized infrastructure. Figure 9.7 shows the self-organizing nature of TC. It is a mechanism which, by acting on the transmit power levels (local choice at the nodes), returns a network with certain desired topology features (global property).

In this subsection, we first describe the motivations for using TC, that is, their potential for reducing node energy consumption and increasing network capacity. We then discuss the ideal features of a distributed TC mechanism and briefly summarize the main approaches presented in the literature.

Before starting, we would like to emphasize that the term TC is used to differentiate it from power control techniques, whose goal is to optimize the choice of the transmit power level for a single, though possibly multi-hop, wireless transmission. Power control thus has a radio channel-wide perspective, while TC has a wider, network-wide perspective.

9.3.2.1 Optimization of Energy Efficiency

A node usually consumes a significant part of its energy to communicate with peer nodes. Reducing the energy consumed to exchange messages is therefore an important issue. The energy spent for sending a message is related to the transmit power used by the sender. The path loss formula relates the transmit power P_t to the received power P_r as a function of the distance d between the sender u and the receiver v of the transmission as follows: $P_r(d) \propto P_t/d^{\alpha}$ where α is the distance–power gradient, which depends on environmental conditions and is usually between two and six.

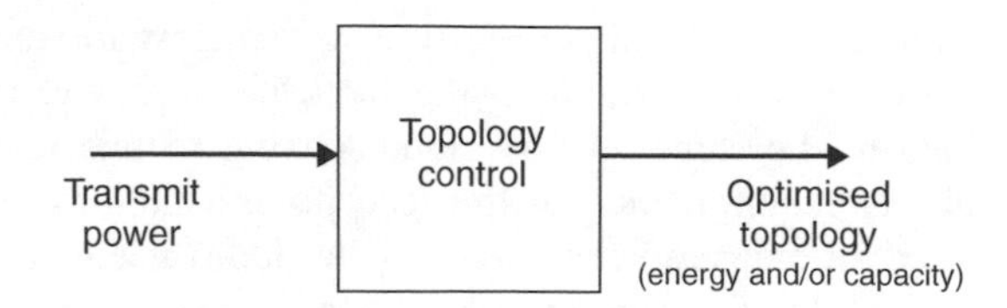

Figure 9.7 Topology control as a self-organizing paradigm

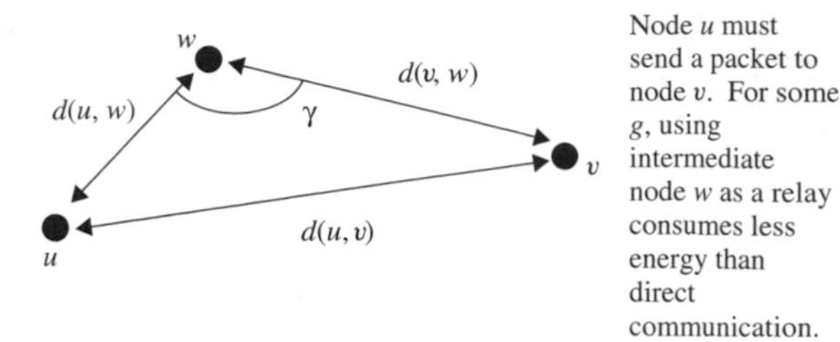

Figure 9.8 Direct transmission and use of a relay w

Assume we have the following situation; node u wants to send a message to node v, which is located within the transmit range. However, there exists node w, with $d(u, w)$, $d(v, w) < d(u, v)$, where $d(x, y)$ is the distance between x and y, as shown in Figure 9.8. Assume nodes u and w are equipped with similar transceivers, and that they can choose an arbitrary transmit power, provided it does not exceed the maximum power. Is it less power consuming overall to transmit the message directly to v, or to use a lower transmit power, and use node w as a relay?

To resolve this, we compare the transmit power used for the single, long-distance transmission, with the sum of the powers used in the two short-distance transmissions. Assume that $\alpha = 2$, and $P_r = 1$ is sufficient for correct message reception. In these hypotheses, the transmit power needed for directly sending the message to v equals $d(u, v)^2$. Under the same hypotheses, the sum of the transmit powers in case of two-hops transmission equals $d(u, w)^2 + d(w, v)^2$. Consider triangle uwv, and let γ be the angle opposite to side uv (see Figure 9.8). By elementary geometry, we have

$$d(u, v)^2 = d(u, w)^2 + d(w, v)^2 - 2d(u, w)d(w, v)\cos\gamma$$

It follows that if $\gamma > \pi/2$ then communicating through the two-hop path is more convenient (from the energy consumption point of view) than direct communication. The situation is reversed when $\gamma < \pi/2$.

The example above has outlined the importance of a careful choice of the wireless channels to use in communications; by properly choosing the channels, energy consumption can be considerably reduced. This is the first motivation for TC, and also one of the design goals – to identify and 'remove' energy-inefficient links from the network topology.

9.3.2.2 Optimization of Network Capacity

Contrary to the case of wired point-to-point channels, wireless communications use a shared medium, the radio channel. The use of a shared communication medium implies that particular care must be paid to avoid concurrent wireless transmissions corrupting each other.

Suppose node u must transmit a packet to node v, which is at distance d. Furthermore, assume there are intermediate nodes $w_1, \ldots, w_k$ between u and v, and that $d(u, w_1) = d(w_1, w_2) = \ldots = d(w_k, v) = d/(k + 1)$ (see Figure 9.9). From the network capacity point of view, is it preferable to send the packet directly from u to v, or to use the multi-hop path $w_1, w_2, \ldots, v$?

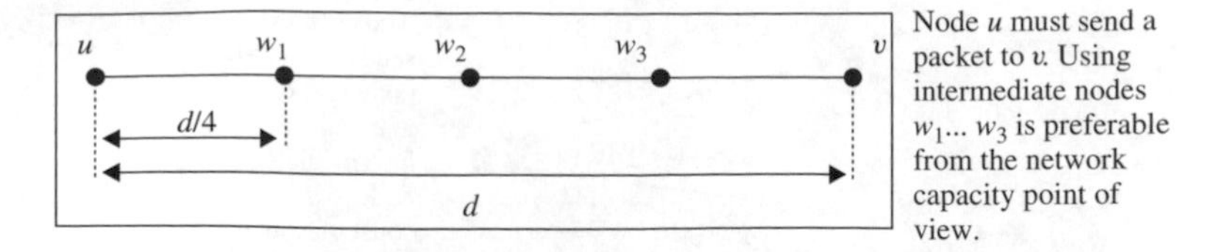

Node u must send a packet to v. Using intermediate nodes $w_1 \ldots w_3$ is preferable from the network capacity point of view.

Figure 9.9 Direct transmission and use of a relays for multi-hop communication

Here again, we need an appropriate interference model. Maybe the simplest such model is the Protocol Model used in [31]. In this model, the packet transmitted by a certain node u to node v is correctly received if $d(v, w) \geq (1 + \eta)d(u, v)$ for any other node w that is transmitting simultaneously, where $\eta > 0$ is a constant that depends on the features of the wireless transceiver. When a certain node receives a packet, all the nodes in its interference region must remain silent in order for the packet to be correctly received. The interference region is a circle of radius $(1 + \eta)d(u, v)$ (the interference range) centred at the receiver. The area of the interference region measures the amount of wireless medium consumed by a certain communication. Since concurrent non-conflicting communications occur only outside each other interference region, this is also a measure of the overall network capacity.

Referring to Figure 9.9, for a direct transmission, the interference range of node v is $(1 + \eta)d$, corresponding to an interference region of area $\pi d^2(1 + \eta)^2$. For multi-hop transmission, we have to sum the area of the interference regions of each short, single-hop transmission. The interference region for any such transmission is $(k + 1)(d/(k + 1))^2(1 + \eta)^2$, and there are $k + 1$ regions to consider overall. By Holder's inequality, we have

$$\sum_{i=1}^{k+1} \left(\frac{d}{k+1}\right)^2 = (k+1)\left(\frac{d}{k+1}\right)^2 < \left(\sum_{i=1}^{k+1} \frac{d}{k+1}\right)^2 = d^2.$$

It can be concluded that from the network capacity point of view, it is better to communicate using short, multi-hop paths between the sender and the destination.

This conclusion is the other motivating reason for a careful design of the network topology; instead of using long-distance links, we can use a multi-hop path composed by shorter links. A second goal of TC techniques is to identify and prune capacity-inefficient links.

9.3.2.3 Ideal Features

In the previous sections, we have provided the motivation to use TC techniques in ad hoc networks. A TC protocol should have the following ideal features:

- It should be fully distributed, asynchronous, and localized. The first two features are obvious for fully decentralized scenarios typical of ad hoc networks. Being localized means that every network node should take decisions regarding its transmitting range on the basis of information provided by neighbour nodes only. Locality in general ensures that exchanging few messages is sufficient to generate the desired network topology. This feature is extremely important in the presence of mobility. If the topology

can be computed locally, the TC protocol can quickly react to node mobility, updating the topology accordingly.
- It should generate a network topology that preserves all the original connections between nodes and avoids energy-inefficient and/or capacity-inefficient wireless links.
- It should generate a network topology that avoids unidirectional links. Avoiding unidirectional links is fundamental to facilitating the integration of TC mechanisms with existing lower and upper layer network protocols.
- It should rely on 'low quality' information. In general, there is a trade-off between the type of information available to the nodes, for example, exact location, or directional information, and the 'quality' of the topology generated. The more accurate the information is, the less energy is consumed and the more network capacity we have. However, the price to be paid in terms of additional hardware on the nodes, or of additional messages to be exchanged to obtain high quality information must be carefully considered. A good TC protocol should be able to provide considerable energy saving and/or capacity increase using 'low quality' information.

9.3.2.4 Existing Approaches

Current approaches to TC can be roughly subdivided into three categories:

- *Location-based TC*: In this approach to TC, it is assumed that every network node knows its exact position, which is typically provided by a low-power Global Positioning Systems (GPS) receiver. By exchanging location information with neighbour nodes, a node can identify inefficient, typically energy-inefficient, links, which are not used in the communications. Examples of location-based TC protocols are discussed in [32, 33]. The advantage of this approach is that inefficient links are accurately identified. An obvious drawback is that it is quite demanding on the nodes, such as the need for a GPS receiver or a similar device. Also, location-based TC protocols are not very suitable for application in mobile networks. Since node positions change continuously because of mobility, location-based protocols might not stabilize.
- *Direction-based TC*: This family of protocols assume that nodes are equipped with directional antennas. The network topology is computed by exchanging directional information with neighbour nodes. Examples of direction-based TC protocols are discussed in [34–36]. As for location-based TC, this approach has the advantage of producing 'good' network topologies, but it has the drawback of being quite demanding on the nodes. As for mobility, the topology generated using directional information is more resilient regarding node migration, especially for group mobility, that is, when we have a set of nodes moving approximately in the same direction.
- *Neighbour-based TC*: This family of protocols is based on the simple idea of connecting each node to some of its closest neighbours. This can be accomplished without requiring any additional hardware on the nodes. However, the topologies obtained using this approach are less efficient than those produced by location- and direction-based protocols. In particular, full network connectivity in some situations might be compromised, because there are situations in which the network is connected when all the nodes transmit at full power, but the topology generated by neighbour-based TC is not connected. Neighbour-based TC requires little message exchange, and it is suitable for

Table 9.1 Main features of the various approaches to TC

	Local information	Worst-case connection	Additional hardware	Optimize energy	Optimize capacity
Location-based	1-hop	Yes	GPS	Yes	No
Direction-based	1-hop	Yes	dir. antenna	Yes	No
Neighbour-based	1-hop	No	No	Yes	Yes

implementation in mobile networks. Examples of neighbour-based TC protocols are discussed in [37–39].

- The main features of the various approaches to TC are summarized in Table 9.1.

9.3.3 Address Self-configuration

The automatic configuration of addresses to nodes in ad hoc networks needs additional functionality compared to the approaches used in traditional networks. There are two major problems; first, IP-based auto-configuration protocols require all nodes to be reachable within a single hop. The multi-hop nature of ad hoc networks complicates the protocol operation, since not all nodes in a sub-network can be reached via a link-level broadcast. The second problem is that two or more ad hoc networks may merge because of their mobility. If so, it is possible that nodes located in different former networks own the same address in the merged network. To cope with this situation, DAD should be used. Using standard IPv6 DAD [6] might cause problems due to a high signalling overhead and the fact that a supplementary process is necessary to detect whenever different sub-networks have merged.

The first problem is addressed by Perkins *et al.* [40]. This expired Internet draft proposes some modifications to standard IP auto-configuration protocols. Each node chooses an IP address randomly and then performs DAD by flooding the network with a message containing the address. If a node has the same address, it defends it by replying using the reverse path. The originating node considers its address to be unique if it does not receive a reply within a certain time. The document also introduces a 'MANET local scope' in addition to the 'link-local' scope. Network merging is not supported. Perkins' concept has been used in [41] for hierarchical address auto-configuration in a hybrid ad hoc/cellular network.

The second problem is addressed by Vaidya [42]. He distinguishes between 'strong' and 'weak' DAD. Unlike strong DAD, weak DAD ensures that packets are routed to the correct destination, but does not imply that this process detects identical addresses immediately. The section shows that strong DAD cannot be guaranteed in ad hoc networks if message delays are not bounded. The author proposes a weak DAD mechanism on the basis of an enhancement of a link-state routing protocol. This mechanism works as follows; each node owns a unique identifier. If a node sends a control packet indicating its link state, it adds this identifier to the packet. Each node keeps track about the links it is connected to, the corresponding nodes it is in relation with, and their identifier. If a node receives a control packet with an address he knows but a different identifier, the node concludes it has detected a duplicate address. The node thus announces the duplicate address and keeps sending the packets to the node that it knows uniquely owned the address previously.

Other approaches for self-configuration in ad hoc networks can be found in [43, 44]. Recently, the charter of a new IETF working group on 'ad hoc network configuration' is under discussion. The goal is to standardize solutions for IPv4 and IPv6 address auto-configuration in two scopes: a local scope in which addresses that are only valid in a particular ad hoc network, and a global scope where addresses are routable in the global Internet.

9.4 Self-organization in Network Management

Centralized network management has some well-known problems. It gives poor scalability with respect to management traffic, processing load on the management station, and execution time. Furthermore, it is not suited for dynamic topologies, where networks, rather than terminals, emerge at the edges of the networks and compose in an 'ad hoc' manner. To cope with this increased dynamics, network management should be based on a distributed paradigm. This section reviews three approaches that go in this direction: (1) policy-based management, (2) pattern-based management, and (3) knowledge planes (This section is based on selected material from [45]).

9.4.1 Policy-based Management

The aim of policy-based management is to control and manage a communication network on a high abstraction layer. In this context, policies are general rules describing the action that needs to be taken in a node upon the occurrence of certain events. A network manager defines these high-level rules independent of the implementation of the nodes. The policies are then distributed in the network and translated to the operation system and hardware-specific processes. Two examples are management of groups of nodes into logical networks and management quality of service (QoS) parameters. A policy-based management is a layered approach, where the policies can be seen as a logical layer of event handling mechanisms ('policy framework'). The following paragraphs give an overview of the architecture and protocols for policy-based network management.

The policy framework architecture defined by the IETF in [46, 47] contains the key functional entities needed for policy-based management. The policy decision point (PDP) is the point where policy decisions are made. The policy decision function can be located at a specialized policy server or within the network device as a local policy module. The policy enforcement point (PEP) is where the policy decisions are actually enforced when the rule conditions evaluate to 'true'. It is responsible for execution of actions and may perform device-specific condition checking and validation. The policy management function provides the interface to the network manager; it comprises the functions of policy editing, rules translation and validation. With the policy editor the administrator can enter, view, and edit policy rules in the policy repository. Policy prescriptions and rules can be of various forms.

The Common Open Policy Service (COPS) protocol [48] is an emerging solution for distributed policy management. It is a simple request–response protocol between PDPs and PEPs for exchanging policy information and decisions. COPS may also be used between bandwidth brokers that essentially act as PDPs for dynamic inter-domain policy exchange. COPS includes a client–server model where the PEP sends requests, updates, and deletes to the remote PDP and the PDP returns the decisions to the PEP. COPS includes the utilization of TCP as transport protocol. The protocol is extensible in that it is designed to take advantage of self-identifying objects and can support diverse PEP-specific information without requiring modifications to the protocol itself. COPS provides message level security for authentication, replay protection, and message integrity. It can reuse existing protocols for security to authenticate and secure the channel between PEP and PDP. The protocol is stateful in two aspects; first, requests from the PEP are installed or remembered by the remote PDP until they are explicitly deleted by the PEP; at the same time, decisions from the remote PDP can be generated asynchronously at any time for a currently installed request state. Second, the PDP may respond to new queries differently because of previously installed request/decision states that are related. Additionally, the protocol allows the PDP to push configuration information to the PEP, and then allows the PDP to remove such state from the PEP when it is no longer applicable.

The Adaptive Resource Control for QoS using an IP-based Layered Architecture (AQUILA) project 'defines, evaluates, and implements an enhanced architecture for QoS in the Internet' [49]. It describes the service level specifications (SLS) negotiation process as an interaction between end-user application toolkits, admission control agents, and resource control agents. The control of the routers providing certain resource control is defined through policies, which has to conform to the SLSs defined at the domain-level.

The objective of the Traffic Engineering for Quality of Service in the Internet at Large Scale (TEQUILA) project is to 'study, specify, implement, and validate a set of service definition and traffic engineering tools' [50]. The goal is to obtain 'end-to-end Quality of Service (QoS) guarantees through careful planning, dimensioning, and dynamic control of scalable and simple qualitative traffic management techniques'. It addresses SLS; protocols for negotiating, monitoring, and enforcing SLSs; and intra- and inter-domain traffic engineering schemes. To increase the flexibility of its architecture, TEQUILA uses a policy-based SLS. In addition, policy-based management offers a flexible, extensible vehicle for the realization of dynamic resource management aspects. The policy handling in TEQUILA is based on COPS.

The IPSec policy architecture [51], standardized within the IETF, provides a solution for secure policy exchange. It can provide a scalable, decentralized framework for managing, discovering, and negotiating the host and network IPSec policies that govern access, authorization, cryptographic mechanisms, confidentiality, data integrity, and other IPSec properties. The IPSec policy architecture defines the mechanisms and protocols needed for discovering, accessing, and processing security policy information in multi-domain scenarios. The architecture accommodates topology and policy changes without the need for manual reconfiguration of clients and security gateways. It proposes a protocol for gateway discovery and security policy distribution. Moreover, it introduces a trust management system and language to resolve and exchange policies.

9.4.2 Pattern-based Management

Pattern-based management [52] is a distributed management approach that uses graph traversal algorithms to control and coordinate the processing and aggregation of management information inside a network. From a network manager's perspective, the algorithm provides different means to diffuse (spread) the computational processes over many nodes. A key feature of the approach is its ability to separate the diffusion and aggregation mechanism from the semantics of the management operation. This is achieved using the concepts navigation patterns and aggregators. A navigation pattern represents a generic graph traversal algorithm that controls the execution flow of a (distributed) management operation. In summary, the main benefits of a pattern-based management approach are as follows: (a) it separates the semantics of the task from its flow control, (b) it enables us to build scalable management systems, and (c) it facilitates management of dynamic networks exhibiting frequent topology changes. The concepts of pattern-based management have been implemented in the platform Weaver [53] as well as in the SIMPSON simulator [52].

A particularly useful navigation pattern is the so-called 'echo pattern' [52]. This pattern propagates to all nodes in a network (expansion phase), and returns the requested result to the originating node (contradiction phase). Among other things, the echo pattern can be used to poll the network for its current state. For example, the management node asks 'What are the top 10 flows in the network right now?' and the nodes report back their top 10 flows during the contraction phase. Furthermore, the intermediate nodes might process the data and in turn report only their top 10 flows, thereby distributing the computation effort over several nodes. Other useful patterns are, for example, the 'progressive wave pattern' and the 'stationary wave pattern' [54]. It is shown how these can be used to update Internet routing protocols and node interaction in peer-to-peer networks. Last but not least, the paper [55] shows how pattern-based network management can be used to support service deployment in active networks.

9.4.3 Knowledge Plane

A new approach to adding some intelligence and self-learning to the network management has been presented in [56]. The authors introduce a 'knowledge plane', whose goal is to eliminate unnecessary multilevel configuration. If one specifies the high-level design goals and constraints, the network should make the low-level decisions on its own. The system should reconfigure itself in a distributed manner according to the changes in the high-level requirements. Figure 9.10 illustrates the logical relation of the knowledge plane and the classical data and control planes within a network domain. Each network element implements the knowledge plane. The nodes interact with each other to keep themselves informed about global (network-wide) states and events. This interaction is also used to negotiate contradictory service levels and requirements.

The knowledge plane must also work in the presence of partial, inconsistent, and possibly misleading or malicious information. It must operate appropriately even if different stakeholders of the Internet define conflicting higher-level goals. To meet these challenges, the authors suggest employing cognitive techniques. In this way, the knowledge plane integrates behavioural models and reasoning processes into a distributed networked environment.

9.5 Graph-theoretical Aspects of Self-organization

The theoretical study of self-organization in complex systems has been a topic of research in science for a long time. Many specific approaches exist in the literature on mathematics, physics, biology, chemistry, medicine, and different fields of engineering and computer science.

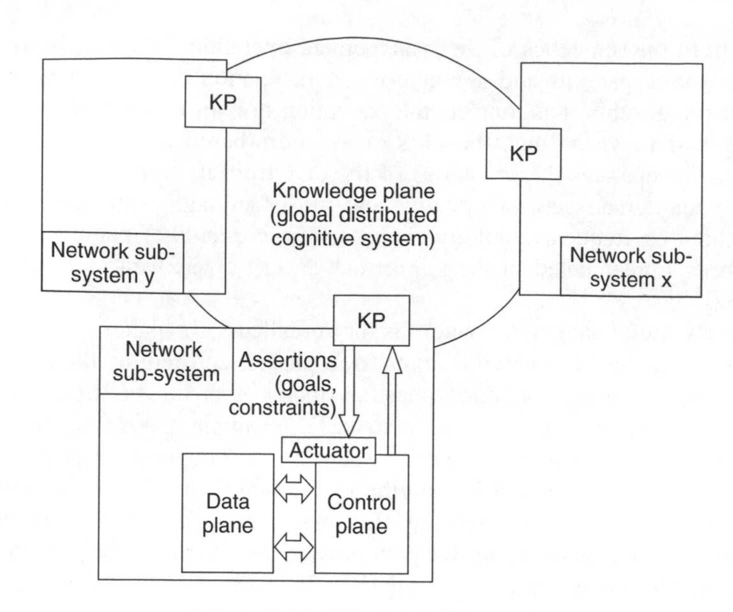

Figure 9.10 The knowledge plane

In many complex systems, graph theory plays an important role in the description and analysis. The system entities are represented by nodes, the interrelations between them by links. For example:

- the brain is a graph of neurons and their interactions;
- the economy is a graph of companies and customers as well as their trade relations;
- the World Wide Web is a graph of Web sites and hyperlinks.

There have been several recent graph-theoretical insights and tools to model self-organizing systems. For example, random graphs have been used to describe the topology of the World Wide Web and peer-to-peer networks, and small-world and scale-free graphs are used to model the Internet. In the following, we explain these concepts and discuss their application in communication networks:

9.5.1 Random Graphs

Until the late 1950s, physical phenomena were usually modelled assuming that interactions among entities can be represented by the regular and perhaps universal structure of Euclidean lattices, where the structure repeats. In 1960, Erdös and Rényi [57] achieved a breakthrough result in graph theory, giving birth to random graph theory. To construct a random graph, we subsequently choose a random pair of nodes and add a link between them, until a given number of links are achieved. Interestingly, this model creates graphs that capture several typical characteristics of real-world complex networks.

An important insight was to show that certain graph properties (such as connectivity, k-connectivity, and existence of isolated nodes) experience a threshold effect (also called *phase transition phenomenon*). For example consider a random graph process in which we start with a given set of nodes and subsequently add links between a randomly chosen pair of nodes. The following threshold effect may occur. While a random graph with a certain number of links might be very unlikely to have a given property (e.g. be connected), a random graph with only a few more links is very likely to have the given property.

Until the late 1990s, random graph theory was the main approach to study complex networks. Most networks in the real world, however, are neither fully regular nor completely random. With the availability of supercomputers and huge storage capabilities, the analysis of real-world complex network became feasible. These investigations resulted in two major discoveries – complex systems experience the 'small-world phenomenon' and are 'scale-free graphs'.

9.5.2 Small-world Phenomenon

The 'small-world phenomenon' goes back to the sociologist Milgram [19], who analysed social networks, that is, the graph of who knows whom. He showed that any two random persons in the United States are six people 'away' from each other. This finding is referred to as 'six degrees of separation' and gained popularity in Kevin Bacon's movie and John Guare's screenplay both called 'Six degrees of separation'.

A small-world graph is mainly characterized by two structural properties:

- The average path length L is short (i.e. most nodes are on average few hops away from each other).
- The clustering coefficient C, defined as the average fraction of pairs of a node's neighbours that are also neighbours of each other, is high.

To define a model for small-world graphs, Watts and Strogatz [58, 59] interpolated between a regular lattice and a random graph. They showed that the 're-wiring' of links between a few randomly chosen node pairs in a regular graph significantly reduces the average path length. This concept is illustrated in Figure 9.11, which shows a regular graph (a), a small-world graph (b), and a random graph (c). As can be seen, the average path length between two nodes of the regular graph is high; in contrast, the average path length of the random graph is low. To create a small-world graph, some links of the regular graph are rewired. To be more precise, a link is rewired with probability p. As

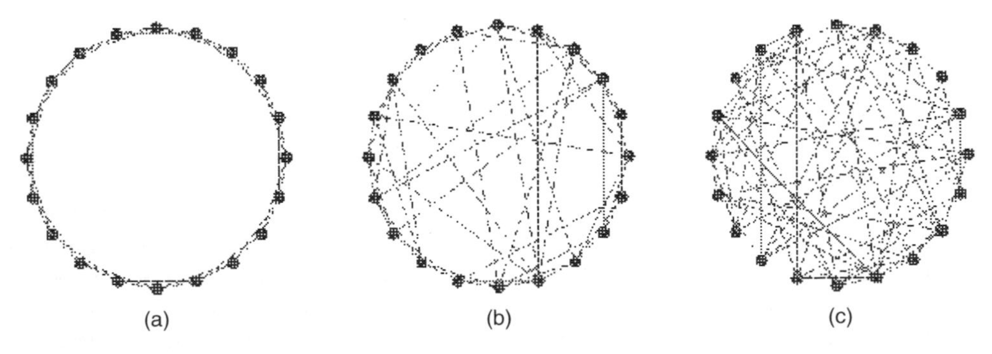

Figure 9.11 (a) Regular ($p = 0$), (b) small-world ($p = 0.5$), and (c) random graph ($p = 1$)

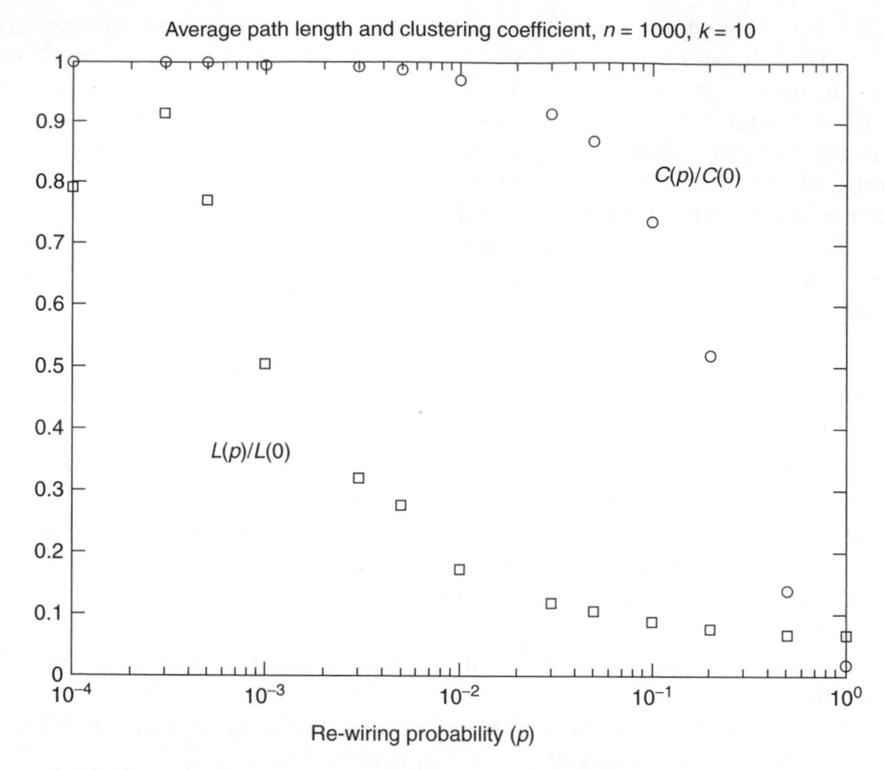

Figure 9.12 Normalized average path length L and clustering coefficient C as a function of rewiring probability p

can be seen, for $p = 0.5$ some irregularities are added, but the average path length is significantly reduced compared to the regular graph.

Figure 9.12 shows the normalized average path length L and clustering coefficient C as a function of rewiring probability p. Already at low probabilities (say $p = 10^{-3}$), the path length L is significantly reduced, while the clustering coefficient C remains high. At medium p (say $p = 10^{-2}$ the value of L is reduced further, while C still remains high. Only at p approaching 10^{-1}, the value of C shrinks as well.

9.5.3 Scale-free Graphs

The Watts and Strogatz's scheme is able to describe the short average path length of a small-world, but it does not model another fundamental property found in real-world complex networks; these networks are scale-free. The notion of 'scale-free' means that the vast majority of nodes have very few neighbours (i.e. a low degree), and only a few nodes have many neighbours (i.e. a high degree), and this degree distribution is independent of the size or scale of the network. In other words, in many complex self-organizing networks, only a few well-connected nodes nicely connect a large number of poorly connected nodes. If a network is scale-free, it is also a small world.

The Watts and Strogatz's scheme does not capture these effects, because it is based on the assumption of a fixed number of nodes and a creation of links with uniform probability.

Real complex network, however, expands continuously by the addition of new nodes and, in addition, the new nodes prefer to attach to nodes that are already well connected. These two principles, growth and preferential attachment, lead to the scale-free property.

Barabási *et al.* [60–62] proposed a model that is based on these two properties to construct an artificial scale-free network. The algorithm starts with a small number of m_0 nodes. At each time step, a new node is added and is linked to m existing nodes, where $m \leq m_0$. The probability that the new node is linked to a certain node v depends on the degree of v. Figure 9.13 illustrates this process for $m_0 = 3$ and $m = 2$. A new node is linked to an already existing node with a probability that is proportional to the degree of the existing node. The total number of nodes is denoted by m.

This model creates a graph that has a small average path length and power-law degree distribution. In fact, it was shown that the combination of growth and preferential attachment has a very important role to create a power-law degree distribution. If the growth property is removed, that is, the number of nodes is fixed, the degree of the nodes follows a Gaussian distribution. If the attachment is not preferential but uniform, the degree follows an exponential distribution. Hence, both properties are needed to create power-law degree distributions and scale-free features. For illustration, Figure 9.14 shows the probability distribution $P(k)$ of the degree k in a scale-free network consisting of $n = 4000$ nodes. Although the number of data points collected is not high, the power-law distribution of the degree is observed.

9.5.4 Application of Graph-theoretical Aspects to Communication Networks

The analysis of graph-theoretical properties of complex self-organization networks has made significant advances during the past few years. Most important, it was shown that

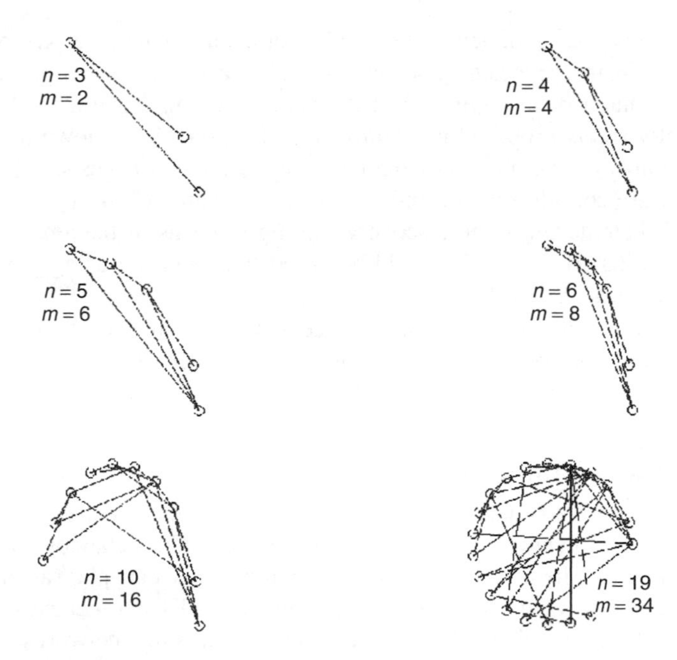

Figure 9.13 Using growth and preferential attachment to create a scale-free graph

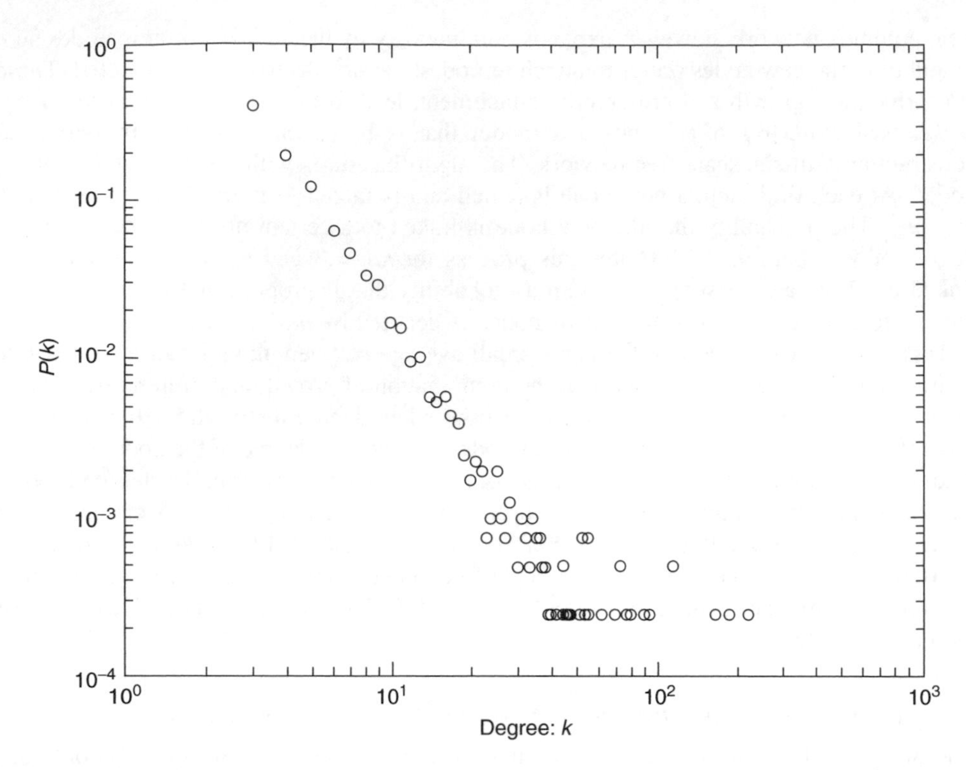

Figure 9.14 Power-law degree distribution for a scale-free network with $n = 4000$ nodes

many complex networks experience the small-world and scale-free phenomena. These insights might help in the design and analysis of self-organizing communication networks. For example, a recent approach that applies the small-world model ideas to ad hoc wireless networks is proposed by Helmy *et al.* [63, 64]. It is shown that by rewiring some links randomly (using physical wires), the network experiences small-world effects. The advantages are straightforward and obvious, for example, the rewired links create shortcuts, which help during route discovery, saving in terms of the amount of overhead compared to flooding. However, it should be noted that rewiring using physical wires is not realistic in wireless networks.

Therefore, it is important to develop new concepts where connectivity is strictly wireless in its manifestation and yet, attempt to make the network behave in a ‘small-world’ fashion. The same concepts could potentially be applied at higher layers.

9.6 Potential and Limitations of Self-organization

Self-organization as we understand it today means that simple behaviour of individual components can lead to surprisingly complex and pattern-like behaviour of large groups of these components. The exact nature of the components can play an important role for overall system behaviour. This property, often termed *emergent behaviour*, has so far not been exploited much in the design of communication networks. The previous sections have shown several examples where self-organization is applied efficiently in communication networks to shift the burden from centralized administrators to a self-organizing mechanism, while attaining the expected system behaviour. For example, the open-content encyclopaedia 'Wikipedia' represents an almost fully self-organized process among human beings, having just very few and simple high-level rules or policies defining the system. This concept also becomes visible in the management of networks, where high-level policies or patterns are employed.

All in all, the application of principles from self-organization has the potential to cope with increasing complexity of communication networks, without a similarly complex human interaction that no one can afford. All layers, from physical to content, are potential areas of application.

But the application of self-organization has its limitations also. First, the evolution of peer-to-peer networking shows that there is trade-off between completely decentralized and more centralized structures for self-organization. Second, as discusses for ad hoc networks, there might be a need for incentives to achieve cooperation, or punishment for non-cooperation. Also Wikipedia has some 'fall back states' if people misbehave. Hence, a crucial aspect is how security, privacy, and trust are maintained when using concepts of self-organization. In these areas, self-organization still is at large an embryonic field.

9.7 Acknowledgement

The following individuals contributed to this chapter: G. Kunzmann, R. Schollmeier (TU Munich, Germany), J. Nielsen (Ericsson, Sweden), P. Santi (IIT/CNR, Italy), R. Schmitz, M. Stiemerling, D. Westhoff (NEC, Germany), A. Timm-Giel (University Bremen, Germany).

References

[1] M. Weiser, "The computer for the 21st century," *ACM SIGMOBILE Mob. Comput. Commun. Rev.*, vol. 3, no. 3, pp. 3–11, 1999.
[2] H. Haken, "*Synergetics: Introduction and Advanced Topics*," Springer, 2004.
[3] S. Dixit and A. Sarma, "*Self-organisation in Wireless World Systems*," WWRF SIG3 charter, Apr. 2004.
[4] R. Droms, "*Dynamic Host Configuration Protocol DHCP*," RFC 2131, Mar. 1997.
[5] S. Deering and R. Hinden, "*Internet Protocol, Version 6 (IPv6) Specification*," RFC 2460, Dec. 1998.
[6] S. Thomson and T. Narten, "*IPv6 Stateless Address Autoconfiguration*," RFC 2462, Dec. 1998.
[7] R. Droms, J. Bound, B. Volz, T. Lemon, C. Perkins, and M. Carney, "*Dynamic Host Configuration Protocol for IPv6 (DHCPv6)*," RFC 3315, July 2003.
[8] B. Haberman, "*Automatic Prefix Delegation Protocol for Internet Protocol Version 6*," Work in progress: Draft-haberman-ipngwg-auto-prefix-02.txt, Feb. 2002.
[9] G. Chelius, "*Using OSPFv3 for Router Autoconfiguration*," Work in progress: Draft-chelius-router-autoconf-00.txt, June 2002.
[10] A. White and A. Williams, "*Unique Identifier Allocation Protocol (UIAP)*," Work in progress: Draft-white-zeroconf-uiap-01.txt, Oct. 2002.
[11] M. Crawford, "*Router Renumbering For Ipv6*," RFC 2894, August 2000.
[12] J. Linton, "*Automatic Router Configuration Protocol*," Work in progress: Draft-linton-arcp-00.txt, Oct. 2002.
[13] E. Guttman, C. Perkins, J. Veizades, and M. Day, "*Service Location Protocol, Version 2*," RFC 2608, June 1999.
[14] S. Czerwinski, B. Zhao, T. Hodes, A. Joseph, and R. Katz, "An architecture for secure service discovery service," in *Proc. ACM Intern. Conf. on Mobile Comp. and Netw.*, MobiCom, pp. 24–35, 1999.
[15] Universal Plug and Play (UPnP). http://www.upnp.org.

[16] S. Cheshire, B. Aboba, and E. Guttman, "*Dynamic Configuration of IPv4 Link-local Addresses*," Work in progress: Draft-ietf-zeroconf-ipv4-linklocal-17.txt, July 2004.

[17] L. Esibov, B. Aboba, and D. Thaler, "*Linklocal Multicast Name Resolution (LLMNR)*," Work in progress: Draft-ietf-dnsext-mdns-37.txt, Oct. 2004.

[18] C. Fraleigh, S. Moon, B. Lyles, C. Cotton, M. Khan, D. Moll, R. Rockell, T. Seely, and C. Diot, "Packet-level traffic measurements from the Sprint IP backbone," *IEEE Network*, vol. 17, no. 6, pp. 6–16, 2003.

[19] S. Milgram, "The small world problem," *Psychol. Today*, vol. 2, pp. 60–67, 1967.

[20] Wikipedia. http://www.wikipedia.org/, 2005.

[21] W. Altmann, R. Fritz, and D. Hinderink, "*TYPO3*," Open Source Press, 2004.

[22] M. Kurzidim, "*Wissenswettstreit: Die kostenlose Wikipedia tritt gegen die Marktführer Encarta und Brockhaus an*," c't – Magazin für Computertechnik, no. 21, pp. 132–139, Heise-Verlag, 2004.

[23] C. E. Perkins, ed., "*Ad Hoc Networking*," Addison–Wesley, 2001.

[24] S. Marti, T. Giuli, K. Lai, and M. Baker, "Mitigating routing misbehaviour in mobile ad hoc networks," in *Proc. ACM Intern. Conf. on Mobile Comp. and Netw.*, MobiCom, pp. 255–265, Aug. 2000.

[25] H. Yang, X. Meng, and S. Lu, "Self-organised network layer security in mobile ad hoc networks," in *ACM Workshop on Wireless Security (WiSe)*, Atlanta, USA, Sept. 2002.

[26] S. Buchegger and J.-Y. L. Boudec, "Performance analysis of the confidant protocol (cooperation of nodes: Fairness in dynamic ad-hoc networks)," in *Proc. ACM Intern. Symp. on Mobile Ad Hoc Netw. and Comp.*, MobiHoc, Lausanne, Switzerland, June 2002.

[27] K. Paul and D. Westhoff, "Context aware detection of selfish node in DSR based ad-hoc network," in *Proc. IEEE Globecom*, Taipei, Taiwan, Nov. 2002.

[28] S. Zhong, J. Chen, and Y. R. Yang, "Sprite: A simple, cheat-proof, credit-based system for mobile ad-hoc networks," in *Proc. IEEE Infocom*, San Francisco, USA, Mar. 2003.

[29] B. Lamparter, K. Paul, and D. Westhoff, "Charging support for ad hoc stub networks," *Elsevier Comput. Commun.*, vol. 26, pp. 1504–1514, Aug. 2003.

[30] N. B. Salem, L. Buttyan, and J.-P. Hubaux, "A charging and rewarding scheme for packet forwarding in multi-hop cellular networks," in *Proc. ACM Intern. Symp. on Mobile Ad Hoc Netw. and Comp.*, MobiHoc, Annapolis, MD, USA, June 2003.

[31] P. Gupta and P. Kumar, "The capacity of wireless networks," *IEEE Trans. Inform. Theory*, vol. 46, no. 2, pp. 388–404, 2000.

[32] N. Li, J. Hou, and L. Sha, "Design and analysis of an MST-based topology control algorithm," in *Proc. IEEE Infocom*, 2003.

[33] X. Li, "*Localized Construction of Low Weighted Structure and its Applications in Wireless Ad hoc Networks*," ACM/Kluwer Wireless Networks (WINET), Vol. 11, No. 6, pp. 697–708, November 2005.

[34] M. Bahramgiri, M. Hajiaghayi, and V. Mirrokni, "Fault-tolerant ad 3-dimensional distributed topology control algorithms in wireless multi-hop networks," in *Proc. IEEE Int. Conf. on Computer Commun. Netw.*, Miami, Miami, Florida, USA, pp. 14–16, October 2002.

[35] Z. Huang, C. Shen, C. Srisathapornphat, and C. Jaikaeo, "Topology control for ad hoc networks with directional antennas," in *Proc. IEEE Int. Conf. Computer Commun. Netw.*, Miami, Miami, Florida, USA, pp. 14–16, October 2002.

[36] L. Li, J. Halpern, P. Bahl, Y. Wang, and R. Wattenhofer, "Analysis of a cone-based distributed topology control algorithm for wireless multi-hop networks," in *Proc. ACM PODC*, 2001.

[37] D. Blough, M. Leoncini, G. Resta, and P. Santi, "The k-neighbours protocol for symmetric topology control in ad hoc networks," in *Proc. ACM Intern. Symp. on Mobile Ad Hoc Netw. and Comp.*, MobiHoc, Annapolis, USA, June 2003.

[38] J. Liu and B. Li, "Mobilegrid: Capacity-aware topology control in mobile ad hoc networks," in *Proc. IEEE Int. Conf. on Computer Communications and Netw.*, Miami, Miami, Florida, USA, pp. 14–16, October 2002.

[39] R. Ramanathan and R. Rosales-Hain, "Topology control of multihop wireless networks using transmit power adjustment," in *Proc. IEEE Infocom*, Tel Aviv, Israel, pp. 206–30, March 2000.

[40] C. Perkins, J. T. Malinen, R. Wakikawa, E. M. Belding-Royer, and Y. Sun, "*IP Address Autoconfiguration for Ad hoc Networks.*" IETF draft, 2001.

[41] J. Xi and C. Bettstetter, "Wireless multi-hop internet access: Gateway discovery, routing, and addressing," in *Proc. Intern. Conf. on Third Generation Wireless and Beyond (3Gwireless)*, San Francisco, USA, May 2002.

[42] N. H. Vaidya, "Weak duplicate address detection in mobile ad hoc networks," in *Proc. ACM Intern. Symp. on Mobile Ad Hoc Netw. and Comp.*, MobiHoc, Lausanne, Switzerland, pp. 9–11, June 2002.

[43] S. Nesargi and R. Prakash, "MANETconf: Configuration of hosts in a mobile ad hoc network," in *Proc. IEEE Infocom*, New York City, USA, pp. 23–27, June 2002.

[44] K. Weniger, "PACMAN: Passive autoconfiguration for mobile ad hoc networks," *IEEE J. Select. Areas Commun.*, vol. 23, no. 3, pp. 507–519, Mar. 2005.

[45] M. Brunner, S. Schuetz, G. Nunzi, E. A. Asmare, J. Nielsen, A. Gunnar, H. Abrahamsson, R. Szabo, S. Csaba, M. Erdei, P. Kersch, Z. L. Kis, B. Kovács, R. Stadler, A. Wágner, K. Molnár, A. Gonzalez, G. Molnar, J. Andres, M. Ángeles Callejo, L. Cheng, and A. Galis, "*Ambient Network Management: Technologies and Strategies (d-8.1)*," Tech. Rep., Ambient Networks project, Dec. 2004.

[46] B. Moore, E. Ellesson, J. Strassner, and A. Westerinen, "*Policy Core Information Model – Version 1 Specification*," RFC 3060, Feb. 2001.

[47] B. Moore, ed., "*Policy Core Information Model (PCIM) Extensions*," RFC 3460, Jan. 2003.

[48] J. Boyle, R. Cohen, D. Durham, S. Herzog, R. Rajan, and A. Sastry, "*The COPS (Common Open Policy Service) Protocol*," RFC 2748, Jan. 2000.

[49] IST AQUILA Website. http://www-st.inf.tu-dresden.de/aquila/, May 2005.

[50] IST TEQUILA Website. http://www.ist-tequila.org/, May 2005.

[51] J. Jason, L. Rafalow, E. Vyncke, "*IPsec Configuration Policy Information Model*," RFC 3585, Aug. 2003.

[52] K. S. Lim and R. Stadler, "Developing pattern-based management programs," in *Proc. IFIP/IEEE Intern. Conf. Manage. of Multimedia and Netw. Serv.*, Chicago, USA, Oct. 2001.

[53] K. S. Lim and R. Stadler, "Weaver: Realizing a scalable management paradigm on commodity routers," in *Proc. IFIP/IEEE Intern. Symp. Integrated Netw. Manage.*, Colorado Springs, USA, Mar. 2003.

[54] C. Adam and R. Stadler, "Patterns for routing and self-stabilization," in *Proc. Netw. Oper. & Manage. Symp.*, Seoul, Korea, Apr. 2004.

[55] M. Bossardt, A. Mühlemann, R. Zürcher, and B. Plattner, "Pattern based service deployment for active networks," in *Proc. Intern. Workshop Active Network Technol. and Appl.*, Osaka, Japan, May 2003.

[56] D. D. Clark, C. Partridge, J. C. Ramming, and J. T. Wroclawski, "A knowledge plane for the Internet," in *Proc. ACM SIGCOMM*, Karlsruhe, Germany, 2003.

[57] P. Erdös and A. Rényi, "On the evolution of random graphs," *Publ. Math. Inst. Hungar. Acad. Sci.*, vol. 5, pp. 17–61, 1960.

[58] D. J. Watts and S. H. Strogatz, "Collective dynamics of 'small world' networks," *Nature*, vol. 393, pp. 440–442, 1998.

[59] S. Strogatz, "Exploring complex networks," *Nature*, vol. 410, pp. 268–276, Mar. 2001.

[60] A.-L. Barabási and R. Albert, "Emergence of scaling in random networks," *Science*, vol. 286, pp. 509–512, Oct. 1999.

[61] A.-L. Barabási, R. Albert, and H. Jeong, "Mean-field theory for scale-free random networks," *Physica A*, vol. 272, pp. 173–187, 1999.

[62] A.-L. Barabási and E. Bonabeau, "Scale-free networks," *Sci. Am.*, vol. 288, pp. 50–59, May 2003.

[63] A. Helmy, "Small worlds in wireless networks," *IEEE Commun. Lett.*, vol. 7, no. 10, pp. 490–492, Oct. 2003.

[64] R. Chitradurga and A. Helmy, "Analysis of wired short cuts in wireless sensor networks," in *Proc. IEEE Intern. Conf. Pervasive Serv.*, Beirut, Lebanon, pp. 19–23, July 2004.

Appendix: Glossary

3GPP	3rd Generation Partnership Project
3GPP2	3rd Generation Partnership Project 2
AAA	Authentication, Authorization and Accounting
ACL[1]	Access Control List
ACL[2]	Asynchronous Connectionless
ADC	Analogue to Digital Conversion
AEP	Application Environment Profile
ALT PHY	ALTernative PHYsical
AmI	Ambient Intelligence
AP	Access Point
API	Application Programming Interface
ARCP	Automatic Reconfiguration Protocol
ARIB	Association of Radio Industries and Businesses
ARQ	Automatic Repeat Request
ARTEMIS	Advanced Research and Development on Embedded Intelligent Systems
AS[1]	Access Stratum
AS[2]	Angle Spread
ASIC	Application-specific Integrated Circuit
ATIS	Alliance for Telecommunications Industry Solutions
ATM	Asynchronous Transfer Mode
AWT	Abstract Windowing Toolkit
BB	Baseband
BD	Block diagonalization
BER	Bit-Error Rate
BF	Beamforming
BICM	Bit-interleaved Coded Modulation
BIOS	Basic Input Output System
BMM	Bandwidth Management Module
BPEL4WS	Business Process Execution Language for Web Services (BEA, IBM, Microsoft)
BPML	Business Process Modelling Language
BS[1]	Bearer Services
BS[2]	Base Station
BT	Bluetooth

Technologies for the Wireless Future – Volume 2 Edited by Rahim Tafazolli

BT-AP	Bluetooth Access Point
BTS	Base Transceiver Station
CALLUM	Combined Analogue Locked Loop Universal Modulator
CAST	Configurable radio with Advanced SW Technology
CCF	Computing and Communication Foundations
CCSA	China Communications Standards Association
CDC	Connected Device Configuration
CDG	CDMA Development Group
CDMA	Code Division Multiple Access
CEPT	European Conference of Postal and Telecommunications
CF	Core Framework
CI-OFDM	Carrier Interferometry OFDM
CIR	Carrier-to-interference Ratio
CIR	Channel Impulse Response
CISE	Computer & Information Science & Engineering
CLDC	Connected Limited Device Configuration
CM[1]	Configuration Management
CM[2]	Connection Management
CMM	Configuration Management Module
CMOS	Complementary Metal-oxide Semiconductor
CN	Core Network
CNS	Computer and Network Systems
CODEC	CODer/DECoder
COM	Component Object Model
COPS	Common Open Policy Service
CORBA	Common Object Request Broker Architecture
CP	Cyclic Prefix
CPE	Common Phase Error
CPU	Central Processing Unit
CSA	Client–Server Architecture
CSI	Channel State Information
CSIK	Complementary SIK
CSM	Common Signaling Mode
CSMA	Carrier Sense Multiple-access
CSW	Computer Software Components
DAB	Digital Audio Broadcast
DAC	Digital to Analogue Conversion
DAD	Duplicate Address Detection
DAML	DARPA Agent Markup Language
DARPA	Defense Advanced Research Projects Agency
Das	Directory agents
DCF	Distributed Coordination Function
DCOM	Distributed Component Object Model (Microsoft)
DECT	Digital Enhanced Cordless Telecommunication
DFE	Decision-feedback Equalization
DFT	Discrete Fourier Transform

DGAPS	Design of Generic and Adaptive Protocol Software
DHCP	Dynamic Host Configuration Protocol
DL	Downlink
DLL	Data Link Layer
DNS	Domain Name Service
DoD	Department of Defense
DOM	Document Object Model
DRM	Digital Rights Management
DS-CDMA	Direct Sequence Code Division Multiple Access
DSP	Digital Signal Processor
DSSS	Direct-sequence Spread-spectrum
DVB	Digital Video Broadcast
EADF	Effective Aperture Description Function
ECC	European Communication Committee
ECO	E-Commerce Project
EIR	Equipment Identity Register
ETSI	European Telecommunications Standards Institute
EVM	Error Vector Magnitude
FAA	Federal Aviation Administration
FCC	Federal Communication Commission
FDD	Frequency Division Duplex
FDOSS	Frequency-domain Orthogonal Signature Sequences
FEC	Forward Error Correction
FFT	Fast Fourier Transform
FHSS	Frequency-hopping Spread-spectrum
FLP	Flexible Linearity Profile
FoV	Field of View
FP	Foundation Profile
FPGA	Field Programmable Gate Array
FPs	European Framework Programmes
FTP	File Transfer Protocol
FuTURE Forum	FuTURE Mobile Communication Forum
-g	Generic
GFD	Generalized Frequency Domain
GGSN	Gateway GPRS Support Node
GIOP	General Inter-ORB Protocol
GLL	Generic Link Layer
GMC	Generalized Multicarrier
GMITS	Guidelines for the Management of IT Security
GP	Generic Protocol
GPP	General Purpose Processor
GPR	Ground Penetration Radar
GPRS	General Packet Radio Service
GPS	Global Positioning Systems
GSEs	Generic Service Elements
GSM	Global System for Mobile Communications

GUI	Graphic User Interface
H/2	HiperLAN 2
HARQ	Hybrid Automatic Repeat Request
HDD	Hybrid Division Duplexing
HDR	High Data Rate
HiperLAN/2	High Performance LAN type 2
HLR	Home Location Register
HO	Handover
HRM	Home Reconfiguration Manager
HSDPA	High-speed Downlink Packet Access
HSUPA	High-speed Uplink Packet Access
HTTP	HyperText Transport (or Transfer) Protocol
HVAC	Heating, ventilation, air-conditioning
HW	Hardware
IAM	Identity and Access Management
ICI	Inter-carrier Interference
ICT	Information Communication Technologies
IDL	Interface Definition Language
IdM	Identity management
IdP	Identity Provider
IEEE	Institute of Electrical and Electronics Engineers
IETF	Internet Engineering Task Force
IF	Intermediate Frequency
IFDMA	Interleaved FDMA
IFFT	Inverse FFT
IGMP	Internet Group Management Protocol
IIOP	Internet Inter-ORB Protocol
IIS	Information and Intelligent Systems
IMD	Intermodulation Distortion
IMEI	International Mobile Equipment Identity
IMS	IP multimedia subsystem
IMSI	International Mobile Subscriber Identity
IMT-2000	International Mobile Telecommunications 2000
IOTA	Isotropic Orthogonal Transform Algorithm
IP	Internet Protocol
IPv6 FORUM	Internet Protocol v6 forum
IR	InfraRed
IRDA	InfraRed Data Association
IrTM	Travel Mobility Special Interest Group
ISA	Instruction Set Architecture
ISDN	Integrated Services Digital Network
ISI	InterSymbol Interference
ISO	International Standards Organization
IST	Information Society Technologies
ISV	Independent Software Vendor
ITAP T9	Intelligent Text prediction

ITU	International Telecommunication Union
ITU-R WP8F	International Telecommunication Union – Radio Sector, Working Party 8F
ITU-T SSG	ITU – Telecommunication Sector, Special Study Group
J2ME	Java 2 Micro Edition
JEDI	Java Event Distribution Infrastructure
JMS	Java Message Service
JNDI	Java Naming and Directory Interface
JOBS	Jointly Opportunistic Beamforming and Scheduling
JSRs	Java Specification Requests
LA	Liberty Alliance
LAN	Local Area Network
LBS	Location Based Services
LDR	Low Data Rate
LED	Light Emitting Diodes
LED DLP	Light emitting Diodes Digital Light Processing
LINC	Linear Amplification with Nonlinear Components
LLC	Logical Link Control
LMSC	LAN/MAN Standards Committee
LNA	Low Noise Amplifier
LO	Local Oscillator
LOS	Line Of Sight
LR	Location Register
LSA	Link State Advertisement
LTE	Long-term Evolution
LTI	Linear Time Invariant
MAC	Medium Access Control
MAN	Metropolitan Area Networks
MBOK	M-ary bi-orthogonal keying modulation
MC	Multicarrier
MC-CDMA	MultiCarrier CDMA
MCM	Modes Convergence Protocol
MCSSS	Multicarrier Spread Spectrum Signal
MDA	Model Driven Architecture
mDNS-SD	Multicast DNS-Service Discovery
ME	Mobile Equipment
MEMS	Micro ElectroMechanical Systems
MExE	Mobile Execution Environment
MIB	Management Information Base
MIDPv2	Mobile Information Device Profile version 2
MIMM	Mode Identification & Monitoring Module
MIMO	Multiple Input–Multiple Output
MIP	Mobile IP
MIPS	Million Instructions Per Second
MISO	Multiple Input–Single Output
mITF	mobile IT Forum

MLD	Maximum-likelihood Detection
MM	Mobility Management
MMS	Multimedia Message Service
MMSE	Minimum Mean Squared Error
MN	Mobile Node
MNOs	Mobile Network Operators
MNSM	Mode Negotiation and Switching Module
MOBIVAS	downloadable MOBIle Value Added Services
MOM	Message-oriented Middleware
MS	Mobile Station
MSC	Mobile Switching Center
MSMQ	Microsoft Message Queue Server
MT	Mobile Terminal
MU	Multiple User
MUI	Multiuser Interference
MVCE	Mobile Virtual Center of Excellence
NB	Narrowband
NBS	Network Bearer Services
NE	Network Element
NFC	Near Field Communications
NGMC	Next Generation Mobile Communication
NLOS	Nonline of Sight
NO	Network Operator
NSF	National Science Foundation
OEM	Original Equipment Manufacturer
OFDM	Orthogonal Frequency Division Multiplexing
OFDM/OQAM	OFDM/OffsetQAM
OFDMA	Orthogonal Frequency-Division Multiple Access
OIL	Ontology Interchange Language
OMA	Open Mobile Alliance
OMG	Object Management Group
OMG DSIG SDO	OMG Domain-SIG for Super Distributed Objects
OOK	On–off Keying
OPEX	OPeration EXpenses
OQAM	Offset-QAM
ORB	Object Request Broker
OS	Operating System
OSI	Open Systems Interconnection
OSPF	Open Shortest Path First
OTA	Over the Air
OW	Optical Wireless Communications
OWL	Web Ontology Language
P2P	Pear to Pear
PA	Power Amplifier
PAN	Personal Area Network
PAPR	Peak to Average Power Ratio

PAS	Power-Azimuth Spectrum
PBA	Parameterizable Basic Architecture
PBP	Personal Basis Profile
PDA	Personal Digital Assistant
PDF	Probability Density Function
PDP	Power-delay Profiles
PDU	Protocol Data Unit
PEP	Policy Enforcement Point
PHS	Personal Handyphone System
PHY	Physical Layer
PIM	Platform Independent Model
PIN	Personal Identification Number
PLMN	Public Land Mobile Network
PN	Public Network
POTS	Plain Old Telephone Services
PP	Personal Profile
PPM	Pulse-position Modulation
PRM	Proxy Reconfiguration Manager
PRP-OFDM	Pseudorandom Postfix OFDM
PSM	Platform Specific Model
QAM	Quadrature Amplitude Modulation
QoS	Quality of Service
QPSK	Quadrature Phase-shift Keying
RAN	Radio Access Network
RAT	Radio Access Technology
RCMAC	Regulated Contention Medium Access Control
RCS	Radio Control Server
RF	Radio Frequency
RFC	Request for Comments
RFID	Radio Frequency Identity module
RLC	Radio Link Control
RM	Reconfiguration Manager
RMA	Reconfiguration Management Architecture
RMI	Java Remote Method Invocation
RMM	Reconfiguration Management Module
RNC	Radio Network Controller
ROM	Read-only Memory
RPC	Remote Procedure Call
RRC	Radio Resource Control
RRM	Radio Resource Management
RSFQ	Rapid Single Flux Quantum
RSMM	Resource System Management Module
RSSI	Receive Signal Strength Indicator
RX	Receive
-s	Specific
SAML	Security Assertion Markup Language

SAP	Service Access Point
SAs	Service Agents
SAW	Surface Acoustic Waves
SC	Single Carrier
SC-FDE	Single Carrier with Frequency Domain Equalization
SCM[1]	Spatial Channel Model
SCM[2]	Subcarrier Modulation
SCOUT	Smart user-centric COmmUnication environmenT
SDF	Service Deployment Framework
SDL	System Description Language
SDM	Software Download Module
SDMA	Space Division Multiple Access
SDOs	Super Distributed Objects
SDP	SW Download Provisioning
SDR	Software Defined Radio
SDRC	Software Download and Reconfiguration Controller
SDRF	SDR Forum
SDS	Service Discovery Service
SDU	Service Data Unit
SEE	Service Execution Environment
SEG	Security Gateway
SIC	Successive Interference Cancellation
SIK	Sequence Inversion Keying Modulation
SIM	Subscriber Identity Module
SIMO	Single Input–Multiple Output system
SINR	Signal-to-Interference and Noise Ratio
SIR	Signal to Interference Ratio
SISO	Single-input, Single-output
SLA	Service Level Assignment
SLP	Service Location Protocol
SLS	Service Level Specifications
SMEs	Small and Medium Enterprises
SMMSE	Successive MMSE
SMS	Short Message Services
SO	Successive Optimization
SOAP	Simple Object Access Protocol
SoC	System on Chip
SPRE	Software Download and Profile Repository
SPUCPA	Stacked Polarimetric Uniform Circular Patch Array
SRA	Software Radio Architecture
SRM	Serving Reconfiguration Manager
S-RNC	Serving RNC
SS	Spread-spectrum
SSID	Service Set Identifier (802.11b)
SS-MC-MA	Spread-spectrum Multicarrier Multiple Antennas
SSO	Single Sign On

STBC	Space–Time Block Codes
STTC	Space–Time Trellis Codes
SW	Software
TC	Topology Control
TCP	Transport Control Protocol
TCSI	Transmit Channel State Information
TDD	Time Division Duplex
TDL	Tapped Delay Line
TDMA	Time Division Multiple Access
THP	Tomlinson–Harashima Precoding
TLA	Top Level Assignment
TOI	Third-order Intercept
TPM	Trusted Computing Platform
TRSA	Terminal Reconfiguration Serving Area
TRUST	Transparently Reconfigurable UbiquitouS Terminal
TTA	Telecommunication Technology Association
TTC	Telecommunication Technology Committee
TX	Transmit
UAs	User Agents
UCD	User Centric Design
UDP	User Datagram Protocol
UE	User Equipment
UI	User Interface
UIAP	Unique Identifier Allocation Protocol
UL	Uplink
ULA	Uniform Linear Array
UML	Unified Modelling Language
UMTS	Universal Mobile Telecommunication System
UP	User Profile
UPnP	Universal Plug-and-play
UP-OFDMA	Unitary-precoded OFDMA
UPS	User Plane Server
USB	Universal Serial Bus
USIM	UMTS Subscriber Identity Module
UTRAN	UMTS Terrestrial Radio Access Network
UW	Unique Words
UWB	Ultra WideBand
UWBWG	Ultra Wideband Working Group
VCO	Voltage Controlled Oscillator
VHE	Virtual Home Environment
VLR	Visitor Location Register
VM	Virtual Machine
VSCRF-CDMA	Variable Spreading and Chip Repetition Factor CDMA
WAN	Wide Area Network
WAP	Wireless Application Protocol
WB	WideBand

WBAN	Wireless Body Area Networks
W-CDMA	Wideband–Code Division Multiple Access
WG	Working Group
WiFi	Wireless Fidelity
WiMAX	Worldwide Interoperability for Microwave Access
WLAN	Wireless LAN
WRC	World Radio Conferences
WSCI	Web Service Choreography Interface
WSDL	Web Services Description Language
WSI	Wireless Strategic Initiative
WSN	Wireless sensor networks
WWRF	Wireless World Research Forum
WWRI	Wireless World Research Initiative
XML	eXtensible Markup Language
ZP	Zero-Padding
ZP-OFDM	Zero-Padded OFDM

Index

Page numbers followed by '*f*' indicate figures and '*t*' indicate tables.

Technologies for the Wireless Future – Volume 2 Edited by Rahim Tafazolli